1st Materials Research Society Indonesia Conference and Congress (MRS-INA C&C 2017)

Innovation of Advanced Materials for A Better World

IOP Conference Series: Materials Science and Engineering Volume 432

Yogyakarta, Indonesia
8 – 12 October 2017

ISBN: 978-1-5108-7510-4
ISSN: 1757-8981

Printed from e-media with permission by:

Curran Associates, Inc.
57 Morehouse Lane
Red Hook, NY 12571

Some format issues inherent in the e-media version may also appear in this print version.

Printed by Curran Associates, Inc. (2019)

For permission requests, please contact the Institute of Physics
at the address below.

Institute of Physics
Dirac House, Temple Back
Bristol BS1 6BE UK

Phone: 44 1 17 929 7481
Fax: 44 1 17 920 0979

techtracking@iop.org

Additional copies of this publication are available from:

Curran Associates, Inc.
57 Morehouse Lane
Red Hook, NY 12571 USA
Phone: 845-758-0400
Fax: 845-758-2633
Email: curran@proceedings.com
Web: www.proceedings.com

TABLE OF CONTENTS

NUCLEAR SCIENCE AND TECHNOLOGY APPLICATION

NANO-MATERIALS. NANO-TECH, THIN FILM AND FUNCTIONAL

POLYMERS, BIO-MATERIALS AND OTHER SOFT MATTERS

THEORETICAL MODELLING AND SIMULATION OF MATERIALS

ENERGY MATERIALS AND DEVICES

NUCLEAR SCIENCE AND TECHNOLOGY APPLICATION

IOP Conf. Series: Materials Science and Engineering **432** (2018) 011001 doi:10.1088/1757-899X/432/1/011001

The Materials Research Society Indonesia Conference & Congress 2017

MRS-INA C&C 2017

SAHID JAYA HOTEL AND CONVENTION
YOGYAKARTA, INDO OCTOBER 8-12, 2017 — INDONESIA

ORGANIZED BY

MATERIAL RESEARCH SOCIETY OF INDONESIA

Accepted, peer reviewed papers from the MRS-INA C&C 2017

Edited by

Evvy Kartini *et.al*
Indonesia

2018

PREFACE

The **Materials Research Society-Indonesia Conference and Congress 2017** (MRS-INA C&C 2017) was held in Yogyakarta on October 8 – 12, 2017. The main theme of MRS-Ina C&C 2017 is ***"Innovation of Advanced Materials for A Better World".*** The conference is organized by Materials Research Society-Indonesia (MRS-INA) with the aim to provide international forum for presenting scientific results in the field of materials science by scientists, academicians, industries and government from national, as well as regional and international.

The pupose of the MRS-INA C&C 2017 is to gather Indonesian Materials Researchers, and to promote international collaboration and cooperation among scientists and academia from regional and developed countries. This should be accomplished by presence of invited world-class speakers for the scientific program. In this event, the fundamentals, innovations and industrial applications of Materials Sciences and Technology will be discussed.

The Committee received over **200 scientific articles** showing a diverse and stimulating program, from more than 40 national and international institutions, covering 15 countries (UK, Taiwan, Sweden, Austria, Germany, USA, Malaysia, Singapore, Thailand, Japan, Bangladesh, China, Australia, Korea and Indonesia). The publication committees were responsible for arranging approximately **150 full peer review** of the manuscripts that will appear in several media of internationla publications, such as Progress in Natural Science: Materials International, IONICS, Journal of Atom Indonesia, Indonesian Journal of Materials Sciences as well as the IOP Conference Proceedings. After selection process, there are 65 articles are to be published in IOP Materials and Engineering.

There are several International Symposia covering subjects in material research, namely (1) Nuclear Science and Technology Application (NST], (2) Metallurgy, Ceramics, Composites and High Temperature Materials (MCT), (3) Nanomaterials, Nanotechnology, Thin Film, Carbon Based Materials and Functional Materials (NTC), (4) Energy Materials and Devices (EMD), (5) Polymers, Biomaterials and other Soft Matters (PBS), (6) Theoretical Modeling and Simulation of Materials (TMS). I truly acknowledged all the publication team and reviewers who have been working very hard on reviewing all those submitted articles.

The conference was remarked by the Minister of Research, Technology and Higher Education of the Republik of Indonesia, and a Plenary Speech from the President of International Union of Materials Research Society (IUMRS), Professor Soo Wohn Lee. The inauguration of MRS-INA as official member of IUMRS was declared by the IUMRS President, and a certificate of membership was given to Prof.Dr. Evvy Kartini, President of MRS-INA.

MRS-INA C&C 2017
Prof. Dr. Evvy Kartini
Chair of the Committee
President of MRS-INA

IOP Conf. Series: Materials Science and Engineering **432** (2018) 011001 doi:10.1088/1757-899X/432/1/011001

ORGANIZING COMMITTEE

CONFERENCE ORGANIZERS

Steering Committee

Prof. Dr. Djarot Sulistio Wisnubroto Chairman of BATAN

Prof. Dr. Ir. Muhammad Anis, M. Met. Rector of Universitas Indonesia (UI)

Dr. Agus Pudji Prasentyono M. Eng.

Senior Advisor to the Minister of Relevance and Productivity Dr. Utama Padmawinata National Research Council (DRN)

Prof. Dr. Ridwan National Research Council (DRN)

International Advisory

Prof. Dr. B.V.R. Chowdary President of MRS-Singapore.

Prof. Dr. Sung Min Choi President of AONSA, KAIST, Korea

Prof. Soo Wohn Lee President of IUMRS, Korea

Prof. Dr. Hsiung Chou President of NCU Taiwan

Prof. Dr. Santi Maensiri The President of MRS-Thailand

Prof. Dr. Toshiji Kanaya Director of MLF J-PARC, Japan

Prof. Dr. Kazuyoshi Yamada Director of IMSS of KEK, Japan

Prof. Dr. Masatoshi Arai The European Spallation Source (ESS) Sweden

Prof. Dr. Takashi Kamiyama KEK, Japan

Prof. Dr. Takashi Sakuma Ibaraki University, Japan

Prof. Dr. Shahidan Radiman UKM, Malaysia

Prof. Dr. Stefan Adams NUS, Singapore

Dr. Jamie Schultz Director of ANSC ANSTO, Australia

Prof. Dr.J. A. Fernandez-Baca Oak Ridge National Laborator, USA

Dr. Maxim Avdeev ANSTO, Australia

Local Advisory

Dr. Azwar Manaf Universitas Indonesia (UI)

Dr. Bambang Sugiono Universitas Indonesia (UI)

Prof. Dr. A. Herman Yuwono Universitas Indonesia (UI)

Prof. Dr. Anne Zulfia Syahrial Universitas Indonesia (UI)

Prof. Dr. Bondan Tiara Sofyan Universitas Indonesia (UI)

Dr. Agus Purwanto Universitas Sebelas Maret (UNS)

Prof. Dr. Zaki Su'ud Institut Teknologi Bandung (ITB)

Prof. Dr. Khairul Basar Institut Teknologi Bandung (ITB)

Dr. Agustinus Agung Nugroho Institut Teknologi Bandung (ITB)

Prof. Dr. Darminto Institut Teknologi Sepuluh Nopember (ITS)

Prof. Dr. Ratno Nuryadi The Agency for the Assessment and Application of Technology (BPPT)

Prof. Dr. Eng. Eniya L. Dewi The Agency for the Assessment and Application of Technology (BPPT)

Dr. Andika Pramono The Indonesian Institute of Sciences (LIPI)

Dr. Sunit Hendrana The Indonesian Institute of Sciences (LIPI)

Dr. Bambang Prihandoko The Indonesian Institute of Sciences (LIPI)

Dr. Muhammad Taufik Universitas Negeri Malang (UNM)

Dr. Triyana Kuwat Universitas Gadjah Mada (UGM)

ORGANIZING COMMITTEE

Committee Chair

Prof. Dr. Evvy Kartini

Conference

Prof. Dr. Akhmad Herman Yuwono

Congress

Dr. Sunit Hendrana

Dr. Agus Taufiq

Dr. Eni Sugiarti, M.Eng.

Program

Prof. Dr. Ratno Nuryadi

Publication

Dr. Abu Khalid Rivai

Dr. Nofrijon Sofyan

Heri Jodi, M.Eng.

Rida Nurul Shelni Rofika, M. Si.

Symposia

Dr. Agus Purwanto

Dr. Agung Nugroho

Dr. Iwan Sumirat

Dr. Azwar Manaf

Dr. Bambang Prihandoko

IOP Conf. Series: Materials Science and Engineering **432** (2018) 011001 doi:10.1088/1757-899X/432/1/011001

Website

Ir. Chairul Hudaya, Ph.D.

Finance

Dra. Noeraida

Exhibition

Drs. Yustinus Purwamargapratala

Logistic

Enggay Sugati, S.E.

Dra. Deswita

Agus Sudjatno, A.Md.

Jadigia Ginting, M.Sc.

Secretariat

Contact Person 1. Dr. Sudaryanto

Contact Person 2. Teguh Panca Putra, Ph.D.

Secretary

Yualina Riastuti Pratiwi

Cipta Panghegar Supriadi

Address Bldg. 71, Kawasan Puspiptek Serpong, Setu, Tangerang Selatan, Banten, INDONESIA, 15314

Phone +62.21.756.0922 ext. 4023

Facsimile +62.21'756.0926

Website http://mrs-ina.org

e-mail cnc@mrs-ina.org

IOP Conf. Series: Materials Science and Engineering **432** (2018) 011001 doi:10.1088/1757-899X/432/1/011001

SCIENTIFIC COMMITTEE

IOP Conf. Series: Materials Science and Engineering **432** (2018) 011001 doi:10.1088/1757-899X/432/1/011001

- Maria Kayra SASKIA (Binus International School, Indonesia)
- Evvy KARTINI (BATAN, Indonesia)

LAYOUT EDITOR

- Rida Nurul Shelni ROFIKA (ITB, Indonesia)
- Yustinus PURWAMARGAPRATALA (BATAN, Indonesia)

PUBLICATION

- Dr. Rivai Abu KHALID M.Eng (BATAN, Indonesia)
- Dr. SUDARYANTO (BATAN, Indonesia)

IOP Conf. Series: Materials Science and Engineering **432** (2018) 011002 doi:10.1088/1757-899X/432/1/011002

Peer review statement

All papers published in this volume of *IOP Conference Series*: *Materials Science and Engineering* have been peer reviewed through processes administered by proceedings editors. Reviews were conducted by expert referees to the professional and scientific standards expected of a proceedings journal published by IOP Publishing.

IOP Conf. Series: Materials Science and Engineering **432** (2018) 012001 doi:10.1088/1757-899X/432/1/012001

Development of lotus-type porous material for heat pipe application using centrifugal slip casting process

S Supriadi[1]*, M Rasis[1], T Heriyanto[1], B Ariantara[1], and N S D Putra[1]

[1] Departement of Mechanical Engineering, Faculty of Engineering, University of Indonesia, UI Depok, Depok 16424 Indonesia

sugeng@eng.ui.ac.id

Abstract. Heat pipe is a heat conductor which has a large conduction capacity using a special fluid-filled pipe as a conductor of heat from the hot end (evaporator) to the cold end (condenser) used for thermal management. In the heat pipe, there is a sintered wick which transfers refrigerant fluid from condenser to evaporator. The objective of this research is to improve capillary performance by making a straight pore called as lotus-type porous material (LTP) using centrifugal slip casting technique. Copper powder is used as wick materials due to its high thermal conductivity. A starch solution is used as binder material to make the copper slurry. Nylon wire is applied as straight pore mold. Freeze drying is applied to curing and demolding. Vacuum sintering is conducted to increase metallurgical bonding between the particle. The results show that centrifugal casting machine is capable to make a wick with LTP structure. On the performance department, capillary pumping is affected by rotation speed, powder size, and powder loading. The optimum parameters were copper powder with 260.54 μm powder size, 50% Powder Loading with 688 rpm rotation speed. The LTP wick is proved to have a higher capillary pumping capacity compared to the conventional wick. Further work is necessary to verify the performance of LTP wick in the heat pipe.

1. Introduction

Heat pipe is the most effective heat exchanger because of its ability to transfer large amounts of heat with the passive work system so that it does not require additional energy to operate it except the device itself [1]. The heat pipe is operated by utilizing latent heat and refrigerant to exchange large quantities of heat with small differences temperature across the area [2, 3]. The latent heat of the refrigerant is utilized to improve the thermal conductivity of the heat pipe. Example of heat pipe application is as a coolant to withstand the heat flux generated due to processor performance on CPU so that processor does not overheat.

Capillary wick is an important component of heat pipe that serves to re-drain condensed vapors to the evaporator quickly. The wick works based on the capillary behavior of the fluid [4]. This capability is one of the keys in determining of how the quality of pipe heat works. There are 3 types of capillary wick which are widely used: sintered powder, groove, and screen mesh. The sintered powder wick has better thermal performance than the screen mesh wick [5]. The sintered powder wick has a higher thermal resistance performance compared to the groove wick [6]. Sintered powder wick has a very small structure that allows the flow of refrigerant works at high capillary pressure, however, it is prone to blockage so that the fluid flow is not able to penetrate the capillary wick [7].

In a 2014 article by Putra *et al.* introduced a capillary wick [8] based on biomaterials in the form of coral reefs. However, being a living creature, the coral reef will attract new problems if it is used for mass producing capillary wick, which eventually could lead into the destruction of underwater ecosystems. The conventional method to fabricate LTP metals utilizes pressurized hydrogen to control porosity and pore size [9]. One significant disadvantage is to rely on pressurized hydrogen gas, which requires the use of high-pressure chamber, so there is a definite need to develop fabrication methods that do not require high-pressure hydrogen [9]. LTP material using solid structure has been utilized with a loop heat pipes to improve the performance of battery thermal management systems so that it has high performance with long service life [10]. However the previous study used solid LTP for wick of the heat pipe. Therefore the objective of this research is to fabricate a hollow LTP for heat pipe using centrifugal slip casting and sintering process.

2. Experimental method

The fabrication of heat pipe by the method of centrifugal casting method that has the characteristics of Lotus-Type Porous is still relatively new. To get a heat pipe that has Lotus-type porous characteristics, the method is divided into three main stages: design centrifugal casting machine, centrifugal casting process, and densification.

The design of centrifugal casting machine using DC motor of 100-120V capacity with a maximum rotation speed reaching up to 888 rpm. Figure 1 shows the design centrifugal slip casting machine. Pulley and belt that serves to transfer the rotation from dc motor to pipe to produce centrifugal force on the pipe. Bearing is installed to restrain rotation axis. Support mold is designed to regulate the cavity structure on the capillary wick by being placed on both ends of the pipe. Figure 2 shows the illustration of support mold.

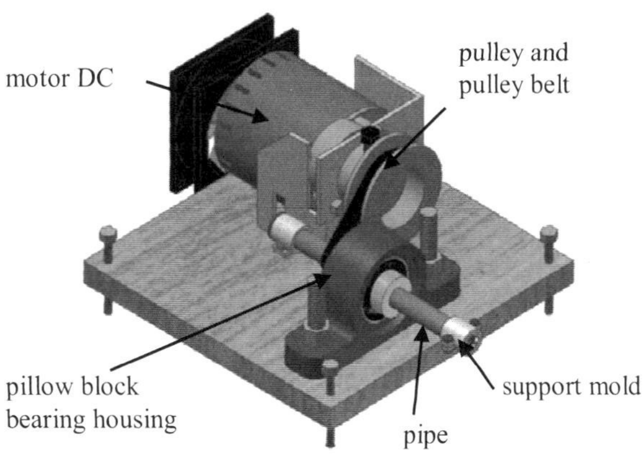

(a)

(b)

Figure 1. Centrifugal Slip Casting Machine: (a) centrifugal slip casting machine design with Inventor app, (b) centrifugal slip casting machine has been made.

Centrifugal casting process stage is the process of pouring copper powder that has been mixed with the binder material. The characteristic diamaters of the used copper powder area 141.52 μm, 192.95 μm, and 260.54 μm respectively. The binder material used is tapioca. The ratio of mixture between copper powder and binder is 55% copper powder and 45% binder [11]. Lotus-type Porous Material (LTP) cavity builders use nylon-sized yarn of 0.12 micrometer. The principle of centrifugal casting is essentially done by pouring molten metal into a rotating mold. Due to the influence of the liquid metal centrifugal force

IOP Conf. Series: Materials Science and Engineering **432** (2018) 012001 doi:10.1088/1757-899X/432/1/012001

will be distributed to the wall of the print cavity and then freeze. This casting process takes approximately 30 minutes for each sample.

Densification stage or sintering process is to enter the green part that has been prepared into the vacuum furnace. Vacuum sintering is conducted to increase metallurgical bonding between the article. This process enters all green part samples into the vacuum furnace and then closes tightly and turns on the vacuum pump to make the atmospheric pressure decrease to -0.1 Mpa after which set up the temperature setting of 850°C [11]. Continuous vacuum is conducted during 60 minutes [11]. The aims to remove by product gases from sintering processes that potentially damage the microstructure of the product.

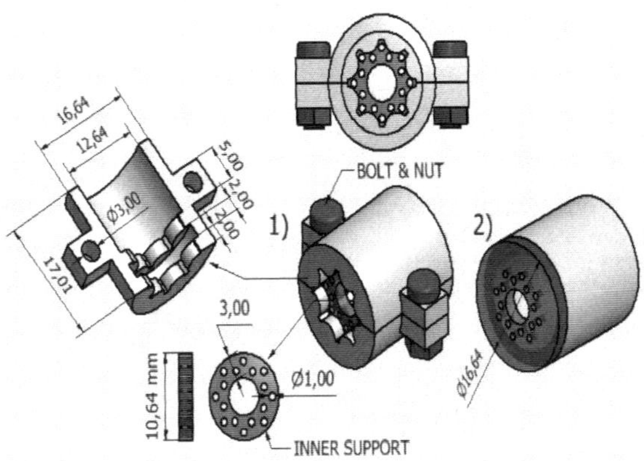

Figure 2. Design support mold.

In this research, capillary data were collected on 14 samples and 5 times for each sample. Data recording is done five times in one second using WinCT software and processed using Microsoft Excel and Origin Lab. Data retrieval on each seal with 3 main variables used as reference are Powder size (PS), Powder Loading (PL) and Pipe Rotation Speed (rpm) are shown in Table 1:

Table 1. Summarized data of 3 main sample variables.

Sample	PS (μm)	PL	RPM
S.1[a]	141.52	40.00%	888
S.2[a]	141.52	50.00%	888
S.3[a]	141.52	55.00%	888
S.4[a]	192.95	40.00%	888
S.5[a]	192.95	50.00%	888
S.6[a]	192.95	55.00%	888
S.7[a]	260.54	40.00%	888
S.8[a]	260.54	50.00%	888
S.9[a]	260.54	55.00%	888
S.10[b]	192.95	40.00%	888
S.11[b]	192.95	50.00%	888
S.12[b]	192.95	55.00%	888
S.13[a]	141.52	50.00%	688
S.14[a]	141.52	50.00%	788

[a] with pore former

[b] without pore former

Capillary pumping is the ability of a hollow material to create a capillary pressure difference [12]. The method used for testing capillary pumping performance is capillary pumping amount real time changing curve [4]. Figure 3 shows a schematic of capillary pump testing.

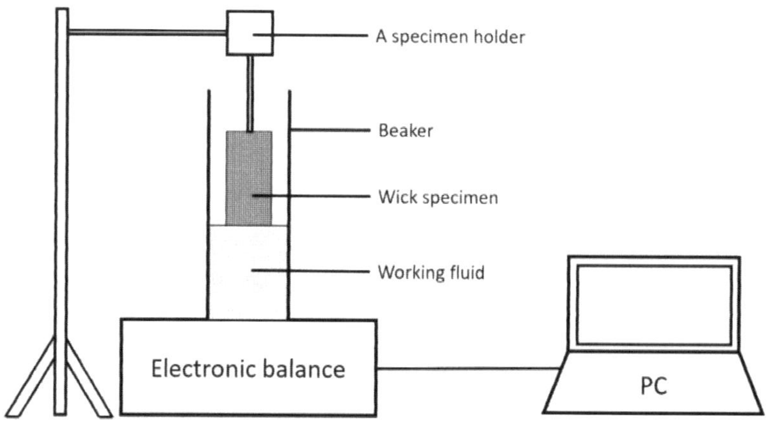

Figure 3. Schematic for measuring the capillary pumping amount changing curve.

Testing begins by filling beaker glass with working fluid which in this test is water. Then beaker glass is placed on an electronic scale that is connected to the computer to transfer data. Electronic scales used are A & D type GF-300. Then, once the computer and scales are ready, the porous wick sample is lowered slowly to the surface of the water. Data on the amount of fluid inhaled will be recorded in the computer which will be used to determine the capillary pumping rate [4].

Li *et al.* [4] states that the change in the capillary pumping amount curve can be described as shown in the equation (1),

$$y = y_0 + A \text{ x } e^{-t/\tau} \tag{1}$$

where y_0 and A in equation (1) are related to the total amount of capillary pumping in hollow material. The time constants in the equation are related to the capillary pumping rate. The time constant can be used as a comparison of capillary pumping rate between hollow material. The higher the speed, it can be said the better the performance [4]. The derivative of equation (1) gives the capillary pumping rate as equation (2).

$$y'(t) = \frac{y_{sat}}{\tau} e^{-t/\tau} \tag{2}$$

In equation (2) the value (y_o/t) is the capillary pumping rate coefficient. In addition Li *et al.* [4] also concluded that along with increasing porosity, the performance of its capillary pumping will increase. When the porosity has the same value, then the material that has a narrower cavity will have better performance.

3. Result and discussion

Heat pipe wick was successfully fabricated using centrifugal slip casting process as shown in Figure 4. The thickness of wick is affected by viscosity. Figure 4 shows the difference in the diameter of the thickness of each sample wick. With the pouring volume of each pipe in the same way, then with an increasingly large powder loading (PL), the viscosity will also rise. Figure 4(d) shows the heat pipe without pore former for comparison with the sample Figure 4(a) Figure 4(e) shows the difference of wick diameter with the same PL and powder size but different rotational speed. Performance test results of capillary pumping of each sample heat pipe shown in Table 2.

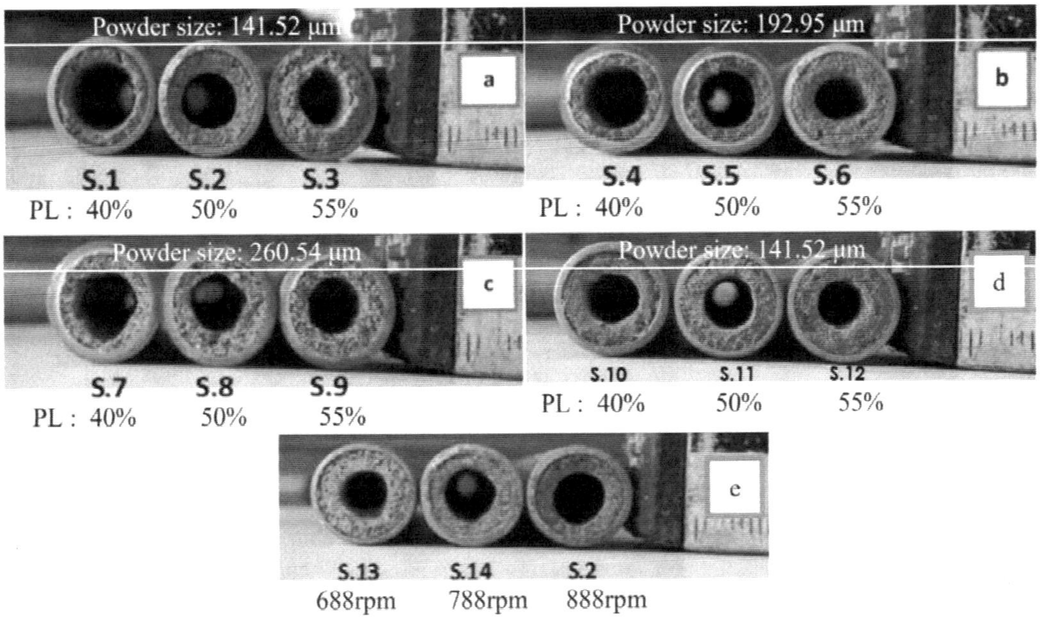

Figure 4. Sample heat pipe with centrifugal casting process: (a) wick with powder size 100 µm, (b) wick with powder size 150 µm, (c) wick with powder size 200 µm, (d) wick without pore former using powder size 150 µm, (e) comparison of rotation speed with powder size 100 µm.

Table 2. Summary of experiment result data.

Sampel	y0	Std. Error	t	Std. Error	y0/t	Std. Deviasi
S.1	0.974	0.00276	1.195	0.01891	0.850	0.20174
S.2	1.032	0.00284	1.107	0.01680	1.249	0.78677
S.3	1.356	0.00367	1.000	0.01679	1.519	0.65120
S.4	1.036	0.00275	0.808	0.01100	1.329	0.73733
S.5	1.107	0.00272	0.759	0.00672	1.514	0.64126
S.6	1.434	0.00377	0.767	0.01263	1.785	0.52146
S.7	1.277	0.00325	0.926	0.01539	1.486	0.48144
S.8	1.392	0.00346	0.806	0.01483	1.797	0.45628
S.9	1.508	0.00365	0.756	0.01590	2.189	0.64268
S.10	0.769	0.00136	1.001	0.01889	1.012	0.31260
S.11	0.830	0.00104	0.875	0.01561	1.211	0.44068
S.12	0.872	0.00201	0.860	0.01733	1.279	0.50358
S.13	1.254	0.00318	0.873	0.01633	1.604	0.65735
S.14	1.221	0.00263	0.951	0.01483	1.353	0.37634

There are three main variables that are used as references to study the relationship between Powder size (PS), Powder Loading (PL) and Pipe Rotation Speed (rpm).

Thickness of the capillary axis is influence by the amount of powder loading. Figure 5 shows that the Capillary Pumping Amount (CPA) capability increased when using larger powder sizes as well as the capillary wick with powder loading 55% with the largest capillary axis thickness. A result indicating that the powder size 260.54 µm and PL 55% remain the best. The larger powder size will produce a larger cavity so that the time required to reach the maximum amount of fluid that can be absorbed by the capillary axis becomes relatively shorter.

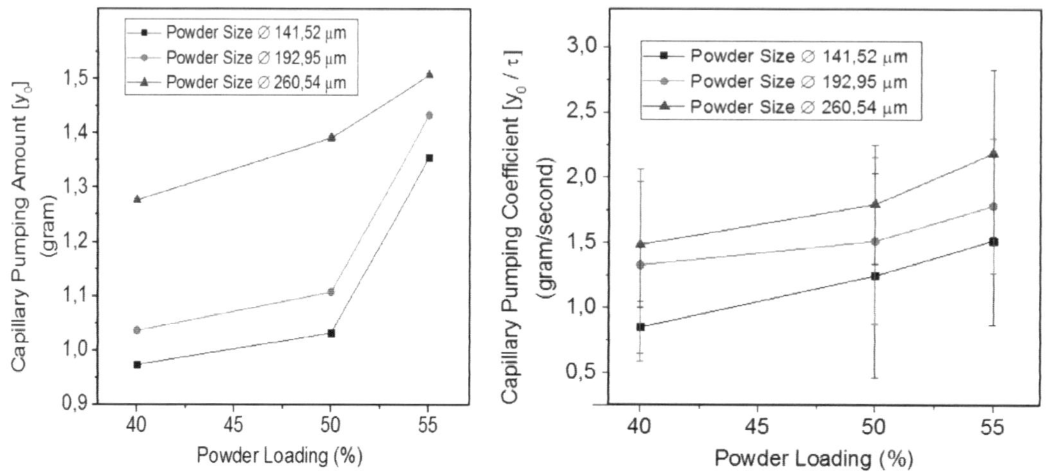

Figure 5. Capillary pumping as a function of powder loading for different powder size.

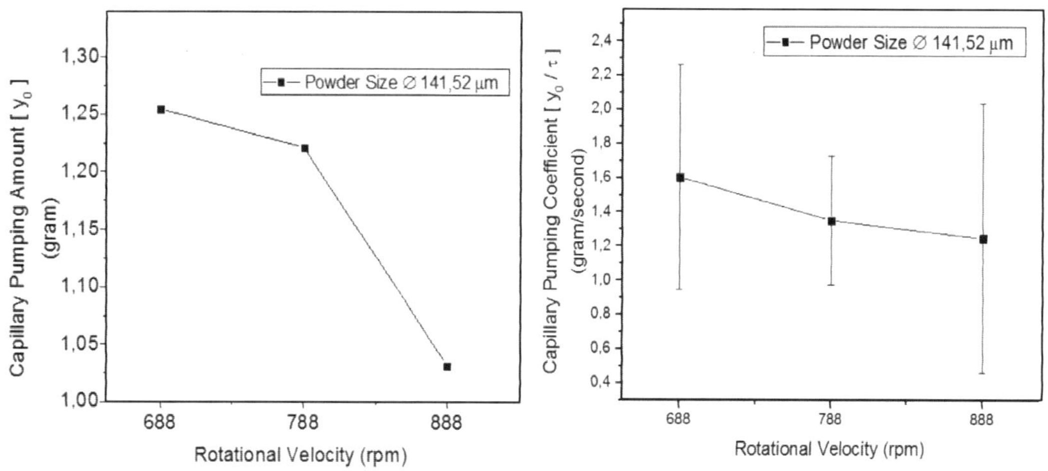

Figure 6. Graph comparison between rotational velocity and capillary pumping.

Variations of different rotation speeds will form capillary wick with different thickness because the force received by the slurry is different is shown in Figure 6. According to the effect of powder size as shown in Figure 5, the result of CPA which is influenced by rotation speed will have the similar trend with different powder size, so the optimum parameters are copper powder with powder size 260.54 μm, powder loading 55% and will be more optimum when done at 688 rpm rotation speed. However, the capillary wick fabricated with a 55% PL composition tends to be more difficult to produce capillary wick with hollow characteristics, because the slurry dries faster due to the higher viscosity of the slurry and makes it more difficult to be evenly distributed on the inside of the pipe during the centrifugal casting process. So the optimum parameter to produce an even more optimal capillary wick is when at 50% PL composition, using powder size 260.54 μm, and 688 rpm rotation speed.

The LTP wick performs higher capillary pumping capacity. The use of yarn as a pore former in the manufacture of Lotus-Type Porous Material on the capillary wick shows improvement in ability when compared with no pore former. Figure 7 shows the performance comparison of the sintered copper powder capillary wick seen in terms of capillary pumping rate coefficient (CPR Coeff.) with the recommended optimum parameter reaching 1.198 gr/s, while the CPR Coeff. research that the authors do, the sample with the optimum parameters obtained is 1797 gr/s, a result that shows improvement in the performance from previous research.

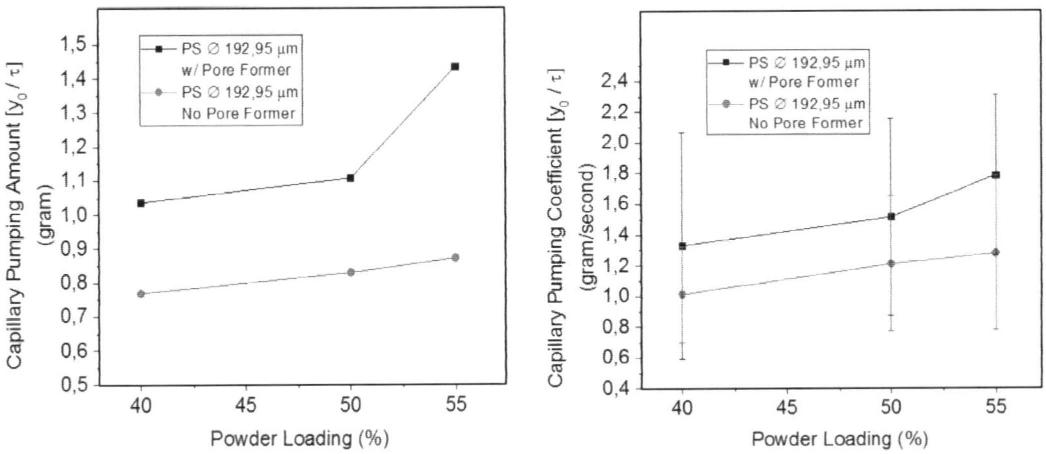

Figure 7. Graph comparison between with pore former and without pore former.

4. Conclusion

Copper LTP wick with hollow structure has successfully fabricated using centrifugal slip casting method followed by sintering process. The mold is used to maintain the structure of the pore former so as not to deform when the centrifugal casting process takes place. The sintering method by re-vacuum process can reduce the occurrence of decarburation or oxidation on the capillary wick. The optimum parameters of centrifugal slip casting machine made are copper powder with powder size 260.54 μm, Powder Loading 50% and will be more optimum when done at 688 rpm rotation speed. Performance maximum capillary pumping rate of LTP ranges from 1.329 gr/s to 1.785 gr/s.

Acknowledgment

The authors would like to thank LPDP Department of Finance of Republic of Indonesia under RISPRO 2014.

References

[1] Putra N *et al.* 2015 *Experimental Thermal and Fluid Science* **66** 13-27
[2] Faghri A 1995 *Heat pipe science and technology* **1** Global Digital Press Washington DC
[3] Faghri A 2012 *Journal of heat transfer* **134(12)** 123001
[4] Li J Zou Y and Cheng L 2010 *Experimental Thermal and Fluid Science* **34** 1403-08
[5] Kumaresan G *et al.* 2014 *International Communications in Heat and Mass Transfer* **57** 208-15
[6] Russel M K *et al.* 2011 *Applied Thermal Engineering* **31(1)** 69-76
[7] Korn F 2008 *Heat pipes and its applications* MVK160 Project Reports
[8] Putra N *et al.* 2014 *International Journal of Thermal Sciences* **76** 128-36
[9] Lee Y and Hyun 2012 *Material Letters* **78** 92-4
[10] Ariantara B Putra N and Supriadi S 2018 *IOP Conf. Ser.: Earth & Env. Sci.* **105** 012045
[11] Supriadi S *et al.* 2016 *Applied Mechanics and Materials* **819** 601-5
[12] Dullien F A L and Brenner H 2012 Porous Media: *Fluid Transport and Pore Structure.*: Elsevier Science

Transformation of orthodontics bracket geometry in metal injection molding process

S Supriadi[1], D Abdussalam[1], T Heriyanto[1], B Irawan[2], B Suharno[3]

[1] Department of Mechanical Engineering, Faculty of Engineering, University of Indonesia, UI Depok, Depok 16424, Indonesia
[2] Department of Dental Material, Faculty of Dentistry, University of Indonesia, Salemba Raya 4 St., Central Jakarta, 10430, Indonesia
[3] Department of Metallurgy and Materials Engineering, Faculty of Engineering, University of Indonesia, UI Depok, Depok 16424, Indonesia

sugeng@eng.ui.ac.id

Abstract. Orthodontic Bracket is a complex component; Therefore, it requires manufacturing process of a high accuracy like metal injection molding. Metal injection molding consists of three main steps. At first, the feedstock which is a mixture of steel powders and polymer is injected into a mold to form the desired design. Then the molded part is unbound to remove the polymer from the part, this process results in an absence of polymer that generates porosity. Furthermore, the unbound part is sintered to densify steel particles through capillarity force, this process is also the cause of part shrinkage that leads to a geometrical change. The bracket consists of three sections namely the geometrical functions that are affected by the sintering process, they are: wing, trunk, and base. The wing section works as a hook to hold the orthodontic ring, the trunk section supports all the bracket sections, while the base section provides surface contact with orthodontic adhesive. Each section of the bracket has its specific function namely to attach object with different geometries and to withstand a certain load. The functions will not work properly if during the sintering process the geometrical changes caused by shrinkage occurs. This paper presents transformation analysis of orthodontic bracket geometric transformation in the MIM process. In the case of wing geometry, the shrinkage of the outer part of the arch is greater than that of the inner part resulting the reduction on its slope from 35,01° to 32,11°. In the case of square geometry, the corners shrink greater than the edge side so that the square getting rounded, especially the bottom surface of the bracket that the radius of convexity reduced from 43.51 mm to 19.23 mm. The effects of shrinkage due to sintering have less impact on the trunk in the middle of the bracket that is equal to 7.23% on x-axis and 4.994% on y-axis while the wings section shrink by 10.98% on x-axis and -6.637% (elongated) on y-axis. In conclusion, the changes of size and shape are different from each section, depend on the section's geometry, the sintering tends to effect more on the outer side of geometry. The disproportion of shrinkage will result in overall shape change in the manufacturing outcome. On the wing of the bracket, there will be a thinning and a slope reducing that will increase the risk of fracture and can not refute the orthodontic rubber well. On the base, occurs the rounding that will reduce the surface contact on the adhesive that attaches the bracket to the teeth. To adapt to the geometrical change, the first design must be larger, and the part's faces must be designed to be more concave.

1. Introduction

The orthodontic bracket is used to align teeth and correct malocclusion of the teeth. It is transferring the force applied by orthodontic wire to the tooth [1]. The orthodontic bracket must be strong enough to apply the load while expected to be as small as possible to provide comfort.

There are several methods to fabricate an orthodontic bracket, such as machining, metal injection molding, investment casting [2]. The orthodontic bracket fabrication has to be economical for mass-production but also able to produce complex geometry with high dimensional tolerance. The machining process is good in producing high complexity shape but bad in mass-producing, while investment casting is the otherwise [3]. Therefore, metal injection molding is chosen to fabricate the orthodontic bracket.

Metal injection molding (MIM) is a process with several steps as shown in Figure 1. The first step is feedstock preparation, where the metal powder is mixed with the polymer. The second step is injecting the feedstock to the molding using a machine with a high temperature. The third step is unbinding, where the injected part is unbound to remove the polymer from the part, furthermore, the absence of polymer will result in porosity. The last step is sintering, where the unbound part is heated to fuse metal particle then make the unbound part denser.

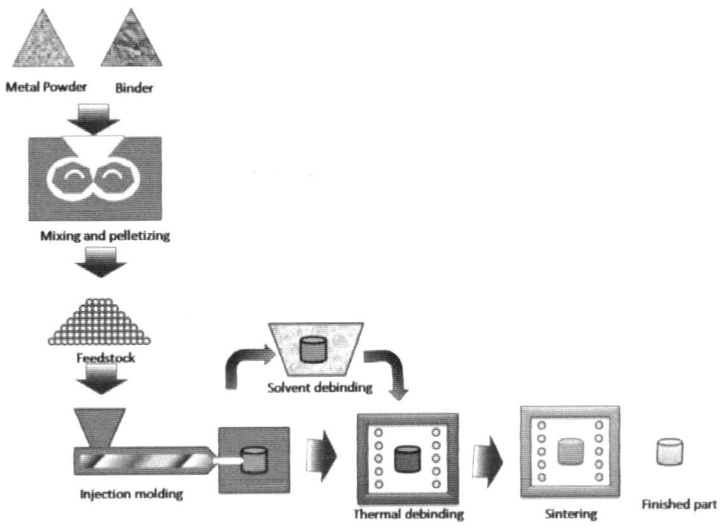

Figure 1. Illustration of metal injection molding process [4].

The long process of metal injection molding is the disadvantage of this manufacturing process. During the sintering, the part is not only shrinking, but also changing in geometry that is probably caused by ununiformly shrinkage all over the part [5].

This phenomenon makes it more complicated to design the mold. This paper focuses on formulating the geometrical transformation from which the result will be used to overcome the difference between the design and the end product.

2. Experimental method

This research was carried out in several steps. The first step was the machine and feedstock preparation. The type of feedstock used in this research is Stainless Steel 17-4PH which has a specific temperature of 200°C. The feedstock was injected using pneumatic injection machine with the force of 5.2 kN/m^2 and injecting-speed of 89 m/min, while the temperature of the barrel was 100 °C, the nozzle was 200 °C and mold was 50 °C. The second step was unbinding, this process consists of solvent and thermal unbinding. The solvent unbinding used agitation method with the temperature of 50 °C, the heating rate of 1 °C/min, and the holding time of 60-minute, during 90-minute duration. The thermal unbinding use the temperature of 510 °C, the heating rate of 1 °C/min, and the holding time of 60-minutes. The third step was sintering, this process occurred under vacuum atmosphere with the temperature of 1360 °C, and the holding time of 1.5 hours [6]. The last step was comparing and analyzing the geometrical differences between the design and the end product.

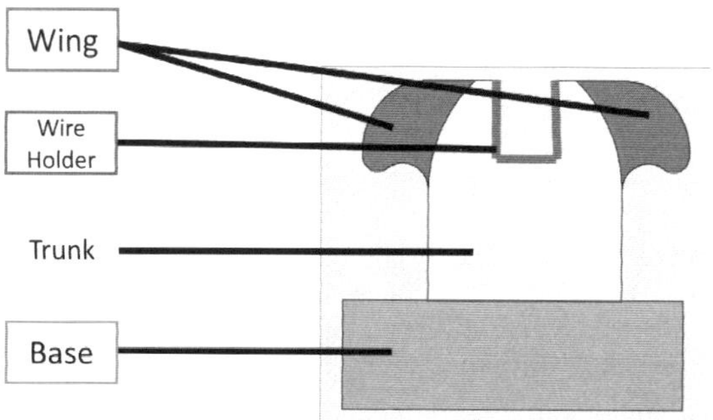

Figure 2. The four sections of orthodontic bracket from front view.

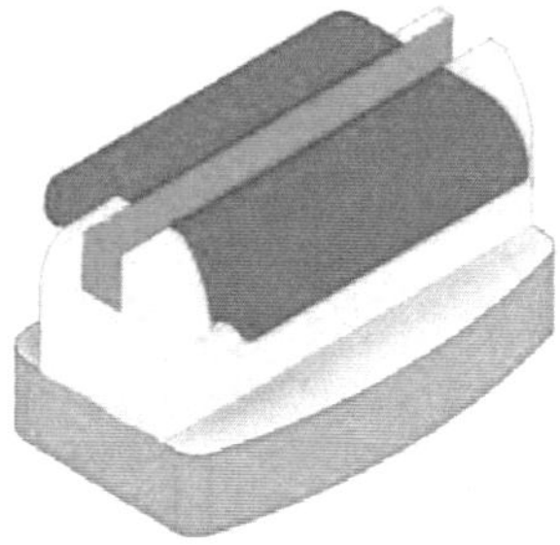

Figure 3. The four sections of orthodontic bracket from isometric view.

For analytical purpose, the bracket was divided into 4 sections based on their geometries and position as shown in Figure 2 and Figure 3. Each of the sections have their specific functions and withstood loads, therefore, they used different calculations to determine the shrinkage occurred and its criticality to the functions. The shrinkage was calculated using a formula in equation (1).

$$Shrinkage = 1 + \frac{size\ after\ sintering}{size\ before\ sintering}\,[\%] \quad (1)$$

The wings section as illustrated in Figure 2 and Figure 3 is thick hooks extruded along the bracket's length, both on the right and the left side with bilateral symmetry. This section works as a hook to hold the orthodontic ring. The lower arc of the wing (as illustrated in Figure 4) is important to hold the rubber on, since the rubber won't hold safely if the curve is too flat. The angle of the hook (as illustrated in Figure 5) is also important to keep the rubber held on. Therefore, the measurement need to be done to prevent unexpected size changes which reduces the wing function to hold the rubber. The angle of hook is measured in degree relative to the trunk.

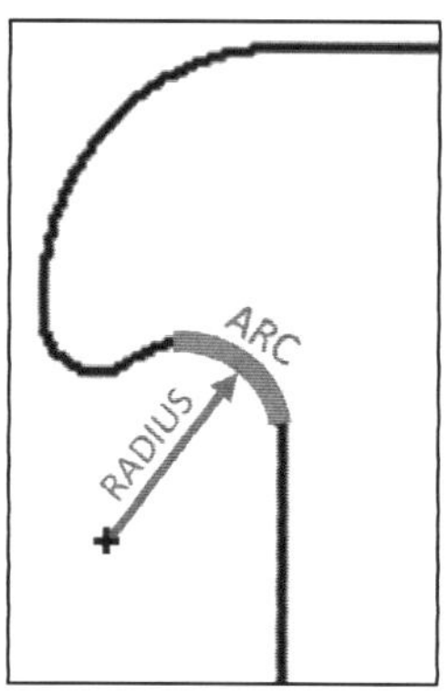

Figure 4. The lower arc of the wing (red) and its radius (blue) measurement.

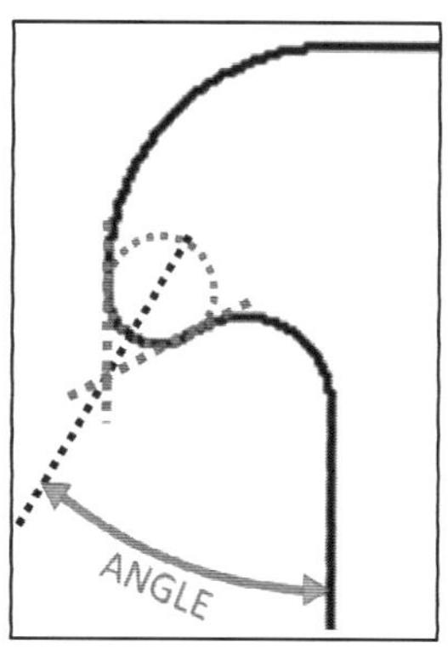

Figure 5. The hook's angle measurement.

The base section as illustrated in Figure 2 and Figure 3 works to provide surface contact to the tooth. The excessive surface rounding on the bottom of the base that can be caused by manufacturing process such as sintering, will reduce the surface contact with orthodontic adhesive and increase the risk of the bracket to detached from the tooth. Therefore, measurement is important to find out how much of the surface rounding has occurred.

As shown in Figure 6, the surface rounding was determined by the radius of roundness that appeared as an arc line on 2-dimensional view. The corner roundness was also measured to determine sintering reactions that occurred in different shapes as shown in Figure 6.

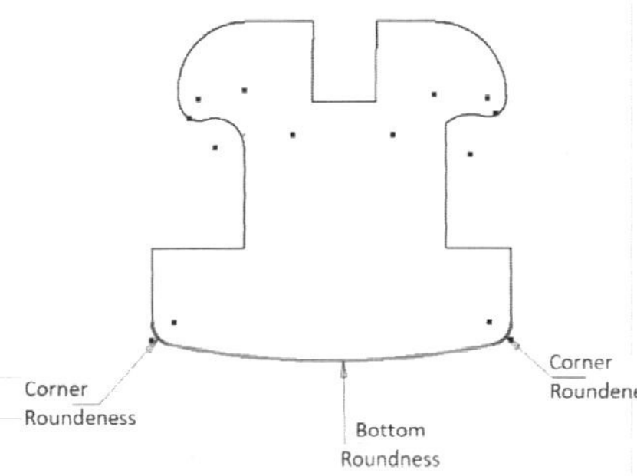

Figure 6. Bracket's base measurement method.

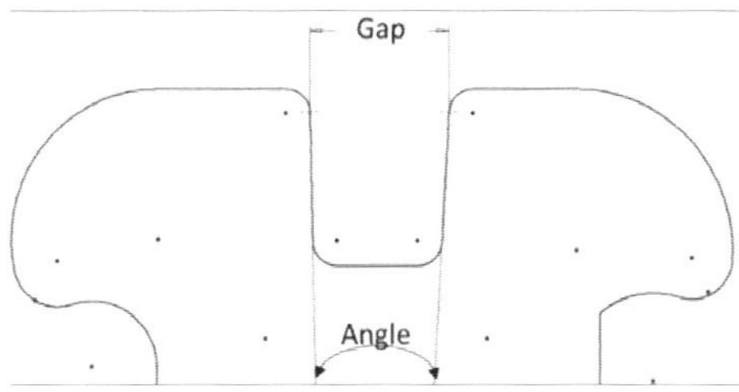

Figure 7. Wire slot measurement method.

The wire slot section as shown in Figure 2 and Figure 3 as is a u-shape cavity that receives the torsional force from the orthodontic wire.

The wire slot was designed to be bigger than the wire to give a space that prevents an unexpected friction with the wire. Therefore, the measurements need to be done on the parameters that indicate a narrowing of the slot, those parameters are the gap and the angle as illustrated in Figure 7.

3. Result and discussion

The results of the measurements are geometry's comparations of each section before and after sintering process. The differences in shrinkage ratio are also used to find the correlation between shape and shrinkage. Furthermore, the results are related with heat distribution on the sintering furnace.

3.1. Overall body shrinkage

To analyze the shrinkage that has occurred in different dimensions, the measurement is conducted by comparing the 5 components of measurement that are shown in Figure 8. From the data provided in Table 1, the most shrunk dimension is the length of the bracket that shrinks 10.95% along the z-axis while the least shrunk is the height of the base that shrinks only 0.85% along the y-axis. Based on the measured data, the longer the geometry was likely to shrink greater than the shorter geometry. Based on comparison between 3-axis dimension, the dimensions along the z-axis are shrinking the most while the dimensions along the y-axis are shrinking the least.

The reliability of transformation of before and after sintering was checked by *Pearson product-moment* correlation. The correlation coefficient reflects the relations of shrinkage and sintering,

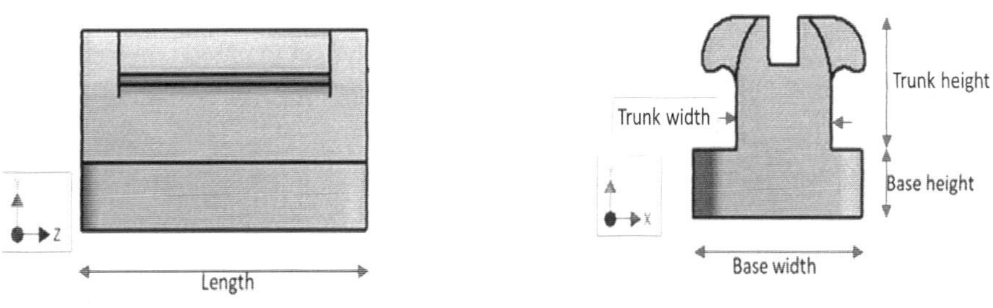

Figure 8. overall body measurement.

since the shrinkage would have nothing to do with the sintering if the correlation has not satisfy the requirement. The correlation check concludes that the sintering effects all measured dimension to be shrunk except a shrinkage in trunk height. The shrinkage of trunk height has a p-value of 0.845917 that exceeds the significant level of 0.10, therefore the geometrical transformation is not related to the sintering.

Table 1. Measuring results from the Figure 8.

	Size (mm)			
Dimension	Design	Before Sintering	After Sinterin	Shrinkage due to
Length	5.400	5.947	5.295	10.95%
Trunk Width	1.825	1.994	1.850	7.24%
Base Width	3.240	3.350	3.179	5.11%
Trunk Height	2.000	2.086	1.982	4.99%
Base Height	1.000	1.175	1.165	0.85%

3.2. Bracket base geometric transformation

The corner rounding, and bottom rounding as illustrated in Figure 9 and Figure 10, are the indicators of the base geometrical transformation. The results of the measurement are shown in Figure 11. After the sintering, the bottom surface of the bracket seems to be more rounded than before.

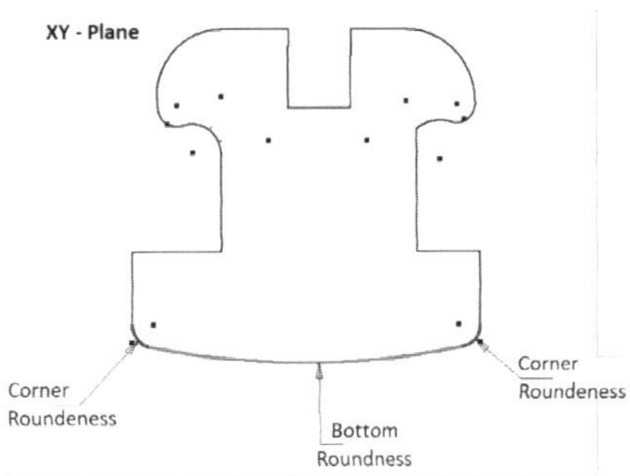

Figure 9. Definition of bottom roundness and corner roundness on xy-plane.

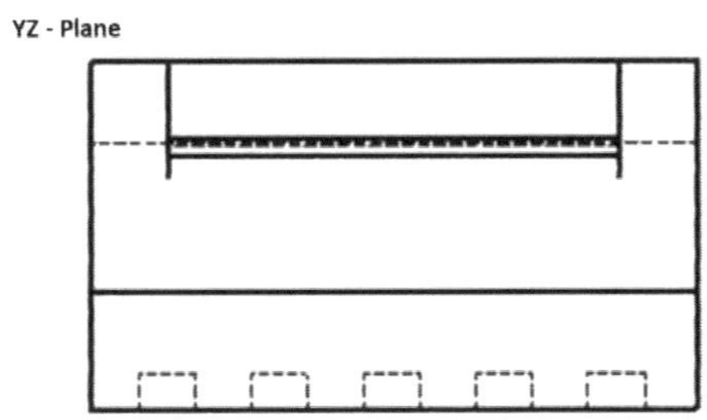

Figure 10. Definition of bottom roundness and corner roundness on yz-plane.

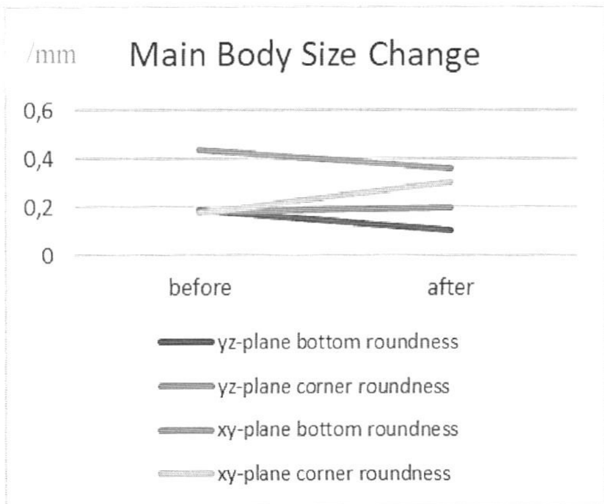

Figure 11. Roundness change in radius.

From the data obtained in Figure 11, the corners roundness radiuses are reduced while the bottom roundness radius is increased. This indicates that the shrinking move along the edge toward the center of geometry which resulting in the rounding on the base surface as the illustration in Figure 12.

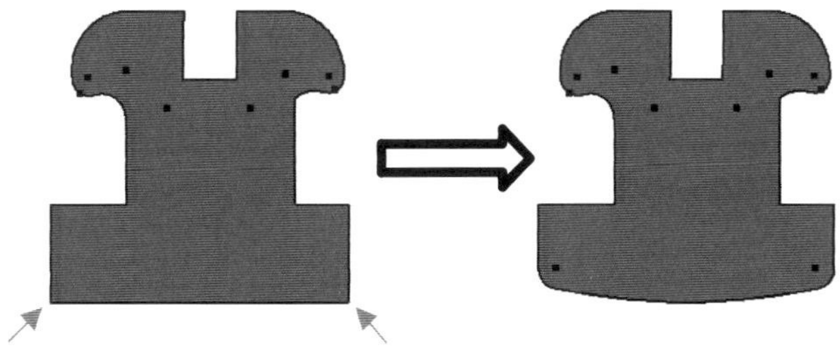

Figure 12. the red arrows represent the shrinking direction on the edge that resulting in the geometric on the right side.

3.3. Wing geometric transformation

The wings width is shrinking while the heights contracted. The wing shrinks along the x-axis for 6.92% and stretches along the y-axis for -6.11%. The shrinkage and contraction are not uniform on all the section of the wing geometry, so the leaning angle measurement is conducted to understanding the pattern. The details are presented in Table 2.

Table 2. Wing geometry shrinkage.

	Shringkage			
Dimension	Left Wing	Right Wing	Average	Percentage
X-Axis	0.034 mm	0.05 mm	0.04 mm	6.92%
Y-Axis	-0.022 mm	-0.07 mm	-0.05 mm	-6.11% (stretch)
Cross-section area	0.01 mm^2	0.019 mm^2	0.014 mm^2	3.4%

During the sintering, the hook slop moves leaning down. The angel measured by inspecting the direction of the arc in the end of the hook to the y-axis.

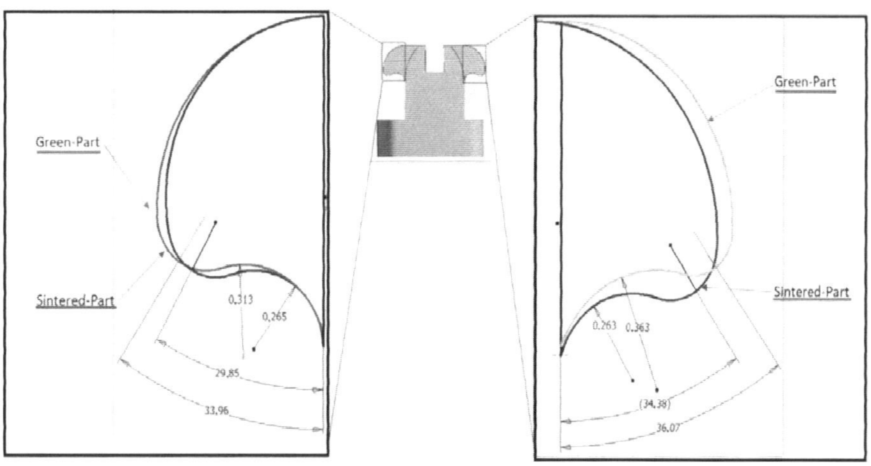

Figure 13. The angle of leaning difference before (green) and after the sintering (blue).

IOP Conf. Series: Materials Science and Engineering **432** (2018) 012002 doi:10.1088/1757-899X/432/1/012002

Based on Figure 13, the angle of the hook is reduced indicate that there is a partial rotation occuring on the wings. The average of the anglular decreased is 3°, but there is uncertainty if the leaning angle is a linear transformation since the geometry is complex. The radius of the lower arc of the wing is decreasing from 0.50 mm to 0.45 mm which means that the orthodontic rubber will safely hold even after the part was sintered.

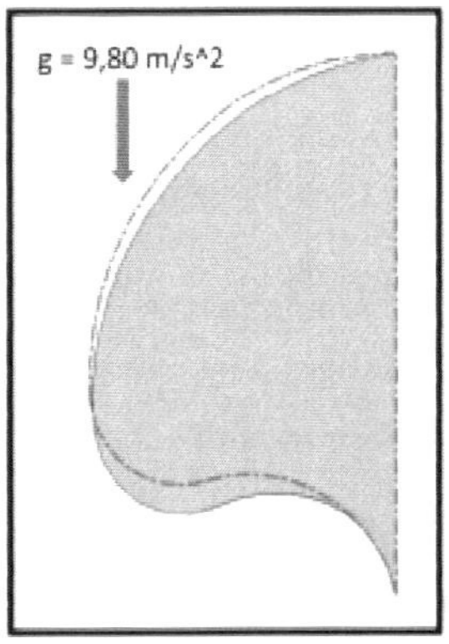

Figure 14. Gravitational influence simulation on the green-part with the gravitation of 9.8m/s^2 and fix constraint to the right side. The red dashed line represents the geometry before gravitation influence and the grey solid after the influence.

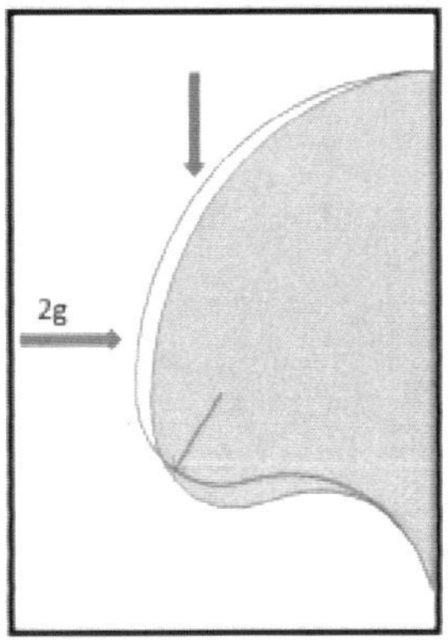

Figure 15. Gravitational influence and additional force to simulate a shrinkage movement. The red line represents the pre-gravitational influence geometry and the grey solid line represents the post-gravity influence geometry.

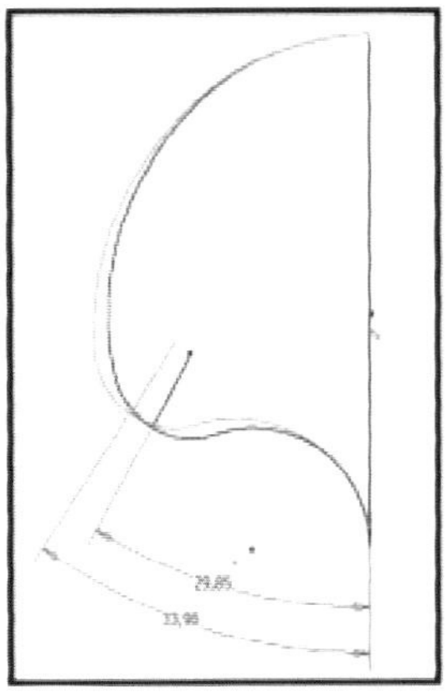

Figure 16. Left-wing geometric transformation before and after sintering.

The furnace heats the sample through radiation during the sintering process. The heat distribution imaging as shown in Figure 17, occurs during unsteady state at the uncertain time. Since the purpose only to estimate which part experience the geometrical transformation earlier, the exact time is considered to be unnecessary. Overheating will turn the phase of the metal particles from solid-state to

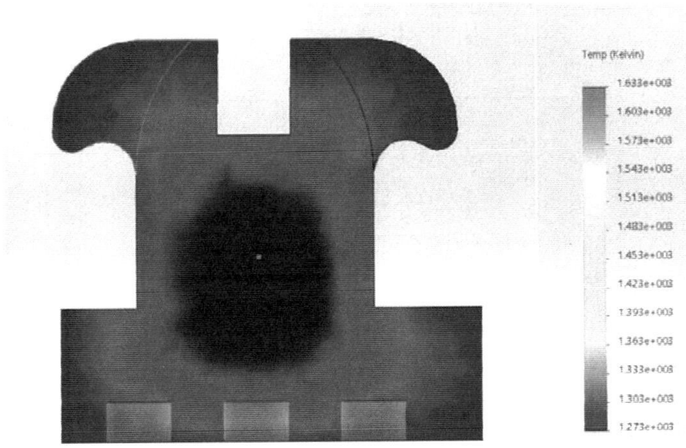

Figure 17. Heat distribution on the bracket cross-section during the sintering process. The simulation is conducted using *Solidworks* thermal analysis.

liquid-phase. The liquid-phase causes the bond between particles to loosen and easier to attract by the gravitational force [7]. The simulation is conducted to compare the geometrical change of the wing caused by sintering process with the geometry change caused by the influence of gravitation. This simulation uses CAD software for measuring bending disposition. By adjusting the parameter, the software will be able to simulate the gravitational influence accurately. Based on the heat distribution simulation on Figure 17, the inner side of the wing attached to the trunk receives less heat. Therefore, the inner side is assumed to not experience any geometrical changing and considered to be a fixed constraint. The force applied on the bending is equivalent to the wing mass that is distributed in all parts of the wing. The result of this simulation is shown in Figure 14. Based on the comparison of Figure 14 and Figure 16, the simulation has not been similar enough to the result of sintering. This means that there is another significant factor that must be calculated.

In the second simulation, the axial force along x-axis toward the trunk is added to simulate the shrinking force. The magnitude is based on the uniform force on the wing plane with weight value of 2 times gravitation. The particles densifying also depends on the heat distribution, therefore the shrinkage assumed to occur less on the inner (right) side of the wing. The result of the second simulation is shown in Figure 15.

As shown above, the second simulation result in Figure 15 is closer to the geometrical change of the sintering process in Figure 16. Based on the comparison, the research concludes that in this sintering process, both shrinkage and gravitation influence is significant to a hook geometric transformation of the wing section.

3.4. Wire slot geometric transformation

The gap of the upper cavity is narrowing by 0.02 mm and the angle is increased by 3° as shown in Figure 18. The u-shape cavity is opening but upper and lower gap is narrowing, since the angle is increasing, the lower gap is narrowing greater than the upper gap. However, the variables between the gap length and angle of before and after sintering are not correlated since the p-value of the gap (0.6855) and the angle (0.4734) is exceed the significant level (0.10).

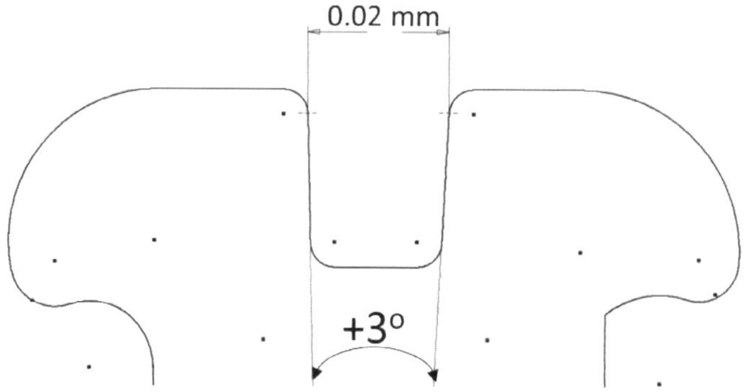

Figure 18. Wire slot measurement method.

4. Conclusion

During the sintering process, the geometries of the orthodontic bracket are transforming. The transformations are different in each section of the bracket, depend on the geometries. The longer geometry tends to shrink more than the shorter geometry based on the shrinking percentage. The edge and the corner side of the geometry tend to shrink more than the face side of the geometry, this phenomenon is related to the heat exposure direction. The gravitational attraction influences the geometric transformation, especially on the geometry with the less support. The wing geometry is leaning down, the base surface geometry is rounding. Since there is a shrinkage from the molded part into the sintered part, the design of the mold must be larger, with the part's face must be more concave to adapt the geometrical change caused by sintering that resulting the desired dimension product. Therefore, for the base section of the mold, the design of bottom surface should be concave with the radius of 43.51 mm along the width and 88.86 mm along the length. For the wing section of the mold, the width and the height respectively are set to be 7.43% and 6.51% bigger then the first design. For the overall body, the length, trunk width, base width, trunk height, base height, respectively are set to be 12.29%, 7.80%, 5.38%, 4.37%, 0.86% bigger than the first design.

Acknowledge

The authors would like to thank for a research grant of PUPT from Ministry of Research, Technology and Higher Education Indonesia.

References

[1] Frank C A and J Nikolai R 1980 *Am. J. Orthod.* **78** 593-609
[2] Tamizharasi S K 2010 *J. of Indian Academy of Dental Specialists* **1** 25-30
[3] Supriadi S *et al.* 2015 *Int. J. of Technology* **4** 613-621
[4] Supriadi S *et al.* 2016 *BICET2016* **1** 103-109
[5] Kumar V *et al.* 2016 *Powder metallurgy* **59** 170-180
[6] Ashari A 2017 *J. of Dentistry Indonesia* **24** 1-7
[7] German R 1985 Liquid Phase Sintering (New York: Plenum Press)

Ex-situ Li^+ doped into HKUST-1 and their application in hydrogen storage

D Ni'maturrohmah[1], W W Lestari[1*], H Suwarno[2]

[1] Research Group of Porous Materials for Sustainability, Department of Chemistry, Faculty of Mathematics and Natural Sciences, Sebelas Maret University, Ir Sutami St. No. 36A, Kentingan-Jebres, Surakarta, 57126, Central Java, Indonesia
[2] Center for Nuclear Fuel Technology, BATAN, Serpong, South Tangerang, 15314, Indonesia

witri@mipa.uns.ac.id

Abstract. The aim of this research is to know the influence of Li^+ doping into $[Cu_3(BTC)_2]$ (BTC: benzene 1,3,5-tricarboxylate) or called as Hongkong University of Technology-1 (HKUST-1) structure through ex-situ modification under the solvothermal condition and testing for hydrogen storage. The HKUST-1 structure remains stable after Li^+ doping with variation 3, 5, and 10 wt.%, according to XRD analysis, however surface area of HKUST-1 changes significantly up to 1040.96 m^2/g after doped with 5 wt.% of Li^+ and gave impact to the hydrogen uptake. At temperature 30°C (1.4 bar), hydrogen uptake of Li^+ @HKUST-1 reached up to more twice than HKUST-1 without Li^+. In addition, the presence of Li^+ also increases hydrogen uptake at high temperature (80°C, 1.4 bar). Hydrogen uptake enhancement after doping with Li^+ could be due to Li^+ supplies additional active site to bind hydrogen molecules.

1. Introduction

Energy is urgently needed for modern society development. One of the promising alternative energies is hydrogen. Hydrogen combustion does not lead to the pollution and results in high energy density, which is almost three times than a petroleum [1]. The barrier in the utilization of hydrogen as an energy carrier is the storage system. Conventionally hydrogen storage used cryogenic tanks, which is neither efficient nor effective if be applied for the vehicle. In 2015, a target set by Department of Energy (DOE) of United State for hydrogen storage system is 5.5 wt.% on gravimetric capacity and 40 g L^{-1} on volumetric capacity [2] under operational temperature (-40)-60 °C with pressure below 100 bar [1, 3].

Recently, several porous materials have been intensively explored, including Metal-Organic Framework (MOFs) as hydrogen storage materials. MOFs are hybrid materials of organic-inorganic, constructed from metal ions or metal-oxide clusters combined with organic ligands form crystalline porous coordination network [4]. MOFs can be used for hydrogen storage due to their porosity and active sites on both of ligand and metal [5]. MOFs store hydrogen effectively within reduced volume resulting in a high density [6].

HKUST-1 $[Cu_3(BTC)_2]$ is a well-known type of MOFs frequently investigated materials in hydrogen storage [4, 7, 8]. In order to enhance the hydrogen uptake, HKUST-1 could be doped with an electropositive metal. Based on the previous report [6], in-situ lithium and sodium ions doping into HKUST-1 can increase surface area and hydrogen uptake. Lithium ions act as additional binding sites to interact with hydrogen molecules [6]. In this present study, we investigate the influence of Li+ doping into HKUST-1 toward its structural changes and the hydrogen sorption properties. HKUST-1 was synthesized under the solvothermal condition and doped with a various amount of lithium ions (3, 5, 10 wt.%) through ex-situ modification in order to improve the hydrogen adsorption capacity of HKUST-1.

2. Materials and methods

2.1 Materials

H_3BTC (95%) and LiCl (8M) were purchased from Sigma Aldrich Germany, $Cu(NO_3)\cdot 3H_2O$ (98%) and ethanol (96%) were commercially provided by Merck Germany. Deionized water was obtained from Bratachem-Indonesia. All reagents were used as purchased without any further purification.

2.2 Synthesis of Li^+@HKUST-1

HKUST-1 was prepared via the solvothermal method based on the previous literature procedures [4]. HKUST-1 was doped with Li^+ through ex-situ modification from LiCl solution (8M) with a variation of 3, 5 and 10 wt.% Li^+ in 1 g of HKUST-1. The solution of LiCl (1.027; 1.71; 3.42 mL) was added with HKUST-1 (0.97; 0.95; 0.90 g) in 24 mL ethanol: deionized water (1:1). The mixture was placed into Teflon vessels covered with stainless-steel autoclaves and heated in an oven MEMMERT (type 67 UN30) at 393 K for 12 h. After cooled, the obtained blue precipitate was filtered, washed with ethanol (96%) two times and dried at ambient temperature for 16 hours.

2.3 Hydrogen storage measurement

General procedures: the material (0.5 g) was activated under vacuum condition at 473 K for 1 h. Hydrogen storage measurement was tested using Sievert system as published by Suwarno *et al.* [9] at 30 and 80 °C with pressure from 0.2 to 1.5 bar. At each pressure, the measurement was conducted for 15 minutes.

2.4 Material characterization

The X-ray powder diffraction (PXRD) data as quantitative analysis were obtained on Rigaku Miniflex 600 Benchtop, using Cu as anode tube, Kα radiation (λ = 1.5406 Å), with the current and voltage under 30 mA and 40 kV (2θ = 5-50°) at a scan rate of 1°/min. The nitrogen sorption isotherms at 77 K were obtained from a Quadrasorp Evo (Quantachrome instruments). The samples were activated at 120 °C for 24 h prior measurement. The morphology and elemental compositions were observed using SEM-EDX FEI Inspect S50.

3. Results and discussion

X-Ray diffractogram of HKUST-1 compared with Li^+ @HKUST-1 are performed in Figure 1. Three basic peaks at 2θ: 6.73° (2 0 0); 9.47° (2 2 0) and 11.65° (2 2 2) are confirmed through similarities with the simulated pattern (CCDC: 112954). The diffraction patterns of HKUST-1 are remained unchanged after doped with Li^+. It may be considered that the presence of Li^+ does not chemically influence the HKUST-1 structure. Moreover, the absence of Li^+ peaks in the diffractogram showed that HKUST-1 atoms are not chemically bonded with Li^+ ions [6]. In addition, the presence of some impurities possibility such as Cu_2O (2θ: 36.43°) and CuO (2θ: 35.5° and 38.7°) are not observed in the obtained materials [10].

The determination of pore type, surface area, and pore volume distribution were performed using N_2 sorption isotherm measurement at 77 K. Figure 2 shows that before and after doping with Li^+, the materials could be classified as microporous materials. Doped 5 wt.% of Li^+ into HKUST-1 increases surface area from 713.28 up to 1040.96 m^2/g with the smaller pore volume up to 0.048 cc/g which assumed that Li^+ ions filled the HKUST-1 pores, as shown in Table 1.

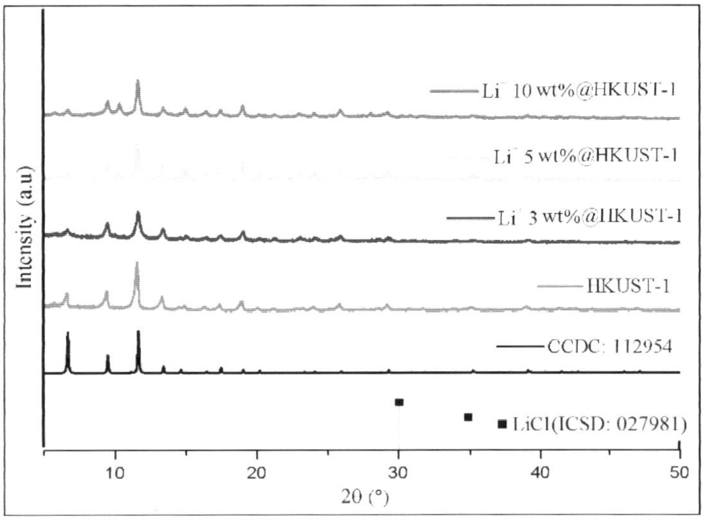

Figure 1. Diffractogram of doped HKUST-1 with Li^+ and undoped HKUST-1 compared to simulated pattern #CCDC 112954.

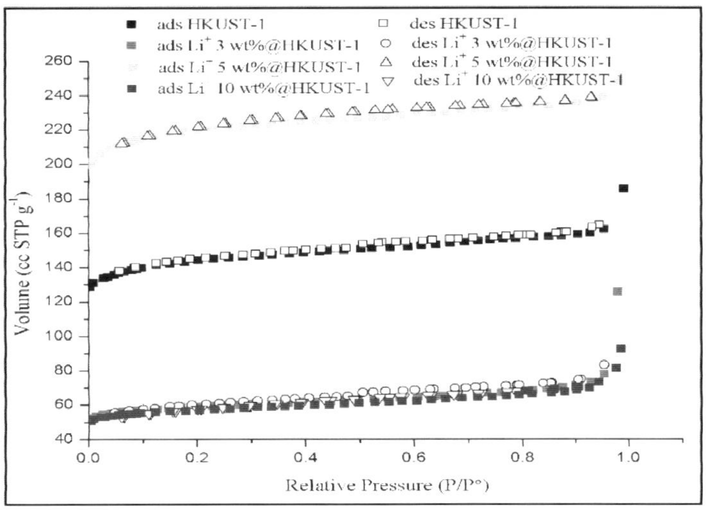

Figure 2. N_2 adsorption-desorption isotherm at 77 K of doped HKUST-1 with Li^+ and undoped HKUST-1.

Table 1. Pore volume and surface area of doped HKUST-1 with Li^+ and undoped HKUST-1.

Sample	Langmuir surface area (m^2/g)	Pore volume (cc/g)
HKUST-1	713.28	0.068
Li^+ 3 wt.%@ HKUST-1	334.49	0.108
Li^+ 5 wt.%@ HKUST-1	1040.96	0.048
Li^+ 10 wt.%@ HKUST-1	314.84	0.059

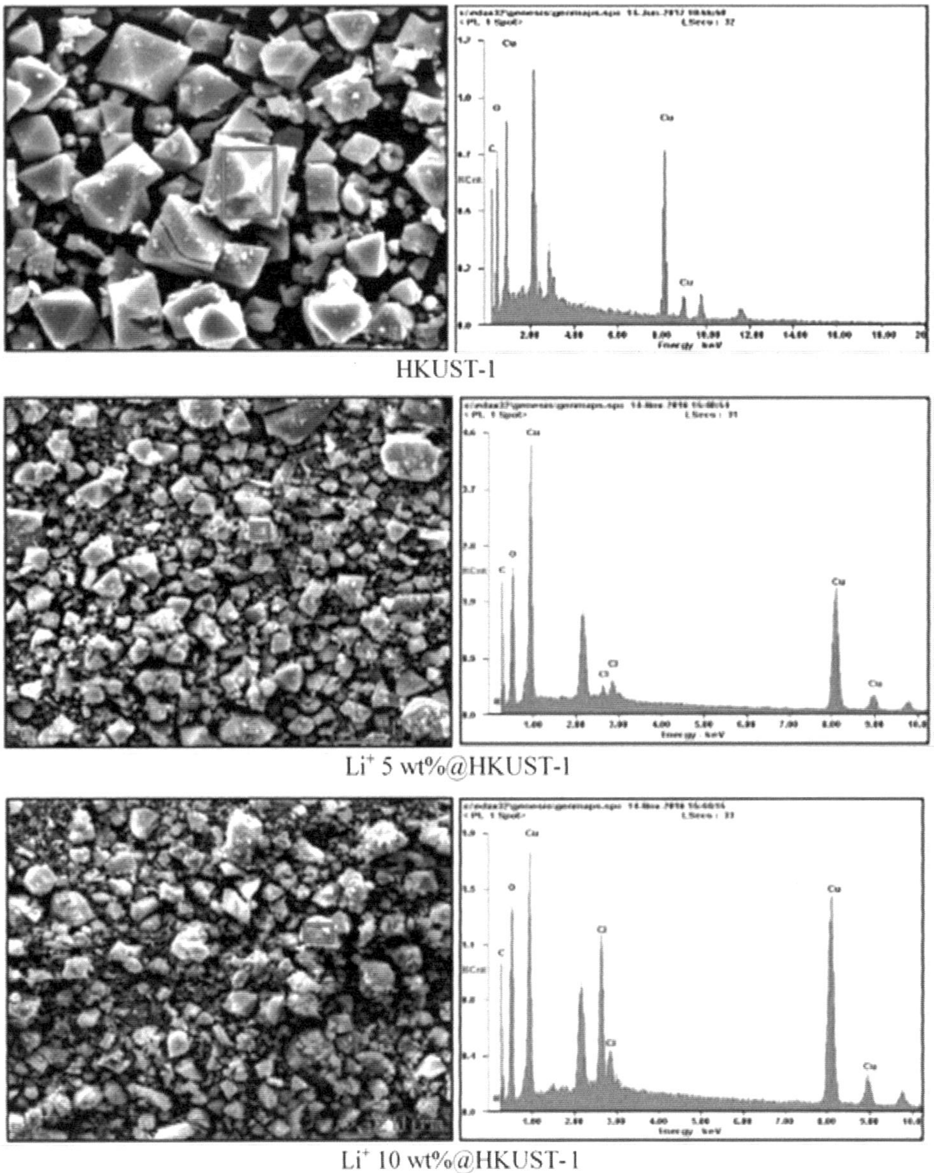

Figure 3. SEM-EDX Images of doped HKUST-1 with Li^+ and undoped HKUST-1.

Table 2. Elements composition of doped HKUST-1 with Li^+ and undoped HKUST-1.

The material	Element (wt %)			
	C	O	Cu	Cl
HKUST-1	33.68	19.40	46.92	-
Li^+ 5 wt.%@HKUST-1	30.13	21.19	47.69	0.99
Li^+ 10 wt.%@HKUST-1	25.79	19.04	49.25	5.91

EDX spectrum and HKUST-1 and Li^+@HKUST-1 morphology were shown in Figure 3. The basic shape of HKUST-1 (in octahedral) remained similar after doped with Li^+, however, the morphology is changed due to stirring process during the treatment which may break the HKUST-1 crystal. In addition, observed Cl elements in Table 2 is an indication that Li^+ successfully doped into HKUST-1 since Li^+ itself could not be observed by EDX analysis. Cl^- ions size are very small, 0.18 nm [11] which may diffuse to the HKUST-1 pores. The presence of Li^+ and Cl^- in HKUST-1 also changed the element composition of HKUST-1 (C, O, and Cu) as listed in Table 2.

The hydrogen storage capacityies at temperature 30 and 80 °C is shown in Figure 4. Hydrogen uptake enhancement of HKUST-1 could be due to Li^+ ions supply additional active site to bind hydrogen molecules. At temperature 30 °C (1.4 bar), the maximum hydrogen uptake is 0.24 wt.%, obtained from Li^+ 5 wt.%@HKUST-1. Additionally, the maximum adsorption capacity at 80°C (1.4 bar) is up to 0.18 wt.% obtained from Li^+ 10 wt.%@HKUST-1. According to the previous study conducted by Anbia and Faryadas [6], in-situ doping about 16.9 wt% of Li^+ gave hydrogen storage capacity at ambient temperature (14 bar) up to 8 mmol g^{-1}. The result indicates that electropositive metal ion embedding to HKUST-1 increases hydrogen uptake due to the increasing surface area and additional sites provided by the metal cations in the HKUST-1 environment. In overall, hydrogen storage capacity at temperature 80 °C is lower than temperature 30 °C. According to Lin *et al.* [4], interaction between hydrogen molecules and materials would be more effective at low temperature.

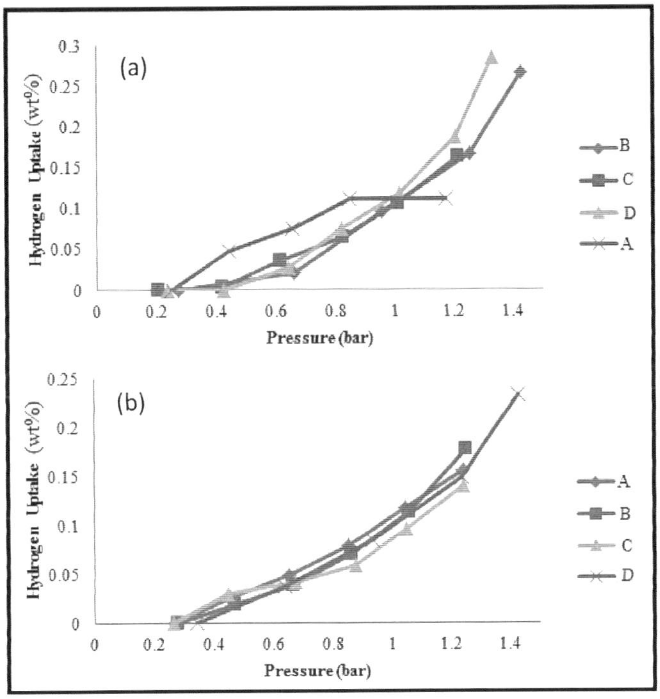

Figure 4. Hydrogen storage capacity of undoped HKUST-1 (A) and doped HKUST-1 with Li^+ (B= 3 wt.%; C=5 wt.%; D=10 wt.%) at temperature (a) 30 °C (b) 80 °C.

4. Conclusion

Ex-situ doping of Li^+ into HKUST-1 was successfully prepared and tested for hydrogen storage. The presence of Li^+ does not chemically change the HKUST-1 structure but enhances hydrogen storage capacity up to 0.24 wt.% (for 5 wt.% of Li^+) at 30 °C 1.4 bar. Due to its high surface area and porosity, simple synthesis, and possible to store hydrogen, Li^+@HKUST-1 can be a good candidate material for hydrogen storage. For the next study, the hydrogen uptake measurement should be conducted with longer time and optimized under higher pressure.

Acknowledgments
The authors gratefully acknowledge to LPPM Sebelas Maret University through the Mandatory Grant 2018 for funding.

References

[1] Xiang Z Cao D 2013 *Journal of Mater. Chem. A* **1** 2691-718
[2] Basdogan Y Keskin S 2015 *Cryst. Eng. Comm.* **17** 261-75
[3] Zhao D Yuan D Zhou H C 2008 *Energy & Environmental Sci.* **1** 222-35
[4] Lin K S Adhikari A K Ku C N Chiang C L Kuo H 2012 *Int. J. of Hydro. Energy* **37** 13865-71
[5] Lee J Li J Jagiello J 2005 *J. of Solid State Chem.* **178** 2527-2532
[6] Anbia M Faryadras M 2015 *J. of Nanostructure in Chem.* **5** 357-64
[7] Prestipino C Regli L Vitillo J G Bonino F Damin A Lamberti C Zecchina A Solari P L Kongshaug K O Bordiga S 2006 *Chem. of Mater.* **18** 1337-46
[8] Wong-Foy A G Matzger A J Yaghi O M 2006 *J. of the American Chem. Soc.* **128** 3494-5
[9] Suwarno H Wiryolukito S Dewa M D K 2015 *J. of Nuclear Mater. Tech.* **10** 96-103
[10] Maiti S Pramanik A Manju U Mahanty S 2016 *Microporous and Mesoporous Mater.* **226** 353-9
[11] Sitharaman B Kissell K R Hartman K B Tran L A Baikalov A Rusakova I Sun Y Khant H A Ludtke S J Chiu W Laus S 2005 *Chem. Communication* **31** 3915-7

Application of nanocatalyst iron oxide (Fe_2O_3) to reduce exhaust emissions (CO and HC)

P Puspitasari[1,2*], C Yazirin[3], L A Bachtiar[4], J W Dika[3], SM S N S Zaharis[5]

[1] Mechanical Engineering Department, Engineering Faculty, Malang State University, Semarang street 5, Malang, East Java, Indonesia
[2] Center of Nano Research and Advanced Materials, Malang State University, Semarang street 5, Malang, East Java, Indonesia
[3] Master Program of Mechanical Engineering, Engineering Faculty, Malang State University, Semarang street 5, Malang, East Java, Indonesia
[4] Bachelor Program, Mechanical Engineering Department, Malang State University, Semarang street 5, Malang, East Java, Indonesia
[5] Industrial Chemical Technology Programme, Faculty of Science and Technology, Universiti Sains Islam Malaysia, Bandar Baru Nilai, 71800 Nilai, Negeri Sembilan

poppy@um.ac.id

Abstract. Fe_2O_3 has been the interest of many studies due to its applications in various scientific and industrial fields including in environment, corrosion, soil science, and exhaust emissions. Fe_2O_3 is a chemical mixture that can be applied as a catalyst to reduce exhaust emissions because it is capable of binding CO and HC contained in motorcycle exhaust gases The growth of motor vehicles results in air pollution and an increase in exhaust emissions consisting of 70.50% CO and 18.34% HC. One effort to reduce exhaust emission levels is by applying a catalytic converter to the exhaust system. This research was an experimental research aiming to investigate the difference in emission level of CO and HC between a motorcycle with and without a catalytic converter. The types of fuel used were RON 88, RON 90, RON 92, and RON 98. The results showed that the application of Fe2O3 as a catalyst could reduce the level of CO significantly in RON 88, i.e. 91.43% CO, and the level of HC in RON 88 and RON 92, i.e. 53.27% and 91.07% HC respectively.

1. Introduction

Air pollution is one form of pollution with an ever-increasing rate due to motor vehicle emissions and exhaust gas that can reduce the quality of air that causes global warming. The traffic growth also causes air pollution which contributes to the deterioration of the air quality. Technological developments in various fields, particularly in the field of land transportation, have resulted in a massive increase in the number of vehicles with different types and brands. The dramatic growth of motor vehicles triggers the occurrence of air pollution and increases exhaust emissions. Emissions discharged from motor vehicles consist of 70.50% CO, 18.34% HC, 8.89% HC, 0.88% SOx, and 1.33% other particles [1]. CO and HC are highly hazardous pollutants from motor vehicles because they can negatively affect human health [2].

Motor vehicles that use fuel oil containing lead are a significant contributor to air pollution. This condition is exacerbated by the depletion of fuel oil that is hitting the country [3]. Compelled by these phenomena, many people are encouraged to look for alternative fuels that are more economical, energy efficient and reduce air contamination by exhaust emissions. Several types of alternative fuels are Premium, Pertalite, Pertamax and Pertamax Turbo. Premium is a fuel with RON 88. Pertalite is a fuel oil from Pertamina with RON 90. Pertalite has several advantages compared to RON 88, one of which is creating a better combustion in the vehicle engine. Pertamax, a gasoline product with RON 92, is superior compared to the formers. The last is Pertamax Turbo which is a fuel with RON 98 and the best among all.

2. Experimental method

The material used is iron oxide (Fe_2O_3) because of its many benefits one of them is its ability to reduce exhaust emissions. The present study included phase characterization and morphology of iron oxide oxide (Fe_2O_3) nanopowder. The iron oxide (Fe_2O_3) data sintered at various durations (1, 2 and 3 hours) were obtained from the characterization tools XRD and SEM. The apparatus consisted of a gas analyzer, a tachometer, a fan, a motorcycle with a four-stroke engine (Honda MegaPro 4, 2013), a spray bottle, an adsorption tube, a spoon, a mortar, a beaker glass, a measuring cup, and a dropper. Furthermore, the specification of Fe2O3 nanoparticles was treated as the controlled variable, the activation treatment on Fe2O3 nanoparticles as the manipulated variable, and the concentration of CO and HC as the response variable. The design of the tool used in this study is as follows.

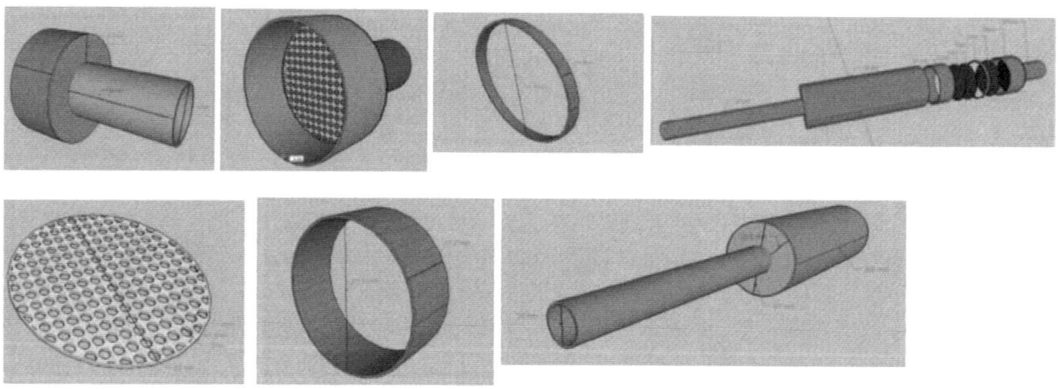

Figure 1. Design of the adsorption tube.

The measurement of CO and HC emissions from activated carbon was carried out in stages, namely (1) igniting the motorcycle (Honda Mega Pro, four-stroke engine, 2013); (2) inserting the gas analyzer probe into the exhaust to determine the level of CO and HC prior to treatment; (3) installing an adsorption tube with a tube length of 17 cm and a diameter of 8 cm containing 3 variations of nanoparticles on the exhaust, then inserting the gas analyzer probe into the adsorption tube to determine the post-treatment levels of CO and HC with variations in RPM; and (4) analyzing the measurement results of CO and HC emissions shown on the gas analyzer screen [4].

3. Result and discussion

3.1 Phase of characterization

The result of XRD testing of nanopowder iron oxide (Fe_2O_3) with 1 hour sintering process variation, the highest peak was positioned 33,1543° 2θ with intensity 66,46 cts and FWHM 0,3936° 2θ. In the variation of 2 hours peak sintering process is positioned 33,2733° 2θ with intensity of 75,35 cts and FWHM 0,3542° 2θ and variation of 3 hours peak sintering process is positioned 33,1949° 2θ with intensity 78,56 cts and FWHM 0.3936 ° 2θ. There is a slight shift in the highest peak between iron oxide (Fe_2O_3) 1 hour, 2 hours, and 3 hours due to increased intensity of iron oxide (Fe_2O_3) during sintering. From the XRD pattern below shows that the crystal structure is rhombohedral. This is evident from the highest peak at the 2nd angle of theta pattern above [5-7].

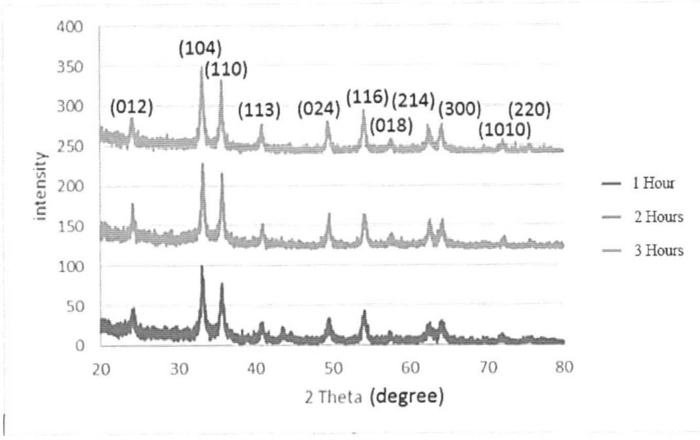

Figure 2. Phase identification of iron oxide (Fe_2O_3) with different sintering times (1, 2, and 3 hours).

Table 1. Intensity, FWHM, d-spacing and crystallite size of Fe_2O_3.

Sample Material iron oxide (Fe_2O_3)	X-Ray Diffraction (correspond to [222] peak)			
	Intensity (counts)	FWHM (rad)	d-spacing (Å)	Crystallite Size (nm)
iron oxide (Fe_2O_3) 1 hours	66.46	6.8723	2.70214	21.05
iron oxide (Fe_2O_3) 2 hours	75.35	6.1844	2.69275	23.40
iron oxide (Fe_2O_3) 3 hours	78.56	6.8723	2.69893	21.05

Seen from the above table it is concluded that the smallest grains are in iron oxide (Fe_2O_3) 1 hour sintering process with a crystal diameter of 21.05 nm and the intensity is 66.46 cts. Although it has the same crystal diameter, but one hour sintering process has a lower intensity than the three hours sintering process because X-rays are irradiated at the time of XRD experiment directly on the item being tested [6, 8, 9].

3.2 Morphological characterization

IOP Conf. Series: Materials Science and Engineering **432** (2018) 012004 doi:10.1088/1757-899X/432/1/012004

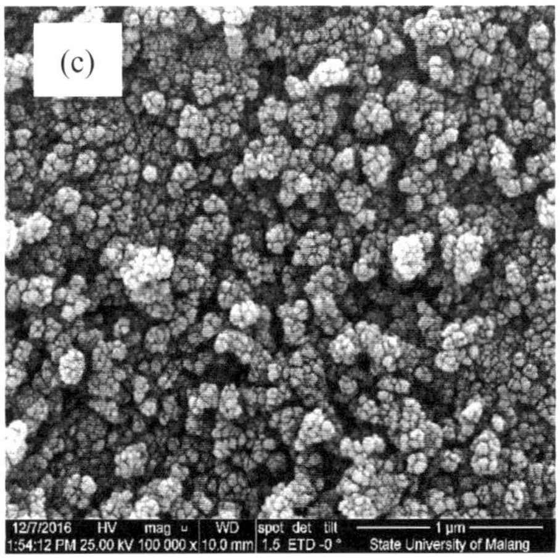

Figure 3. (a) Morphology of iron oxide (Fe_2O_3) sintered for 1 hour (b) morphology iron oxide (Fe_2O_3) sintered for 2 hours (c) morphology iron oxide (Fe_2O_3) sintered for 3 hours at a magnification of 100.000.

Scanning electron microscope (SEM) test aims to analyze and compare the morphology of iron oxide material (Fe_2O_3). In the result of identification of iron oxide material (Fe_2O_3) with variation of sintering time of 1 hour, 2 hours, and 3 hours showed at magnification 100 K all material has a homogeneous nanostructure in the form of nanosphere [10, 11].

3.3 Result of measurements with and without using a catalytic converter

Table 2. Result of research on a motorcycle without a catalytic converter.

No	Rotation	CO (%)				HC (ppm)			
		RON 88	RON 90	RON 92	RON 98	RON 88	RON 90	RON 92	RON 98
1	1500	3,12	1,66	1,47	0,92	141	76	100	19
2	1750	3,2	1,71	1,58	0,97	135	67	95	18
3	2000	3,38	1,73	2,09	1,37	121	61	94	15
4	2250	3,4	1,76	2,16	1,36	100	57	94	11
5	2500	3,45	1,95	2,2	1,48	98	51	77	10
6	2750	3,5	2,36	2,24	1,53	94	46	77	8
7	3000	3,53	2,36	2,48	1,6	92	44	72	6
8	3250	3,38	2,31	2,31	1,74	87	39	66	3
9	3500	3,22	2,31	2,26	1,79	77	38	64	3
10	3750	3,2	2,34	2,15	1,66	73	32	61	0
11	4000	3,07	2,38	2,12	1,5	66	30	55	0
12	4250	2,82	1,98	2,02	1,41	59	18	48	0
13	4500	2,68	1,97	1,61	1,38	51	17	39	0
14	4750	2,98	1,99	1,18	1,15	51	16	32	0
15	5000	3,34	1,94	0,9	0,69	57	14	31	0
16	5250	3,43	1,7	0,41	0,6	56	10	18	0
17	5500	3,67	1,2	0,27	0,59	54	10	14	0
18	5750	3,75	0,98	0,54	0,53	56	12	18	2
19	6000	3,86	0,97	0,69	0,51	60	12	20	4

Table 2 showed the level of CO and HC at engine speed of 1500 rpm to 6000 rpm with 250 rpm increments using fuel ron 88, ron 90, ron 92, and ron 98 without catalytic conventers. The condition showed the factory standard condition of Honda MegaPro.

Table 3. Result of research on a motorcycle with a catalytic converter.

No	Rotation	CO (%)				HC (ppm)			
		RON 88	RON 90	RON 92	RON 98	RON 88	RON 90	RON 92	RON 98
1	1500	3,27	3,74	4,86	3,16	108	175	150	78
2	1750	3,47	3,85	4,76	3,18	127	173	138	72
3	2000	3,76	4,01	4,73	3,38	136	186	134	70
4	2250	3,98	4,21	4,69	3,49	123	184	105	70
5	2500	3,87	4,57	4,50	3,52	95	143	98	68
6	2750	3,78	4,83	4,57	3,68	76	122	87	54
7	3000	3,69	4,91	4,58	3,37	46	119	72	45
8	3250	3,59	5,01	4,28	3,58	28	111	48	38
9	3500	3,44	4,75	4,16	3,36	22	105	37	28
10	3750	3,31	3,94	3,83	3,27	18	87	28	20
11	4000	2,83	3,87	3,65	3,20	11	65	21	11
12	4250	2,73	3,35	3,50	3,12	8	59	18	9
13	4500	2,68	2,80	3,48	3,05	6	43	12	5
14	4750	2,55	2,54	3,47	2,87	6	38	11	3
15	5000	2,18	2,11	3,41	2,54	3	23	8	0
16	5250	2,09	2,07	3,39	2,37	1	15	7	0
17	5500	2,10	2,06	3,28	2,18	0	8	1	0
18	5750	2,11	2,11	2,89	1,94	0	4	1	0
19	6000	2,15	2,17	2,72	1,97	0	5	3	1

Table 3 showed the level of CO and HC at engine speed of 1500 rpm to 6000 rpm with 250 rpm increments using fuel of RON 88, RON 90, RON 92, and RON 98 with catalytic converter Fe_2O_3. This condition as comparator the result of CO and HC level with and without the use of catalytic converter.

3.4 Measurement result of carbon monoxide (CO) emissions in RON 88

Under standard conditions (without the addition of Fe_2O_3 as a catalyst), the emission of CO in RON 88 had an average of 3.3147 %vol. With the addition of Fe_2O_3 as a catalyst, the emission of the exhaust gas had an average of 3.0305 %vol. The results of both treatments also showed that the Sig. level was 0.000, indicating that $\alpha < 0.05$. In other words, there was a significant difference in the average emission of both treatments. Data on the concentration of CO in RON 88 are presented in Figure 4.

As shown in Figure 4, the treatment without using a catalytic converter produced a CO level of 3.12 %vol at 1500 RPM, and it increased to 3.53 %vol at 3000 RPM. Then it dropped to 2.68 %vol at 4500 RPM and rose again to 3.86 %vol at 6000 RPM. Upon the application of Fe_2O_3 as a catalyst, the CO concentration was 3.27 %vol at 1500 RPM, peaked at 2250 RPM (3.98 %vol), and then decreased dramatically at 3750 RPM (2.09 %vol). At 6000 RPM, it gradually increased again to 2.15 %vol. Therefore, the authors determined how much influence of a catalytic converter applied to the four-stroke gasoline motorcycle engine by giving a percentage of the difference in CO level between the engine using and without using a catalytic converter at the same RPM. The authors took the sample with 5250 RPM because it had the lowest decrease in CO level upon the application of a catalytic converter on Honda Mega Pro motorcycle. The percentage difference is 0.0343ppm - 0.0209%ppm = 0.0134%ppm, or 60.93% if converted into a percentage (With Catalytic Converter/Without Catalytic Converter x 100 = 60.93%) [12, 13].

IOP Conf. Series: Materials Science and Engineering **432** (2018) 012004 doi:10.1088/1757-899X/432/1/012004

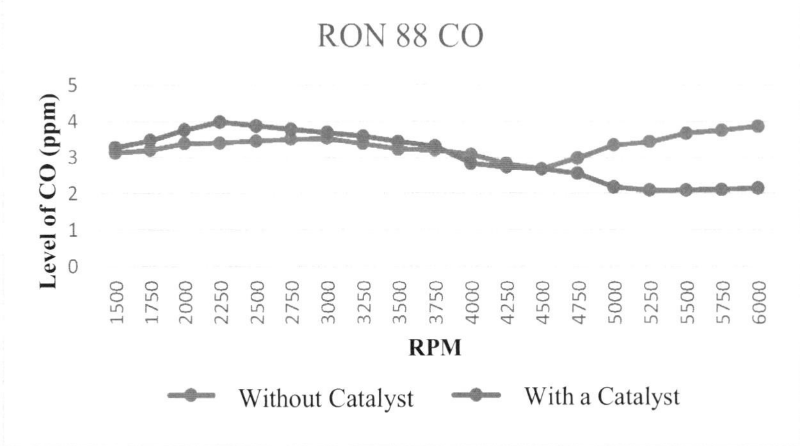

Figure 4. Comparison graph of CO level with and without a catalytic converter in RON 88.

3.5 Measurement result of hydrocarbon (HC) emissions in RON 88

Under standard conditions (without the addition of Fe_2O_3 as a catalyst), the emission of HC in RON 88 had an average of 80.4211 ppm. With the addition of Fe_2O_3 as a catalyst, the emission of the exhaust gas had an average of 42.8421 ppm. The results of both treatments also showed that the Sig. level was 0.004, indicating that there was a significant difference in the average emission of both treatments. Data on the concentration of HC in RON 88 are presented in Figure 5.

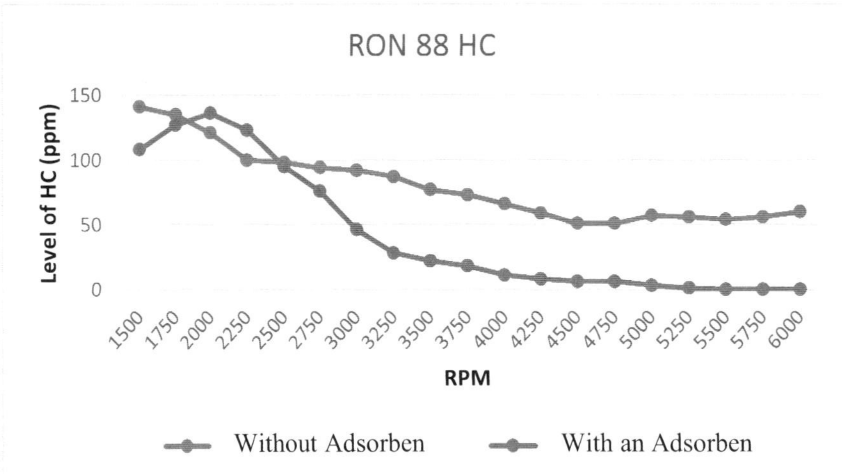

Figure 5. Comparison graph of HC level with and without a catalytic converter in RON 88.

Based on the above diagram, upon the treatment without a catalytic converter, the recorded level of HC level at 1500 RPM was 141 ppm and gradually decreased to 51 ppm at 4500 RPM. In the application of Fe_2O_3 as a catalyst, the level of HC at 1500 RPM was 108 ppm, then increased to exceed the standard of RON 88 without using adsorbent at 2000 RPM i.e. 136 ppm, and gradually decreased to 0 Ppm from 5500 RPM to 6000 RPM. The authors took the sample with 5250 RPM because it had the lowest decrease in CO level upon the application of a catalytic converter on Honda Mega Pro motorcycle. The percentage difference is 56%ppm - 1%ppm = 55%ppm, or 1.79%, if converted into a percentage (With Catalytic

Converter / Without Catalytic Converter x 100 = 1.79%). This result indicates that a catalytic converter is efficient to bind vehicle exhaust.

3.6 Measurement result of hydrocarbon (HC) emissions in RON 92

Under standard conditions (without the addition of Fe_2O_3 as a catalyst), the emission of HC in RON 92 had an average of 56.5789 ppm. With the addition of Fe_2O_3 as a catalyst, the emission of the exhaust gas had an average of 51.52363 ppm. The results of both treatments showed that the Sig. level was 0.004, indicating that there was a significant difference in the average emission of both treatments. Data on the concentration of HC in RON 92 are presented in Figure 6.

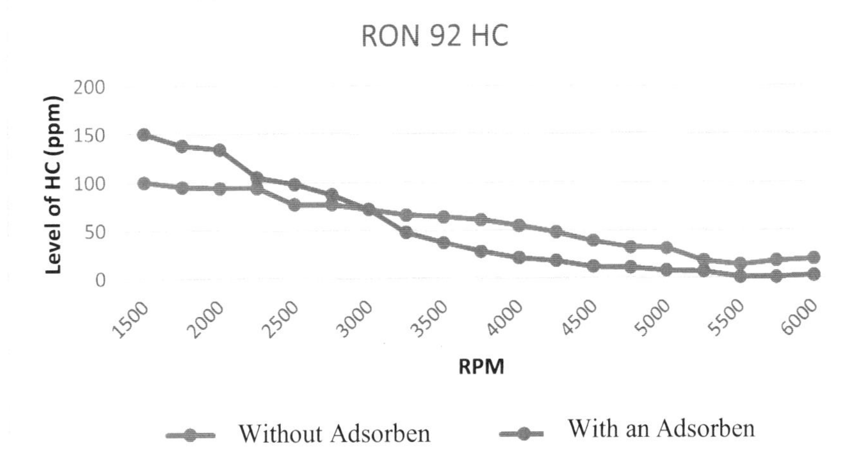

Figure 6. Comparison graph of HC level with and without a catalytic converter in RON 92.

Based on the above diagram, upon the treatment without a catalytic converter, the recorded level of HC level at 1500 RPM was 100 ppm, then gradually decreased to 14 ppm at 5500 RPM, and rose to 20 ppm at 6000 RPM. In the application of Fe_2O_3 as a catalyst, the level of HC at 1500 RPM was 150 ppm, then gradually decreased to 1 ppm at 5500 RPM, and increased to 3 Ppm at 6000 RPM. The lines representing the concentration of HC in both treatments intersected at a high rotation, i.e. 3000 RPM.

From the two treatments, there was a tendency of decreasing HC level, which was proportional to the increase in engine speed. This was due to the low RPM (1500 RPM) and thus the low temperature around the walls of the combustion chamber. Also, the fuel combustion process was not complete because the amount of air used in combustion did not meet the ideal combustion ratio, hence high level of HC. The authors took the sample with 5500 RPM because it had the lowest decrease in HC level upon the application of a catalytic converter on Honda Mega Pro motorcycle. The percentage difference is 14%ppm – 1%ppm = 13%ppm, or 7.14%, if converted into a percentage (With Catalytic Converter / Without Catalytic Converter x 100 = 7.14%). This result indicates that a catalytic converter is efficient to bind vehicle exhaust [12-14].

4. Conclusion

Based on the research and discussion on the application of ferro oxide (Fe_2O_3) as a catalyst in RON 88, RON 90, RON 92, and RON 98 fuel to reduce exhaust emissions (CO and HC), we conclude that:

1. Applying Fe_2O_3 as a catalyst is only effective to reduce CO emissions from motor vehicles using RON 88 fuel, but is not effective for those vehicles using RON 90, RON 92, and RON 98 fuel.
2. Applying Fe_2O_3 as a catalyst is significantly effective to reduce CO emissions from motor vehicles using RON 88 fuel, RON 90, and RON 92, but is not effective for those vehicles using RON 98 fuel.

References

[1] Wardhana W A 2001 *Imp. Of Env. Pollut.* Yogyakarta ANDI (In Indonesia)
[2] Kusuma W G 2003 Flue Gas Emm. Reduc. in Motors, Cars, Outboard Motors and Moving Vehicle Machine *J. of Tech.* (In Indonesia)
[3] Iskandar A 2006 *Estimated Use of Fuel for Land Transport Agency of Transport. Agency of the Ministry of Trasport. Of the Republic of Indonesia* Jakarta (In Indonesia)
[4] Ismiyati D Marlita and Saidah D 2014 *J. Manag. Transp. Logistics* **1(3)** 241–8 (In Indonesia)
[5] Lin M Tng L Lim T Choo M Zhang J Tan H R and Bai S 2014 *J. Phys. Chem. C* **118(20)** 10903–10
[6] Wang F Qin X F Meng Y F Guo Z L Yang L X and Ming Y F 2013 *Mater. Sci. Semicond. Process* **16(3)** 802–6
[7] Cuong N D Khieu D Q Hoa T T Quang D T Viet P H Lam T D Hoa N D and Van H 2015 *Mater. Res. Bull.* **68** 302–7
[8] Puspitasari Y Poppy N 2011 *Appl. Sci.* **11(7)** 7
[9] Yahya N B Puspitasari P Koziol K K K and Pavia G 2011 *J. Nano Res.* **16** 119–30
[10] Mohammadikish M 2014 *Ceram. Int.* **40(1)** 1351–8
[11] Babay S Mhiri T and Toumi M 2015 *J. Mol. Struct.* **1085** 286–93
[12] Jayanti I S Eka N and Hakam M 2006 *J. of Mechan. Engin. PPNS* (In Indonesia)
[13] Syahrani A 2006 *J. SMARTEK* **4** 260–6 (In Indonesia)
[14] Rami N 2014 *8th Int. Conf. Mater. Sci.* CSM8-ISM5 55 373–82

Nanocatalyst manganese oxide (MnO) for exhaust emissions (CO and HC) reduction

P Puspitasari[1,2*], M I N Sasongko[3], G A Kurniawan[3], Andoko[1] and D I Tsamroh[3]

[1] Mechanical Engineering Department, Engineering Faculty, Malang State University, Semarang Street 5, Malang, East Java, Indonesia
[2] Center of Nano Research and Advanced Materials, Malang State University, Semarang Street 5, Malang, East Java, Indonesia
[3] Master Program, Postgraduate Program, Malang State University, Semarang Street 5, Malang, East Java, Indonesia

poppy@um.ac.id

Abstract The growing number of motor vehicles, particularly the exhaust gas as a result of incomplete combustion of fuels, has a very adverse impact on the environment. One effort to reduce exhaust emission levels is by applying a catalytic converter to the exhaust system. MnO is a chemical mixture that can be applied to a catalytic converter to reduce exhaust emissions because it is capable of binding CO and HC contained in motorcycle exhaust gases. This research was an experimental research aiming to investigate the difference in emission level of CO and HC between a motorcycle with and without a catalytic converter. The types of fuel used were RON 88 and RON 92. The results showed that the application of MnO as a catalyst could reduce the level of CO significantly in RON 88, i.e. 24.36%, and the level of HC in RON 88 and RON 92, i.e. 82.08% and 38.14 % respectively.

1. Introduction

Technological developments in various fields, particularly in the field of land transportation, have resulted in a massive increase in the number of vehicles with different types and brands. Emissions discharged from motor vehicles consist of 70.50% CO, 18.34% HC, 8.89% NOx, 0.88% SOx, and 1.33% other particles [1]. The combustion process converts fuel energy into motion. Energy transformation is derived from the combustion of fuel. According to the theory of complete combustion, a combustion reaction is represented as follows:

$$C8H18 + 12{,}5O2 \rightarrow 8CO2 + 9H2O + Energy \quad (1)$$

In real conditions, air contains ± 21% O2 and ± 79% N2. On top of that, complete combustion can only occur in a laboratory. Thus, in practice, the following combustion reaction takes place:

$$C8H18 + 12.5\,(O2 + 79/21N2) \rightarrow 8CO2 + 9H2O + 2.5\,(79/21N2) + Energy \quad (2)$$

On the basis of the equation (2), one mol of fuel requires 12.5 mol of air to produce 8 mol of CO2, 9 mol of H2O, 12.5 (79/21) mol of N2 and energy [2].
Emissions generated from vehicle exhaust are the result of incomplete combustion. The chemical reaction of incomplete combustion in the engine combustion chamber is as follows:

$$C8H18 + O2 + N2 \rightarrow CO + CO2 + HC + NOx + SO2 + Pb + O2 + \text{Other particles} \quad (3)$$

MnO used as a catalytic converter to reduce motor vehicle emissions reacts directly with the exhaust gases in the exhaust system and has the following chemical reaction:

$$MnO2 + CO \rightarrow MnO + CO2 \text{ [3]} \quad (4)$$

The above reaction shows that MnO is able to reduce exhaust emissions generated by motor vehicles. In some studies, a catalytic converter is attached to a vehicle's exhaust tip. In general, the study material consists of two main parts: the inside and outside of the catalytic converter.

The use of a catalytic converter to reduce exhaust emissions produced by motor vehicles is constantly encouraged. Most catalytic converters for motor vehicles are pellet-type and monolithic-type catalysts and made of noble metal materials, such as palladium (Pd), platinum (Pt) and rhodium (Rh). These metals have high volatility (tendency to vaporize), are easily oxidized, and break easily at 500-9000°C, resulting in a reduction in catalyst activity.

Materials known as oxidation catalysts are platinum, plutonium, nickel, manganese, chromium and other metal oxides, while some metals used as reduction catalysts are iron, copper, nickel alloys and other oxide materials. In addition, several metals are oxidizable substances and reduction catalyst materials. From larger to smaller crystallite size, those metals are Pt, Pd, Ru> Mn, Cu >> Ni> Fe> Cr> Zn [4]. This type of catalytic converter can reduce exhaust emissions (CO, HC, NO_x) by 16% to 80% [5]. Therefore, optimizing the ability of catalytic converters to reduce exhaust emissions is necessary [6].

The focus of this research was to determine the effect of MnO as a catalyst on RON 88 and RON 92 fuel to reduce exhaust emissions (CO and HC), since RON 88 contained of 88% of iso-octane and 12% of n-heptane. Content of 12% n-heptane caused RON 88 had low maximum boiled point, while RON 92 contained of 92% iso-octane and 8% of n-heptane. MnO selection as catalyst, due to MnO was material oxide thus decrease exhaust emission of fuel combustion [7, 8]. The others properties of MnO was greatly used as adsorbent, thus that material could be utilized as catalyst [9].

2. Methodology

MnO that used as adsorbent was nano powder of MnO, after synthesis process by using sol-gel method. Synthesis process passed through several steps; they were weighing, stirring, crushing, and sintering. MnO was placed between three layers of glasswool, each of which weighs 0.5 grams; 0.33 grams of MnO were filled in between layers. Overall, the amount of MnO used was one gram. As proposed by [7], 1 gram of reactant substance is used in a catalyst to absorb other substances [7].

Next, the three layers of glasswool were inserted into a catalyst chimney which was connected to the motorcycle exhaust shown in Figure 1(a) which is where the catalyst is located. Then, the effect of the catalyst on the motorcycle exhaust emissions was examined.

Each layer of glasswool was flanked by two hollow sieves made of iron, a total of four. Thus, two layers of glasswool were made. On the first layer, glasswool was placed in the middle of the iron sieve, while on the second layer it was placed on the edge of the iron sieve, shown in Figure 1(b). As a result, there was no exhaust gas that could pass through the glasswool layers containing MnO.

The catalyst chimney was used to store MnO inserted in the glasswool layers as an absorbent (catalyst) for CO and HC emissions. This tube was made of galvanized iron with the design as follows:

A gas analyzer was used to examine the exhaust emissions generated from a motorcycle with and without a catalyst converter by comparing the amount of CO and HC emitted. The data result obtained from gas analyzer that be able to analyze the result of CO which was the emission from fuel combustion [10-12].

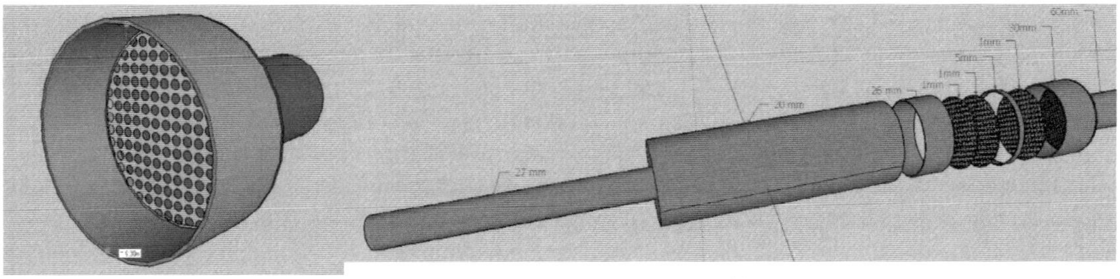

(a) (b)

Figure 1. Design of adsorption tube, (a) glasswool head place and (b) full design.

3. Results and discussion

3.1 Phase Characterization

The phase identification was performed using the X-Ray Diffraction (XRD) aimed to compare the crystalline phases in the materials and powders, and analyze the properties of the structure (crystallite size, phase composition, crystal orientation, and crystal defect) in each phase [13, 14].

The crystallite size of MnO materials can be determined using the following Scherer Equation [15, 16]:

$$d = \frac{K . \lambda}{\beta \cos \theta} \tag{5}$$

where:

- d: Crystallite Diameter
- K: Constant = 0,89-0,9
- λ: Wavelength = 1,5406 Å
- β: Full-Width Half Maximum (FWHM).

The results of calculation using the above equation are presented in Table 1. It shows that the crystallite size of the synthesized MnO powders sintered at different times (30, 60, and 90 minutes) was relatively small compared to the raw MnO powders which had the size of 143,44 nm. The comparison of all MnO samples is also shown in the graph of intensity and °2Th, is shown in Figure 2.

Table 1. Intensity, FWHM, d-spacing and crystallite size of MnO.

Sample Material	X-Ray Diffraction (correspond to [222] peak)			
	Intensity (counts)	FWHM (rad)	d-spacing (Å)	Crystallite Size (nm)
Manganese Oxide (MnO) raw	264.33	1.03	2.23	143.44
Manganese Oxide (MnO) 30 minutes	72.00	0.17	2.71	47.30
Manganese Oxide (MnO) 60 minutes	436.62	2.40	2.71	68.30
Manganese Oxide (MnO) 90 minutes	516.91	1.71	2.72	84.55

The transition of MnO from grain to crystallite size is demonstrated in the graph above; there was a very significant change between the raw MnO and the synthesized MnO with various sintering times (30, 60 and 90 minutes). The transition occurred as a result of the synthesis process with variations in sintering time. The shift of crystalline intensity (as shown in Figure 2) indicated that during the sintering process there was development in MnO nanopowders. The shifting and increasing of peak occurred in sintering process while conducting synthesis process of MnO, thus, the crystal intensity shifted (as showed at Figure 2 [17]. The crystalline intensity is a peak that indicates an increase in the purity level of MnO nanopowders. Figure 2 shows that the change or shift from the MnO peak is obvious. The peak that underwent such a shift is peak [222]. Peak [222] was like the fingerprint of MnO. MnO has several peaks, namely [111], [200], [220], [311], and [222], respectively [9, 18, 19]. Peak [222] represented the peak of MnO, since the result of XRD analysis showed the peaks at Figure 2 [9, 18]. Shifting and peak intensity containing one of the above MnO peaks indicate that the purity level of MnO is increasing. The similarity of position of the crystalline intensity indicated that the crystalline intensity was a single phase. Single phase also showed that the cubic structure centered in the MnO nanopowders already contained electrons in each atom and also a large number in the lattice. As a result, the electrons in the atoms of MnO nanopowders can be used effectively as a catalyst or as the main material in the production of batteries; in fact, the movement of the MnO atoms containing electrons reacts and spreads very quickly.

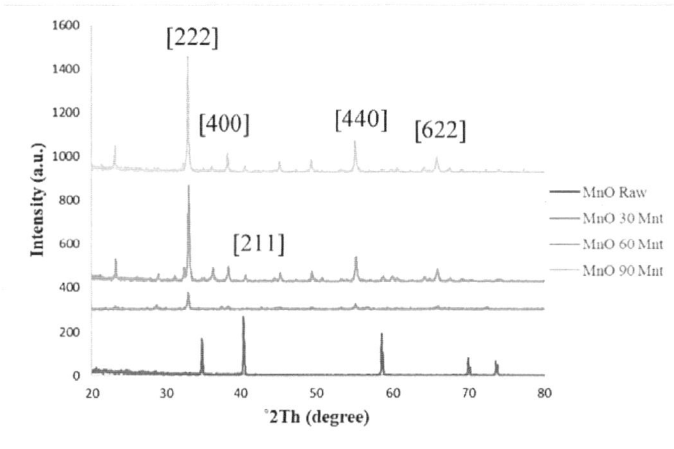

Figure 2. Phase identification of MnO with different sintering times.

Table 1 shows that the variation in crystallite size was influenced by the duration of the crushing process, whereas the difference in crystalline purity or intensity was affected by the variations in sintering time. The longer the sintering time is carried out (according to the atomic development limit of MnO), the higher the purity level. The best purity and size of the three samples with various sintering times occurred at the sample sintered for 60 minutes the crystalline intensity or purity level was 436.62 counts, and the size was 68.30 nm. It can be concluded that the use of sol-gel method a simple method without many treatments in the synthesis process, can produce MnO nanopowders with great purity level and very small crystallite size compared to the raw MnO powder of micron size. The results of the XRD test are highly relevant because of the beam of X-ray is shot directly at the object being tested.

3.2 Morphological characterization

SEM was used to analyze and compare the morphology of MnO materials [20, 21]. The analysis aimed to examine whether there was a change in the morphology of MnO after undergoing different synthesis processes. Morphological changes that occurred in the synthesized MnO with variations in sintering time as the results of SEM are shown in Figure 3.

The different morphologies of MnO can be seen from the results of SEM at a magnification of 100k. The results showed morphological changes in MnO due to variations in sintering time. The changes may be caused by the temperature of 600 °C, as stated by [17] that drying and shrinkage occur due to the sintering process. Drying and shrinkage occurred due to molecular reactions occurring at room temperature, gel formation and drying solution. Under these conditions the mixed materials have many characteristics morphologies and manganese oxide (MnO) which corresponds to the sintering temperature range, starting from 600 ºC to 1000 ºC.

(a) (b)

Figure 3. Morphology of manganese oxide (MnO), (a) without synthesis and (b) synthesis with 60-minute sintering.

Figure 3 shows the raw MnO morphology-which is not synthesized. SEM is used to check the MnO standard morphology at 100k magnification. The results show that the raw MnO is triangular in shape. The difference in grain shape on the surface shows that agglomeration occurs when particles are deposited on the membrane of the MnO grains. MnO powders were tested using SEM to determine the initial form of MnO powder prior to synthesis. The different things occur in MnO synthesized with various sintering times (30, 60, dan 90 minutes), the initial shape of MnO (triangular with little dots) turned into spheres with agglomeration (grouping of granules in some places). This was influenced by the sintering temperature (600 ºC) and time (30 minutes), and the duration of the crushing process. The transition of MnO from micro to nano indicates that the synthesis had been successfully done.

The SEM results at 100k magnification (as shown in Figure 3) show that MnO morphological changes also occur as a result of 60-minute sintering. There were no significant changes in grain form compared to 30 min sintered MnO. However, morphological differences are indicated by the number of spheres in which large agglomerations occur. Due to sintering process and duration of crushing thus the agglomeration enlarge but the diameter of powder became smaller [22]. It can be concluded that the transition of MnO from micro to nano was successful. This could occur due to the duration of sintering, the uneven crushing, and the duration of crushing.

3.3 The addition of MnO as a catalyst could reduce carbon monoxide (CO) emissions in RON 88

Data on the concentration of CO in RON 88 are presented in Figure 4. It shows, the use of MnO catalyst resulted in lower CO emission levels than that without a catalyst. In principle, the maximum adsorption process occurs only in catalysts having nanoparticle size. As proposed by [8], the smaller the nanoparticle size, the wider the contact, hence a high absorbency [8]. Rami Nader, *et al.* (2014) also explained that in the form of nanoparticles, atoms have the ability to make direct contact with other substances better than larger particles [23]. The application of MnO to RON 88 showed quite significant results. Without the use of MnO catalysts, RON 88 produced high CO emissions, i.e. 3.86%vol, while with MnO catalysts, the highest CO concentration reached only 2.735%vol.

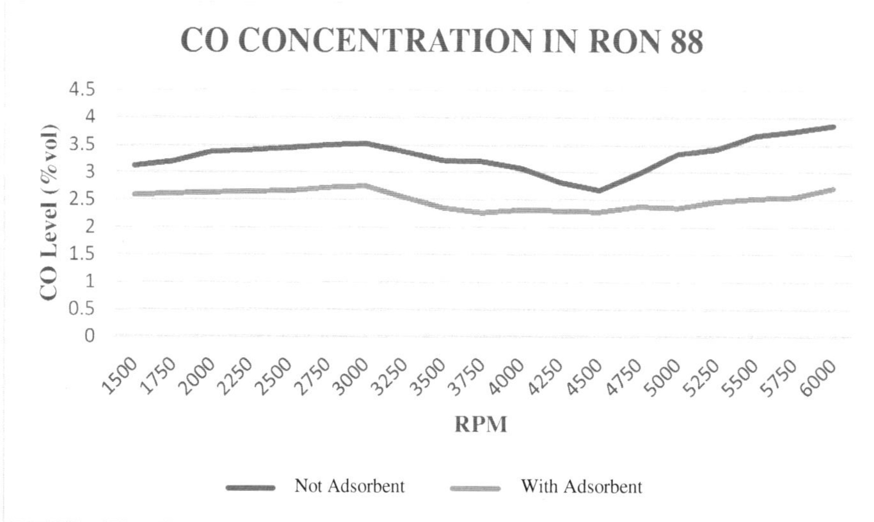

Figure 4. Comparison graph of CO level with and without a catalytic converter in RON 88.

Based on the mean value, the level of CO in the treatment with the addition of MnO catalyst was lower than that in the non-catalyst treatment, and both treatments had a significant difference. Thus, it can be concluded that the addition of MnO as a catalyst in RON 88 results in lower CO emission levels. This demonstrates that MnO as a catalyst affects vehicles with RON 88 fuel positively it can significantly reduce CO emissions.

3.4 The addition of MnO as a catalyst could reduce hydrocarbon (HC) emissions in RON 88

Data on the concentration of HC in RON 88 are presented in Figure 5. The motorcycle-generated hydrocarbons come from fuels that break down due to a heat reaction which then turns into other HC

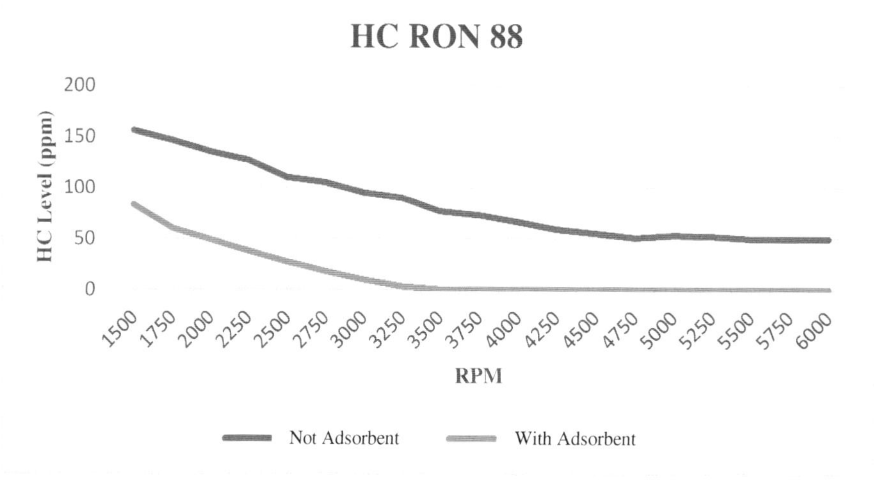

Figure 5. Comparison graph of HC level with and without a catalytic converter in RON 88.

clusters emitted along with the exhaust gas:

$$C8H18 \rightarrow H + C + H \quad (6)$$

Hydrocarbons are crude fuels which are unburned during combustion processes in a combustion chamber. They are the crude fuels left on the cylindrical wall after combustion and emitted. They are also unburned gas in the combustion chamber after misfiring when the engine is accelerated or decelerated [24].

The maximum adsorption process occurs only in nanoparticle-sized catalyst materials. The microparticle-MnO catalyst used in the treatment in this study was fairly capable of showing significant differences compared with the treatment without using a catalyst. In fact, the HC emissions in the treatment without a catalyst were high with the highest HC of 156 ppm, whereas that with the use of MnO catalyst had the highest HC of only 83 ppm.

In sum, it is known that the addition of MnO as a catalyst in RON 88 results in lower HC emission levels. This shows that the use of MnO as a catalyst in vehicles with RON 88 fuel can significantly reduce HC emissions.

3.5 The addition of MnO as a catalyst could reduce hydrocarbon (HC) emissions in RON 92

Data on the concentration of HC in RON 92 are presented in Figure 6. At a high rotation rate (5750 RPM), the HC produced by both treatments had the same level of concentration *i.e.* 18 ppm. It happened due to the fact that the higher the temperature, the lower the level of HC.

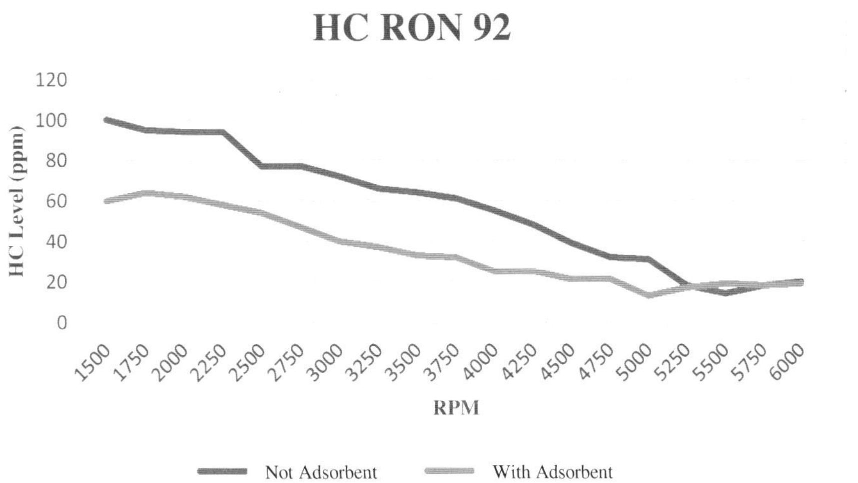

Figure 6. Comparison graph of HC level with and without a catalytic converter in RON 92.

Based on the mean value, the level of HC in the treatment with the addition of MnO catalyst was lower than that in the non-catalyst treatment, and both treatments had a significant difference. Thus, it is known that the addition of MnO as a catalyst in RON 92 results in lower HC emission levels. This indicates that MnO as a catalyst could significantly reduce the HC emissions from vehicles with RON 92 fuel.

4. Conclusion

Based on the research and discussion on the application of manganese oxide (MnO) as a catalyst in RON 88 and RON 92 fuel to reduce exhaust emissions (CO and HC), we conclude that:

1. Using MnO as a catalyst is only effective to reduce CO emissions from motor vehicles using RON 88 fuel (24.36%), not those using RON 92 fuel since they used microparticle-sized MnO—the most effective size is catalysts in nanoparticles. Also, since RON 88 fuel has a low maximum boiling point, MnO could react faster to bind the exhaust emissions.

2. Using MnO as a catalyst is only effective to reduce HC emissions from motor vehicles using RON 88 (82.08%) and RON 92 fuel (38.14%), not those using RON 90 and RON 98 fuel.
   MnO could react well in RON 88 and RON 92 because they produced high HC levels and had a low maximum boiling point

Acknowledgement
The authors would like to thank Ministry of Research, Technology and Higher Education for Applied Research Grants Leading Universities 2017 No. 3.4.8/UN32.14/LT/2017.

References
[1] Wardhana W A 2001 *Impact of Environmental Pollution* Yogyakarta: Pub. ANDI
[2] Anonim *No Title*
[3] Schubert N H and Ullrich 2004 *Synthesis of Inorganic Materials, 2nd, Revised and Updated Edition* Weinheim: Wiley-VCH Verla
[4] D D A 1970 *Catalytic Hand Book* New York
[5] Irawan R B Purwanto P and Hadiyanto H 2015 *Proc. Environ. Sci* **23** Ictcred 2014 86–92
[6] Irawan H Bagus Purwanto and Hadiyanto P 2014 *Int. Conf. Trop. Coast. Reg. Eco-Development 2014(ICTCRED 2014)* **23** 86–92
[7] Rahmayani S M and Fatimah 2013 *J. Tech. Chem. USU* **2** (In Indonesia)
[8] Munandar A 2016 *J. Chem. and Envir. Engin.* **1** 24–31
[9] Wu R Qu J and Chen Y 2005 *Water Res* **39(4)** 630–8
[10] Emiroğlu A O 2017 *Atmos Pollut Res* **8(1)** 149–53
[11] Cho C P Pyo Y D Jang J Y Kim G C and Shin Y J 2017 *Appl. Therm. Eng.* **110** 18–24
[12] Sharaf J 2013 *Int. J Eng. Res. App.* **3(4)** 947–60
[13] N A M Z and N A A P Puspitasari and N Yahya 2011 *App. Sci.* **11(7)** 7
[14] Wu X 2016 *Int. J. Hydrogen Energy* **41(36)** 16087–93
[15] Yahya N B Daud H Tajuddin N A Daud H M Shafie A and P Puspitasari 2010 *J. Nano. Res.* **11** 25–34
[16] Yahya N B P Puspitasari K K K Koziol and Pavia G 2012 *J. Nano. Res.* **16(2011**) 119–30
[17] Cammarata R C 1996 *Mater. Sci.* **580**
[18] Pandey B K Shahi A K and Gopal R 2013 *App. Surf. Sci.* **283** 430–7
[19] Li P 2013 *Ceram. Int.* **39(7)** 7773–8
[20] Cahyana A and Marzuki A 2014 *Proc. Math. Sci. Forum 2014* 23–6 (In Indonesia)
[21] Corneal L M Masten S J Davies S H R Tarabara V V Byun S and Baumann M J 2010 *J. Memb. Sci.* **360(1–2**) 292–302
[22] Theodore L 2006 *Nanotechnology: basic calculations for engineers and scientists* Wiley-Interscience 2006
[23] Rami N 2014 *8th Int. Conf. Mater. Sci. CSM8-ISM5* **55** 373–82
[24] Syahrani A 2006 *J. SMARTEK* **4** 260–6 (In Indonesia)

IOP Conf. Series: Materials Science and Engineering **432** (2018) 012006 doi:10.1088/1757-899X/432/1/012006

A study of the compression molding method for manufacturing a UHMWPE liner

F S Wicaksono[1], **R Ismail**[1,2] **and J Jamari**[1]

[1] Laboratory for Engineering Design and Tribology. Departement of Mechanical Engineering, Diponegoro University, Prof. Sudharto St. UNDIP Tembalang Semarang 50275, Indonesia
[2] Center for Biomechanics, Biomaterial, Biomechatronics and Biosignal processing (CBIOM3S) 5th Floor Central Laboratory UNDIP, Prof. Sudharto St. UNDIP Tembalang Semarang 50275, Indonesia

fabiansinggih@gmail.com

Abstract. Hip and knee joints are some examples of the vulnerable joints in the human body. However, due to internal or external factors, the symptoms of osteoarthritis disease may occur, so the replacement of the natural joints with an artificial ones becomes crucial. The most common material used for this part is one type of polymer, namely the ultra-high-molecular-weight polyethylene (UHMWPE) for the liner bearing, a component in an artificial hip or knee joint. Usually, the liner component of the artificial joint prosthesis is manufactured from machining process from a bar of UHMWPE material, which is made using ram extrusion process. This research aims to conduct a manufacturing process of UHMWPE liner using compression molding process, changing the powder UHMWPE into solid UHMWPE. The present method developed aimed to reduce the pressure and the amount of energy required during the compression molding process. The hardness and density tests were conducted for measuring several indicators related to the success of the solidification process. It was found that the resulting solid UHMWPE conformed to international standards for manufacturing the joints liners and polymer bearings for joint prosthesis. The best parameters resulting from the experiments were the UHMWPE hardness of 9.37 VHN and the density of 0.94 g/cm^3. After a good manufacturing process resulting from this research, the ASTM international standards for joint liners and polymer bearing for joint prosthesis will easily be fulfilled in production in in Indonesia.

1. Introduction

Joints are one of the parts of human body that are vulnerable to damage. Joints have the possibility of damage or dysfunction. There are two types of damage, namely: first, damages due to accidents as a result of over imposition that exceeds the limits of weight that can be held by the joint itself; and second, bones getting porous as it ages. The damage and abnormality of joint functions encourage the creation of artificial joints. In the medical world, the most commonly used artificial joints are the artificial hip and knee joints. Compared to other joints of the human body, the hip and knee joints are most vulnerable to damage [1].

Previous projects have researched hip joints using numerical analysis and finite element method [2-5]. Presently, our researchers focus on the manufacturing of the part in hip or knee joints. The part that becomes our focus is manufacture a part that is often affected by wear, namely, the liner. Among ways and methods to produce the liner, the researchers choose the compression molding method.

The compression molding cycle process is consolidation of resin powder into a rod- or net-shaped implants is a necessary step for the final shape machining of UHMWPE joint implants. The process of consolidation in UHMWPE requires proper designation of pressure, temperature and time. Changes in these three molding variables can impact the mechanical properties of the entire UHMWPE implants. For example, mechanical performance of UHMWPE is dependent on the degree of fusion of the powder particles during compression and can be enhanced by heating the polymer powder at temperatures above 220 °C [6].

In Indonesia, the manufacture of hip or knee joints is rare, not equitable and also don't have International Standard compared with the others nations. Otherwise the size and quality too are far away from International Standard or ASTM because most of it used material can easily found in Indonesia is not equitable. For example, one part of the joint is made of AISI 316L,while most Indonesian industry products are ferromagnetic, which it not allowed for use as implant.

Based on the aforementioned background, a research on how to process UHMWPE powder into solid UHMWPE is needed in order to help the medical world in Indonesia. Up to now, Indonesia mostly imported the solid UHMWPE from abroad, andthere is no focus on producing solid UHMWPE. In this study, the authors focus on developing compression molding method for manufacturing solid UHMWPE.

2. Method

2.1. Literature study
The first step was to conduct a literature study on standard product specifications of the liner for the hip or knee joints that use polymer have international standard for manufacturing implant process. Afterward, UHMWPE material was purchased and compression molding tools were designed. The next step was the processing of powder UHMWPE specimens into a solid one. Based on the specimens that had been produced, evaluation of measurements was conducted to find out whether the compression molding tool was capable during the manufacturing process. The next stage was the evaluation of test results data on specimens, to determine whether the specimens met the standards.

2.2. Material specification
UHMWPE resin powder that the researchers used had a molecular weight of 7×10^6 g/mol and had the same peak condition in XRD analysis as other brands of UHMWPE resin powder.

2.3. Specimen preparation

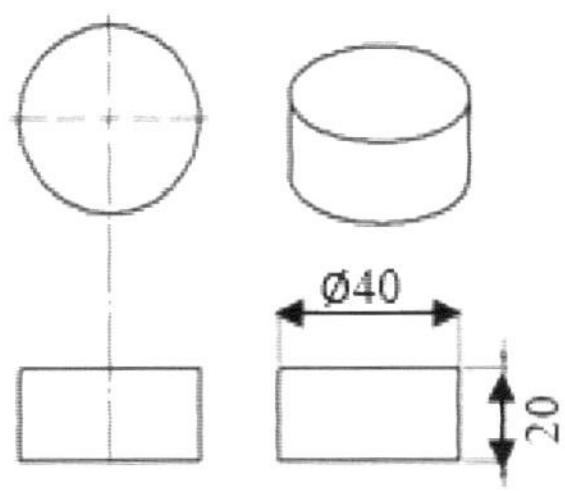

Figure 1. Design of the specimen.

Figure 2. Molding tools set.

Samples or specimens had been prepared from resin powder UHMWPE and formed into solid cylinders that were obtained by compression molding method with the molding tool. The pressure and temperature during product were varied. We conducted the best variations for performing the compression molding to the resin powder UHMWPE. The 3D design of the specimen is pictured in Figure 1 and the molding tools set is in Figure 2.

2.4. Compression molding method

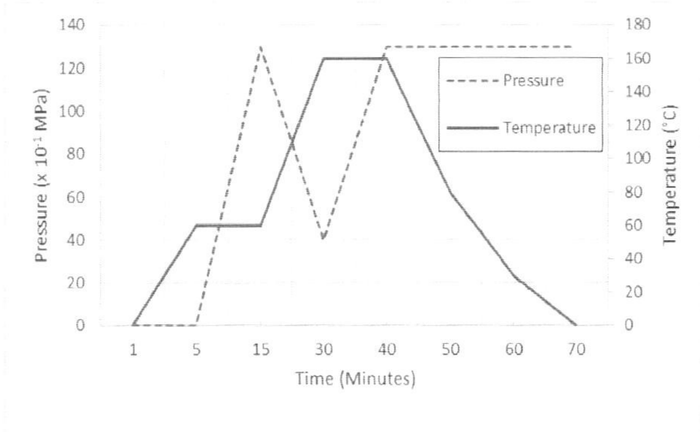

Figure 3. Compression molding method.

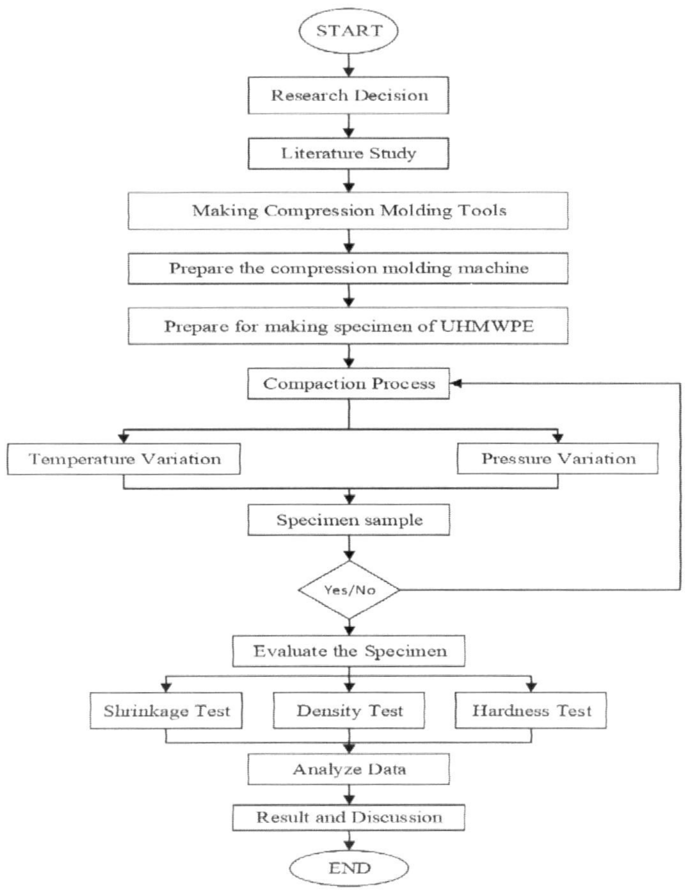

Figure 4. Flowchart of this research.

IOP Conf. Series: Materials Science and Engineering **432** (2018) 012006 doi:10.1088/1757-899X/432/1/012006

Compression molding method chosen in this research is the best and effective way to produce liner using UHMWPE. The compression molding method chosen was varied for the case of 'lower' and 'higher' point compared to the melting point of UHMWPE, to know the effects on the result. Before the final result has been making, the resin powder was pre-compressed in a mold under the pressure of 10 MPa and at the temperature of 40 °C. Then, we use the selective variations of pressure and temperature but with different temperature in final condition to see the best sample of UHMWPE after doing the compression molding method. Finally, the solid cylinder UHMWPE removed from the mold and then was gradually cooled to room temperature (See Figure 3). The phase of this research step by step until measurement the specimen can be seen in Figure 4. After doing the compression molding method, finally the solid cylinder UHMWPE removed from the mold anf then was gradually cooled to room temperature.

3. Results

In this study, the variation of temperature for process compression molding method that was used is 110 C, 140 °C, and 160 °C and the pressure was 70 kgf (9.7 MPa), 100 kgf (13.8 MPa), and 130 kgf (18.03 MPa) in Figure 5.

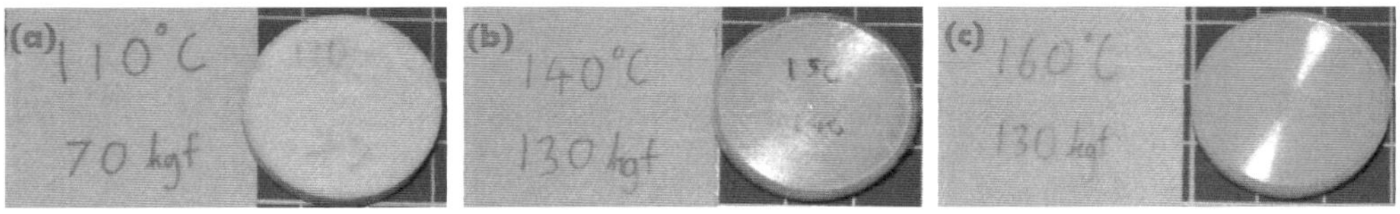

Figure 5. The specimen (a), (b), (c), are examples of after process the compression molding method with the different temperature and pressure.

3.1. Hardness test

Hardness is one of the indicators of the quality of the produced polymer. The hardness test was conducted using Vickers HM-200 Mitutoyo. Three best samples were prepared molded with differrent pressure and temperature is shown in Table 1. Results showed the hardness of specimen 1 through specimen 3 is above 5 VHN. Samples 140 °C 13.8 MPa are almost 10 VHN. This data is compared with previous research shows increasing hardness value, that similar to the hardness value showed in Kurtz research [1].

Table 1. Hardness test results.

Specimen Variation	Hardness (VHN)			
	Test 1	Test 2	Test 3	Average
110 °C 9.7 MPa	6.5	6	5.9	6.13
140 °C 13.8 MPa	8.8	9.6	9.7	9.37
160 °C 18.03 MPa	7.8	9.4	9.9	9.03

3.2 Density test

The density of the material that we processed needed to be tested. Results from the density test showed whether the material met the standards. We tested the sample specimen with palm oil. The density test was conducted using the Vibra Density Meter. The density values shown in Table 2 indicates that the manufacture of UHMWPE that is carried out by compression molding process above melting point matches the standard mentioned in Kurtz [1].

Table 2. Density test results.

Specimen Variation	Density (g/cm^3)
110 °C 9.7 MPa	0.92
110 °C 13.8 MPa	0.92
110 °C 18.03 MPa	0.92
140 °C 9.7 MPa	0.94
140 °C 13.8 MPa	0.94
140 °C 18.03 MPa	0.94
160 °C 9.7 MPa	0.94
160 °C 13.8 MPa	0.94
160 °C 18.03 MPa	0.94

3.3 Shrinkage test

Polymers have a special condition after melting and resolidification called *shrinkage*. This is one of the anomalous effect exhibited by polymers after processes. Table 3 is shown the latest condition of UHMWPE specimen after cooled by room temperature on day 3.

Table 3. Shrinkage test results.

Specimen Variation	Day 1 Vo (mm^3)	Day 3 Vi (mm^3)	% Shrinkage Volume
110 °C 9.7 MPa	7771.50	389.29	5.01%
110 °C 13.8 MPa	7771.50	322.84	4.15%
110 °C 18.03 MPa	7771.50	342.62	4.41%
140 °C 9.7 MPa	7771.50	272.70	3.51%
140 °C 13.8 MPa	7771.50	264.02	3.40%
140 °C 18.03 MPa	7771.50	274.34	3.53%
160 °C 9.7 MPa	7771.50	286.95	3.69%
160 °C 13.8 MPa	7771.50	270.02	3.47%
160 °C 18.03 MPa	7771.50	273.25	3.52%

4. Discussion

All the results of this experimental research provide evidence that compression molding process is effective for making of liner components in artificial hip or knee joints. The most optimal variation for making solid UHMWPE using compression molding process conducted in UNDIP consists of a temperature 160 °C and a load of 130 kgf (18.03 MPa).

The density of the specimen after being processed at a temperature above melting point 138 °C already matched the standard density of 0.94 g/cm^3 according to density properties of UHMWPE by Kurtz [1]. The average hardness value is 9.03 VHN has already exceeded the standard on journal Slouf *et al.* [7], which is 4.64 VHN but still below the hardness value shown Kurtz [1] which is 10 VHN for pure UHMWPE. Thus, the density and hardness value of acetabular liner specimens used in this research already meet the international standard for hip or knee joint liner.

Shrinkage effect of the polymer occurs after UHMWPE processes above melting point, the shrinkage effect stop at precentage ± 3%. The hardness value of making solid UHMWPE it should not be too hard or too soft because bad effects causes when it has become a liner. If too soft, the deformation form comes from impact load which makes part of this liner stuck in the other part or may can not accept the burden of which is supposed to receive the part. And if it is too stiff, it will result in the rest of the unwanted wear giving rise to a new problem.

5. Conclusion

UHMWPE surely influenced by heat dan pressurre, when the processing is done and UHMWPE is heated above the melting point, size of the density of the UHMWPE powder raw will be different from the UHMWPE with shaped solid form. The mechanical property of UHMWPE with different pressure that was disgned 9.7 MPa, 13.8 MPa, and 18.03 MPa in this paper, the UHMWPE samples processed at 13.8 MPa obtain the best mechanical properties.

Shrinkage is one of the polymer properties after reach the melting point. The shrink factor is determined by measuring before procesess, after procesess, and day by day until the shrinkage precentage stop. Subtracting the length of the part from the length of the cavity, and dividing that result by the length of the core in molded part. Then measuring the samples after procesess by the volume samples.

References

[1] Kurtz S M 2009 UHMWPE biomaterials handbook Elsevier Amsterdam
[2] Saputra E Budiwan I Ismail R Jamari J and van der Heide E 2014 *Sci.and Engin.* **66(3)** 53-8
[3] Anwar I B Santoso A Saputra E Ismail R Jamari J and Van der Heide E 2017 *Adv. Pharmaceutical Bull.* **7(2)** 335
[4] Ismail R Saputra E Tauviqirrahman M Legowo A B Anwar I B and Jamari J 2014 *App. Mechanics and Mater.* **493** 426-31
[5] Anwar I B Saputra E Ismail R Jamari J Van der Heide E Ghazali M J B and Abdollah M F B 2015 *Proc. of Malaysian Int. Tribology Conf.* 56-7
[6] Wang S and Ge S 2007 *Wear* ***263*****(7-12)** 949-56
[7] Slouf M Vackova T Nevoralova M and Pokorny D 2015 *Polymer Testing* **41** 191-7
[8] Wu L Jing D and Ding J 2006 *Biomaterials* **27(2)** 185-91
[9] Firouzi D Youssef A Amer M Srouji R Amleh A Foucher D A and Bougherara H 2014 *J. of the Mechanical Behav. of Biomed. Mater.* **32** 198-209
[10] Khalil Y Kowalski A and Hopkinson N 2016 *Additive Manufacturing* **10** 67-75
[11] Huang Y F Xu J Z Li J S He B X Xu L and Li Z M 2014 *Biomaterials* **35(25)** 6687-97

A comparison between mechanical properties of UHMWPE from ram extrusion process and UHMWPE from compression molding process for a hip joint liner

M Wahyudi[1*], Y E Putra[1], S Arrohman[1], J Jamari[1], R Ismail[1,2]

[1] Laboratory for Engineering Design and Tribology, Departement of Mechanical Engineering, Diponegoro University, Prof. Sudharto St. UNDIP Tembalang Semarang 50275, Indonesia
[2] Center for Biomechanics, Biomaterial, Biomechatronics and Biosignal processing (CBIOM3S), 5th Floor Central Laboratory UNDIP, Prof. Sudharto St. UNDIP Tembalang Semarang 50275, Indonesia

mujibwahyudi@gmail.com

Abstract. The compression molding (CM) forms the liner from UHMWPE powder into solid form whereas the machining process forms the liner from a bar of UHMWPE made from ram extrusion (RE) process. Our research goal is to determine and compare the mechanical properties of UHMWPE resulting from compression molding process and ram extrusion. The compression molding is experimentally proceed in UNDIP laboratory whereas the UHMWPE from ram extrusion (RE) process is imported from an European country. The mechanical properties test consists of tensile test and hardness test. The final test results of the tensile strength: CM and RE have yield strength value of 23.66 MPa and 25.25 MPa, respectively. Ultimate tensile strength of CM and RE are 35.36 MPa and 29.70 MPa, respectively. The elongation of CM is 214.75% whereas RE is 120.45%. The material hardness is 15.1 VHN for CM and 8.9 VHN for RE. The results in this study show that the CM specimens have better ultimate tensile strength, elongation, and hardness than RE specimens.

1. Introduction

Ultra-high molecular weight polyethylene (UHMWPE) has been extensively used as a successful load bearing material in orthopedic applications due in part to its superior properties of biocompatibility, good wear resistance, chemical stability, and lubricity which fit the biological and mechanical requirements [1]. Despite the recognized success and worldwide acceptance of total joint arthroplasty, wear is a major obstacle limiting the longevity of implanted UHMWPE components. Increased attention has been devoted to the understanding of factors infuencing the mechanical behavior of UHMWPE. As with any semicrystalline polymer, the mechanical properties of UHMWPE are inextricably linked to its chemical structure, molecular weight, crystalline organization, and thermal history. All of these factors, in turn, affect the morphological, chemical, and mechanical processes which may infuence [2]. To meet the technical requirements of UHMWPE before being used as an acetabular liner in the artificial hip joint component, the UHMWPE processing technique becomes an important matter to be considered and studied. UHMWPE manufacturing methods commonly used today are compression molding and ram extrusion. Many researchers have conducted studies on the development of compression molding and ram extrusion in order to continuously improve the mechanical and tribological properties of UHMWPE.

In compression molding process, high melting temperature of consolidated UHMWPE was effective in improving its toughness by using properly selected melting temperatures and durations. The chain scissioning and improved chain entanglements through granule boundaries were two major causes for the improvement of the mechanical properties.The improved plasticity and ductility by high temperature melting did not significantly compromise the wear rates [3]. To improve wear resistance, many researchers have conducted the croslinking method against UHMWPE, but cross-linking also restricts relative chain movement in the amorphous regions and hence decreases toughness [4-7].

In paralel, equal channel angular extrusion (ECAE) method produces a highly entangled material with better mechanical properties than radiation cross-linked material. Entanglement density in ECAE processed UHMWPE was significantly greater than in control materials, although there was no significant relationship found between entanglement density and processing conditions. Still, mechanical properties in ECAE processed materials were found to vary with processing conditions, suggesting underlying material structure changes [8].

To meet the high demand for enhanced tribological and mechanical properties of ultra-high molecular weight polyethylene (UHMWPE), the incorporation of nano-reinforcements has generally been applied. UHMWPE nanocomposites reinforced with graphite nanoplatelets (GNPs). The modification of GNP surfaces with organosilane was found to effectively disperse the GNP in the polymer matrix and to improve interfacial interactions between phases. As a result, simultaneous improvements in wear resistance and storage modulus have been realized for the UHMWPE nanocomposites with silanized GNPs, up to 980% and 170%, respectively [9].

Some of the aforementioned studies are the development of compression molding and ram extrusion methods. Both of these methods continue to grow due to the demands of UHMWPE production requirements with good mechanical and tribological properties. However, there is no recent study that discusses compression molding and ram extrusion comparisons in terms of the mechanical properties of the products produced. In this study, both methods are re-analyzed. The latest side of the study is no longer using conventional compression molding methods, but has used direct compression molding (DCM) methods for the manufacture of test specimens. The advantage of this DCM method is that it can directly convert UHMWPE powder into acetabular liner product. This cannot be done by the method of ram extrusion.

2. Experimental method

2.1. Compression molding

Historically, the UHMWPE powder has been converted by compression molding since the 1950s because the industries in the area around Ruhrchemie already had experience with this processing technique. At first, the semifinished material was distributed under the trade name RCH-1000/ Hostalen GUR 412. During the 1970s, the compression molded UHMWPE was manufactured and distributed by Ruhrchemie/Hoechst and later distributed by the company Europlast specifically for orthopedic applications under the trade name CHIRULEN, which denoted that the material was produced in Germany by a dedicated press using established processing parameters. Thus, references in the orthopedic literature to RCH-1000 and CHIRULEN apply to compression molded UHMWPE, which was similar to contemporary GUR 1020. Note, however, that GUR 412 would have contained calcium stearate, whereas GUR 1020 does not [1].

An example of a compression molding press, currently installed at MediTECH in Vreden, Germany, is shown in Figure 1. This particular press, originally designed by Hoechst in the 1970s for production of CHIRULEN, molds two 1 m by 2 m sheets in a single press cycle. One UHMWPE sheet is pressed between the upper and middle platens, and the second is produced between the middle and lower platens. The platens are oil heated and hydraulically actuated from below. The heating and load- ing systems are

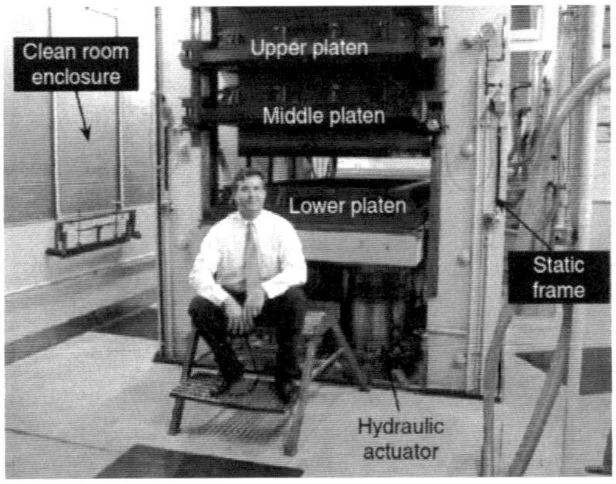

Figure 1. Compression molding press (along with the author, for scale) for production of 1 m by 2 m sheets of UHMWPE [1].

all computer controlled. Finally, the entire press is contained in a clean room to reduce the introduction of extraneous matter into the sheet. Due to the relatively low thermal conductivity of UHMWPE, the duration of the molding cycle will depend upon the particular geometry of the press and the size of the sheet to be produced, but the processing time can last up to 24 h. The long molding times are necessary to maintain the slow, uniform heating and cooling rates throughout the entire sheet during the molding process [1].

2.2. *Ram extrusion*

In contrast with compression molding, which originated in Germany in the 1950s, ram extrusion of UHMWPE was developed by converters in the United States during the 1970s. Historically, a wide range of UHMWPE resins have been extruded over the past three decades. From the early 1970s to early 1980s, ram-extruded Hi-Fax 1900 resin from Hercules Powder Company (Wilmington, Delaware, USA) was commonly supplied to bulk form converters. In the early 1960s Hoechst's GUR 412 resin was made available to converters. Although GUR 412 tended to have a lower extraneous particle count than 1900, it was also more difficult to process using ram extrusion due to its lower average molecular weight and lower melt viscosity. In the mid 1980s, converters began ram extruding using the higher average molecular weight Hoechst GUR 415 resin, which also had a lower extraneous particle count relative to 1900 and was easier to extrude than GUR 412.

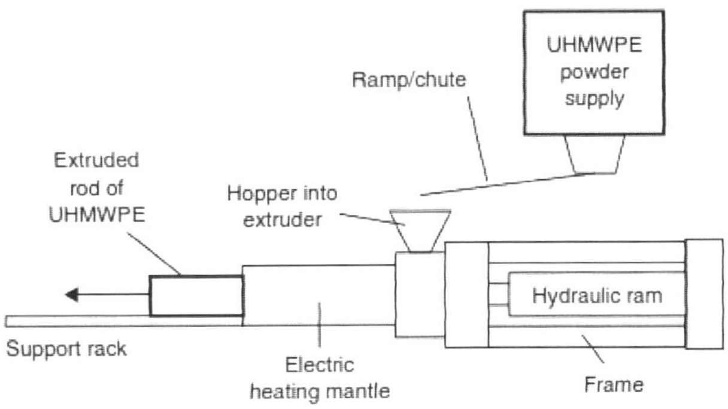

Figure 2. Schematic of a ram extruder [1].

During the 1980s and early 1990s, extruded rods of GUR 412, 415, and (more rarely) 1900 CM (a calcium stearate containing grade of 1900 resin) were all used in orthopedics. Thus, it is difficult to generalize about the resin types used in implants without tracing the lot numbers used by a particular manufacturer. The generic schematic of a ram extruder is illustrated in Figure 2. UHMWPE powder is fed continuously into an extruder. The extruder itself consists essentially of a hopper that allows powder to enter a heated receiving chamber, a horizontal reciprocating ram, a heated die, and an outlet. Within the extruder, the UHMWPE is maintained under pressure by the ram, as well as by the back pressure of the molten UHMWPE. The back pressure is caused by frictional forces of the molten resin against the heated die wall surface as it is forced horizontally through the outlet. Beyond the outlet, the rod of UHMWPE is slowly cooled in a series of electric heating mantles [1].

A wide range of rod diameters is achievable (up to 12 inches in diameter) using extrusion, but the rate of production depends on the rod size due to the increased cooling times needed with the larger rod diameters. Rod sizes ranging from 20 to 80 mm in diameter are most often used in orthopedic applications. Typical production rates are on the order of mm/minute [1].

2.3. Materials

For UHMWPE material from Ram Extrusion (RE) result, obtained by buying from Europe and already in the form of solid cylinder. Images of UHMWPE from Ram Extrusion process can be seen in Figure 3.

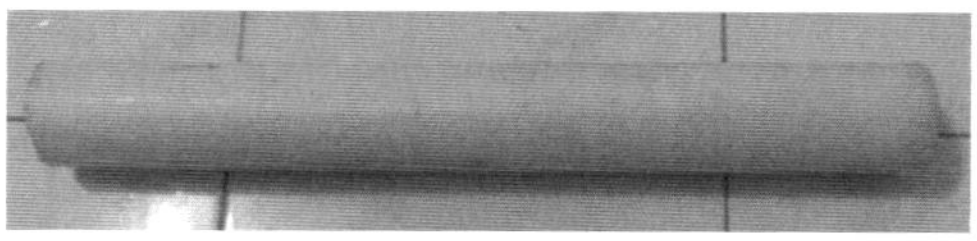

Figure 3. UHMWPE from ram extrusion process.

For UHMWPE material of Compression Molding (CM) result, obtained by compression molding process of UHMWPE powder using compression molding machine at Laboratory for Engineering Design and Tribology, Diponegoro University. The physical form of the compression molding machine facilities contained in the Laboratory for Engineering Design and Tribology, Diponegoro University can be seen in Figure 4.

Figure 4. Compression molding machine at Laboratory for Engineering Design and Tribology, Diponegoro University.

2.4. Method

2.4.1. Tensile test preparation. For UHMWPE material from Ram Extrusion (RE) result, a tensile test specimen is made using a machining process with a specimen dimension in accordance with JIS K-7113 standard. The number of test specimens used was 2 specimens. Tensile test specimens for RE product can be seen in Figure 5.

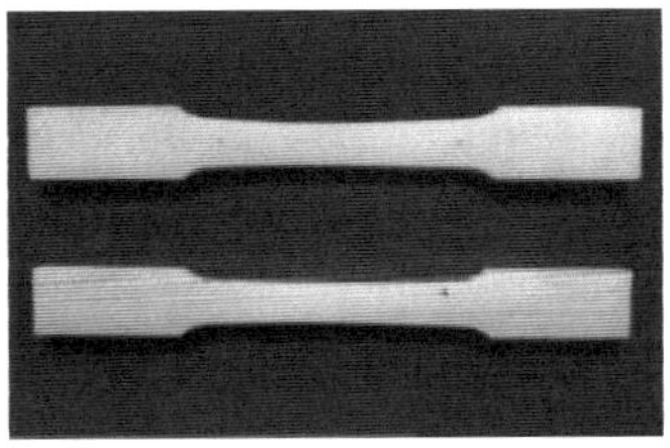

Figure 5. Tensile test specimen for RE product.

For UHMWPE material from Compression Molding (CM) result, tensile specimens were made using compression molding process with dimension according to ASTM D-638 Type IV standard. The compression molding process takes place at a certain pressure. Figure 6 shows the compression molding process. This process is to convert UHMWPE powder form to UHMWPE solid form. The orange color line in Figure 6 represents the compressive force variation (kgf) given during the compression molding process. While the blue color line represents the amount of temperature (^{0}C) used in the compression molding process. In this cycle, there are two special treatments in the melting phase and in the recrystallization phase. This treatment is done by holding the compressive force and temperature at specified temperature as determined in Figure 6. The compression molding machine made by Laboratory for Engineering Design and Tribology, Diponegoro University has been equipped with special software that allows the machine to operate automatically and and which one could easily monitor.

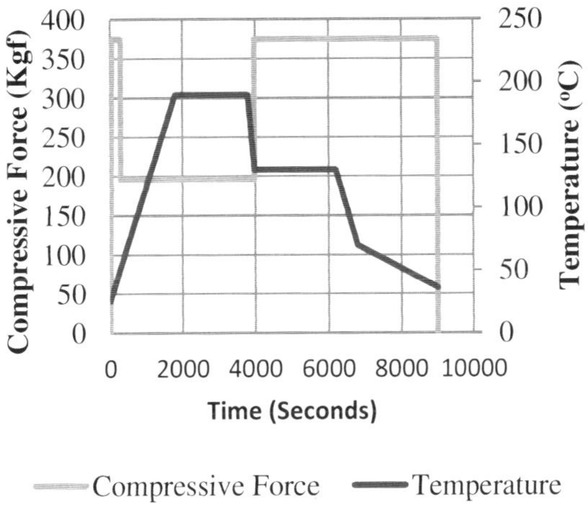

Figure 6. The parameter of compression molding process.

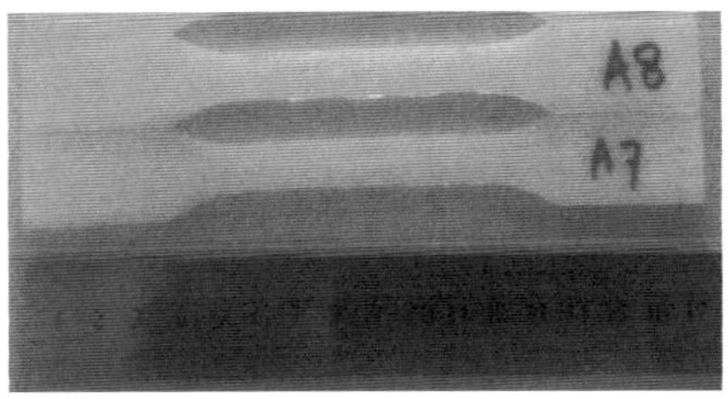

Figure 7. Tensile test specimen for CM product.

The results of the compression molding process in Figure 6 will produce UHMWPE in the form shown in Figure 7. This UHMWPE product will be used as a tensile test specimen with dimensions in accordance with ASTM D-638 Type IV standard. The number of specimens used for the tensile test process is the number of two test specimens.

Although using different test specimen standards between JIS-7113 on RE specimens and ASTM D-638 Type IV on CM specimens. However, the tensile test results can still be scientifically justified because JIS 7113 and ASTM D-638 Type IV standards have the same function as standard tensile testing for plastic materials, which in this case are UHMWPE materials. Differences in the use of test standards are intended to enrich the reference of researchers to not only fix on one standard test alone.

2.4.2. Micro hardness test preparation. This hardness testing process aims to determine the UHMWPE material hardness character. In this test, micro-hardness measurement method is used. For micro hardness test specimens, each RE product or CM, tested three specimens. Of the three test specimens it is expected to represent the value of micro hardness in the RE product or CM product which will be used to analyze the UHMWPE mechanical properties further.

3. Result and discussion

3.1. Tensile test

To find out the result of tensile test of CM product and RE product, we can see it in Figure 8. The result of tensile test consist of three vales, i.e., value of yield strength, ultimate tensile strength, and elongation value. The value shown in Figure 8 are the average values of two specimens in each product. The RE product has a higher yield strength than the CM product. That is 25.25 MPa for RE product and 23.67 MPa for CM product. Contrariwise, the CM product has a higher ultimate tensile strength than the RE product at 35.36 MPa for the CM product and 29.70 MPa for the RE product. The last parameter of this tensile test is the elongation value. It appears that CM products have greater elongation values than RE products at 214.75% for CM products and 120.45% for RE products.

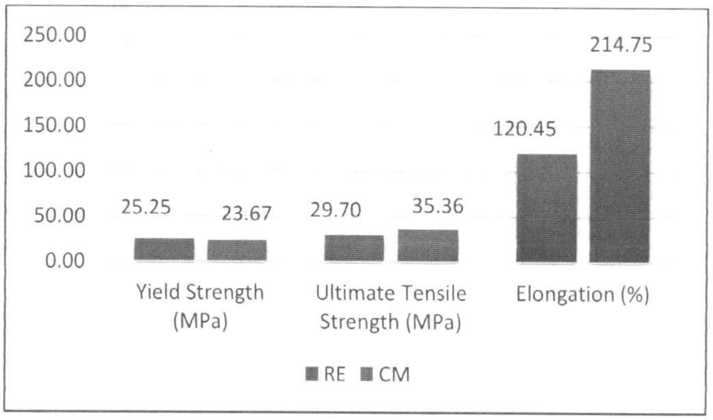

Figure 8. Tensile test result.

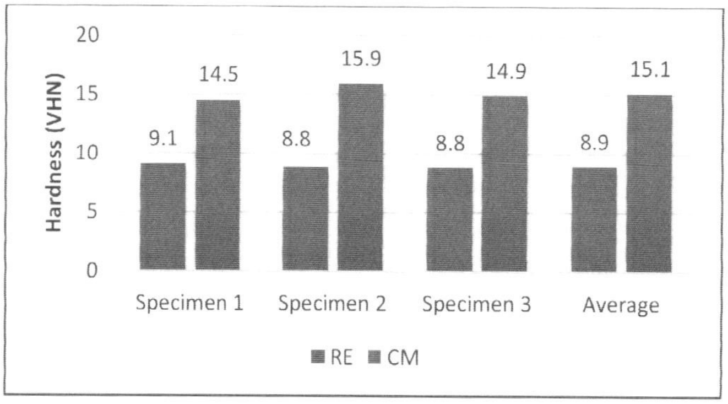

Figure 9. Hardness test result.

3.2. Hardness test
To find out the results of the CM product hardness test and the RE product, we can see it in Figure 9. This figure provides information to us about the value of each material hardness test. The CM product specimens have hardness values of 14.5 VHN; 15.9 VHN and 14.9 VHN. Whereas the RE product specimens have hardness values of 9.1 VHN; 8.8 VHN and 8.8 VHN. The pattern that can be seen for the whole test was that on all specimens for hardness testing, CM products are superior to RE products. The mean hardness values for these two products are 15.1 VHN for CM products and 8.9 VHN for RE products.

3.3. Discussion
Our research goal is to find out and compare the mechanical properties of UHMWPE resulting from compression molding process and ram extrusion. We can see in Figure 8 of the test results of RE and CM specimens. The value of yield strength, there is a difference but not significant. While the significant difference occurs in the ultimate value of tensile strength and elongation. It is known that a higher crystalline quantity can inhibit crack initiation and slow down crack propagation by encouraging crack deflection around lamellae structures [10]. The property of preventing crack and its propagation is one result of the higher value of ultimate tensile strength and elongation of a material. Therefore, the sinificant difference between the ultimate tensile strength and elongation values of CM and RE specimens can be due to the crystallinity value of the CM specimen larger than the RE specimen. Different results are also seen from the values of hardness of specimens CM and RE. The CM specimen has an average hardness value of 70% greater than the RE specimens. This is closely related to the difference in microstructural properties of both CM and RE specimens. Although this study showed better mechanical properties of CM specimens than RE specimens, it can not be generalized that the product of the compression molding (CM) process will always be better in mechanical properties compared to the ram extrusion product (RE). Differences in mechanical properties occur due to many factors that affect both resin types, manufacturing methods and UHMWPE sterilization processes.

The processing of UHMWPE has evolved considerably over the last half century and is uniquely tied to changes in the microstructural features of this polymer. Such evolution manifests as trade-offs in the clinical performance of this material. Crosslinking improves the wear resistance of UHMWPE but with a compromise in fatigue fracture resistance along with susceptibility to oxidation and fatigue wear *in vivo*. Annealing crosslinked materials above and below the melt can retain wear resistance and improve oxidative resistance, but influence crystalline quality and quantity that drives down fatigue crack propagation resistance. Enhancing crystalline quantity can restore this fatigue resistance, but increases the modulus and accompanying contact stresses that can render an implant more susceptible to delamination and pitting *in vivo*. Future iterations of this polymer should consider all aspects of microstructural changes, especially those that affect fatigue wear and fracture, to promote UHMWPE durability under greater and longer-lasting loading cycles and extend the lifetime of modern TJRs [10].

In direct compression molding (DCM), sometimes also called net shape compression molding, the manufacturer of the polyethylene insert effectively converts the resin to a finished or semifinished part using individual molds. One advantage of DCM is the extremely smooth surface finish obtained with a complete absence of machining marks at the articulating surface. In addition, higher processing pressures may be attained, if desired, because the projected surface area of each individual part mold is relatively small compared to the area of large molds used to compression mold sheets [1]. Starting from this study we will continue to develop the method of manufacturing compression molding or more precisely direct compression molding machine to convert UHMWPE powder into acetabular liner, without going through the process of milling or machining. Certainly still guided against the research that has been done by many researchers.

4. Conclusion
This study shows that the mechanical propertiy of the CM specimens are better than the RE Specimens, especially in terms of ultimate tensile strength, elongation and micro hardness. Although

this study showed better mechanical properties of CM specimens than RE specimens, it can not be generalized that the product of the compression molding (CM) process will always be better in mechanical properties compared to the ram extrusion product (RE). Differences in mechanical properties occur due to many factors that affect both resin types, manufacturing methods and UHMWPE sterilization processes. Direct compression molding (DCM) becomes the right solution for the moment in terms of time and cost efficiency in manufacturing acetabular liner components.

Acknowledgment
This research is supported by Laboratory for Engineering Design and Tribology, Diponegoro University.

References
[1] Kurtz S M 2009 *UHMWPE Biomaterials Handbook* London: Academic Press **2** 13-14
[2] Kurtz S M Muratoglu O K Evans M and Edidin A A 1999 *USA: Elsevier* 1660
[3] Fu J Ghali B W Lozynsky A J and Oral E 2010 *Boston: Elsevier*
[4] Sirimamilla P A Furmanski J and Rimnac C M 2012 *USA: Elsevier*
[5] Slouf M Vackova T Nevoralova M and Pokorny D 2014 *Prague: Elsevier* 191-7
[6] Bellare A Dorfman R Samuel A and Thornhill T S 2016 *Boston: Elsevier* 493-8
[7] Wang H Xu L Li R Hu J Wang M and Wu G M 2016 *Beijing: Elsevier* 41-9
[8] Reinitz S D Engler A J Carlson E M and Van Citters D W 2016 *USA: Elsevier* 623-8
[9] Liu T Eyler A and Zhong W H 2016 *Pullman: Elsevier* 17-20
[10] Ansari F Ries M D and Pruitt L 2015 *USA: Elsevier* 337

IOP Conf. Series: Materials Science and Engineering **432** (2018) 012008 doi:10.1088/1757-899X/432/1/012008

Effect Of cobalt addition on Fe-20Cr-5Al-1Y_2O_3 ODS alloy fabricated by arc plasma sintering (APS) process

B Suharno[1], D E Jodi[1], N R Pratama[1], A Dimyati[2]

[1] Department of Metallurgy and Materials Engineering, Faculty of Engineering University of Indonesia, Depok 16424, Indonesia
[2] Center for Science and Technology of Advanced Materials. BATAN Puspiptek Area, Serpong, Tangerang 15314, Indonesia

prof.suharno@gmail.com

Abstract. In this research, microstructure, phases and hardness of Fe-Cr-Al based Oxide Dispersion Strengthened (ODS) alloy with addition of cobalt (Co) produced using arc plasma sintering (APS) method were investigated. ODS alloy has an important role particularly in high temperature applications, since it has the ability to maintain their strength at high service temperature. On the other hand Co is well known to increase creep resistance of many alloys, which could in turn increases their lifetime in high temperature environment. To produce ODS alloy, usually high cost and long processing time are required. To overcome these problems, a new sintering method based on arc plasma was used in this study, which has been successfully applied in previous works. The alloys composed of Fe-20Cr-5Al-1Y_2O_3 with addition of 0.5, 1, 1.5 and 2 wt% cobalt were first prepared by mechanical alloying (MA) comprising high-energy milling for 8 hours and subsequent isostatic pressing at 20 ton-force/in^2 prior to sintering by APS method up to 28 minutes. The sintered samples then characterized by means of Scanning Electron Microscopy (SEM) equipped with an Energy Dispersive X-Ray Spectrometer (EDX) to observe the microstructure, the X-Ray Diffraction Technique (XRD) to investigate the formed phases, and Vickers microhardness test to examine hardness value of the alloys. The results show that APS method is technically and economically feasible for fabricating ODS alloys since this process significantly reduces time and energy required. Variation of cobalt in the alloy obviously does not cause observable microstructure difference. However, the alloy with 0.5 wt% cobalt exhibited single FeCrAl matrix phase, while at higher cobalt concentration additional AlCo phases were found in the FeCrAl matrix, which can be assumed to improve the high temperature usage of the alloy. Cobalt addition also increased hardness value of these alloys.

1. Introduction

Oxide dispersion strengthened (ODS) alloys are promising material mainly for high temperature and corrosive applications, i.e. nuclear reactor and power generation plant [1, 2], because of the presence of oxide particles that are dispersed on grain and grain boundaries that could inhibit creep during high temperature applications [3]. The presence of chromium and aluminum in Fe-20Cr-5Al alloy help to improve its oxidation resistance due to the formation of alumina scales, whereas the addition of 1 wt% Y_2O_3 improves both the mechanical strength and the high temperature properties, and also lowers the minimum aluminum concentration required for the formation of protective alumina scale on alloy surface [4, 5]. The addition of cobalt in alloys, generally improves tensile strength, but it also helps improving fatigue and creep resistance [6].

ODS alloys are commonly manufactured using powder metallurgical technique and subsequent sintering in a conventional heating furnace, which is constrained by the high cost and it is also very time consuming. To overcome these issues, a new method for sintering ODS alloy and other high temperature alloys based on plasma generated by an electrical arc has been developed in Center for Science and Technology of Advance Materials (PSTBM) BATAN which is known as Arc Plasma Sintering. APS method for producing Fe-Cr with variation of Cr content and with addition of ZrO_2 has been proved to be an effective sintering method with less time and energy required [7]. This method also has been previously studied for the fabrication of silicon carbide, which resulted in an increase of mechanical strength followed by lower energy and less time required for its sintering process [8].

This study's primary objective is to investigate the effect of cobalt content variation in the range of between 0.5 and 2 wt% on the microstructure, phase formation and mechanical property of the Fe-20Cr-5Al-1Y_2O_3 ODS alloy after sintering by the APS method.

2. Experimentals

Four samples of ODS Fe-20Cr-5Al-1Y_2O_3 alloy were fabricated from Fe with a purity of 99.9%, Cr of 99%- and Al of 99% purity, all starting materials are powder, with cobalt addition which was varied 0.5, 1, 1.5, and 2 wt% firstly by means of powder metallurgy. The initial powder size was measured and presented in Table 1. The powder mix was mechanically alloyed using high-energy mill HEM E3-D for 8 hours with ball to powder ratio of 2:1 in order to reduce powder size and to homogenize the powder mixing. The powder was then pressed using uniaxial pressing method with pressure of 20 ton-force/in^2 to produce sample coin as green material. Sintering process is performed using APS method up to 28 minutes at 12 Volt and 80 Amperes to consolidate the samples. The morphology and composition of the alloys were observed by JEOL SEM-EDX. Investigation of formed phase was further investigated by RIGAKU-SMARTLAB X-Ray Diffraction (XRD), and mechanical properties were measured using Vickers microhardness (HV) tester using 300 gf load and 10s indentation time.

Table 1. Initial powder size of Fe, Cr, Al, Y_2O_3, and Co.

Powder	Fe	Cr	Al	Y_2O_3	Co
Average Powder Size	5 μm	5 μm	20 μm	50 nm	5 μm

3. Results and discussion

3.1. Microstructures

The microstructures of the specimens are observed using optical microscope (OM) and SEM, which are shown in Figure 1 and 2. Based on the microstructure observation, the addition of cobalt in the ODS alloy alloys seems to give no significant difference. The sharp-angle grains in the microstructure in all samples were mainly caused by the impact force that occurred during high energy mechanical alloying process, in which the powder was work hardened and fractured as the result of fatigue mechanism, and the powder turn to flaky shape with sharp angles [9, 10].

There is also no significant microstructure and grain size changes with the cobalt content. The average grain size for each ODS alloy sample with different cobalt content is shown in Table 2, the grain size has obviously been increased significantly compared with initial powder used in this study.

In addition, the increasing cobalt content also does not yield any significant changes in the number and size of porosities. All of the specimens exhibit localized and reduced diameter in porosity size and shape, which is obviously caused by the compaction and the sintering process [11, 12].

To observe the relationship between cobalt content and its dispersion in the alloys, SEM and EDX measurement were used. EDX examination was performed on the sample with 0.5 and 2 wt% cobalt addition, the results are shown in Figure 3 and Table 3.

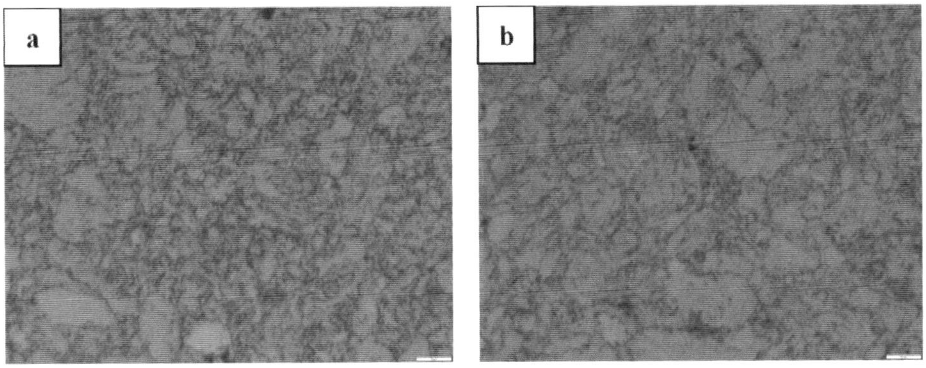

Figure 1. OM micrographs of the arc plasma sintered Fe-20Cr-5Al-1Y_2O_3 ODS alloy observed at 1000x magnification, with addition of (a) 1 and (b) 1.5 wt% Co.

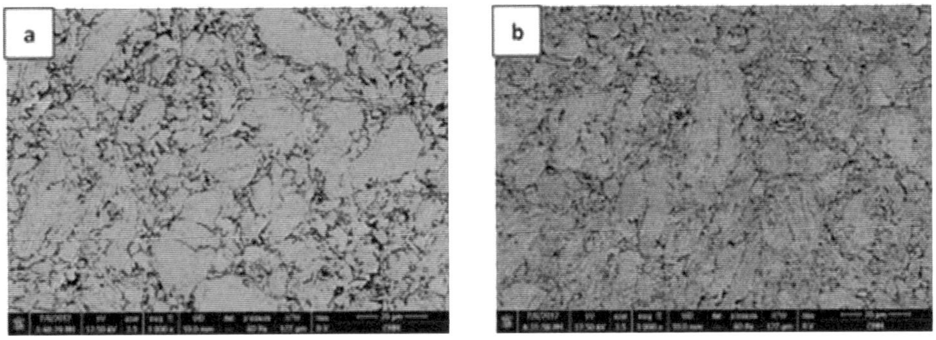

Figure 2. SEM micrographs of the arc plasma sintered Fe-20Cr-5Al-1Y_2O_3 ODS alloy with addition of (a) 0.5 wt% Co (b) 2 wt% Co.

Table 2. Average grain size for each ODS alloy sample after arc plasma sintering process.

Composition	0.5 wt% Co	1 wt% Co	1.5 wt% Co	2 wt% Co
Average grain size	12.86 μm	13.14 μm	13.77 μm	13.37 μm

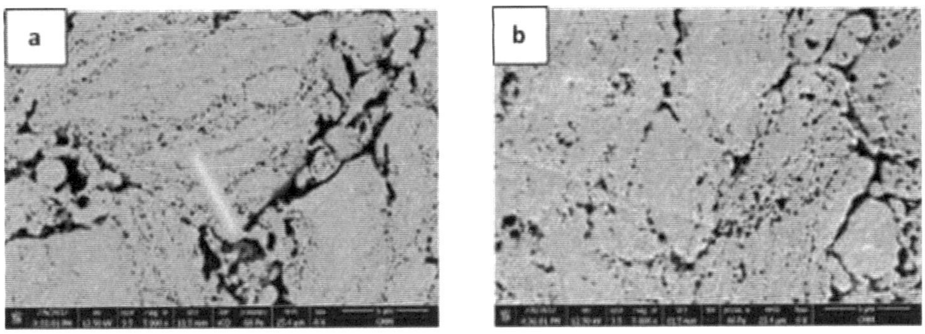

Figure 3. SEM images with the area of EDX examination on Fe- 20Cr- 5Al- 1Y_2O_3 ODS alloys samples with addition of (a) 0.5 and (b) 2 wt% Co, indicated with arrows.

From EDX results as shown in Table 3, in general, it is similar to the main composition used in the alloys, with the main elements consisting of Fe and Cr. A unique feature was found in the Fe-20Cr-5Al-

$1Y_2O_3$-0.5Co specimen, shown by the yellow arrow, the phase consists of Y and Al in higher composition compared to EDX result in general position which is Fe-rich, and makes the color of this phase tends to be darker. The location of this yttrium-containing phase itself is on the edge of the grain, which is caused by the original trait of Y_2O_3 that is located on the grain boundary that would help to impede the creep movement [13]. Also, from the EDX result, there is a presence of oxygen in the microstructure, which could be the oxide phase that was formed during the sample preparation before EDX, and oxygen may react with chromium and formed Cr_2O_3. This formed oxide also suppress the densification process by providing a barrier to atomic diffusion between particles, resulting in lower density value and lower mechanical strength [14, 15]. Further investigation about the formed phase will be described in the next section.

Table 3. Chemical composition of the arc plasma sintered Fe-20Cr-5Al-$1Y_2O_3$ ODS alloys with addition of 0.5 and 2 wt% Co measured by EDX.

Alloy	Fe (%)	Cr	Al	Co	O
0.5 Co	70.81	17.78	4.91	1.44	5.06
2 Co	70.9	20.2	2.4	4.1	2.4

3.2 Phase formation

X-ray diffraction (XRD) is used to identify phases that exist in the specimens. Results are shown in figure 4. From the XRD results, for alloy with the addition of 0.5 wt% cobalt, there are 3 main diffraction peaks at the 2θ angular positions of 44.56°, 64.84°, and 82.01°, which are identified as ferrite phase that is rich in iron, chromium, and aluminium, since these peaks closely resemble this phase's Miller Indices and also yields the lattice parameter of 2.87 A°. The main reason of the formation of this ferrite phase is because the main elements of this specimen is α-Fe and Cr, which cover 93.5% of total composition in the specimen. Also, the addition of 0.5% cobalt could be assumed to be too low to form any intermetallic phase of cobalt. For alloy with 2 wt% cobalt addition, besides the three main peaks that are found in alloy with 0.5 wt% cobalt, there are also three small peaks are observed at 2θ of 31.10°, 55.35°, and 73.69° respectively, which could be identified as AlCo phase, since the angular positions closely correspond to the AlCo reflection angles, as well as its Miller Indices number, and also its lattice parameter with a value of 2.862 A°. This intermetallic compound has the ability for high temperature applications, since this compund has high melting temperature, moderate oxidation resistance, and the ability to maintain its crystal structure up to its melting temperature [16].

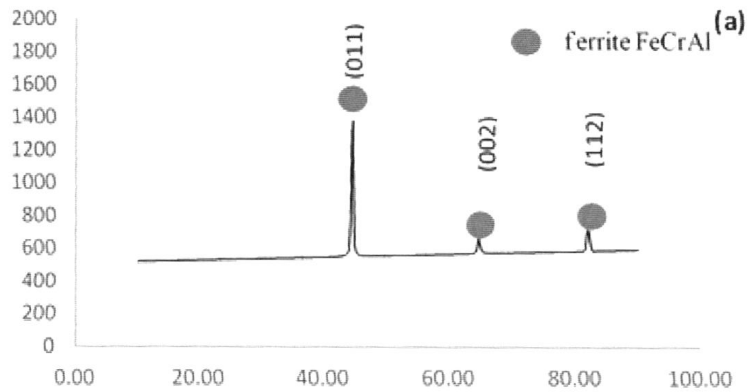

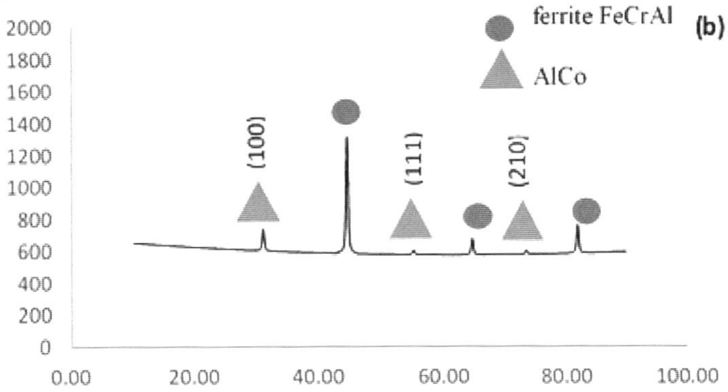

Figure 4. XRD spectren of ODS Fe-20Cr-5Al-1Y_2O_3 alloy with addition of (a) 0.5 and (b) 2 wt% Co.

Although theoretically AlCo phase is formed when the composition is around Al-65%Co to Al-73%Co, this AlCo phase could also be formed with a lower amount of cobalt when processed using mechanical alloying method, due to composition change of the powder mixture during milling process and also the enhanced solubility of AlCo phase by milling [17]. Also, in this investigation, the oxide phase that is predicted to be present on the alloys is predicted to be too low in composition, since there is no oxide phase that is detected in the microstructure. Further research is needed to understand the effect of AlCo phase on the ODS Fe-20Cr-5Al-1Y_2O_3 alloy and also the correlation between AlCo phase and Y_2O_3 particles in high temperature applications.

3.3 Mechanical properties

Figure 5 shows the Vickers microhardness results after arc plasma sintering process as a function of cobalt content. Hardness value increases with an increase in cobalt addition, which mainly affected by the addition of cobalt in ODS alloys. This result is similar to the results obtained in other previous experiments, and the reason is because Co helps to decrease the stacking fault energy and increase the plastic deformation resistance, although this behavior tends to significantly affective for face-centered cubic metals [18]. The increase of hardness in these alloys couldn't be fully associated with the presence of AlCo phase in the microstructure, but the hardness value of these alloys could reach higher number if the sintering process had been done in fully argon or vacuum atmosphere, since there's no oxide phase that would present in the specimens that could decrease the hardness value. Changes in the hardness value also affect the mechanical strength, since both of these aspects are related to each other, and increase in hardness value should also increase the mechanical strength of these alloys.

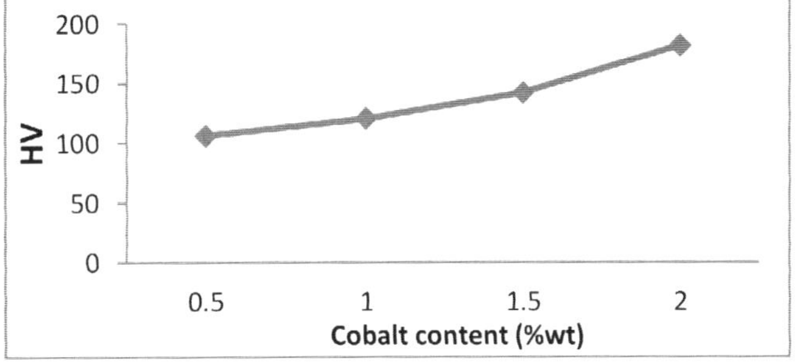

Figure 5. Vickers microhardness after arc plasma sintering vs. cobalt content.

4. Conclusion

This research has successfully fabricated alloys of ODS Fe-20Cr-5Al-1Y_2O_3 with cobalt addition using arc plasma sintering method, which is proved to reduce fabrication time, cost, and energy significantly. The addition of cobalt in 0.5 wt% in this alloy form FeCrAl main phase, and as cobalt content increase in around 2 wt%, AlCo starts to form in Fe-Cr matrix that could help this alloy in high temperature application, although further research are needed to observe the actual effect. Increase in cobalt also increase the hardness and mechanical strength value of these ODS alloys.

Acknowledgement

The authors acknowledge financial support from Directorate of Research and Community Service University of Indonesia (DRPM UI) and technical support from Indonesian National Nuclear Energy Agency (BATAN).

References

[1] Sakasegawa H *et al* 2009 *J. Nucl. Mater.* **384(2)** 115–8
[2] Capdevila C Serrano M Campos M 2014 *Mater. Sci. Technol.* **30(13b)** 1655–7
[3] Hayashi T Sarosi P M Schneibel J H Mills M J 2008 *Acta Mater.* **56(7)** 1407–16
[4] Revie R W Uhlig H H 2000 *Electrochemical Society Uhlig's corrosion handbook*
[5] Yang F 2012 *J. Nanomed. Nanotechnol.* **4(1)** 1–3
[6] Sharov B P Zikeev V N 1977 *Met. Sci. Heat Treat.* **19(1)** 265–8
[7] Bandriyana Sujatno A Salam R Sugeng B Dimyati A 2017 *IOP Conf. Ser. Mater. Sci. Eng.* **176**
[8] Kim J J Venkateswaran V Phoenix R C 1987 *Plasma arc sintering of silicon carbide United States Patent 4,676,940, Jun. 30*
[9] Suryanarayana C 2011 *Progress in Mater. Sci.* **46(1–2)** 1–184
[10] Oka H Watanabe M Ohnuki S Hashimoto N Yamashita S Ohtsuka S 2014 *J. Nucl. Mater.* **447(1-3)** 248–53
[11] Sheppard T McShane H B 1980 *Powder Metall.* **23(3)** 427–32
[12] Kurgan N 2013 *Mater. Des.* **52** 995–8
[13] Elzey D M Arzt E 1988 *Superalloys 1988* 595–604
[14] Ye H Liu X Y Hong H 2008 *Mater. Lett.* **62(19)** 3334–6
[15] Sung H J Ha T K Ahn S Chang Y W 2002 *J. Mater. Process. Tech.* **130–131** 321–7
[16] Takano Y Yoshinaka M Hirota K Yamaguchi O 2001 *J. Am. Ceram. Soc.* **84** 2445–7
[17] Sui H X Zhu M Qi M Li G B Yang D Z 1992 *J. Appl. Phys.* **71(6)** 2945–9
[18] Helis L Toda Y Hara T Miyazaki H Abe F 2009 *Mater. Sci. Eng. A* **510–511(C)** 88–94

IOP Conf. Series: Materials Science and Engineering **432** (2018) 012009 doi:10.1088/1757-899X/432/1/012009

Development of Fe-20Cr-5Al -1Y_2O_3 ODS alloy with addition of Ni by means of Arc plasma sintering (APS) method

A Dimyati[1], N R Pratama[2], D E Jodi[2], B Suharno[2]

[1] Center for Science and Technology of Advanced Materials, Indonesian National Nuclear Energy Agency Puspiptek Area, Serpong 15315, South Tangerang, Indonesia
[2] Department of Metallurgy and Materials Engineering, Faculty of Engineering University of Indonesia, Depok 16424, Indonesia

arbi.dimyati@gmail.com

Abstract. APS has two main advantages over the conventional method since it reduces the energy consumption and process time significantly. In this work, ODS alloys composed of Fe-20Cr-5Al-1Y_2O_3 with addition of 0.5, 1, 1.5, and 2 wt% nickel were produced. These alloys are firstly fabricated via powder metallurgy method consisting of mechanical alloying by high energy milling for 8 hours and uniaxial isostatic pressing with 20 ton-force/in^2 pressure, and subsequent sintering in APS for 28 minutes. The alloy samples were investigated using Scanning Electron Microscope (SEM) equipped with X-Ray Diffraction Spectroscopy (EDX) to reveal the microstructure and chemical composition, X-Ray Diffraction (XRD) to analyze the phases which occur after sintering, and Vickers micro hardness testing to examine the hardness value of specimens. The results show that APS method in processing ODS alloy has given two main technical advantages over the conventional method: first, low energy need, and second, short process time. The addition of nickel leads to the formation of the microstructure consisting of three main phases, the (Fe,Ni) or kamacite, NiAl and ferrite matrix. The presence of NiAl is recognized as essential since it can improve oxide particles within the alloy matrix and along grain boundaries to impede dislocation motion particularly at high temperature (creep). Nickel addition also shows increasing hardness value of the ODS alloys.

1. Introduction

Oxide Dispersion Strengthened (ODS) alloy is a kind of alloy which consists of high temperature matrix phases such as Fe-Cr, Fe-Al, Fe-Cr-Al, Ni-Cr or Ni-Al and small addition of oxide particles, which give the alloy sufficient resistance to corrosion or oxidation at high temperature applications, such as nuclear applications [1-4]. ODS alloy is fabricated by mechanical alloying method using milling process, in which cold welding, fracturing and re-welding can occur. Both cold welding and re-welding are actually the same process, the difference is located in the milling stages. Cold welding occurs when milling is in progress because of the collision between particles of powder at room temperature (without heating process) while re-welding happens as powder inside HEM-E3D tool collide with each other and make the particles of powder separated (fractured) but at the end it will *merge again* (re-weld process) [5, 6]. Then, the mixed powder is pressed, as the result of the powder stick to each other (more compact/denser). It will make the sintering process easier, which means less energy consumption, because the gaps between particles are close that will maximized the necking process. After that conventional sintering process was used but requires long processing time to reduce the porosity significantly and consumes a high energy. As an alternative way producing ODS alloys is sintering with plasma in APS which is developed in the Center for Science and Technology of Advanced Materials (PSTBM)-BATAN. It has been proved to reduce sintering energy and time significantly.

In this study, a new ODS alloy based on FeCrAl with addition of nickel was developed using APS method. Nickel is assumed to act as a strengthening element in super alloys and improves their creep and oxidation resistance. Microstructure, mechanical properties, and phase formation were investigated using electron microscopy and X-ray diffraction to get detailed insight into the effectiveness of addition Ni in ODS alloy.

2. Experimentals

ODS alloy samples are fabricated from powder materials with compositions and particle sizes as shown in Table 1.

Table 1. Chemical composition of ODS alloy samples.

Element	Weight percentage	Size
Fe	Bal	5 µm
Cr	20 %	5 µm
Al	5 %	20 µm
Y2O3	1 %	50 nm
Ni	0.5 % , 1% , 1.5% , 2%	10 µm

The materials were mixed and milled using high energy milling (HEM-E3D) from Nanotech Indonesia for 8 hours in order to reduce the particle size and for homogenization, followed by isostatic pressing 20 ton-force/in^2 using AKS 3030 pressing tool to produce green body as sample coins. The sintering process was conducted using APS method at 12 V and 80 Amps DC current in Argon atmosphere up to 28 minutes. JEOL SEM-EDX was used to observe the microstructure of the samples at magnification as high as 5000x, and to identify the chemical composition. Determination and identification of new phases which occur in the sample are investigated using RIGAKU-SMARTLAB X-Ray Diffraction (XRD). The result was then analyzed and refined with Xpert HighScore software.Vickers micro hardness test was performed to examine the relation of nickel addition to alloy hardness. The test was conducted with 300 grams load and 10 seconds indentation time.

3. Results and discussion

3.1. Microstructures

SEM observation of the sample alloys containing 0.5 and 2 wt% Ni after sintering in APS for 28 minutes are shown in Figure 1.

SEM images of the samples containing 0.5 and 1 wt% Ni in Figure 1 generally show an alloy with large and small grains containing porosities (appear as black) spread in grains and grain boundaries. The fraction of porosities to observable area is about 10%. This is quite good, which indicates that although the APS method uses a relatively short process time and pressing is absent during sintering, it can significantly reduce the size and amount of porosity, which is in agreement with the investigation of Kumar *et al.* [7]. The reason for this is the difference in diffusion rate between the phase formation and the necking process during sintering [8].

The effect of nickel variation on the microstructure of ODS alloy sample cannot be identified in the Figure 1. No change of the microstructure and the number of the porosity on the specimens was observed. This can suggest that the powder metallurgy preparation process prior to sintering has successfully homogenized the powder mix and make the samples more compact after pressing.

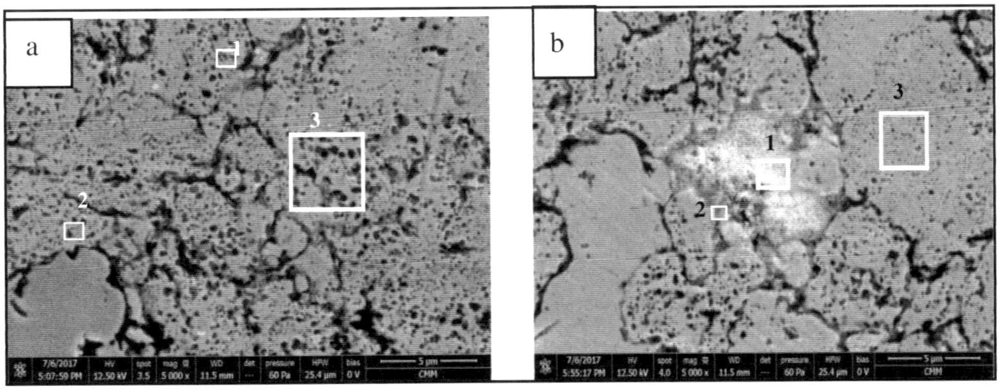

Figure 1. SEM micrographs of Fe-20Cr-5Al-1$Y_2O_3$with addition of (a) 0.5 and (b) 2 wt% Ni sintered in APS for 28 minutes.

Table 2. Chemical composition of ODS alloy measured by EDX at position as indicated in Figure 1.

ODS Steel	No	Fe	Cr	Al	Ni	Y	O
	1	26.8	10.3	9.7	-	30.6	22.5
Ni 0.5%	2	28.7	58.8	2.4	-	0.7	2.7
	3	64.9	17.7	7.3	-	1.2	8.8
	1	22.8	7.1	2.7	-	16.2	51.2
Ni 2%	2	23.2	9.3	9.6	-	20.8	37.2
	3	69.4	19.3	3.5	3.2	4.6	-

Plasma sintering was quite efficient at consolidating the samples after compaction process due to very high temperature, and will improve heat transfer (free electron movement) together with ionization process in plasma [9, 10]. As the effect of the high temperature of the plasma, the green body more easily consolidated, forming necking in relatively short time compared to in conventional sintering process.

From the EDX results as presented in Table 2, in general Fe-rich phase form the alloy matrix, which appears as grey area. This phase can be recognized as ferrite since the main composition of this alloy consists of Fe-Cr-Al which is categorized as ferritic steel [11].

Another reason why the main phase formed in ODS steel is ferrite is the base elements such as Fe and Cr. Both elements are well known as ferrite formers. These elements can form Fe-Cr matrix which is a ferrite phase. Also, a recent study stated that if the addition of Cr in weight percentage reaches more than 12%, it will be categorized as ferritic steel [12]. A distinct feature found in Figure 1(a) point 2, where there is smooth and spotless region that represents Cr-rich phase which is also categorized as ferrite phase. Higher Cr content in a steel can make an alloy have dominant ferrite phase because Cr element is one of the ferrite formers [13]. Another unique feature there is also found in Figure 1(b) is the presence of white region that represents Y-rich phase. This Y-rich phase will make a compound together with O to become Y_2O_3 and this oxide particle will prevent dislocation motion at high temperature, providing resistance to creep [14]. This oxide particle is very stable at elevated temperature since it has a high melting point [15]. Nickel was identified in the microstructure of ODS steel with 2% wt of Ni and this element will be a compound with Al to create NiAl intermetallic compound. According to EDX result, with the increasing of nickel content, it will form more intermetallic compound and can affect the resistance to high temperature [16]. To further identify phase in the specimen, X-Ray Diffraction is used to analyze formed phase.

3.2 Phase formation

Type of phases which were present in the samples were analyzed using X-Ray Diffraction, which results are shown in Figure 2.

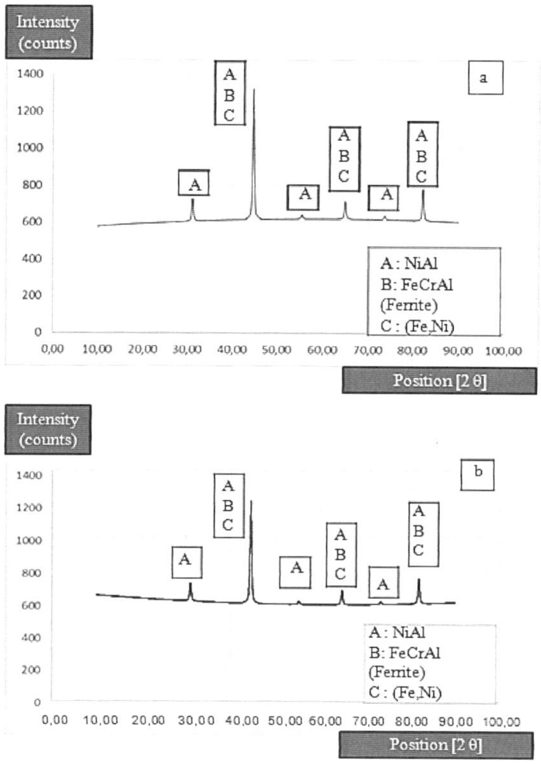

Figure 2. XRD patterns of ODS alloy with addition of (a) 0.5 wt% Ni and (b) 2 wt% Ni.

Based on Table 3, In the XRD patterns of ODS alloy with addition of 0.5 wt% Ni, there are 6 diffraction peaks at angles 2θ of 44.6°, 64.9°, 82.2°, 98.8°, 116.2°, and 137° with Miller indices of (110), (200), (211), (220), (310), and (222). Based on the reference, type of phase which are present is (Fe,Ni) or kamacite and has cubic crystal [17]. Another phase present is FeCrAl or ferrite. The diffraction peaks of this phase have angles 2θ of 44.6°, 65°, 82.3°,98.9°, 116.3°, and 137.1° with Miller indices of (110), (200), (211), (220), (310), and (222) [18].

The ferrite phase is present because ODS alloy with the main composition of Fe-Cr-Al is categorized as ferritic steel [11]. According to the result of XRD analysis, there are also 6 diffraction peaks at angles 2θ which are 30.9°, 44.3°, 55°, 64.4°, 73.2°, and 81.6° with Miller indices of (100), (110), (111), (200), (210) and (211). Based on the reference, type of phase which formed is intermetallic compound of NiAl and has cubic crystal [19]. The reason that Ni is added to ODS steel in this work is to observe whether nickel aluminide phase can be formed in this alloy, as this phase can improve the resistance of oxidation at high temperature and creep due to high temperature application [20]. NiAl possesses excellent high temperature strength and resistance to oxidation and corrosion because of the long range ordered structure of this phase, and this structure impedes the mobillity of atoms [16]. This NiAl intermetallic compound can form small precipitates, and together with oxide particle (Y_2O_3), it helps to preserve the atomic order from dislocation, especially in high temperature. Also, this intermetallic compound acts as strengthening constituent in nickel based super alloy [16, 19].

Table 3. Result of XRD Pattern for addition of 0.5% Ni.

No	Addition Element	Diffraction Peaks (2θ)	Miller Indices	Phase
1	Ni 0.5%	44.6 °	(110)	Kamacite
		64.9°	(200)	
		82.2°	(211)	
		98.8°	(220)	
		116.2°	(310)	
		137°	(222)	
2	Ni 0.5%	44.6°	(110)	FeCrAl (Ferrite)
		65°	(200)	
		82.3°	(211)	
		98.9°	(220)	
		116.3°	(310)	
		137.1°	(222)	
3	Ni 0.5%	30.9°	(100)	NiAl
		44.3°	(110)	
		55°	(111)	
		64.4°	(200)	
		73.2°	(210)	
		81.6°	(211)	

There is no significant change between ODS alloy with addition of 0.5 wt% Ni and addition of 2 wt% Ni. Based on the results of XRD analysis in the ODS alloy with 2 wt% Ni, the phases formed are similar. There are also three type of phases, which are NiAl, (Fe,Ni) or kamacite, and FeCrAl (ferrite), although increase in nickel addition would increase the amount of NiAl phase that will give advantage in the improvement of resistance to corrosion or oxidation at high temperature [21]. Further research of oxidation or corrosion resistance to high temperature should be done to determine the effect of NiAl phase.

3.3 Hardness

Vickers micro hardness testing was used to know the relation of nickel addition to the hardness value of ODS alloy. The result is shown in Figure 3.

Based on the graph above, the increase of Ni percentage to ODS alloy will result in the increase of hardness value of the samples. The results of the hardness data are 109.37 HV for addition of 0.5 wt% Ni, 142.43 HV for addition of 1 wt% Ni, 157.17 HV for addition of 1.5 wt% Ni, and 159.9 HV for addition of 2 wt% Ni. According to references, the addition of Ni to steel alloy will increase the strength of ferrite phase, accompanied with the improvement of toughness and hardness [22]. The increase of nickel weight percentage to ODS alloy induces improvement of hardness value. Based on the result of XRD analysis, with the increasing nickel content, more intermetallic compound of NiAl formed. This intermetallic compound has similar properties to both ceramic and metal, and it is usually also used as strengthening constituent in nickel-base super alloy for high temperature application [23]. Hardness value of the samples with NiAl phase formed can be related since NiAl can act as strengthening phase to ODS alloy. The more NiAl formed in ODS alloy, the higher hardness value of specimens is achieved. The conclusion is increasing of nickel content to ODS alloy improve the hardness value of the samples.

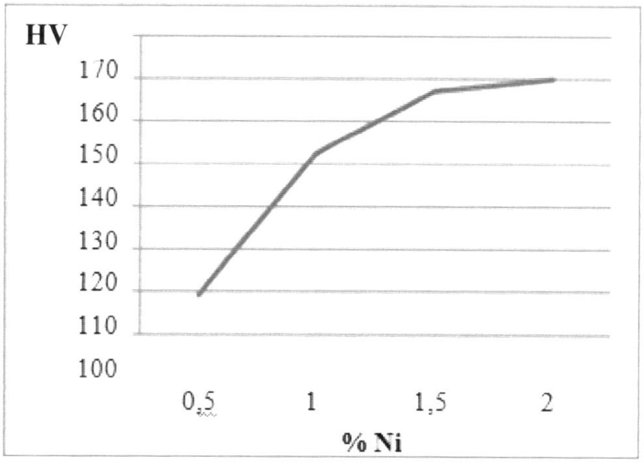

Figure 3. Vickers micro hardness vs nickel content result.

4. Conclusion

According to microstructure of the sintering process result, there is still porosity which lays across the samples. This is a usual phenomenon since sintering process only reduces the size and amount of porosity. Then, there are three phases which are present in the samples, FeCrAl (ferrite), NiAl and (Fe,Ni). One of the reasons for this research was to achieve the NiAl phase which can support the function of oxide particle in ODS alloy for the resistance of oxidation and corrosion also creep in high temperature environment. As the percentage of Ni in the ODS alloy increases, more NiAl will form, and the oxidation resistance becomes better. However, further research should be done to know the effect of this NiAl phase. The result of Vickers micro hardness testing indicates that with the increase of Ni addition to ODS alloy, the hardness value of samples improves because more NiAl formed in the samples which act as strengthening constituent.

Acknowledgment

The authors wish to thank Directorate of Research and Community Service University of Indonesia (DRPM UI) for the financial support and BATAN for technical support, also Mr. Rohmad Salam for APS work assistant.

References

[1] Hoelzer D T Bentley J Sokolov M A Miller M K Odette G R and Alinger M J 2007 *J. Nucl. Mater.* **367–370 A** 166–72
[2] Noh S Choi B Han C and Kang 2013 *Nucl. Eng. Tech.* **45(6)** 821–6
[3] Li M Zhou Z He P Liao L Xu Y and Ge C 2010 *Fusion Eng. Des.* **85(7–9)** 1573–6
[4] Silalahi M Dimyati A Harjanto S Untoro P and Suharno B 2014 *Int. J. of Tech.* 169–82
[5] Suryanarayana C Ivanov E and Boldyrev V V 2001 *The sci. and tech. of mechan. Alloy.* **306** 151–8
[6] Suryanarayana C 2001 *Prog. Mater. Sci.* **46(1–2)** 1–184
[7] Kumar S Zinzuwadia M Dabhade V and Prakash U 2012 *Comparative study of microstructural development during sintering and powder forging of ODS steels*

[8] Höganäs A B Sweden H and Publications 2012 *Sintering*
[9] Pfender E 1967 *Pure Appl. Chem.* **48(1)** 199–213
[10] Vardelle M Vardelle A Fauchais P and Boulos M I 1988 *AIChE J.* **34(4)** 567–73
[11] Bandriyana Sujatno A Salam R Sugeng B and Dimyati A 2017 *IOP Conf. Ser. Mater. Sci. Eng.* **176**
[12] Sundararajan G Vijay R and Reddy A V 2013 *Curr. Sci.* **105(8)** 1100–6
[13] Maalekian M 2007 Technische Universität Graz Christian Doppler Laboratory for EarlyStages of Precipitation The Effects of Alloying Elements on Steels (I) October
[14] Pasebani S Dutt A K Burns J Charit I and Mishra R S 2015 *Mater. Sci. Eng. A* **630** 155–69
[15] Susila P Sturm D Heilmaier M Murty B S and Sarma V S 2011 *Mater. Sci. Eng. A* **528(13–14**) 4579–84
[16] Yang J 1989 *Mater. Sci. and Engineer. A* **107** 81–91
[17] Keller L Rask J and Buseck P 1986 *ICDD Grant-in-Aid* Arizona
[18] Swanson *et al.* 1955 *Natl. Bur. Stand.* (U.S.) Circ. 539
[19] Baker I 1922 *ICDD Grant-in-Aid* Hanover New Hampshire
[20] Chen T and Wu J 2016 *Applied Sci.* 1–9
[21] Dey G K 2003 *Sadhana* **28** 247–262
[22] Higgins R A 1993 *Part 1 Applied Physical Metallurgy Engineering*
[23] Darolia R Walston W S and Nathal M V 1996 *NASA Lewis Research Center*

IOP Conf. Series: Materials Science and Engineering **432** (2018) 012010 doi:10.1088/1757-899X/432/1/012010

Effective dose analysis of extremely low frequency (ELF) magnetic field exposure to growth of *S. termophilus, L. lactis, L. acidophilus* bacteria

Sudarti[1*], T Prihandono[1], Yushardi[1], Z R Ridlo[1], A Kristinawati[1]
[1] Department of Physics Education, University of Jember, Jember, Indonesia

sudarti.fkip@unej.ac.id

Abstract. The Extremely Low Frequency (ELF) magnetic field, which is a component of a non-ionizing radiation, at low intensity has been shown to increase cell proliferation. This fact is expected to lead to a technological breakthrough in *S. thermophilus*, *L. lactis*, and *L. acidhopilus* bacteria activation which is needed in the production process of cream cheese. The main issue considered here in the production process of cream cheese is the growth of bacteria which often cannot be optimal, causing it to taste bad. This study is an initial step to determine the effective dose of exposure to ELF magnetic fields on the *S. thermophilus*, *L. lactis*, and *L. acidhopilus* bacteria growth. This research used the exposure of ELF magnetic field with intensity of 100 μT and 300 μT to those three types of bacteria with exposure durations of 5, 15, 25, 35, and 45 minutes. Measurements of the amount of bacteria were performed 1 hour, 2 hours, and 16 hours after exposure to ELF magnetic field. The results of this study indicated that the growth of three types of bacteria exposed to ELF magnetic field with an intensity of 100 μT was significantly faster than the control one, but the other one which were exposed to 300 μT ELF magnetic field was not different from the control one. The analysis results provided evidence that the rate of optimal growth occurred in three types of bacteria after being exposed to 100 μT ELF magnetic fields for 5 minutes. Therefore, it could be concluded that exposure to ELF magnetic field with an intensity of 100 μT for 5 minutes is an effective dose for the activation of *S. thermophilus*, *L. lactis*, and *L. acidhopilus* bacteria growth.

1. Introduction

Extremely Low Frequency (ELF) magnetic field is a component of ELF electromagnetic waves, proven to have the ability to penetrate almost all types of matter. It is a form of non-ionizing radiation, low-energy, and non-thermal energy. This characteristic provides a very high advantage that ELF magnetic field radiation is not destructive because it does not cause ionizing effects or thermal effects on the atoms that interact with it. In addition, the ELF magnetic field is easy and cheap to obtain, we are always around of ELF magnetic field, considering that in every current flow in electricity cable and equipment around us will arise the ELF magnetic field. Therefore, the utilization of ELF magnetic field becomes the focus of the researcher's attention, especially in the food industry. The result of Sudarti's research, 2016, showed that ELF magnetic field radiation is potentially used for *Salmonella thypimurium* sterilization in the gado-gado dish, showing that exposure to 646.7 μT for 30 minutes was able to decrease *Salmonella typhimuriun* amount 56% growth [1], while Sudarti and Kristiani A, 2016, proved that exposure to 100 μT magnetic field for 5 minutes was able to lower the water content in the process of making cream cheese [2]. Tessaro LW, *et al.*, 2015, also proved that exposure to ELF magnetic field of 250 μT can increase the growth of *Staphylococcus* bacteria cultures [3].

This study provides inspiration to utilize ELF magnetic field as a breakthrough in technology of bacterial culture activation of lactic acid bacteria that is very useful for the process of cheese production. Given the difficulty in the process of making cream cheese is generally produced cheese is still less tasty, bitter or sour, one of the factors is the proliferation of lactic acid bacteria in the process of less optimal pengmentation. As an alternative solution, this research will utilize Extremely Low Frequency (ELF) magnetic field for activation of lactic acid bacteria proliferation. Lactic acid bacteria used in this study is a mixture of bacteria *Streptococcus thermophilus* (*S. thermophilus*), *Lactococcus. lactis* (*L. lactis*), and *Lactobacillus acidophilus* (*L. acidophilus*).

The ELF magnetic field is the ELF electromagnetic wave component generated around the electric current.The Gauss law of electric field can be expressed as:

$$\nabla \cdot \mathbf{D} = \rho$$

If the assumption of propagation in a vacuum is used, with $\rho = 0$ and $J = 0$, then Maxwell's equation has the following modification with result of formulation written as:

$$\nabla \times \mathbf{H} = \frac{\partial \mathbf{D}}{\partial t}$$

For the propagation of electromagnetic waves in free space, Maxwell's equations are rearranged to be expressed explicitly on time and coordinates.

$$\nabla^2 \mathbf{E} = \mu\epsilon \frac{\partial^2 \mathbf{E}}{\partial t^2}$$

$$\nabla^2 \mathbf{B} = \mu\epsilon \frac{\partial^2 \mathbf{B}}{\partial t^2}$$

In a material, the speed of light is less than c. It can be grouped into materials depicted by refraction index, the ratio of the speed of light in vacuum to velocity in a medium

$$n = \frac{c}{v} = \sqrt{\frac{\epsilon\mu}{\epsilon_0\mu_0}}$$

The wave propagation around the strings that move the energy with the wave is proportional to the square of the wave amplitude. Electromagnetic theory explains that energy density (J/m^3) is associated with electromagnetic waves as

$$U = \frac{(\mathbf{D} \cdot \mathbf{E} + \mathbf{B} \cdot \mathbf{H})}{2}$$

Using relation of $\mathbf{D} = \epsilon\mathbf{E}$ and $\mathbf{B} = \mu\mathbf{H}$ if applied in medium of propagation:

$$U = \frac{1}{2}\left(\epsilon E^2 + \frac{B^2}{\mu}\right) = \frac{1}{2}\left(\epsilon + \frac{1}{\mu c^2}\right) E^2$$

In vacuum place:

$$U = \epsilon_0 E^2 = \frac{B^2}{\mu_0}$$

In addition to propagating, the waves also carry energy. The change in energy density (energy per unit of sectional area, each time unit) transferred is represented by the Poynting vector [19]. John Henry Poynting demonstrates that the presence of both electric fields and magnetic fields at the same time in place produces a flow in the energy field, called the Poynting theorem and the Poynting vector described

$$\mathbf{S} = \mathbf{E} \times \mathbf{H}$$

The unit of the Poynting vector is J/(m^2.sec) using the wave field to describe some portion of the vector. **S** involves a quadratic relationship on **E**, it is important to use the actual form of **E**

$$I = |\langle \mathbf{S} \rangle| = \frac{1}{T}\int_{t_0}^{t_0+T} \mathbf{A}\, cos^2\,(\omega t - \mathbf{k}\cdot\mathbf{r} + \phi)dt[20]$$

The magnetic field on a transmittable carrier is written:
Gauss's law

$$\mu_0 I = \int_0^{2\pi a} B.dl$$

$$\mu_0 I = B.dl|_0^{2\pi a}$$

$$= B(2\pi a) - B(0)$$

$$B = \frac{\mu_0 I}{2\pi a}$$

Based on the formula explains that the intensity of the ELF magnetic field that arises is directly proportional to the electric current flowing in the conductor. If a living being is in an ELF magnetic field, the ELF medal directly interacts with the cell membrane, capable of changing the cell membrane potential, causing the calcium cannal to open, and there is a Ca^{2+} influx that results in an increase in intracellular calcium. Increased intracellular calcium may activate cell propiferation [4]. The impact of exposure to the ELF magnetic field is not linear, but depends on the characteristics of the sample. Exposure of more than 500 μT has a damaging effect. The results of the Sudarti's study (2015) shown that exposure to ELF magnetic fields of less than 300 μT does not cause cellular damage effects, while exposure to ELF magnets of 500 μT suppresses bacterial growth [5].

The intensity of exposure to the ELF magnetic field used in this study was 100 μT and 300 μT with time exposure are 5 minutes, 15 minuntes, 25 minuntes, 35 minuntes, and 45 minutes.

2. Experimental method

This study was an experimental study, with exposure of magnetic field treatment on cultures of *S. thermophilus*, *L. lactis*, and *L. acidophilus* in agar-agar medium. The design of this study used a complete randomized design. This research was conducted at Advanced Physics Laboratory and FMIPA Microbiology Laboratory Building, University of Jember which started from July to December 2016. The ELF magnetic field was produced by current transformer (CT). The voltage regulation (V) was kept minimal (close to zero) and the magnitude of the current adjusted to set the resulting exposure of the ELF field 100 μT and 300 μT. The intensity of exposure ELF magnetic field in a homogeneous CT exposure space, the CT image as the source of the ELF magnetic field is presented in Figure 1.

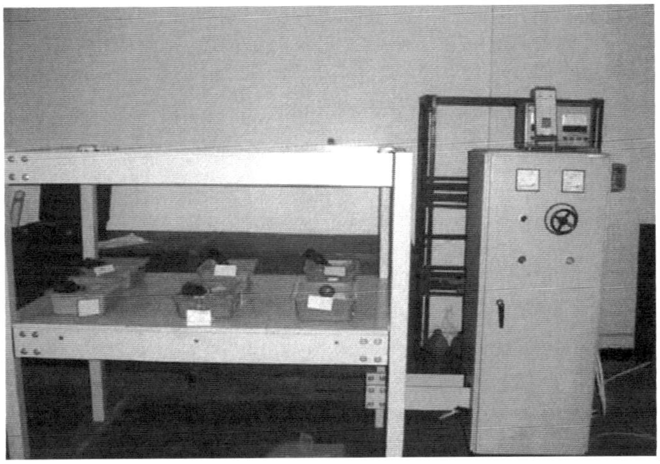

Figure 1. Current Transformer (ELF Magnetics Source).

The substance of this research is a culture of lactic acid bacteria consisting of three species of bacteria, namely *S. thermophilus*, *L. lactis*, and *L. acidhopilus* bacteria, which were cultured in MRSA (deMannRogosa Sharpe Agar) media. The process of exposure of the ELF magnetic field to the bacteria culture was carried out at the beginning of the breeding process. The group exposed to the ELF magnetic field consisted of a group exposed to ELF 100 μT and 300 μT magnetic fields, each exposed for 5 minutes, 15 minutes, 25 minutes 35 minutes, and 45 minutes. The control group was placed in a place not exposed to an ELF magnetic field from an artificial source. The exposure by current transformer was done at the Physical Education Laboratory of FKIP University of Jember.

Examination of number of *S. thermophilus*, *L. lactis*, and *L. acidophilus* bacteria in control group was done shortly after bacterial isolate was adapted in culture medium (K-0), one hour later when bacteria undergoing adaptation process and called Lag phase (K- 1), the next 2 hours when bacterial proliferation enters Log phase (K-2), and the next 16 hours when bacteria enter the stationary phase (K-16). The examination of the bacterial count of the treatment group was done one hour (E-1), two hours (E-2), and 16 hours (E-16) after exposure to ELF 100 μT and 300 μT magnetic fields for 5 minutes, 15 minutes, 25 minutes, 35 minutes, and 45 minutes in units of CFU (colony forming unit).

3. Results and discussion

Breeding of lactic acid bacteria *S. thermophilus, L. lactis*, and *L. acidophilus* bacteria in the control group described the bacterial proliferation condition as naturally depicted in Figure 2.

Figure 2 shown that the proliferation of *S. thermophilus*, *L. lactis*, and *L. Acidhopilus* bacteria in natural conditions without intervention. In general, the breeding of these three bacteria began to increase at the hour to hour 1 and hour 2, but at the 16th hour decreased. Breeding of bacteria at 1 o'clock shows the bacterial response or adaptation phase; this condition is called lag phase. Breeding of bacteria at 2 o'clock shows logarithmic proliferation rate, this phase is called log phase. The proliferation of bacteria at the 16th hour begins to decrease, so this phase is referred to as the stationary phase. Analysis of the number of *S. thermophilus*, *L. lactis*, *L. Acidhopilus* at the 1st hour (lag phase) after exposure to ELF 100 μT magnetic field for 5 minutes, 15 minutes, and 25 minutes are presented at Figure 3.

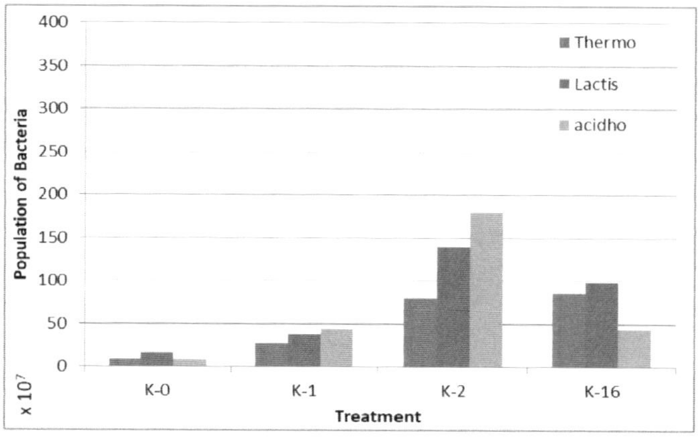

Figure 2. Graphic of Natural Bacteria Growth (control group).

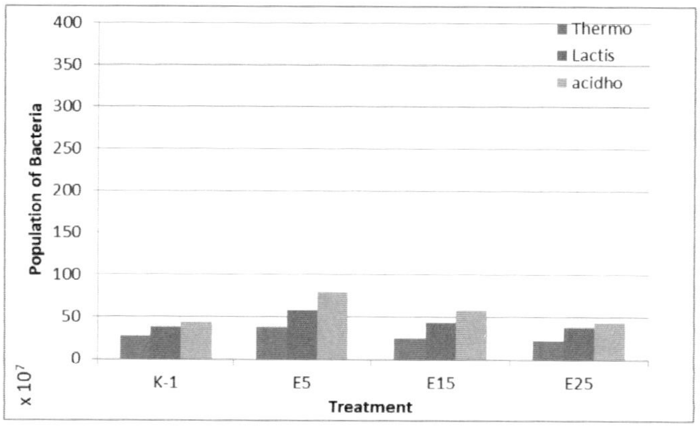

Figure 3. Growth of lag phase bacteria after exposure by ELF magnetic field 100 μT.

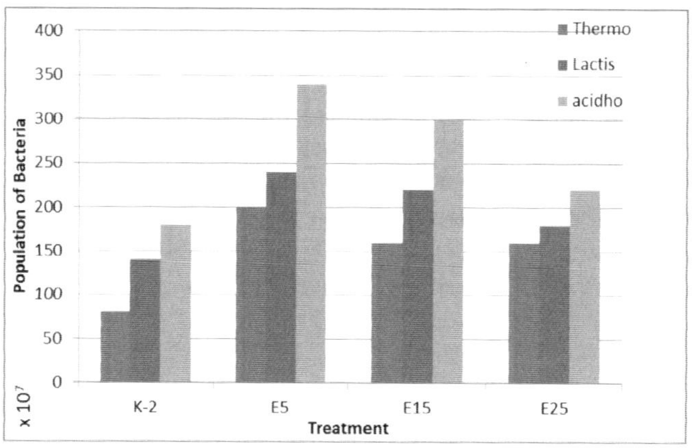

Figure 4. Growth of log phase bacteria after exposure by ELF magnetic field 100 μT.

Breeding of bacteria at 1 o'clock (lag phase), after exposure of ELF 100 μT magnetic field for 5 minutes, 15 minutes, and 25 minutes showed that, the total number of three types of bacteria reached the optimum value occurring in the experiment group of ELF 100 μT for 5 minutes, but the total number of three types of bacteria decreased after exposure to magnetic field 100 μ for 15 minutes and 25 minutes. Further analysis of bacterial count data *S. thermophilus, L. lactis,* and *L. Acidhopilus* in the second clock (log phase) after exposure to ELF 100 μT magnetic field for 5 minutes, 15 minutes, and 25 minutes are presented in Figure 4.

The proliferation of three types of bacteria at log phase (2 o'clock) after exposure to ELF 100 μT magnetic field showed that the number of the three types of bacteria reached the optimum value in the group exposed to ELF 100 μT field for 5 min, and decreased after exposure to 100 μT magnetic field for 15 minutes and 25 minutes but still higher than the control. Furthermore, comparative analysis of bacterial proliferation *S. Thermophilus*, *L. lactis*, and *L. Acidhopilus* after exposure to a 100 μT and 300 μT ELF magnetic field with a 5-minute, 15 minute, and 25-minute exposure in lag phase versus control (K), is presented in Chart 4.4 and the log phase is presented in Figure 5.

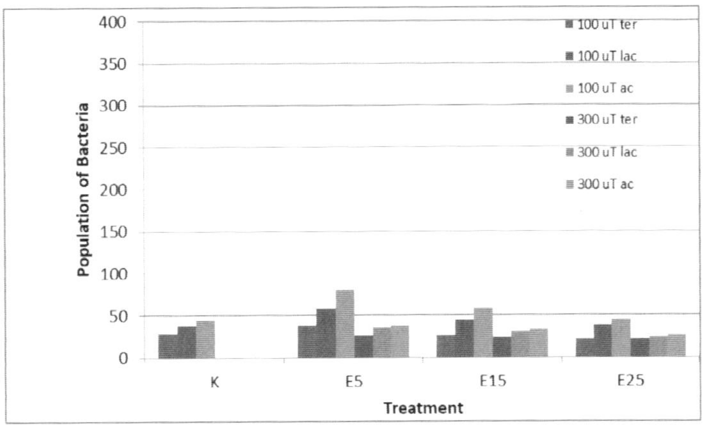

Figure 5. Growth of bcateria lag phase after exposure by ELF magnetic field 100 μT and 300 μT.

The *S. thermophilus* appears to increase only in the group exposed to 100 μT ELF magnetic field for 5 minutes, but the number of *S. thermophilus* bacteria appears to decrease in the group exposed to the ELF magnetic field 100 μT for 15 minutes, 25 minutes, and groups exposed to ELF 300 μT magnetic field. Proven proliferation of three types of *S. thermophilus*, *L. lactis*, and *L. acidhopilus* on lag phase bacteria achieve the optimum value occurs in the bacterial group after exposure to ELF magnetic field with intensity of 100 μT for 5 minutes. Overall, however, the proliferation of three types of bacteria exposed to an ELF magnetic field with an intensity of 300 μT for 5 minutes, 15 minutes, 25 minutes proved to suppress the proliferation of *S. Thermophilus*, *L. lactis*, and *L. acidhopilus* bacteria. Furthermore, comparative analysis of breeding of three types of log phase bacteria after exposure to ELF magnetic field between 100 μT and 300 μT with exposure time of 5 minutes, 15 minutes, and 25 minutes compared to control (K), presented in Figure 6.

Exposure to ELF 100 μT magnetic field for 5 minutes, 15 minutes, and 25 minutes significantly increased bacterial proliferation of *S. thermophilus, L. lactis* and *L. acidhopilus* log phase, but exposure to ELF 300 μT magnetic field for 5 minutes, 15 minutes, and 25 minute significantly suppress the breeding of the three bacteria. The number of bacteria after the 300 μT magnetic field was exposed for 5 minutes, 15 minutes, and 25 minutes lower than the control. Breeding patterns of the three bacteria after exposure to the ELF magnetic field were consistent, with the rate of bacterial proliferation of *L. acidhopilus* being the fastest, and the lowest rate of reproduction occurred in the *S. thermophilus* bacteria.

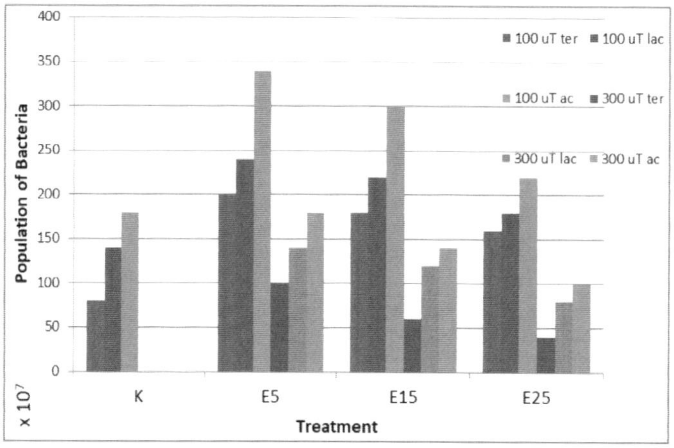

Figure 6. Growth of log phase bacteria after exposure by ELF magnetic field 100 μT and 300 μT.

Based on the descriptive analysis, the proliferation of *S. thermophilus*, *L lactis* and *L. acidhopilus* bacteria in log phase reached the optimum value in the group exposed to ELF 100 μT magnetic field for 5 min. Subsequent analyses were performed on the proliferation of all three bacteria at the 16th hour (stationary phase) after exposure to a 100 μT ELF magnetic field for 5 minutes, 15 minutes, 25 minutes, 35 minutes, and 45 minutes as presented in Figure 7.

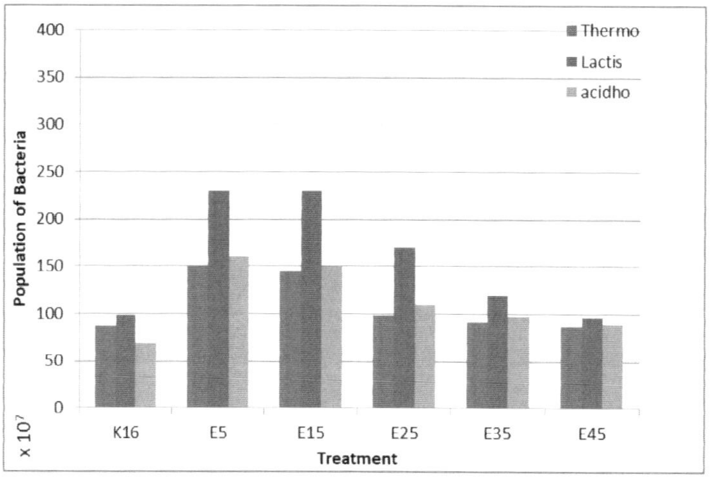

Figure 7. Growth of Stationer phase bacteria after exposure by ELF magnetic field 100 μT.

Proven proliferation of three types of lactic acid bacteria at the stationary phase reached the optimum value in the group exposed to 100 μT intensity ELF magnetic field for 5 minutes and 15 minutes. Furthermore, the total number of three types of bacteria decreased after exposure to ELF 100 μT magnetic field for 25 min, decreased after exposure for 35 min, and decreased again until the position was no different with control after exposure of ELF magnetic field for 45 min. Based on the results of bacterial analysis of *S. thermophilus*, *L. lactis*, and *L. acidhopilus* after exposure to the ELF magnetic field, the proliferation of the three types of bacteria in the Lag phase and Log phase achieved optimal values in the group exposed to the 100 μT ELF for 5 minutes. However, the proliferation of these three bacterial strains at the 16th hour achieved optimal values in the group exposed to 100 μT ELF magnetic field for 5 minutes and 15 minutes. It can therefore be concluded that the effective dose of the ELF magnetic field which can trigger the proliferation of *S. thermophilus, L. lactis,* and *L. acidophilus* bacteria optimally is the exposure to the ELF magnetic field at an intensity of 100 μT with an exposure duration of 5 min.

The bacteria *Streptococcus thermophilus*, *Lactococcus lactis*, and *Lactobacillus acidophilus* comprise a group of lactic acid bacteria can be used as a starter in the process of cream cheese. The important role of the bacterial mixture in the cheese-making process can be to produce lactic acid, have the ability to break down proteins, the ability to produce CO_2 gas, and provide the aroma of cheese. The addition of *Lactobacillus acidophilus* bacteria alone proved to produce higher carbohydrates in the soft cheese making process than the addition of mixed *Lactobacillus acidophilus*. The bacteria *Lactobacillus acidophilus* and *Lactococcus lactis* are able to inhibit the growth of pathogenic bacteria, especially *E. coli*, while actor *S. thermophilus* has an important role especially in the formation of textures and flavors in producing cheese.

The ELF magnetic field has the ability to penetrate a variety of materials including biological materials or living things. The ELF magnetic field directly interacts with the cell membrane, exposure to the ELF magnetic field will force the ion and molecules in the cell and cause cell membrane conductivity changes. According to Gobba and Malagoli (2003), the plasma membrane is a magnetic field interaction medium that affects the enzyme activity and the signal path of the transduction. Changes in the conductivity of the cell membrane will cause the opening of calcium canals. This can lead to an increase in intracellular calcium, and it is this condition that plays an important role in the process of cell cycle as a process of cell proliferation.

Several studies have shown that exposure to the ELF magnetic field affects intracellular calcium alteration. Luo FL *et al.*, 2014, these findings indicate that ELF-EMF exposure specifically influences the intracellular calcium dynamics via a calcium channel-independent mechanism [5]. While Golbach LA *et al.*, 2016, concluded that low-frequency magnetic fields (LF MF) may influence the calcium homeostasis in cells in vitro, thus any potential clinical implications await further investigation. Furthermore, Martynyuk V *et al.*, 2015, clarify that however, the 50-Hz field strengthened the basic intracellular calcium concentration in smooth muscle cells (SMC) in a time-dependent manner, whereas the 8-Hz field alone supports calcium levels. Alabovsky VV *et al.*, 2016, that a single exposure to 171 MHz electromagnetic field (180 V / m, 0.04 mW / kg) that enhancement of the Na^+/Ca^{2+} exchange towards removing Ca^{2+} from the cardiomyocytes electromagnetic field exposure is a result of Ca^{2+} extraction from the sarcoplasmic reticulum and the increase of its intracellular level. This proves that the exposure of ELF magnetic fields causes biological effects through intracellular calcium alteration mechanisms, and it is evident that exposure to low-intensity ELF (less than 300 μT) ELF magnetic fields is not mutagenic and does not cause cellular damage, but enhances cell proliferation. This has been proved by Belyaev I, 2011, that ELF-MF, under specific conditions of exposure, acted as a non-toxic but cell-growth stimulating agent [20]. Verschaeve L, *et al.*, 2016, shown that a 100 μT magnetic field (50 Hz) does not damage DNA and hence is not mutagenic in this assay and that there was also no influence on the DNA damaging capacity of the used mutagens [6].

Verschaeve L, *et al.*, 2011, that ELF-MFs (50 Hz, 100 and 500 μT, 1 and 2 h exposure) do not induce SOS-based mutagenicity in *S. typhimurium* bacteria and do not show any synergetic effect when combined with chemical mutagens [7]. Oncul *et al.*, 2016, these results show that ELF-EMF, 50 Hz, 1 mT (1000 μT) for 2 h affects the crucial physicochemical processes in both Gram-positive and Gram-negative bacteria which need further research [8]. Bayir E *et al.*, 2015, the cultures of bacteria in broth media were exposed to ELF-EMF showed a statistically significant decrease compared to their controls in colony forming capability, especially at long exposure times. An exposure to 4 mT-20 Hz ELF-EMF of 6 h produced maximum inhibition of CFU compared to their controls for both microorganisms (95.2% for *S. aureus* and 85% for *E. coli*) [9].

In addition, Martirosyan V, *et al.*, 2013, It has been shown that EMF exposure has pronounced stimulation while at 8 Hz, 0.4 mT (400 μT), 30 min it has inhibited cell proliferation [10].Zhang J *et al.*, 2016, results showed both biomass and mannatide production increased significantly at MF induction 0.4, 0.6, and 0.9 mT and decreased at both 1.2 and 1.5 mT. Magnetic field (MF) inductions, exposure times, and exposure periods varied in a range of 0-1.5 mT, 0-16 h, and six periods of incubation time, respectively [11].

Result shown by Huwiler SG, *et al.*, 2012, Moreover, short-term exposure (8 min) to the sinusoidal continuous and power line intermittent signal neither affected bacterial growth nor showed evidence for reliable changes in transcription. This experiments did not indicate that the different tested MFs (50 Hz, 1 mT) affected the transcription of *E. coli* [21].

Research by Lugina C, *et al.*, 2008, the results indicate that an exposure to 50 Hz EMF (0.1, 0.5, 1.0 mT), acts as a stressing factor on bacteria which can represent a suitable model to investigate acute and chronic effects related to ELF-EMF exposure [30]. Exposure ELF-EMF (50 Hz, 0.5 mT ELF-EMF for 6 h) induces a decrease in growth rate and morphological changes for both Gram-negative and Gram-positive bacteria [18]. Research by Petecchia L *et al.*, 2015, The PEMF effect is primarily associated to early enhancement of intracellular calcium concentration, which is proposed here as a reliable hallmark of the osteogenic developmental stage [28]. In addition, Ma Q, *et al.*, 2016, suggest that ELF-EMF (50 Hz, 1 mT) exposure for 1, 2, and 3 days with 4 hours per day promotes the neuronal differentiation and neurite outgrowth [14].

Normally cell proliferation is strongly influenced by cell metabolism that is strongly influenced by intracellular calcium ion activity. Cell cycle processes require intracellular calcium activity higher than ordinary conditions, therefore increased intracellular activity of calcium in conditions according to the needs of the cell cycle process will have an impact on increased cell proliferation. However, if an increase in intracellular calcium exceeds the need for a cell cycle process will result in cell death or weakened cell function and impact on cellular proliferation.

The results of this research show that exposure of ELF magnetic field with intensity 100-150 uT intermittently 8 hours/day for 7 weeks and 14 weeks have a significant effect on intracellular calcium increase, thus causing the increase of Germinal apoptosis in Bulb/C mice. Subsequent research of Sudarti, et al, 2014, proves that exposure to ELF magnetic field with intensity of 676 uT can inhibit the proliferation of *Salmonella typhimurium* to 57%. While exposure to Magnetic Field ELF with intensity of 500 μT for 30 min in the process of fermentation of glutinous tape can inhibit microbial proliferation in real 35% at 48 hours [15].

On the contrary, the research results of Sari REYW, Prihandono T, Sudarti, 2015, proved that exposure to ELF magnetic field with intensity 300 μT for 60 minutes can accelerate the growth process of tomato ranti plant. The results of Andika and Sudarti's study, 2016, poved that exposure to ELF magnetic field with intensity of 100 μT for 5 min significantly decreased pH and moisture content in cream cheese making process [16].The results of this study indicate that exposure to ELF magnetic field with intensity of 100 μT for 5 minutes proved to significantly increase the proliferation of *S. thermophilus, L. lactis,* and *L. acidhopilus* bacteries.

It can be stated that exposure to ELF magnetic field with intensity of 100 μT for 5 min is estimated able to interact directly with bacterial membrane, thus causing increase of intracellular calcium as needed in cycle cell process so that mamicu process of cell or bacteria proliferation. However, exposure to the ELF 100 uT magnetic field in more than 25 minutes proved to weaken the bacterial proliferation process, it is estimated that exposure to the ELF 100 uT magnetic field over 25 minutes increases intracellular calcium beyond the requirement in the cycle cell process, resulting in the weakening of the cycle cell process ultimately impact on the inhibition of bacterial proliferation process.

This means that if the cell is placed in an ELF magnetic field, then the cell membrane's electrical potential will be affected.The ELF magnetic field has the properties to penetrate cell membranes, thus affecting uncontrolled cellular intra-cell transport mechanisms. Description of the biological effects caused by exposure to the ELF magnetic field, based on the possibility of a permeability effect on the ion channels contained in the membrane. This can have an impact on cell growth that causes biological changes in organisms [17]. Panagopoulos and Karabarbounis (2002), stated that the relative magnetic permeability value of the tissue is 1, meaning that the magnetic flux density in the cell is almost similar to the flux density outside the cell.

Some possibilities for the impact of cell growth by exposure to the ELF magnetic field are: 1) the formation of free radicals in cells exposed to the magnetic field, 2) the change in the concentration of ions under exposure to the magnetic field [17]. Blank and L. Soo (2001) suggest that any change in membrane transport activity is detected by the movement of molecules and ions across the plasma membrane. This movement can cause changes in metabolic activity.

Based on the theoretical study that has been done, the Maxwell III equation states that the magnetic field changes can produce induced currents. The presence of exposure to the magnetic field creates changes in the movement of Ca^{2+} ions in the extracellular crossing of the cell membrane while the induced current by the ELF magnetic field will increase the rate of movement of Ca^{2+} ions through the magnetic flux density region. Fields exposed to magnetic fields will produce strength in Ca^{2+} ions to move and are actively bound to the protein channel and affect the channel gate opening conditions. The vibrations due to the ELF flux density line density at a time will exceed some critical values, resulting in rotation or movement of ions which can give false signals to the gate gate of the channel on the cell membrane, and cause errors in the electrochemical balance of the cell membrane and its continuity to the whole cell function [17]. Ca^{2+} ion is an ion in cells that have high sensitivity to exposure to ELF magnetic field, this is because Ca2 + ions are classified as paramagnetic materials and have a positive susceptibility price. The properties of a paramagnetic material can be affected by the magnetic field (termagnetization). The form of magnetic field influence on the material is the spin of electrons found in the material that was originally randomly directed by the magnetic field.The potential change of the cell membrane will at some extent be able to open Ca^{2+} canal, this condition will fall on the transport of Ca^{2+} ions and under certain conditions affect the increase of Ca^{2+} influk. The transport activity of Ca^{2+} influk is needed in the growth process of the cell or Cell Cycle, so that in the appropriate conditions it will trigger cell proliferation, but on the condition of the high enough Ca^{2+} influk transport activity will result in cell death. Exposure to high-intensity ELF magnetic fields (greater than 600 μT) will have an impact on cell death, but at low intensity (less than 300 μT) have an impact on increased cell proliferation.

Increased growth of *S. thermophilus*, *L. lactis*, and *L. acidophilus* bacteria will result in a decrease in pH value in the medium. Increasing the rate of bacterial growth will effect the increased production of lactic acid, thereby effecting the decrease in pH value of the medium.

4. Conclusion

Based on data analysis and discussion of the results of this study, can be formulated some conclusions that the effective dose of exposure to ELF magnetic field affecting the speed of bacterial proliferation of *S. thermophilus*, *L. lactis*, and *L. acidhopilus* is exposure at intensity of 100 μT for 5 min.

Based on the results of the discussion of the results of this study, can be formulated some suggestions as follows.

a. Need to further research the implementation of effective dose of exposure to ELF magnetic field in the process of cream cheese ripening.
b. It is necessary to further study the effective dose of exposure to ELF magnetic field to the death of pathogenic bacteria in cream cheese making process.

Acknowledgment

We gratefully acknowledge the support of the 3th Research group Physics Education from FKIP – University of Jember.

References

[1] Sudarti 2016 *Int. Conf. on Food, Agriculture, and Natural Resources FAN Res* **12** 201
[2] Sudarti Krisnawati A 2016 *Int. Sem. IBSC 2016 Conducted at: FMIPA University of Jember*
[3] Tessaro LW Murugan N J Persinger MA 2015 *Microbio. Res.* **172** 26-33

[4] Sudarti 2004 *J.l of Sci. and Tech.* **3** 24-29
[5] Sudarti 2015 *Proc. of Nat. Conf. FKIP University of Jember* **1** 157-63
[6] Afiati F Yopi Maheswari RRA 2014 *J. Tech. and Food Indust.* **25(1)**
[7] Maheswari P Wiryawan IKG Maduningsih GL 2008 *Microbio.* **2** 124-30
[8] Shimizu H Suzuki Y 1995 *Nippon Eiseigaku Zasshi* **50(5)** 919-1031
[9] Luo FL Yang N He C Li HL Li C Chen F Xiong JX Hu ZA Zhang J 2014 *Env. Res.* **135** 236-46
[10] Golbach LA Portelli LA Savelkoul HF Terwel SR Kuster N de Vries RB Kemenade V 2016 *Environ Int* **92** 695-706
[11] Belyaev I 2011 *Mutat. Res.* **722(1)** 56-61
[12] Verschaeve L Wambacq S Anthonissen R Maes A 2016 *Mutat. Res. Genet. Toxicol Env. Mutagen* **795** 31
[13] Verschaeve L Anthonissen R Grudniewska M Wudarski J Gevaert L Maes A 2011 *Bioelectromagnetics* **32** 580
[14] Oncul S X Cuce E Aksu B Inhan A 2016 *Int. J. Radiat. Biol.* **92** 42
[15] Bayır E Bilgi E Şendemir A Hameş E 2015 *Electromagn. Biol. Med.* **34** 14
[16] Martirosyan V Baghdasaryan N Ayrapetyan S 2013 *Electromagn. Biol. Med.* **32** 291
[17] Zhang J Xu C Wan Y Gao M 2016 *Bioelectromagnetics* **37** 331
[18] Huwiler SG Beyer C Fröhlich J Hennecke H Egli T Schürmann D Rehrauer H Fischer H 2012 *Bioelectromagnetics* **33** 488
[19] David J 1999 *Intro. to Electrodyn.* New Jersey: Prentice-Hal Inc
[20] Robert D 1990 *Modern Optics* USA: John Wiley & Sons Inc
[21] Ma Q Chen C Deng P Zhu G Lin M *et al.* 2016 *PLoS One* **11** 3
[22] Kristian RS Sudarti and Abdul G A 2015 *J. of Phys. Edu.* **4** 1-8
[23] Andika K Sudarti 2016 *Prosidia Int. Conf. IBSC* **1** 52
[24] Gaafar *et al.* 2006 *Romanian J. Biophysic* **16** 283-96
[25] Panagopoulos D Karabarbounis A 2002 *Biochem. and Biophys. Research Comm.* **27** 95-102
[26] Inhan-Garip A Aksu B Akan Z Akakin D Ozaydin AN San T 2011 *Int. J. Radiat. Biol.* **87** 1155
[27] Petecchia L Sbrana F Utzeri R Vercellino M Usai C Visai L Vassalli M Gavazzo P 2015 *Sci. Rep.* **14** 5
[28] Lugina C Rosella G Emanuela D Soraya D B Mara D G Iole R Oriana T Maria A Marriggio 2008 *Bioelectromagnetics* **29** 302

IOP Conf. Series: Materials Science and Engineering **432** (2018) 012011 doi:10.1088/1757-899X/432/1/012011

Quality characterization of HVOF thermal spray coating with NiCr matrix composite for protection application of coal fired boiler tubes

M Waldi[1], E A Basuki[1], B Prawara[2]

[1] Metallurgical Engineering Department, Faculty of Mining and Petroleum Engineering – Bandung Institute of Technology, Ganesha 10 St., Bandung 40132, Indonesia
[2] Research Center for Electrical Power and Mechatronic – Indonesian Institute of Sciences (LIPI), LIPI Area, Cisitu 21/154D St., Bandung 40132, Indonesia

waldi05@gmail.com

Abstract. In line with the Indonesian government's policy on national electrification plans by developing a 35,000 MW power generating facility which has been started since 2014, the need arises for high temperature resistant materials for application in the program. Moreover, materials dedicated to engineering equipment such as boilers as a steam producer to drive the coal-fired power plant turbine is required to be compatible in extreme condition. The operational challenge of materials at high temperatures becomes serious, especially as the use of low quality coal as fuel in the combustion system creates degradation problems on tube surfaces such as erosion and corrosion thereby limiting boiler tube life. Thermal spray coating technologies for high temperature operations have been developed with the aim of protecting boiler tubes from aggressive environments, improving thermal efficiency, and reducing time losses from damage. High velocity oxy fuel (HVOF) technology is one of the recommended methods to be used in boiler tubes at coal-fired power plant (PLTU). The following report discusses the HVOF thermal spray coating deposition of metal-matrix composites with combination of feedstock NiCr as metal matrix which has -37μm of size particles with ceramic metal including chromium carbide and alumina as reinforcement on SUS 304 substrate.

1. Introduction

The economic growth rate of Indonesia and the rapid development of multipurpose technology to facilitate the needs of household and industry in the last decade has led to the increasing global demand for electricity from time to time. Based on the 2014 Indonesian Energy Outlook [1] exposition, Indonesia was predicted to experience a population increase from 255 million in 2015 to 335 million people by 2050. Furthermore in 2016, the electrification ratio was 91.35% [2]. Although the trend of installed capacity has been an increase of 4% per year since 2011, it has not been able to attain the ideal target of electrification ratio of 100%. Nevertheless, the construction of power generation facilities will still be upgraded for at least another decade.

The existing power plant facilities in Indonesia in 2015 have an installed capacity of 55,528.10 MW produced by several types of power plants, namely coal-fired power plants (abbreviated PLTU in Indonesian), gas-fired (PLTG), steam-powered and gas-fired (PLTGU), diesel powered (PLTD), hydroelectric (PLTA), geothermal (PLTP), wind-powered (PLT Bayu), solar-powered (PLTS), and coal gasification powered (PLTGB) plants, where the highest share of electricity supply is generated by PLN of 38,314.23 MW and the rest is supplied by private parties or also known as Independent Power Producer (IPP) of 17,213.87 MW [3]. The highest supply is produced by three large types of power plants, namely PLTU, PLTGU, and PLTD, in this particular order, followed by other power plants such as PLTA and PLTG. Thus, to pursue the electrification target of 100%, an additional supply of 35,000 MW is still needed.

With the priority for use of power plants to support electricity supply throughout the region based on feasibility studies [4, 5], the availability of coal, and the consideration of electricity production efficiency, the development of coal-fired power plants is still a solution to accommodate the 100% electrification. The significant constraint encountered in the operation of the coal-fired power plant is the degradation of the components of the boiler tubes at high temperature conditions combined with the destructive extreme environmental conditions, in the form of high temperature corrosion, erosion, wear and oxidation on the surface.

The frequent shutdown of operations due to component turnover decreases the generator's performance. Therefore, the selection criteria for boiler tube materials should be compatible with high operating temperature to support efficiency. In addition to protecting the substrate material against such extreme environments, a compatible metallic layer is required [6].

The application of thermal spray coating (TSC) is the solution to the need for high temperature resistant material that is also to corrosive, oxidative and erosive environments, as well as to creep resistance. TSC provide important properties such as: optimal thickness of up to 500μm, increased component life, good adhesive between coatings and substrates, relatively low production prices, thermal shock resistant, resistance and to corrosion and erosion. In conclusion, TSC acts as a structural jacket against combustion load [7].

Thermal spraying is generally defined as a method of forming metallic or non-metallic coatings resulting from melted or semi-melted conditions caused by energy hereinafter using high velocity oxy fuel (HVOF). The energy is used to melt feedstock materials (derived from powder form) fed through a high speed spray gun and propelled towards the surface of the metal substrate by utilizing gas or atomization jets to form layers with lamellar structures. Referring to the term coating material, the material may be metal, oxide, or carbide. The thermal spray event occurs due to a combination of thermal energy, and the melt and semi-melt kinetic energy of the feedstock from the heat that formed by the interaction of fuel and oxygen, and by an electric source.

One of the most significant issues in boiler operation is erosion damage, the depletion of material due to the impingement of high velocity solid particles and impact on the material surface at the certain angles [8]. The coal used in Indonesia power stations has large amounts of ash (about 10-50%) which contain abrasive mineral species such as hard quartz (up to 15%) which increase the damage.

HVOF sprayed NiCr-Cr_3C_2 coatings are one of the most important candidates for protection of materials from high temperature erosion and have been successfully used to protect pulverized coal fired boiler tubes [9-11]. Stainless steel is used as substrate in the present research. It has higher strength at elevated temperatures so is often used for structural and pressure-containing applications at temperatures above about 500 °C and up to about 800 °C [12]. Austenitic stainless steel contains 16% of chromium and 6% nickel. NiCr powder is selected as bond coat and metal matrix for top coat deposition, chromium carbide and combination of chromium carbide and alumina are used as reinforcement particles in metal matrix composite (MMC). NiCr selection is based on its properties that could withstand in the elevated temperature condition, and hot corrosion issue. Further Martides *et al.* in 2017 has reported their works about the influence of finer size of NiCr matrix that gave improvement to coating properties [13].

Chromium Carbide (Cr_3C_2) was selected as a reliable protective coating to MMC materials that capable to resist oxidation in range 650 – 1200 °C [8]. This material has a good wear resistance which is chemically inert and stable at elevated temperatures. In addition, high purity alumina (white alumina) has good electrical insulation and thermal conductivity, and applied in many applications that require resistance to abrasive and erosive loads.

The present work is aimed to develop a NiCr metal matrix composite (MMC) coating using HVOF. NiCr powder, and Cr_3C_2 with some addition of Al_2O_3 were used as metal matrix and reinforcement particles, respectively. Another purpose of this work is to give valuable information to established

2. Materials and methods

2.1. Materials

The substrate austenitic stainless steel SUS 304 is obtained from the inventory of the Research Centre for Electrical Power and Mechatronics, LIPI, Bandung - Indonesia. The material has been established to have the best performance for the working conditions of the ultra-supercritical pulverized coal (USC) power plant [14, 15]. Table 1 shows the verification result of hardness using Metkon Duroline M Microhardness Tester and chemical composition using Metavision 1008i spectrometer to check as-received material meet the specification defined.

NiCr powder was obtained from Oerlikon Metco (Metco 43C-NS) and employed as bond coat material, and the selection process through sieve shaking process was performed to obtained finer powder distribution size to 400 mesh (-37 μm) for metal matrix of top coat deposition. Reinforcement particles such as Cr_3C_2 were obtained from Oerlikon Metco (Woka 7102), and Al_2O_3 was obtained from Sulzer Metco (Metco 105NS).

Coating feedstock was made into two blending formula with each total weight 1 kg as follows: (1) $60\%NiCr_{[400mesh]} + 30\%Cr_3C_2 + 10\%Al_2O_3$, and (2) $60\%NiCr_{[400mesh]} + 40\%Cr_3C_2$. Hereinafter, formula 1 called as 3 components and formula 2 called as 2 components. After all material were prepared and weighed according to the formula, the material were blended respectively in V-Type blending machine within 64 rpm for as long 16 hrs. The appearance of blending result is shown in Figure 1 and Figure 2.

Table 1. Verification of tensile strength SUS 304, and its chemical composition analysis by.

Hardness	Elements (%-wt)						
(HVN)	C	Si	Mn	P	S	Cr	Ni
209.96	0.0208	0.4087	1.6706	0.0158	0.0055	19.1362	8.8657

(a) (b)

Figure 1. SEM micrographs of NiCr matrix composites powder with Cr_3C_2 and Al_2O_3 reinforcement; (a) feedstock; (b) EDS spectra of MMC 3 components feedstock.

(a) (a)

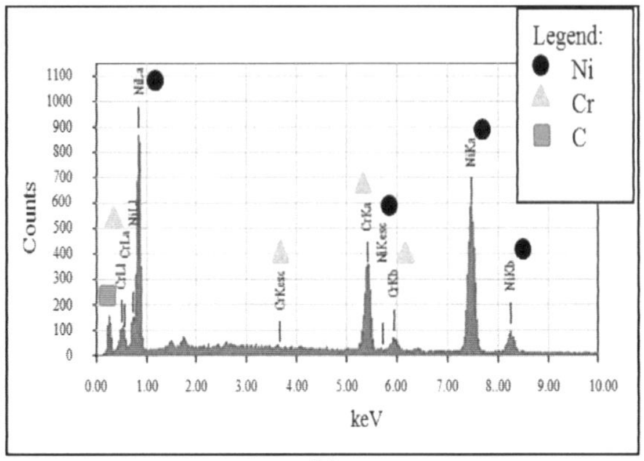

(b)

Figure 2. SEM micrographs of NiCr matrix composites powder with Cr_3C_2 reinforcement; (a) feedstock; (b) EDS spectra of MMC 2 components feedstock.

2.2. HVOF Thermal spraying process

The thermal spraying process is carried out using TECKNOTHERM HVOF 2007 thermal spray machine and spray gun Hipojet-2700 with constant parameters for each MMC formulas, Table 2 describes the HVOF process parameters employed during the. The thermal spraying process is applied on to the SUS 304 plate specimens with size 30 mm x 20 mm X 3mm to be prepared for general characterization including SEM, XRD, hardness, and roughness, and the other is solid cylinder Ø25.4 mm X 38.5mm for bond strength test purpose.

Grit blasting was performed to clean the specimen's surface from contamination and to activate surface condition in order to maximize coating adhesiveness. This latter occurs through the formation of rough textures on the substrate surface to facilitate mechanical interlocking of the splats. Mechanical interlocking plays an important role in the mechanism of cohesion when the splats fill and solidified on the rough surface. The process was employed white aluminium oxide (600 – 850 mesh) and the blasting process was conducted inside the blasting chamber.

Table 2. Parameter process of HVOF thermal spray.

Parameter	Unit	Value
Air pressure	bar	6.2
Pressure O_2	bar	8
N_2	bar	5
Propane	bar	5.5
Flow rate O_2	liter/min	271
N_2	liter/min	8
Propane	liter/min	62.4
Powder feeder rotation	rpm	5
Stand off distance	mm	200
Spray angle	^{o}	90
Substrate preheating	^oC	100

2.3 Characterization

The morphology of the powders and the coating microstructure were observed by means of a integrated scanning electron microscope (SEM) and energy dispersive spectroscopy (EDS) system (JEOL JSM-IT300LA). Observations were carried out on a polished cross-section normal to the surface. The metallographic preparation was done by coarse to fine grinding followed by polishing with 1 μm alumina suspension. The phase characterization of the coating deposit after HVOF thermal spray was conducted by X-ray diffraction (XRD) using Rigaku SmartLab X-ray Diffractometer.

The porosity within the coating microstructure was evaluated by analysis software ImageJ Ver. 1.50i (http://imagej.nih.gov/ij), the software identified the dark pores (cavities) by color mapping. The work by Sadeghimeresht *et al.* has shown that image analysis can reproducibly identify and measure pores, cracks, etc. within thermal spray coatings with a high level of confidence [16].

Surface roughness for each coatings were measured by stylus Kosaka Lab AS-1700α stylus model profilometer equipped with Kosaka Lab SE1700α recorder.

The microhardness was measured on a polished cross-section surface of the HVOF coating by using a Metkon Duroline M microhardness tester with a load selection 300 gf and dwelling time is 10 seconds in accordance with ASTM E92. Seven data point were acquired across the substrate to coating with three times indentation per each sampling points. Each data were collected and averaged for each hardness value.

Tensile bond strength was performed using Hung Ta tensile tester tipe HT-8503 capacity 2 ton and additional test fixture as recommended by ASTM C633. One untreated piece of cylindrical sample face was grit blasted by white alumina oxide and cleaned with compressed air then attached by adhesive Cytec FM1000 and assembled to another cylindrical sample piece that already coated. Five pairs of the specimens were held in V-BLOCK and remain compressed by threaded holder and tightened to 87 kg.cm, heated in the furnace to 180 °C held for 2 hr and cooled inside the furnace. Applied tensile load to each test specimen at a constant rate of cross-head travel between 0.013 mm/s to 0.021 mm/s until rupture occurs.

3. Results and discussion

Figure 1 reveals the blending of the 3 components feedstock performed on a V-Type blending machine for 16 hours at a speed of 64 rpm in which visible components of NiCr, Cr_3C_2, and Al_2O_3 have been well distributed. NiCr component shown in Figure 1 appears to be incomplete, this is caused by collisions with reinforcing components during the blending process, in which the Al_2O_3 powder has a harder property than the NiCr matrix. Qualitatively the EDS spectra verifies the constituent elements of the MMC 3 components, in which the spectrum charts identified the existence of Ni, Cr, Al, and C.

Figure 2 in which the NiCr matrix is still relatively intact after blending process. The observation by using SEM shows that NiCr matrix and Cr_3C_2 particles have been well distributed and qualitatively EDS spectra identified the constituent elements in MMC 2 components, such as Ni, Cr, and C. The blending result of MMC 3 components and MMC 2 components found no contamination of foreign objects that could be reduced the quality of feedstock and the validity of TSC process results. The results obtained after the thermal spray coating process showed an evenly distributed layer, the physical difference between TSC results with MMC 3 components is darker than those produced by MMC 2 components (see Figure 3). This is due to the Al_2O_3 factor that gives the darker character to the exterior appearance of the coating. It is also noted from measurement that coating roughness (Ra) of MMC 3 components and MMC 2 components were 6.2284 μm and 7.3872 μm respectively.

The thickness of the coating has been measured through the SEM instrument and desired thickness composition of 50-75 μm for the bond coat layer and 125-150 μm for the top coat layer. Measurements were performed at least three times at different locations and average of thickness measurement of bond coat and total thickness for each specimens are: 67 μm and 262 μm (MMC 3 components); 72 μm and 368.4 μm (MMC 2 components).

Figure 4 reveals SEM micrograph TSC deposition of MMC 3 components with back scattering feature which capable to distinguished splats, un-melted particles, porosity, and oxide. Bond coat formation is evenly distributed on the substrate with varying thickness across the section that caused by the repeated and accumulation of spray process. Every single bond coat that built and solidified on the substrate can be observed as similar as grain boundaries. Un-melted particles identified in MMC 3 components are quite significant and distributed evenly within the coatings. The presence of un-melted particles is strongly influenced by the parameters during the thermal spray process such as low combustion temperatures and lack of heat input given to the substrate, so it did not have enough time to melt some powder feedstock at the time it propelled out toward the substrate with a high velocity. The non-constant particle rate also greatly affects the melting of the feedstock powder when it travelled from feeder towards the substrate surface, so that the powder was solidified earlier before the substrate strikes the surface. A small part of the oxide also appears to form on the coating structure.

Figure 3. The as-coated result of MMC 3 components and 2 components using TECKNOTHERM HVOF 2007 thermal spray machine and spray gun Hipojet-2700.

Another factor that affects the formation of un-melted particles is the accumulation of spraying conditions to meet the desired thickness targets. The previous coating deposit layer is subject to re-melt from the molten coating powder that propelled by the spray gun, but the heat received by the deposit does not melt entirely and solidified faster than the new coating layer. The coatings have a dense structure with the porosity 1.43% randomly distributed in the coating. Cross-sectional micrograph shows the coating microstructure consists of NiCr matrix (bright area) and chromium carbide (dark area).

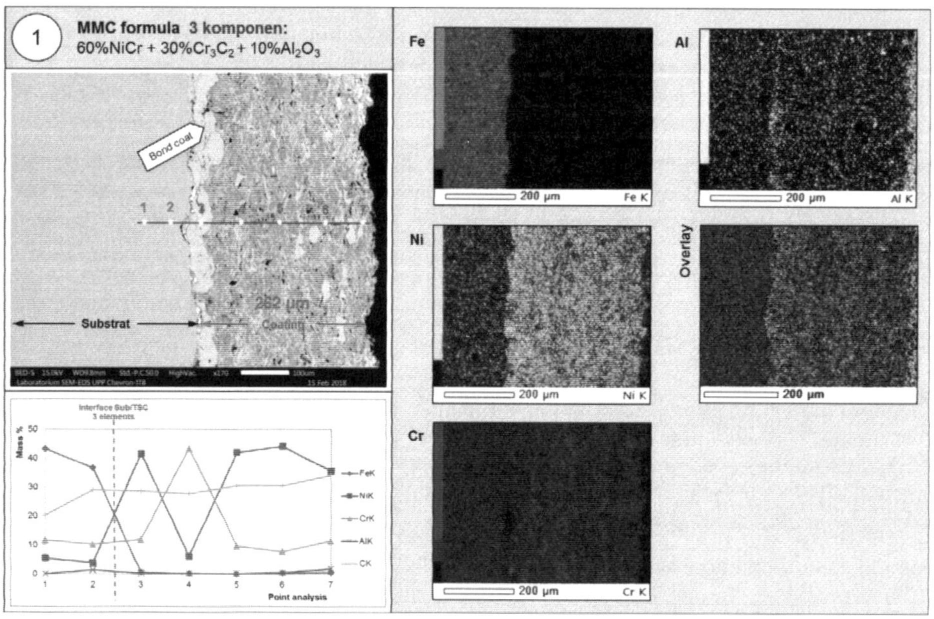

Figure 4. SEM micrograph of HVOF spray MMC 3 components, constituent element mapping, and point analysis.

SEM micrograph of MMC 2 components coating that shown in Figure 5 reveals the different distribution of *splats*, *un-melted particles*, porosities, and oxides with MMC 3 components coating. It is noted that the population of un-melted is lower than the MMC 3 components. The molten droplets that were sprayed continuously were generally spherical and become splats when impacting the surface; thus the shape spreads and fills space in a flat shape, rapidly solidified, and then evolved with the final lamellar morphology.

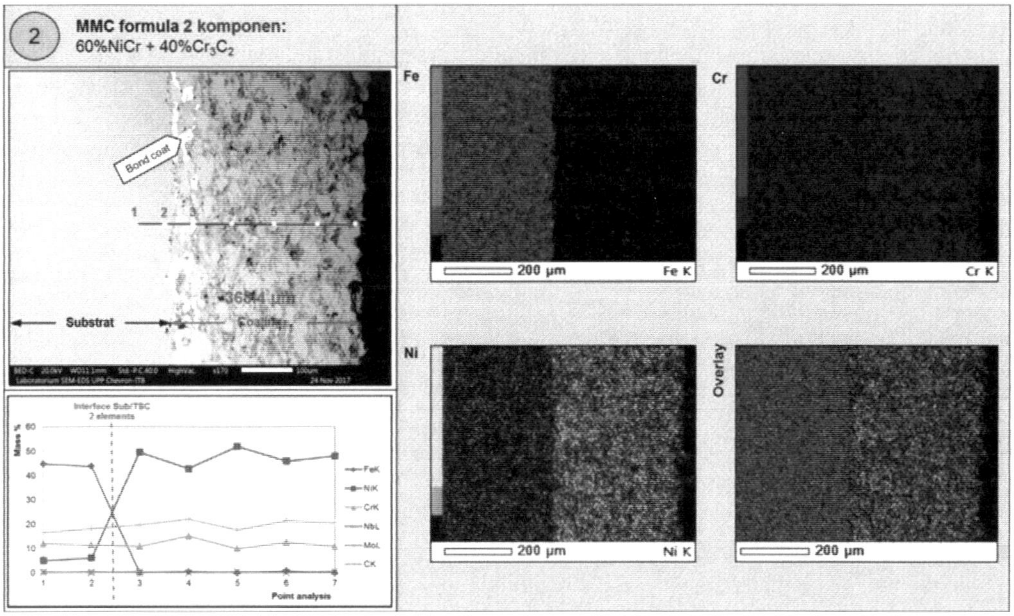

Figure 5. SEM micrograph of HVOF spray MMC 2 components, constituent element mapping, and point analysis.

Splats, signified with lamellar morphology are characteristic of thermal spray coating process. However, only a limited number of un-melted particles can be observed in the microstructures. The coating also possesses some voids and oxide inclusions that are typical characteristics of the HVOF sprayed coating. It is noted that the porosity is 1.29% and is randomly distributed in the coating.

Both in the Figure 4 and Figure 5 have been found that rich of Ni and Cr, and contain small amount of Al within the MMC 3 components coating. It is also shown that Fe has been found to be restricted to the substrate only, which shows that inter-diffusion of the substrate constituents to the coating has not been taken place. This statement is also confirmed by the point analysis performed with the EDS feature to seven points sampling across substrate to coating section.

For both samples, XRD analyses were carried out to identify the different phases present in the coatings. Figure 6 shows the X-ray diffraction pattern of the as coated MMC 3 components, it is seen in the peak that Ni is dominant phase in NiCr based coatings, Cr_3C_2 and Al_2O_3 were also detected and indicated in the microstructure as carbide and oxide. Another phase is $NiCr_2O_4$ that probably formed due to reaction of NiO and Cr_2O_3 during the thermal spray process and had influence to reduce the coating hardness, see point analysis of hardness in Figure 8.

The X-ray diffraction pattern of the specimen MMC 2 components coating is displayed in Figure 7. A significant peak characteristic identify substantial amount of Ni element. The Cr_7C_3 major intensities of peaks as the decarburization result of Cr_3C_2 in the coating, this one similar to the works that has been reported by Hong, *et al.* [17] and Guilemany, *et al.* [18], even though it was also present in the initial powder. It is understood from XRD characterization that existence of Cr3C2 phase provides an increase in hardness of MMC 2 components coating.

Hardness is the significant mechanical property of the coatings. Microhardness distribution data has been acquired to characterize the hardness change along the cross section. The graphic of microhardness shown in Figure 8 and Figure 9 were formed "S" profile. As shown on both profile the microhardness of point 1 exhibited nearly uniform and the value even quite similar to non-sprayed substrate.

At point 2, nearby substrate and coating interface shows little higher value of microhardness, it was both attributed to the work hardening during grit blasting prior to coating [19], and high velocity of molten droplets [20]. At point 3 from both figures shown the increment trend of microhardness, this was associated with bond coat buildup of NiCr powder on the substrate, then at point 4, the increased hardness value at the interface was attributed to the formation of chromium carbide at this region [19]. As indicated in Figure 8 at point 5 the microhardness falls from 397.8 HVN to 315.95 HVN and stay steady to the top region of the coating. In the contrary, Figure 9 from point 4 to 7 the microhardness distribution was relatively stable, only slightly falling to 481.57 at point 6 and rising up again to 538.24 HVN at point 7.

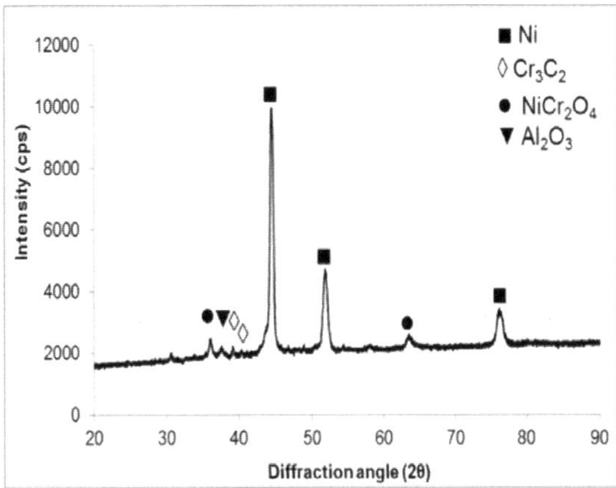

Figure 6. X-ray diffraction pattern of MMC 3 components coating.

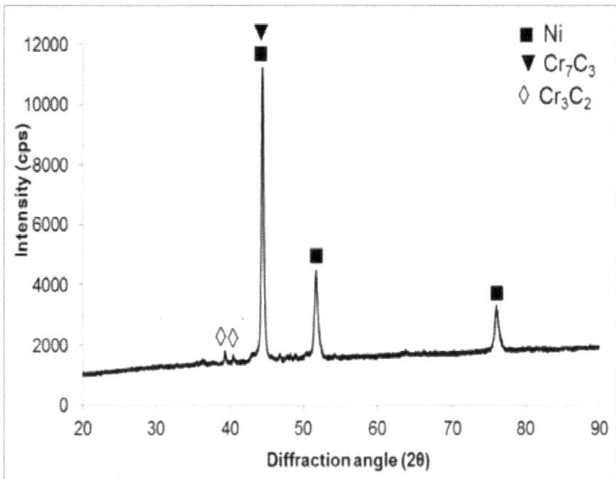

Figure 7. X-ray diffraction pattern of MMC 2 components coating.

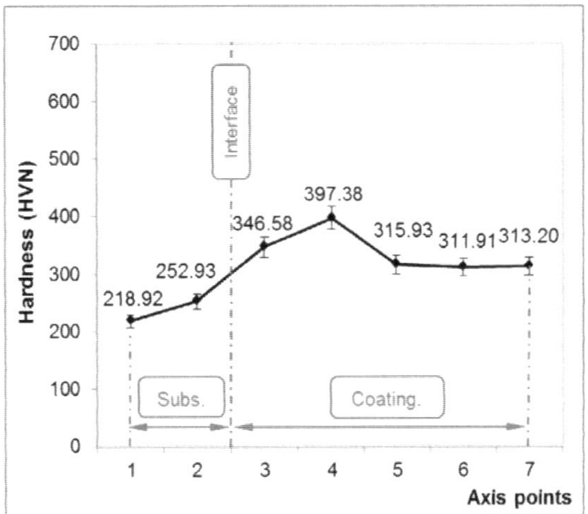

Figure 8. Microhardness profile of MMC 3 components coating.

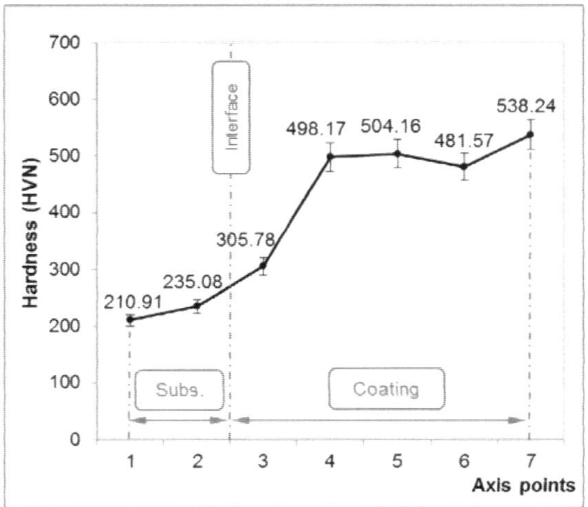

Figure 9. Microhardness profile of MMC 2 components coating.

As shown in Table 3 along with the summary of characteristics of coating quality, the tensile bond strength of MMC 3 components, and MMC 2 components were 5,778 psi and 6,570 psi where both results were greater than acceptance threshold for TSC application on boiler tubes material (> 5,000 psi), MMC 2 components coating was much stronger than the MMC 3 components coating that signified better adhesion property. One thing that has been agreed that adhesion was strongly dependent on the roughness of the substrate surface [21]. The results indicated that adhesion is much stronger between metal/similar metal carbide (NiCr/MMC 2 components) than between metal/cermet (NiCr/MMC 3 components), because of the easy diffusion between similar metal atoms [22]. From the appearance given in the Figure 10 the fracture mode of MMC 3 components shown adhesive fracture since the fracture has separated the whole bond coat off the substrate, and in the other hand as seen in the Figure 11 coating failure of MMC 2 components after the test shown cohesive fracture where the delamination of some top coat section from the bond coat is observed.

Table 3. Summary of characteristics data of HVOF thermal spray coating quality.

MMC Coating	Coating thickness		Hardness value			Porosity level (%)	Roughness index (μm)	Bond strength (Psi)
	Bond coat (μm)	Tot. thickness (μm)	Substrate (HVN)	*BC – BC / TC (HVN)	Average top coat (HVN)			
3 components	67	262	218.92 – 252.93	346.58 – 397.38	313.68	1.429	6.2284	5,778
2 components	72	368.4	210.91 – 235.08	305.78 – 498.17	507.99	1.289	7.3872	6,570

* BC: bod coat; BC / TC: interface between bond coat and top coat.

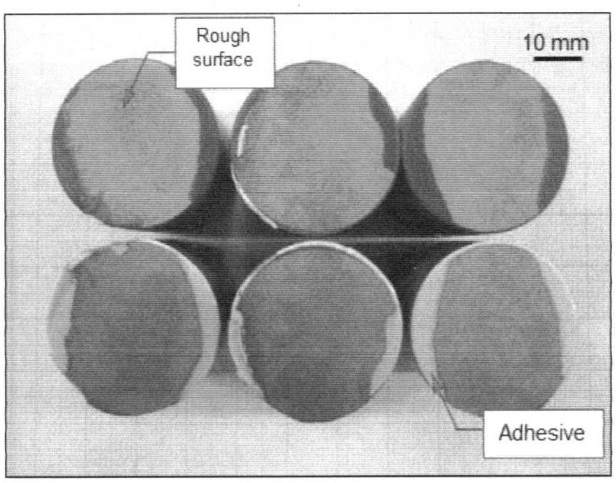

Figure 10. Coating fracture mode of MMC 3 components is adhesive failure, bond coat were detached more than 80% from substrate.

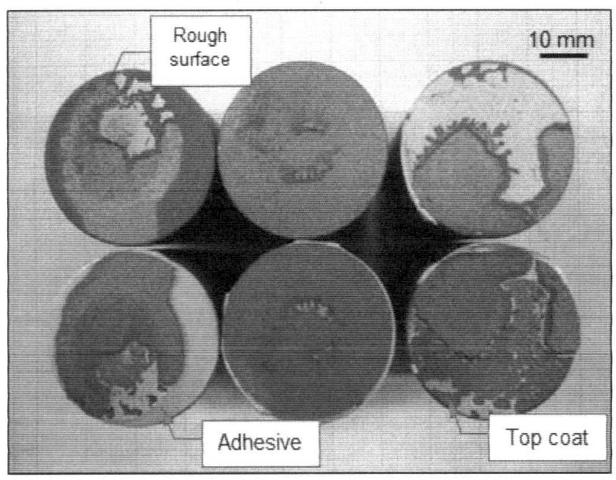

Figure 11. Coating fracture mode of MMC 2 components is cohesive failure, top coat deleamination from its bond coat.

4. Conclusion

This works has produced the following findings about quality characterization of HVOF thermal spray coating with NiCr matrix composite for protection application of coal fired boiler tubes.

1. MMC 3 components and MMC 2 components coating were successfully deposited by HVOF spray technique. The properties reveal uniform and adherent coating, and free from surface and cross sectional cracks with porosity less than 2%.
2. The presence of high carbide in the coating may be the reason for the high microhardness values in coatings as compared with the substrate. Except in MMC 3 components coating the presence of $NiCr_2O_4$ may be the reason for low microhardness.
3. Tensile bond strength test of both MMC 3 components and MMC 2 components were greater than the acceptable threshold. The results indicated that adhesion is much stronger between

metal/similar metal carbide (NiCr/MMC 2 components) than between metal/cermet (NiCr/MMC 3 components), because of the easy diffusion between similar metal atoms.

Acknowledgements
The authors would like to thank all technicians and researchers at Research Centre for Electrical Power and Mechatronics who participated in this research work, sub-program development of material, energy and manufacturing engineering.

References

[1] National board of energy 2014 *Indonesia Energy Outlook in 2014* Ministry of natural resource and mineral of Indonesia (In Indonesia)
[2] Directorate General of Electricity 2017 *Electrification Ratio within the Last Five Year* Ministry of Natural Resource And Mineral of Indonesia Access date 24-Mar-2017 (In Indonesia)
[3] Directorate General of Electricity 2016 *Electrical statistic in 2015* Ministry of Natural Resource and Mineral of Indonesia Ed. 29 access date 24-Mar-2017 (In Indonesia)
[4] Prasetyo P D 2015 *Sepuluh Nopember Institute of Technology undergraduate paper* (In Indonesia)
[5] Muharom A 2015 *Sepuluh Nopember Institute of Technology undergraduate paper* (In Indonesia)
[6] Sivakumar R and Mordike B L 1989 *Surface and Coat. Tech.* **37(2)** 139-60
[7] Nguyentat T Dommer K T and Bowen K T 1992 *Proc. Int. Thermal Spray Conf. and Exposition* (ASM) 321-5
[8] Mruthunjaya M Parashivamurthy K I and Devappa 2016 *Int. J. of Mechan. Engin. and Tech.* **7(3)** 53-62
[9] Kamal S Jayaganthan R and Prakash S 2009 *J. of Alloys and Compounds* **472(1-2)** 378-89
[10] Wang B Q and Luer K 1994 *Wear* **174(1-2)** 177-85
[11] Wang B Q and Verstak A 1999 *Wear* **233** 342-51
[12] Di Gianfrancesco A (Ed.) 2016 *Materials for ultra-supercritical and advanced ultra-supercritical power plants* Woodhead Publishing
[13] Martides E Prawara B Ardy H Junianto E and Priyono B 2017 *Mater. Sci. Forum* **889** 30-5
[14] Kimura H Sato T Bergins C Imano S and Saito 2011 *Hitachi review* **60(7)** 365
[15] Purgert R M 2013 *Proc. Int. Conf. on Sustain. Growth Strategies for Fuel and Effic. (New Delhi, India)*
[16] Sadeghimeresht E Markocsan N and Nylen P 2017 *Surf. and Coat. Tech.* **318** 365-73
[17] Hong S Wu Y Wang Q Ying G Li G Gao W and Guo W 2013 *Surf. and Coat. Tech.* **225** 85-91
[18] Guilemany J M Espallargas N Suegama P H and Benedetti A V 2006 *Corros. Sci.* **48(10)** 2998-3013
[19] Staia M H Ramos E Carrasquero A Roman A Lesage J Chicot D and Mesmacque G 2000 *Thin Solid Films* **377** 657-64
[20] Sundararajan T Kuroda S and Abe F 2004 *Mater. Transactions* **45(4)** 1299-305
[21] Higuera Hidalgo V Belzunce Varela J Martínez de la Calle J and Carriles Menéndez A 2000 *Surf. Engin.* **16(2)** 137-42
[22] Cho T Y Yoon J H Cho J Y Joo Y K Kang J H Zhang S and Kwon S C 2009 *Surf. and Coat. Tech.* **203(20-21)** 3250-3

The synthesis of NiCrAlHf powders with mechanical alloying method for thermal sprayed coating

R D Desiati[1], E Sugiarti[1], S Ramandhany[1], S J Winarno[2]

[1] Research Center for Physics, Indonesian Institute of Sciences (LIPI), Serpong 15314, Indonesia
[2] Physics Division, Faculty of Science and Technology, State Islamic University of Jakarta 15411, Indonesia

rese001@lipi.go.id

Abstract. NiCrAl is a material commonly used for coating in order to increase corrosion or oxidation resistance of materials applied at high temperatures. The addition of hafnium (Hf) as reactive elements to NiCrAl coating are found to prolong the life of the industrial gas turbine blade and increase its performance. NiCrAlHf powders used as the base material were mixed using planetary mill device for 36 hours with a speed of 25 Hz or 1500 rpm. Characterization of NiCrAlHf powder was investigated before and after milling process using Scanning Electron Microscopy (SEM) and X-ray diffraction (XRD). The results showed that the synthesized powders were agglomerated. The obtained particle size of 9.64 μm, 19.97 μm, 39.27 μm, 19.93 μm, and 22.8 μm was observed before milling process, after 9, 18, 27, and 36 hours milling process, resepectively. The phase formed on the synthesized NiCrAlHf powder after 36 hours milling process included Ni and Cr with crystallite size of Ni 25.4 nm and Cr 8.8 nm calculated by Scherrer method. In addition the peak profile of the Al phase gets more amorphous after 27 hours.

1. Introduction

The NiCrAl coating is an interface layer for thermal barier coating that systems commonly used to protect components made of superalloys for high temperature usage. In NiCrAl bond coat systems, reactive elements are required as stabilizers and enhancement for protective binding to extend the life of the gas turbine blade. The commonly used reactive elements are from rare earth element, such as hafnium, due to its high melting point and good corrosion resistance. In addition, hafnium is also primarily used to improve mechanical properties of the coating. Addition of reactive elements is also proven to decrease the formation of oxide scales, thus prolonging the lifetime of coating [1].

Thermal barrier coating that is composed with nano structured bonding has shown improved performance, especially at oxidation behavior at high temperature condition, compared with conventional micro structure bonding [2]. One of the known methods to form nano structured material is mechanical milling [3]. Ball milling as one of mechanical milling method has been proven to successfully prepare a variety of alloy powders exhibiting supersaturated solid solutions, quasi-crystals, amorphous phases and nano intermetallic compounds. The fundamental principle of size reduction in mechanical attrition devices lies in the energy parted to the sample during impacts between the milling media [4].

This study aims to introduce synthesis method of nano-crystallite NiCrAlHf powders through mechanical process and investigate the morphology of NiCrAlHf powders before and after milling process.

2. Experimental procedures

Nickel, chromium, aluminum and hafnium powders are weighed with composition as follows:

Table 1. Composition of powders (gram).

Element	Ni	Cr	Al	Hf
Mass of powders (gr)	137.2	48	14	0.8

The amount of reactive element used (Hf) was 0.4% of the overall mass or 0.8 gram, because the reactive element of $\geq$ 1% or $\leq$ 0.1% can accelerate the removal of the protective oxide layer when it's applied. The powders were mixed and mechanically processed with planetary ball mill (SFM-1 desk-top type) with a milling speed of 1500 rpm for 36 hours. Ball to powder ratio is specified into 2:1.

To obtain the effect of milling time to crystallite size, every 9 hours of milling, mechanically processed powders were sampled and characterized. Morphology of particles are identified using scanning electron microscope (JEOL JIB 4610f), while phases of particles are identified with X-ray diffractometer (Rigaku SmartLab) with a wavelength of CuKα. X-ray diffraction result then will be analyzed through Scherer's method to obtain the crystallite size.

3. Results and discussion

3.1 Scanning electron microscopy (SEM) characterization

Figure 1 shows the morphology of milled powder for 0 hour until 36 hours, with a step of 9 hours. From these images, an increase in particle size can be observed in accordance with milling time. At first the powders are observed to be in various shapes depending on the base powders with an average size of 9.64 μm. After the first 9 hours of milling process, the powders agglomerated to form a homogenous shape and increase in size to 19.97 μm. After 18 hours of milling, the agglomerated powders started to agglomerate further and gain an increase in size to 39.27μm. Agglomeration is caused as an effect of mechanical alloying when powders particles collide with enough energy to form interaction between the powders [5]. However, after 27 hours of milling the powders did not agglomerate further and having a breakdown to the size of 19.93 μm and later increase in size again to 22.8 μm after 36 hours of milling. This breakdown phenomenon may be caused by low binding energy of crystallites in the agglomerated form, therefore continued milling process cause the powder particles to be dispersed again. Figure 2 shows the measurement of particle size from NiCrAlHf powders using image-J software before and after milling process until 36 hours.

3.2 X-ray difraction (XRD) characterization

Phase analysis of the sampled powder was characterized with X-ray diffraction and the results are shown in Figure 3. From the graph it can be observed that before milling process, phases of elementary powder used in this experiment are obtained. However, after milling process, a decrease in peak intensity of detected phases can be observed. It indicated that crystallite size was reduced.

From the figure, it can also be observed that milling time plays an important part in the decrease of crystallite size and deform some of the material. Aluminum peak are not observed after 36 hours of milling time indicate that there is a change in the crystal shape that cause it to deform into an amorphous phase [6]. A study shows that mechanical alloying process with nickel and chromium presence cause by a deformation of crystal shape into amorphous [7, 8].

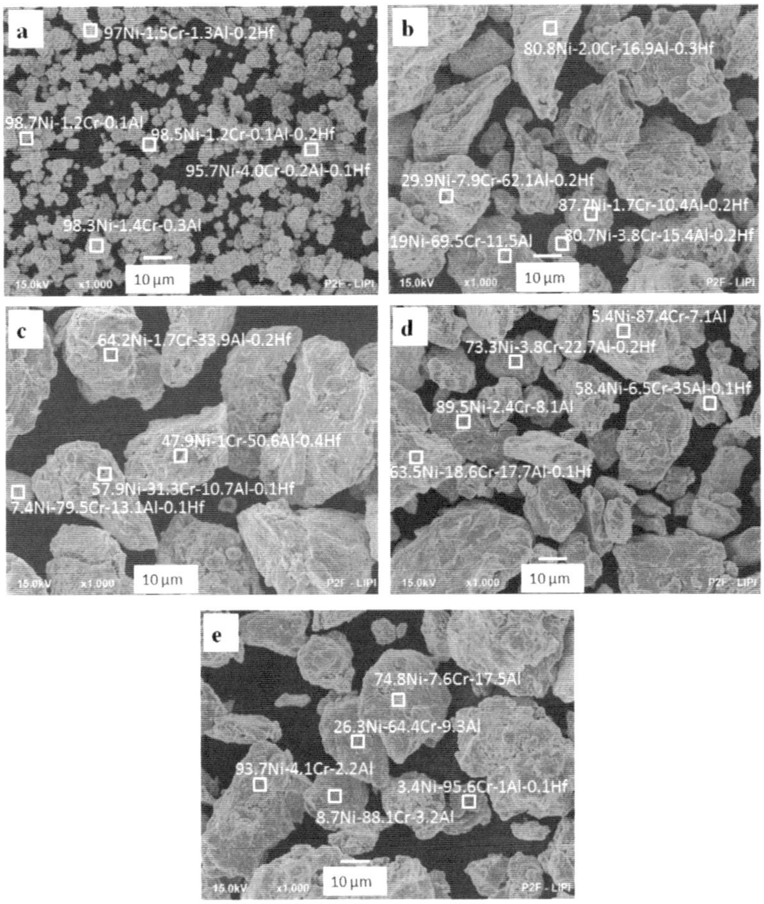

Figure 1. SEM image of NiCrAlHf powder (a) before milling, after milling process of (b) 9 h (C) 18 h, (d) 27 h, and (e) 36 h.

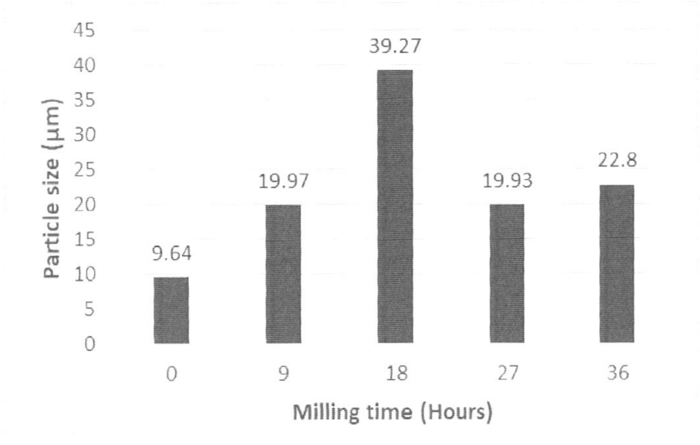

Figure 2. Graphic particle size of NiCrAlHf powders using image-J software at various milling times.

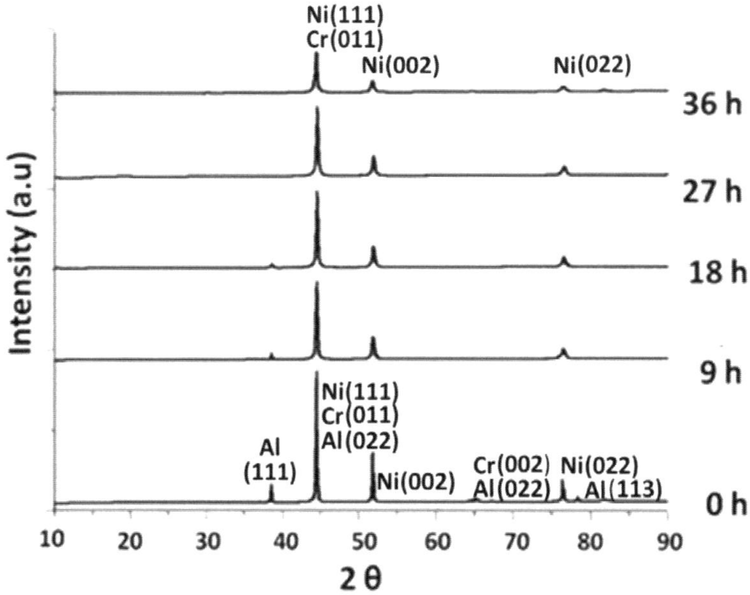

Figure 3. X-Ray difraction pattern analysis of NiCrAlHf powder.

Further analysis of the XRD result with Scherrer's method was done to analyze the crystallite size of processed powder [9]. The method uses calculation shown in equation 1 to estimate crystallite size of powders.

$$D = K\lambda / B \cos \theta \tag{1}$$

Where:
D = Crystallite size (nm)
K = Constant (0,9)
λ = X-Ray wavelength (nm)
B = FWHM at 2θ (rad)

Results of crystallite size calculation are shown in Table 2. According to the results it can be observed that mechanical process of the powders reduces crystallite size of powders, even though there is an increase in particles size due to agglomeration. Milling process duration caused the material size to be decreased further. It can also be observed that the crystallite size of powders is in the scale of nanometer, thus it should be ideally used as a starting material for coating process.

Table 2. The effect of milling times on crystallite size calculated by Scherer method.

Milling Times (hours)	Crystallite size (nm)		
	Ni	Cr	Al
0	61.3	28.5	79.1
9	30.8	13.3	54
18	31.9	12.0	37.6
27	28.1	9.2	
36	25.4	8.8	

4. Conclusions

From the experiments there are several conclusions that could be drawn (**i**) Mechanical alloying process using planetary ball mill can be used to mix Ni-Cr-Al-Hf powders to an alloy for bond coat material; (ii) Milling process for 36 hours caused the powders to form agglomeration with an average size of 22.8 μm; (iii) After 36 hours of milling, phase identification shows that the alloyed powders are mainly constituted of nickel and chromium phases, while other phases are deformed into an amorphous shape; (iv) Using the Scherrer's method, a decrease in crystallite size was able to be proven after 36 hours of mechanical alloying process.

Acknowledgments

The author would like to give thanks to Excellent Activities LIPI, INSINAS IRPI Grant No.04/INS-2/PPK/E/E4/2017 and Toray Science Technology Research Grant for financial support of this work. The authors would also like to thank to researchers in the Group of High Resistant Materials - Research Center for Physics for valuable discussions and support during this work.

References

[1] Hou P Y 2011 *Mater. Sci.* **696** 39
[2] Zhang Q Jiu C Cheng X Li Yang G J and Lui S C 2009 *Surf. Coat. Tech.* **211** 228
[3] Kim G E Addona T Richer P Jodoin B Al-Mathami A and Brochu M 2010 *Thermal Spray Tech.* **112** 29
[4] Hussain M S and Daroonparvar M 2013 *Adv. Phys .Theor. Appl.* **13** 2225
[5] Pratama F B and Widyastuti 2012 *Teknik ITS* **1** 97
[6] Liu K W Mucklich F Pitschke W Birringer R Wetzig K 2001 *Sci. Eng.* A **313** 187
[7] Bidabadi A R S Enayati M H Dastanpoor E Varin R A and Biglar M 2013 *Alloys Compd.* **581** 91
[8] Sheu H Hsiung L Sheu J 2009 *Alloys Compd.* **469** 483
[9] Abdullah M and Khairurrijal 2008 *J. Nanosch. and Nanotech.* **1** 28 (In Indonesia)

Chloride solution leaching of platinum from Indonesia gold ore artisanal mining

S Harjanto[1*], A Lazuardiyani[1], M Y Salam[1], M Taris[1], and F W Pratama[1]

[1] Department of Metallurgical and Materials Engineering, University of Indonesia, UI Depok Campus, Depok 16424, Indonesia

harjanto@metal.ui.ac.id

Abstract. Platinum has been growing in economic importance in recent years. Limited occurrence of large scale deposits and ever-increasing demand have encouraged the utilization of deposits with low grade ore. Efficient yet small scale applicable method of extraction is needed to utilize those deposits. Chloride solution based leaching facilitates excellent condition for platinum dissolution by significantly reducing its reduction potential and through formation of stable aquochlorocomplexes. This paper reports optimization of chloride solution based platinum leaching from Indonesian artisanal ore under variation of lixiviant concentration, leaching time, pulp ratio, and leaching temperature. The Taguchi method was used to optimize the process. Hydrochloric acid was chosen as primary source of chloride ion and H_2O_2 were added as oxidizer. Parameters were arranged in L-9 orthogonal arrays and optimized through signal to noise ratio (S/N Ratio). Analysis of variance was utilized to identify the influence of each parameters to overall leaching process. The suggested optimized condition was: HCl concentration= 12 M, leaching time= 6 h, pulp ratio = 0.1, and temperature= 75 °C. It has been identified that time was the most influential parameter.

1. Introduction

Platinum and other platinum group metals (PGM) are precious metals with high and growing economic importance. Primary producer of platinum and other platinum group metals are South Africa, Russia, Canada, Zimbabwe, United States, and Colombia. South Africa controls over 95% of global deposit of PGM or about 63 000 ton PGM ores [1]. In 2016 global platinum supply deficit reached 861 troy ounces, which was almost 10% of the entire demand of 8677 troy ounce [2].

The extraction of Platinum and other PGM is traditionally carried out through pyrometallurgical routes of matte-smelting-refining technique from high grade ore. Environmental cost is the primary problem of this technique as it produces large volume of pollutants [3]. On the other hand, hydrometallurgical routes provide a promising alternative both in economic and environmental senses as platinum dissolves and form stable complex in several media such as chloride, cyanide, halide, and hypochlorite [4, 5].

Among several other available leaching media chloride media based leaching facilitate excellent condition for platinum dissolution by significantly reduce its reduction potential and through the formation of stable chloridometalate complexes [6]. In addition, this leaching method does not require pressures higher than the atmospheric pressure [3, 4]. Platinum dissolution in chloride media is based on reactions shown in equation (1) and (2).

$$Pt + 6Cl^- \leftrightarrow PtCl_6^{2-} + 4e^- \qquad E° = -0.744V \qquad (1)$$

$$Pt + 4Cl^- \leftrightarrow PtCl_4^{2-} + 2e^- \qquad E° = -0.73V \qquad (2)$$

Several other researchers have successfully leached platinum and other PGM metals by using chloride solutions. Leaching of platinum from industrial catalyst using combination of 60% H_2SO_4 and 0.1 M NaCl has been successfully conducted yet the leaching time was quiet long [4]. Primary platinum ore from South Africa has also been successfully leached using 6M HCl and chlorine gas [7]. Study of different type of chloride media has proven that combination of HCl and H_2O_2 is the most effective lixiviant to dissolved platinum from spent automotive catalyst, however high concentration of HCl were needed [5]. Reaction of platinum dissolution after addition of H_2O_2 is as listed in equation (3).

$$Pt + 4HCl + 2Cl^- \leftrightarrow H_2PtCl_6 + H_2O \qquad (3)$$

$$\text{Anodic} \quad : Pt + 6Cl^- \leftrightarrow PtCl_6^{2-} + 4\,e^- \qquad E^O = -0.744V \qquad (4)$$

$$\text{Cathodic} \quad : H_2O_2 + 2H^+ + 2e- \leftrightarrow 2H_2O \qquad E^O = 1.76V \qquad (5)$$

Based on the discovery of platinum element in some of gold ore artisanal Mining found in Indonesia, the extraction of platinum contained in Indonesian gold ore is become increasingly attractive. However, to maximize the economic gain of the extraction of platinum contained in these ore, an environmentally friendly process with high efficiency is needed. In order to develop such process, this study carried out single-step leaching of platinum from Indonesian artisanal ore by using HCl and H_2O_2 to examine the influence of several leaching parameter to the recovery of platinum. Evaluated parameters were temperature, HCl concentration, pulp ratio, and time. Taguchi method was implemented in experiment design and was used to identify the influence of each parameter involved in the recovery of platinum.

2. Experimental method

2.1. Materials

Indonesian artisanal gold ore was used in this study. The received sample has been roasted and nitrogen pre-treated. Sample was ball milled and sieved. Sample with size of less than 74 μm was used as feed of the leaching process. Platinum content in the sample was known to be 16 ppm. All of the reagent used in this study were of analytic grade. The reagents used were HCl (37%, Merck), and H_2O_2 (30%, Merck). Demineralized water was used to diluting the solutions.

2.2. Methods

Leaching procedure was carried out in 250 mL Erlenmeyer flask. Magnetic stirrer and hot plate were used to agitate the leaching solution and adjust its temperature during leaching process. Watch glass was used as a stopper on the top of the Erlenmeyer flask during all experiment; it was put loosely, preventing the change in pressure in the Erlenmeyer flask. Fine cotton filters (GE Healthcare) with pore size of about 2.5 μm was used to filter the leaching solution at the end of each experiment. Filtration was done at ambient temperature after the solution cooled down. Filtrated solution was then diluted for analysis. Elemental analysis was carried out using AAS (Perkin Elmer). The recovery was calculated based on dissolved platinum in the filtered solution compared to the mass of platinum contained in received sample.

2.3. Taguchi method

In order to optimize the process parameters and minimize the number of experiments needed, this study utilize Taguchi Robust Parameter Design or Taguchi Method. The Taguchi method is a statistical method developed by Taguchi and Konishi to optimize processes involving various parameters with minimal amount of experiment [8]. Through this method relative influence of a particular parameter to the overall process can be characterized. Taguchi method uses orthogonal array that governs the configurations of parameters used in each experiment therefore that the number of experiments can be minimized. On a study with four parameters and three level for each parameter involved, traditional method requires 3^4 or 81 experiments, while through utilization of this method only 9 experiments are necessary [9]. This method utilizes Signal-to-noise ratio (S/N) to analyze the output in graphical form [10]. Signal-to-noise ratio is a logarithmic function that combines the contribution of the mean value and the variation of the sample value into a single value [11]. In this study, larger-the-better S/N ratio were used, since the objective of the study is to achieve the highest recovery of platinum. The larger-the-better S/N ratio formula is shown in equation (6).

$$S/N_{\text{larger-the-better}} = -10\log\left(\frac{1}{n}\sum_i\left(\frac{1}{y_i^2}\right)\right) \quad (6)$$

Table 1. Taguchi's L9 orthogonal array design.

Exp.	Parameters			
	Temperature (°C)	HCl Concentration (Molars)	Pulp Ratio	Time (h)
1	25	3	0.1	2
2	25	6	0.25	4
3	25	12	0.5	6
4	50	3	0.25	6
5	50	6	0.5	2
6	50	12	0.1	4
7	75	3	0.5	4
8	75	6	0.1	6
9	75	12	0.25	2

Through this method relative influence of each parameter involved can also be characterized through ANOVA analysis. Formulas involved in ANOVA analysis is shown in equations (7), (8), and (9). Several researchers have proven the robustness and usefulness of this method in field of extraction [7-10].

$$mi = \left(\frac{1}{Ni}\right)\sum S/N \quad (7)$$

$$\text{Sum of squares (SOS)} = N_i\sum_{i=i}^{i=j}\left(mi - \langle mi\rangle\right)^2 \quad (8)$$

$$\text{Parameter influence} = \frac{SoS}{\left\{DoF \times \sum\left(\frac{SoS}{DoF}\right)\right\}} \quad (9)$$

Four parameters (temprature, HCl concentration, pulp ratio, and time) in three level were considered in this study. Taguchi's L9 orthogonal array design was utilized to dictate the combination of each parameter for each experiment. The design of L-9 orthogonal array for the study is presented in contained in received sample. contained in received sample.

3. Results and discussion

3.1 Ore characteristic

The XRD spectrum of the ore is shown in Figure 1. It is shown that the most prominent peeks are belongs to quartz (SiO_2) and iron oxide (Fe_2O_3). XRF analysis as listed in Table 2 confirms that quartz is the main component of the ore. The presence of Fe_2O_3 indicate the occurrence of FeS on pre-roasted ore, as Fe_2O_3 is the final product of FeS roasting [12]. From these information, it is clear that the ore used was came from quartz-gold veins which is the common type of gold deposit found in Indonesia. The presence of high Fe concentration in the ore was a problem since Fe also dissolved in chloride solution, causing inefficiency of lixiviant usage. Another metallic element contained in the ore also possesses similar threat.

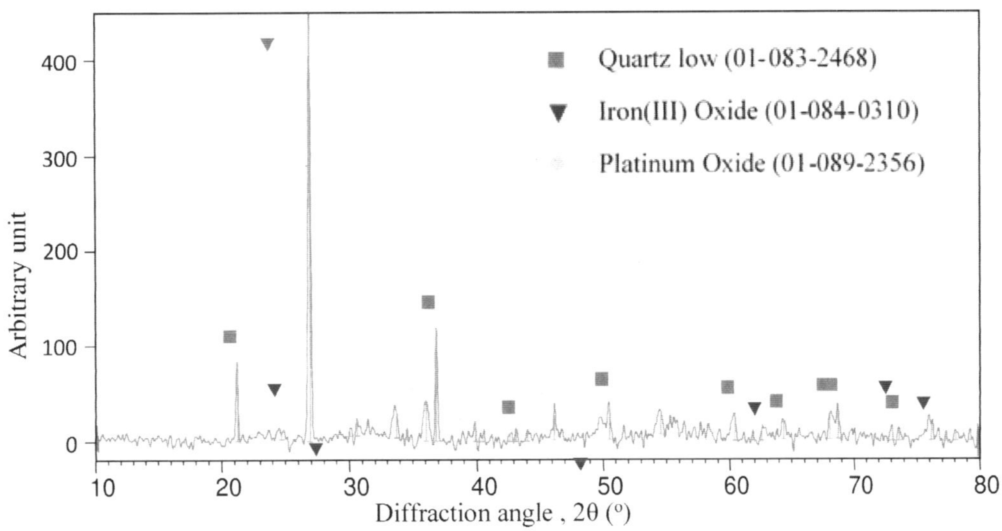

Figure 1. XRD patterns of ore sample used in the experiments.

3.2 Process optimization

The recovery and S/N ratio of each experiment are given in Table 3. The effects of each parameter on the recovery of platinum are listed in Table 4. It is shown in Table 4 that leaching time was the most influential parameter (96.37%) while temperature was the least influential parameter (0.3794%).

Table 2. XRF Analysis of five major elements of the ore.

Element	Atomic Mass (z)	Concentration (wt.%)
Iron	26	43.05
Silicon	14	14.17
Aluminum	13	3.751
Sodium	11	0.685
Titanium	22	0.4188

It can be noted that compared to leaching time, other parameters are insignificant. Figure 2 illustrate the S/N ratio of each parameter. Taguchi's S/N ratio predicts that the optimum leaching condition (corresponding to highest S/N) for platinum extraction is at temperature of 75 °C, pulp ratio of 0.1, HCl concentration of 12 molars, and at 6 hours long.

The leaching experiments were carried on temperature range from room temperature to 75 °C. Figure 2 shows that extraction of platinum increased with increasing temperature. Platinum dissolution kinetics becomes more favorable at higher temperature. However, ANOVA analysis has shown that temperature effect on platinum recovery was insignificant (0.3794%).

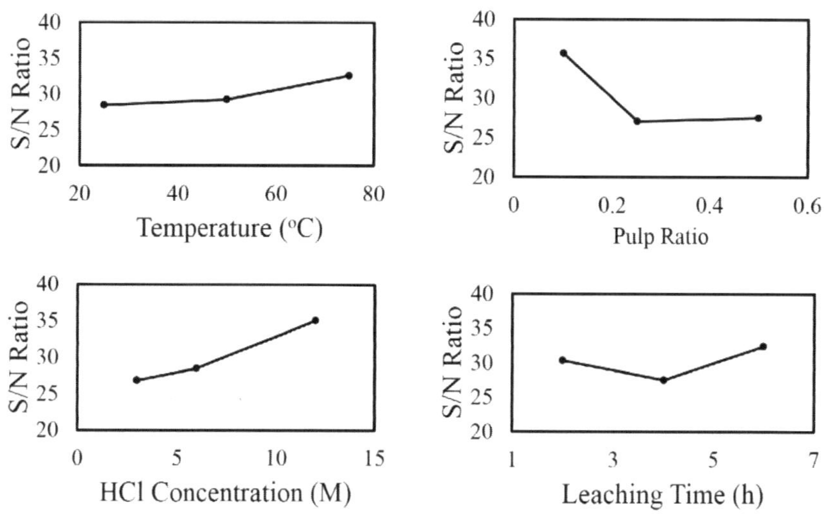

Figure 2. S/N Graph for platinum extraction.

HCl was chosen based on previous study that shows combination of HCl and H_2O_2 is the best lixiviant to dissolve platinum compared to several other lixiviant [5]. S/N ratio of HCl concentration on Table 3 shows platinum recovery was directly proportional to HCl concentration used. ANOVA analysis, as shown in Table 4, revield that HCl concentration increase effect on platinum recovery was only 1.435%. Such minimal effect explains the reason that several other researchers could achieve effective leaching of platinum in relatively low HCl concentration.

Table 3. Recovery and S/N ratio of each experiment.

Experiments	Recovery (%)	S/N Ratio (dB)
1	36.85	31.04
2	11.58	21.22
3	46.01	33.11
4	18.74	25.36
5	18.59	25.35
6	73.36	37.23
7	18.00	24.26
8	89.05	38.93
9	54.92	34.79

Pulp ratio was defined as ratio of feed mass (g) to volume of lixiviant (ml) used. S/N graph of pulp ratio which is shown in Figure 2 shows that pulp ratio 0.1 has highest S/N ratio (35.73) while pulp ratio of 0.25 and 0.5 have relatively comparable S/N ratio (27.13 and 27.57). Low pulp ratio means more lixiviant available to every gram of feed which also means more cost on lixiviant. ANOVA analysis has shown that pulp ratio is the second most influential parameter of the study with effect of 1.8%.

Table 4. Effect of each parameter on the recovery of platinum.

Parameters	Level	Yield			
		mi	$\overline{mi}$	SOS	%Effect
	25	28.46			
Temperatur	50	29.31	30.14	29.60	0.3794
	75	32.66			
	3	26.89			
[HCl]	6	28.50	30.14	112.0	1.435
	12	35.04			
	0.1	35.73			
PR	0.25	27.13	30.14	141.0	1.806
	0.5	27.57			
	2	91.18			
Time	4	27.57	50.41	7518	96.37
	6	32.46			

Time was the most influential parameter of this study as reviled by ANOVA analysis (Table 4). With effect of 96.37% it renders other parameters insignificant. S/N ratio graph for time effect on leaching of platinum shown in Figure 2 did not shows a perfect directly proportional line. The drop on S/N ratio from 2 hours to 4 hours was caused by decomposition of H_2O_2. While at the early hours of the leaching H_2O_2 helped oxidation of platinum and speeds up dissolution process, decomposition of H_2O_2 reduce this effect on latter hours of leaching. Nonetheless, H_2O_2 has never decomposed completely, therefore platinum dissolution was still occurred at latter hours of the leaching although in slower pace of platinum dissolution.

4. Conclusion

Study on leaching of platinum content of Indonesia artisan ore have been carried out. Taguchi S/N ratio suggest that the optimum condition for platinum extraction is at temperature of 75 °C, pulp ratio of 0.1, HCl concentration of 12 molars, and at 6 hours long. The most influential parameter in this study was leaching time with influence of 96.37%, followed by pulp ratio with 1.806%, HCl concentration with 1.435%, and the least influential parameter was temperature with influence of only 0.3794%.

Acknowledgment

The authors wish to acknowledge The Directorate of Research and Community Service of The University of Indonesia for financial support of this study through PITTA Research Grant No. 840/UN2.R3.1/HKP.05.00/2017 of budget year 2017.

References

[1] United States Geological Survey *Mineral Commodity Summaries: Platinum-Group Metals* 2016
[2] Jhonson Matthey *PGM Market Report May* 2016
[3] Mpinga C Eksteen J Aldrich C and Dyer L 2015 *Miner. Eng.* **78** 93

[4] Mahmoud M 2003 *JOM* **55** 37
[5] Harjanto S Cao Y Shibayama A Naitoh I Nanami T Kasahara K Okumura Y Liu K and Fujita T 2006 *Mater. Trans.* **222** 701
[6] Nogueira C A, Paiva A P, Oliveira P C, Costa M C and Rosa da Cosa A M 2014 *J. Hazard. Mater.* **278** 82
[7] Green B Smit D Maumela H and Coetzer G 2004 *J. South Afr. Inst. Min. Metall.* **104** 323
[8] Athreya S and Venkatesh Y 2012 *IRJES* **1** 13
[9] Dasgupta K Singh D Sahoo D Anitha M Awasthi A and Singh H 2014 *Sep. Purif. Tech..* **124** 74
[10] Aderhold J Davydov V Y Fedler F Klausing H Mistele D Rotter T Semchinova O Stemmer J and Graul J 2001 *J. Cryst. Growth* **222** 701
[11] Robinson J T and Anderson-Cook C M 2011 Robust Parameter Designs *Wiley Series in Probability and Statistics : Design and Analysis of Experiments, Special Designs and Applications* 3 ed K Hinkelmann Hoboken: Wiley chapter 13 443-69
[12] Marsden J O and House C I 2006 *Chemistry of Gold Extraction* 19-68

IOP Conf. Series: Materials Science and Engineering **432** (2018) 012014 doi:10.1088/1757-899X/432/1/012014

Polypropylene/clay nanocomposites prepared in an internal mixer: optimization of processing conditions to improve flexural modulus

Zaiby[1], A S Yogi[2], L J Ariadne[1] and U Onny[2]

[1] Department of Physics, Faculty of Mathematics and Natural Sciences, University of Indonesia, Depok, 16424, Indonesia
[2] Center of Polymer Technology. BPPT Puspitek Area, Setu, South Tangerang, 15314, Indonesia

zaiby94@gmail.com

Abstract. Polypropylene (PP)/clay nanocomposites are usually processed by melt mixing. In this method, mixing conditions are important variables to improve nanocomposite properties. Some studies reported the effects of processing on mechanical properties of PP/clay, but there is no clear explanation on optimum conditions. This study aims to predict the optimum conditions of PP/clay nanocomposite prepared by an internal mixer using response surface methodology (DoE). The effect of mixing on flexural modulus and nanocomposite structures were analyzed. Temperature, rotation speed, and mixing time were varied. To improve interfacial bonding, polypropylene-graft-maleic anhydride (PP-g-MA) was added as a compatibilizer. PP/clay formulation was fixed at 88 wt% of PP, 9 wt% of PP-g-MA, and 3 wt% of clay. The results show that the optimum modulus was predicted at 222 °C, 83 rpm and 5 minutes, giving 2085 MPa or 18% improvement compared to control sample. XRD diffractograms showed that nanocomposite peaks shifted to lower angles, suggesting the presence of some intercalated structures that supported the resulting increase in modulus.

1. Introduction

Polypropylene (PP) is one of the biggest commodity polymers in the world. Despite the vast applications of PP, the mechanical properties of PP are not good enough for some applications as building material. Therefore, fillers such as clay are needed to improve PP mechanical properties [1]. The clay is naturally a hydrophilic material, which has different polarity to PP. Therefore, a compatibilizer such as polypropylene-graft-maleic anhydride (PP-g-MA) is needed to facilitate the dispersion of clay layers in PP [2, 3].

The dispersion of clay layers also depends on the preparation process. Internal mixer is one of commonly used process that could facilitate the dispersion clay layers on nanocomposite [4]. The process parameters in internal mixer which can affect nanocomposite performance are process temperature, rotation speed, and time [5, 6]. The rotation speed will produce shear tension. A combination of low temperature and high speed rotation can produce a higher shear tension which can facilitate the dispersion process. However, a high shear tension can make the composite degraded [7]. Therefore, the optimum process condition is required.

Optimum process condition can be achieved by varying process parameters. Some studies showed that 190-240 °C processing temperature [6, 8, 9], 60-100 rpm rotation speed [8, 10, 11], 5-15 minutes mixing time [8, 11, 12] are the ranges that provide a chance of getting an optimum values for the mechanical properties. However, these varieties did not show optimum point directly. Therefore, a method, such as response surface methodology (DoE), to analyze optimum point is needed.

In this research, the effect of processing condition in PP/clay nanocomposite was studied. Various process temperatures, rotation speeds, and mixing times were used in our study. The optimum condition of PP/clay composite was analyzed using response surface methodology (DoE), based on flexural modulus results. The presence of intercalation or exfoliation of clay was also analyzed by X-ray diffraction.

2. Experimental method

2.1. Materials

PP (5169 MAS 2158) with 1.8 g/10min of MFI produced by P.T Politama was used as a matrix. PP-g-MA (Epolene Wax G 3015P) from Eastman Chemical Company were used as a compatibilizer, Southern Clay product, O-MMT clay (Cloisite 20A), was used as filler.

2.2. Preparation of nanocomposites

The nanocomposites were prepared using an internal mixer (Haake Rheomix 600) with various settings as planned in DoE (Table 1.). The clay was dried at 80 °C overnight. After that, the clay, PP-g-MA, and PP were dry mixed, followed by drying at similar conditions. The control sample, neat PP, was processed at 80 rpm, 10 minutes, and 220 °C, according to mid-setting shown in Table 1. The formula was fixed at 88/9/3 of wt% for PP/PP-g-MA/clay as resulted in the best mechanical property [13].

Table 1. Processing conditions used in an internal mixer.

Runs	Temp (°C)	Speed (rpm)	Time (minutes)	Runs	Temp (°C)	Speed (rpm)	Time (minutes)
Control	220	80	10	8	220	60	5
1	230	80	5	9	230	60	10
2	220	60	15	10	230	100	10
3	220	100	5	11	220	80	10
4	230	80	5	12	220	100	15
5	230	80	15	13	220	80	10
6	210	80	15	14	210	80	5
7	210	60	10	15	210	100	10

The specimen was produced by using compression molding, Collin P300P. According to some trials in our laboratory, all samples were compressed twice to produce smooth area. The conditions were shown in Table 2.

Table 2. Settings applied in compression molding.

Parameters	1st Compression					2nd Compression				
	1	2	3	4	5	1	2	3	4	5
Temperature (°C)	205	205	205	205	40	205	0	195	0	40
Pressure (bars)	0	1	1	0	0	0	0	0	0	1
Time (minutes)	10	5	5	0	15	10	0	3	0	15

3. Testings

Mechanical tests were performed according to flexural properties (ASTM D 790), using Universal Testing Machine, Shimadzu AGS 10 kNG. The support span was 25.4 mm. The flexural speeds were set according to samples thickness as explained in the standard. All testing was conducted at 23 ± 2 °C, 50 ± 5% RH. The average of flexural modulus from five specimens was calculated for all samples.

Structural analysis of nanocomposites was done for selected samples namely PP, clay, optimized sample. The analysis was conducted using an X-ray diffractometer (XRD), Rigaku miniflex 600 with Cu Kα radiation. XRD scans were conducted at λ = 1.54 Å, with scan rate of 2.4°/min. The spectra were evaluated in the 2θ range from 2° to 10°.

4. Statistical analysis

The completed flexural modulus data was analyzed and interpreted using Minitab 17 software. The data was regressed using polynomial equation as follows:

$$y = \beta_0 + \sum \beta_i x_i + \sum \beta_{ii} x_i^2 + \sum \beta_{ij} x_i x_j \quad (1)$$

Where y is the model response, x_i and x_j are independent variables, and βo, βi, βii, and βij are the coefficients for the constant, linear, quadratic terms, and interactions, respectively. Coded units (-1, 0, +1) were used in the calculation.

5. Results and discussion

5.1. Flexural modulus

Table 3 shows flexural modulus for all samples. In general, the average flexural modulus ranges from 1512 to 2248 MPa at various conditions. The average modulus of all nanocomposites is 2042 ± 168 MPa while mid-setting is 2096 ± 15MPa. The lowest is 1512 ± 55 MPa, while the highest is 2248 ± 180 MPa, giving -14% to 27% change on modulus. The improvement obtained in this research is higher than previous results reported by Yu Dong *et al.* (±20%) and H. Md. Akil *et al.* (±25%) [4, 14]. This suggests that mixing in this research was done properly.

Table 3. Flexural modulus of samples.

Sample	Temperature (°C)	Rotation Speed (rpm)	Time (minutes)	Average Flexural Modulus (MPa)
Control	220	80	10	1764 ± 86
1	230	80	5	2248 ± 180
2	220	60	15	2078 ± 71
3	220	100	5	2091 ± 99
4	220	80	10	2090 ± 158
5	230	80	15	1993 ± 76
6	210	80	15	2034 ± 121
7	210	60	10	1512 ± 55
8	220	60	5	2110 ± 167
9	230	60	10	2119 ± 57
10	230	100	10	2169 ± 105
11	220	80	10	2113 ± 205
12	220	100	15	2108 ± 94
13	220	80	10	2084 ± 184
14	210	80	5	1900 ± 87
15	210	100	10	1974 ± 131
Average modulus of all nanocomposites				2042 ± 168
Average modulus of mid-setting				2096 ± 15

5.2. Experimental design

Statistical model was developed from flexural modulus of nanocomposites. The model coefficient and p-value is shown in Table 4. It is not showing all terms as significance consideration. The most significant factor is the quadratic effect of temperatures, giving p-value 0.034 suggests curvature region in the range of applied temperatures. P-values of rpm and time are higher than 0.05 for 95% confidence level suggests that these factors are not significant to influence the modulus.

Table 4. Model coefficient and p-value.

Term	Coefficient	P Value
Constant	2077	0.000
Temperature	80	0.087
RPM	52	0.241
Time	-52	0.241
Temp*Temp	-154	0.034
RPM*RPM	-90	0.174
Temp*RPM	-93	0.148

The effect of interacted factors is represented by surface plot shown in Figure 1 (a). There is no significant effect of mixing time in this research (p-value = 0.241 or > 0.1 for 90% confidence level). For this reason, further analysis will be done according to one setting of mixing time, namely at mid-setting.

Figure 1 (a) shows that the modulus increases by higher temperature and rpm in the range of low to medium settings. However, it decreases at higher conditions. This suggests that there might be a degradation mechanism occurred during mixing at extreme high conditions [15] or entropy loss during polymer penetration into clay galleries due to very low polymer viscosity [16]. Meanwhile, as temperature increases might reduce viscosity and facilitate diffusion mechanism, while rpm produces higher shear [6], the improvement on modulus at medium temperature and rpm might be caused by diffusion and shear that work together to disperse the platelets.

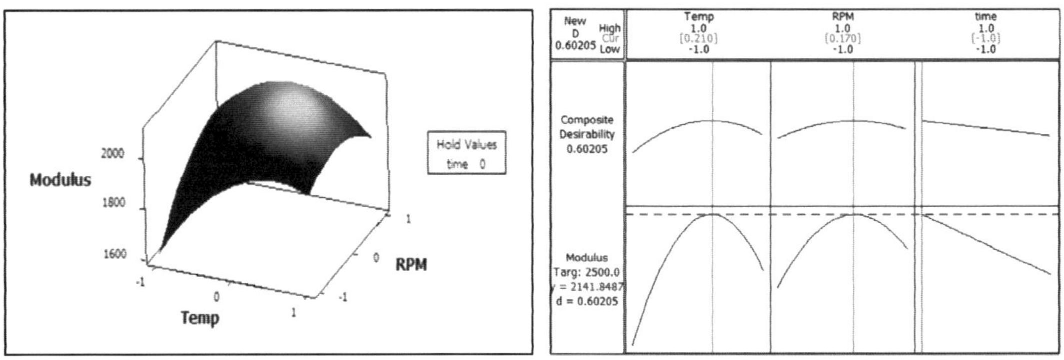

(a) (b)

Figure 1. (a) Surface plots at mid-setting of mixing time, (b) response optimizer analyzed plot.

The optimum conditions to produce the best modulus were predicted by response surface analyzer plot (Figure (b)). The plot shows that the optimum conditions are predicted at (0.21; 0.17; -1.00) in coded unit or 222 °C, 83 rpm, and 5 minutes in actual conditions. Applying these conditions would result in a flexural modulus of 2142 MPa. To validate the model and the predicted result, a verification experiment was carried out at optimized conditions. The flexural modulus value of verification sample was 2085 MPa. The value of the optimum prediction value as well as the verification result was lower than the maximum modulus (2248 MPa). This might be caused by the variations during the processing. However, the ranges of the maximum, optimum predicted, and verification are less than 1 standard deviation of modulus (s.d = 168 MPa) for all nanocomposite samples. This suggests that the model is capable of predicting the modulus.

5.3. XRD analysis

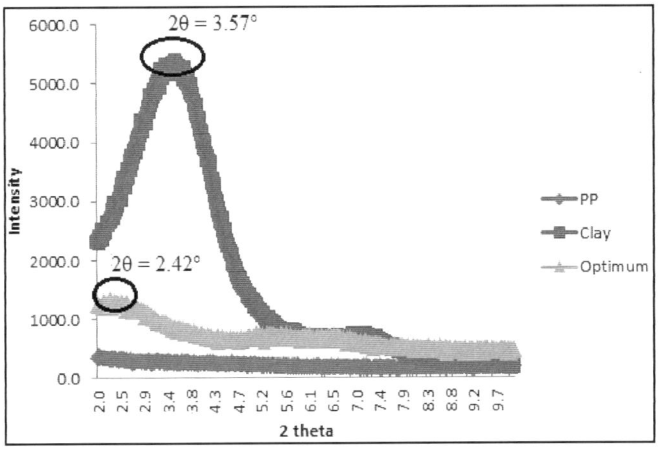

Figure 2. XRD diffractrograms for selected sample.

Figure 2 presents XRD diffractograms of control, clay, and optimized nancomposite samples. The results showed that the clay peak of optimized nanocomposite sample shifted to the lower angle compare to peak of clay. It is shifted from $2\theta = 3.57°$ ($d_{001} = 2.47$ nm) for original clay to $2\theta = 2.42°$ ($d_{001} = 3.65$ nm), giving 48% increase of interlayer spacing. This supported the modulus results and proved that modulus improvement was achieved due to an increase of clay interlayer spacing.

6. Conclusion

Polypropylene (PP)/clay nanocomposites were prepared by an internal mixer. The effects of mixing conditions, namely temperature, rotation speed, and mixing time on flexural modulus were analyzed. The optimum conditions of PP/clay nanocomposite were predicted using response surface methodology (DoE). Nanocomposite with 2085 MPa flexural modulus (18% improvement compare to PP) was generated at 222 °C, 83 rpm, and 5 minutes as the optimum conditions. This improvement might be caused by existence of intercalated clay layers. This hypothesis was confirmed by XRD diffractograms that showed clay interlayer d-spacing increase from 2.47 nm to 3.65 nm.

Acknowledgment

The authors would like to express their gratitude to the University of Indonesia for financial support from HIBAH PITTA and Center of Polymer Technology for facility supports. The authors are also grateful to their families and friends for their supports during this research.

References

[1] Modesti M Lorenzetti A Bon D and Besco S 2005 *Polymer* **46** 10237-45
[2] Hong C K Kim M J Oh S H Lee Y S and Nah C 2008 *J. of Indust. & Engin. Chem.* **14** 236-42
[3] Chrissopouloua K and Anastasiadis S H 2011 *European Polymer Journal* **47** 600-13
[4] Lertwimolnun W and Vergnes B 2005 Polymer **46** 3462-71
[5] Mittal V 2015 *Synthesis Techniques for Polymer Nanocomposites* Abu Dhabi: Wiley-VCH
[6] Kubisova H Merınska D and Svoboda P 2010 *Polym. Bull Springer* 533-41
[7] Musil J and Zatloukal M 2011 *Chem. Engin. Sci.* **66** 4814-23
[8] Lai S M Chen W C and Zhu X 2009 *Composites: Part A* **40** 754-65

[9] Sharma S K and Nayak S K 2009 *Polymer Degradation and Stability* **94** 132-8
[10] Hong C H Lee Y B Bae J W Jho J Y Nam B U Nam G J and Lee K J 2005 *J. of Appl. Polymer Sci. Wiley InterScience* 2375-81
[11] Shariatpanahi H Sarabi F.Mirali M Hemmati M and Mahdavi F 2009 *J. of Appl. Polymer Sci. Wiley InterScience* 922-6
[12] Martın Z Jimenez I Gomez M A Ade H W Kilcoyne D A and Cruz D H 2009 *J. Phys. Chem. B, American Chem. Soc.* 11160-5
[13] Lopez D G Picazo O Merino J C and Pastor J M 2003 *European Polymer J.* **39** 945-50
[14] Dong Y and Bhattacharyya D 2010 *Mater. Sci. and Engin. A* **527** 1617-22
[15] Li J Ton-That M T Leelapornpisit W and Utracki L A 2007 *Polymer Engin. and Sci. Wiley InterScience* 1447-58
[16] Wang Y Chen F B Wu K C and Wang J C 2006 *Polymer Engin. and Sci. Wiley InterScience* 289-302

Failure analysis of S50C carbon steel applied for fine boring transmission system

L Anggraini[1]

[1] Department of Mechanical Engineering, Faculty of Engineering, President University, Jababeka Education Park, Cikarang, Bekasi 17550, Indonesia

lydia.anggra@president.ac.id

Abstract. The purpose of this research was to analyze the failure of the axle of fine boring transmission system shaft engine which was subjected to JIS G4051 S50C carbon steel. In industrial machinery, fatigue failure is often found on the work machine. The transmission system uses a bearing on each side of the shaft so that the shaft rotation becomes smoother and the shock load at the beginning of the rotation becomes smaller. This analysis was carry out by comparison of metallographic results such as fractography by SEM, identification of chemical composition by OES, mechanical properties by tensile and Rockwell test with loading 150 kgf, and calculation of fracture toughness. The analysis obtained from the research showed that, the average of fracture toughness evaluated by hardness and tensile crack were 26 and 77 MPa.m$^{1/2}$, respectively. Therefore, by fracture toughness calculation results and crack propagation in microstructure results, it is confirmed that that the shaft does not meet the JIS Key Slot standard, and these factors would contribute to the occurrence of failures on the transmission shaft.

1. Introduction

Failure of the work system on the transmission is often due to the magnitude of the rotational speed that is passed from the main motor drive [1–3]. The transmission system serves as the successor of power and rotation from the main motor to the other transmission shafts to get the spindle rotation speed as desired [4, 5]. During the operation process, transmission components such as shafts, bearings, gears, and pegs are often damaged by corrosion and fatigue failures. To prevent failures such as fatigue, the transmission system uses pads on each side of the shaft so that the shaft rotation becomes smoother and the shock load at the beginning of the turn becomes smaller. The main causes of failure are the high coefficient of friction on the rotating component, heat transfer, and corrosion, among others [6–8]. The torque load that occurs in the transmission system exceeds the permitted torque load so that the failure happens on the components transmission system [7]. This research was conducted to determine the mechanism and causes of failure in the transmission system based on failure theory. The failure analysis on the transmission system components is done by testing the mechanical properties of the material. The results of this research are expected to reduce the failure rate of transmission system components and can determine preventive maintenance on the machine so that the breakdown maintenance does not occur in the production process.

A transmission system acts as a successor and converts the rotation of the main motor to the successor shafts of rotation by using several main components to work such as shaft, gear, pegs, and bearings [9]. Because the components are a constituent part of a transmission system, the load due to the rotation that occurs in the component is very high. The components of this transmission system often encounter several problems such as failure [9]. The failure that occurs in the transmission system breaks the shaft. The analysis performed is from the results of the hole being processed to be very coarse and the result of the measurement of roughness exceeds the expected roughness tolerance. So that the machine is must stop its production for inspection and repair. The examination found that there was a fault that occurred in the axis precisely on the stake house (keyway). With the fracture at the stake house, the rotation of the main motor could not be continued to the maximum and cause imperfect spindle rotation until the roughness of the results of the process. The objectives of this research are to analyze the failure that occurs on the broken transmission shaft, to analyze the main causes of failure, the forces and stresses that occur on the shaft, also the type of fracture, and the fault mechanisms that occur.

The failure analysis was subjected to S50C carbon steel which is equivalent to JIS G 4051. The composition standard of S50C carbon steel is shown on the following Table 1.

Table 1. Composition standard of S50C.

Steel	C	Mn	P (%)	S (%)	Si	Cr (%)	Ni (%)	Cu (%)
			max	max		max	max	max
S50C	0,47–0,53	0,5-0,8	0,035	0,035	0,17 0,37	0,25	0,25	0,25

2. Experimental method

As the preparation, before the surface fractographic observation of the fracture on the shaft was cleaned by blowing method to avoid the dirt on the fault surface such as dust, oil, and others. The observation of this fractography used a Zeiss EVO MA10 scanning electron microscope (SEM). This microscopic test was divided into several stages of the process, namely the preparation stages. At this stages, specimens to be tested were cut first to facilitate the test. The part to be observed was in the core area of the material so that the cutting of the material is done perpendicular to the broken post plot. The hardness testing used Rockwell test with indenter in the form of diamond cone with angle 120° with 150 kgf loading in accordance with the standard test on JIS Z 2245. The specimens for hardness test were cut into spherical bars with each specimen's size measuring 25 mm. Afterward, the specimens were mounted on test machine with maximal firmness at both end of specimens. The tensile testing used the Intron 3328. In order to confirm the repeatability of results, as many as seven pieces of specimens were prepared for each hardness and tensile test.

Figure 1. Specimens and its dimension.

3. Results and discussion

The testing of the material composition used a spectrometer test device with OES (optical emission spectroscopy) system as shown in Table 2. The requirements of the specimens to be tested was that it should have a flat surface wherein the test specimen would pass through the surface combustion stage resulting in complete combustion. In the A001 transmission axle system, the part that received the largest load was located at the bottom corner of the stake home. Due to the large load obtained from the torque of the main motor rotation, a high concentration of stress on that part would occur. This caused the beginning of cracks because the shaft material could not resist or absorb the high stress or tension that occurs repeatedly.

Table 2. Spectrometer test result of S50C.

Item		C	Si	Mn	P	S	Cr
Std S50C JIS G4051		0.47-0.53	0.15-0.35	0.6-0.9	0.030 max	0.035 max	0.2 max
Gear shaft	Act spark 1	0.51	0.21	0.62	0.027	0.0010	0.136
	Act spark 2	0.51	0.21	0.63	0.026	0.0015	0.137
	Act spark 3	0.51	0.21	0.63	0.027	0.0013	0.136

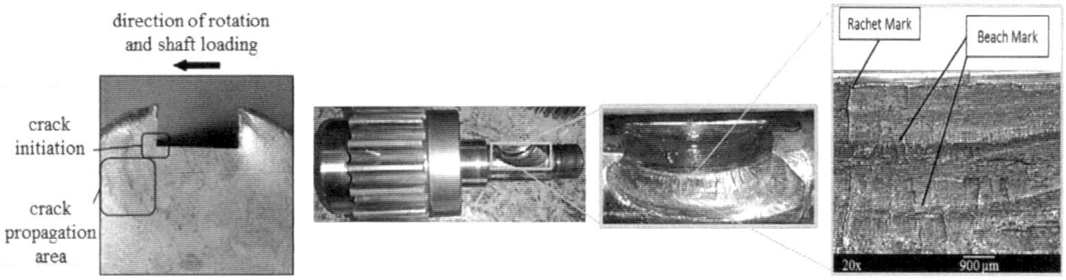

Figure 2. Direction of shaft loading.

In the stake home area, the shaft was flowed with a certain depth so that the shaft part which holds the load from the initial spin that occur less than half the diameter of the shaft. Then the next step was to observe on the part of the shaft that has broken failure to analyze the initial part of the crack that causes the fracture, the pattern of the fracture propagation and the final area of the fault. In Figure 2, some formation that occurs due to fracture can be seen, that is: Sign machine on the post groove, cracking origin, fatigue propagation zone, plastic deformation in the zone of final fracture, and longitudinal sign. An observation by using scanning electron microscope was done on the surface fracture pattern from the beginning of crack initiation, propagation from crack and final rupture to surface of the broken shaft can be seen. The observations were performed with an enlargement of less than 20× in order to obtain an overall breaking surface. The beginning of the crack was affected in the machining process signals. In asddition, the longitudinal marks on the post-plot that exist prior to the occurrence of the fracture were caused by the events occurring when the peg was inserted into the stake groove, thus leaving small scratches on the stake. In addition, there was also a ratchet mark and beach mark that showed high stress at the angle of the stake groove.

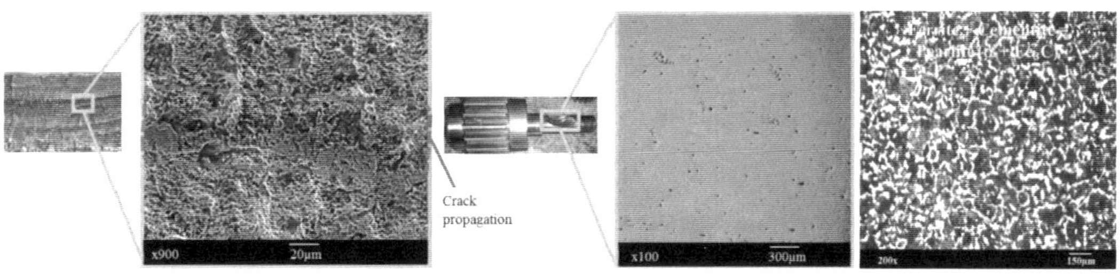

Figure 3. Direction of shaft loading and its microstructure.

The microstructure testing evaluated conformity with the material standard JIS G 4051 S50C also the initial formation process of the material whether the material gets heat treatment during the manufacturing process can be also be known. This test was conducted to obtain an overview of the material structure using etchants of 5 g $CuCl_2$, 40 ml HCl, 30 ml ethanol, and 30 ml of aquadest. Examination of samples on microstructure observation is found in three parts of the specimens which can be in broken transmission porous material with 200× phase enlargement of the material in the form of pearlite which is a combination of two ferrite (α) and cementite (Fe_3C) phases. The pearlite phase was formed by the heating process of the material followed by slow cooling to the room temperature.

Hardness testing on S50C shaft material was performed to determine the hardness of the material itself and the toughness value of the shaft that has breakage failure. The test was carried out on the postal cutter piece of the post which had failed. The test was performed on the surface of the shaft piece and was loaded perpendicularly from the cut surface of the shaft. The test was performed by using Rockwell C hardness test tool with a loading of 150 kgf. The result of hardness testing can be seen in the table where the average test result obtained is 22 HRC, which means the hardness value of the shaft material is uniform and the surface hardening process was absent. From Figure 3, the results of surface observation after a hardness testing can be seen where, on each side of the indenter, cracks occurred; the results can later be used to analyze toughness of the material based on the length of cracks that occurred. The experiment used a S50C shaft material formed from an A001 transmission shaft spare part and latched in accordance with JIS Z 2201 specification standard.

Fracture toughness is an indication to determine what value of maximum stress is allowed to minimize the propagation of pre-existing defects. According to T.L. Anderson [10–13], existing cracks will spread when the elastic strain energy released is at least equal to the energy required to create a new surface crack. Therefore, if the stress that occurs on the shaft exceeds the toughness value of the shaft material, then it can cause a new crack, thus leading to the breakage of the shaft. Based on hardness and tensile tests, there are several equations [14–17] for calculating the value of the fracture toughness. From the several equations obtained, the standard fracture toughness, K_{Ic}, of the S50C medium carbon steel range from 12 – 92 MPa.m$^{1/2}$.

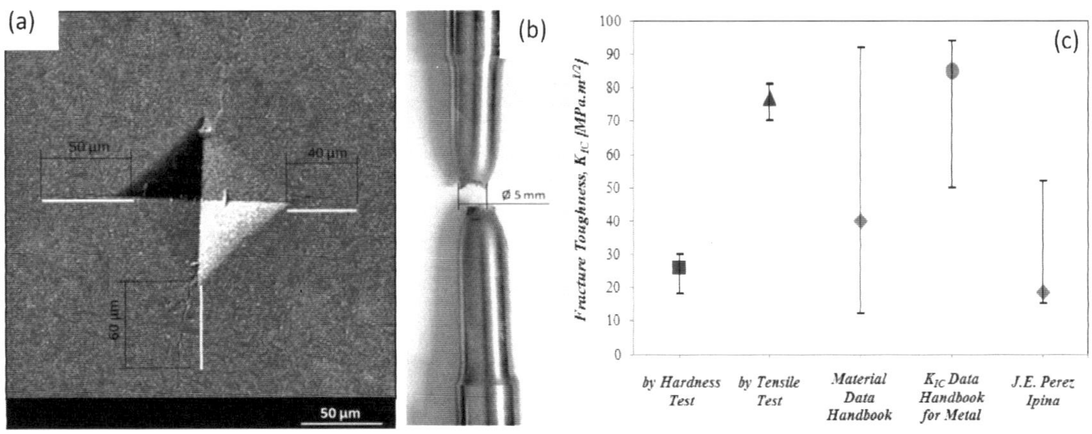

Figure 4. Crack length obtained by (a) hardness indentation and (b) tensile test (c) graph of fracture toughness obtained from calculation compared by some references [14–20].

The average of fracture toughness obtained by hardness and tensile crack are 26 and 77 MPa.m$^{1/2}$, respectively, as shown in Figure 4. The error bars indicate the minimum and maximum for both K_{Ic} obtained by hardness and tensile test, which are 18 – 30 MPa.m$^{1/2}$ and 70 – 81 MPa.m$^{1/2}$, respectively. The differences of fracture toughness obtained by hardness and tensile test are considerable because of the crack length. Based on the macrostructure, it was found that the failure that occurred was a fatigue failure due to the form of propagation from the fracture. The resultant criterion of the fracture surface occurring satisfies from the fatigue of the form because of the high-stress concentration at the stakeholder in the beginning of the fracture. The stages of fracture in the transmission shaft began from the occurrence of plastic deformation or deformation of the post home and the initial crack initiation at the point of the stake home, followed by the crack propagation due to repeated loading, and after the material was not able to resist and reduce the stress that occurred then the material would be broken (final rupture). This is evidenced by the results of microscopic material testing through visual observation or by using an SEM. The fractography evaluation also found similar results on the surface of the shaft that indicate a final rupture. The fracture surface observed that there are fracture propagation and the initial failure occurring due to the stress from the rotation passed by the shaft through the stake housing.

From the microstructure evaluation, the formation of pearlite structure shown, consisting of two phases consists of ferrite (α) and cementite (Fe_3C), was obtained. This explains the fact that the occurrence of heat treatment at the stage of material formation by heating and then gradually cooling at room temperature so that carbon bonds form pearlite structure. In the test results of the chemical composition of the transmission, shaft material meets the standards of the JIS G 4051 S50C material. With the resultant chemical composition of the three tests being within the limits of the percentage composition tolerance standard are C (0.47 – 0.53), Si (0.15 – 0.35), Mn (0.6 – 0.9), P (0.03 max), S (0.035 max) and Cr (0.2 max). All material compositions fall within the standard range of JIS G 4051 S50C material.

Mechanical testing of the material by conducting hardness testing found that the material received a uniform heat treatment as evidenced by the result of the hardness of the shaft material with the loading perpendicular to the shaft diameter obtained a hardness of 22 HRC at all test points. From tensile test sample formed in accordance with JIS Z 2201 standard with 10 mm diameter, it was found that the maximum material stress was 74 kg/mm^2 at 5832 kg with a yield point 62 kg/mm^2 at 4800 kg loading and 24% material length increase. From the test and analysis obtained, it was showed that the shaft failure was in the form of final rupture. As previously mentioned, the main cause of failure was fatigue fracture caused by high stress concentrations at one point and repeated loading. The area that experienced the highest stress was at the stake part of the house, the stake home area is the part that gets high stress at the angle of the stake groove. From the three methods that have been done in the form of analysis on the transmission component material, the design analysis of the transmission components and also of the forces acting on the transmission of shaft material should not fail because the force and the load acting on the shaft do not exceed the yield stress and permit stress axis. In the material, changes due to the environment that affect the structure of the material are also found. Deviations are occurs at the stake home portion where the size of the stake home exceeds the allowed tolerance.

4. Conclusions

The fine boring transmission shaft system of S50C carbon steel has been analyzed the failure experienced by the axle component is caused by fatigue failure. Mechanism of failure of the shaft started with the beginning of crack initiation, crack propagation, and the occurrence of final rupture. Fracture type in the form of a ductile where the material undergoes plastic deformation first then cracks occur until the occurrence of fracture. The occurrence of fracture was cause by the high-stress concentrations in the post hole home that receive a large load. Material transmission shaft components are in accordance with JIS G 4051 S50C standard. However, the size of the post groove formed on the shaft does not meet the JIS Key Slot standard (JIS1301-1976). Therefore, these factors would contribute to the failure occurs on the transmission shaft.

References

[1] Davis H E Troxell G E Wiskocil C T 1955 McGraw-Hill Book Company New York
[2] Lechner G Naunheimer H 1999 Springer Science & Business Media
[3] David C David E F Toshio K Nicholas V 2015 John Wiley & Sons

[4] Gibbs A H Gibbs D P 1994 U.S. Patent 5 306
[5] Dieter JR E G 1987 McGraw-Hill Book Company USA
[6] Kalpakjian S Steven S 2006 Pearson Education South Asia Pte Ltd. Singapore
[7] Smallman R E Bishop R J 1999 Linacri House Oxford
[8] Budynas R G Nisbett J K 2011 McGraw-Hill Book Company New York
[9] Jones D R H 2011 Elsevier Science Ltd. Kidlington
[10] Anderson T L 2017 CRC Press
[11] Dodds R H Shih C F Anderson T L 1993 *Int. J. of Fracture* **64(2)** 101–33
[12] Anderson T L Dodds R H 1991 *J. of Testing and Evaluation* **19(2)** 123–34
[13] Anderson T L Stienstra D Dodds R H 1994 ASTM International
[14] Konopík P Džugan J Procházka R Comtes 2013 FHT Inc. **955** Industri 334–1 Dobřany Republik Ceko Uni Eropa
[15] JIS Z 2201 1998 *Test Pierce for Tensile test for Metallic Material Japanese Standards Association*
[16] O'dowd N P Shih C F 1992 *J. of the Mechanics and PhyS. of Solids* **40(5)** 939–63
[17] Anderson T L Dodds R H 1993 ASTM International
[18] Ipiña J P Centurion S M C Asta E P 1994 *EngiN. Fracture Mechanics* **47(3)** 457–63
[19] Yawny A A Ipina J P 2003 *J. of Testing and Evaluation* **31(5)** 413–22
[20] Bertolino G Perez-Ipina J E 2006 *J. of Mater. Process. Tech.* **179(1)** 202–6

IOP Conf. Series: Materials Science and Engineering **432** (2018) 012016 doi:10.1088/1757-899X/432/1/012016

Analysis of the remaining life of API 5L grade B gas pipeline in the flare gas recovery unit

M Y M Sholihin[1], H S Kusyanto[2], B Soegijono[3]

[1] Department of Mechanical Engineering, Pancasila University Srengseng Sawah St., Lenteng Agung, Jakarta 12640, Indonesia
[2] Material Science Department, Physics Faculty, University of Indonesia, Depok 16424, West Java, Indonesia
[3] Material Science Department, Physics Faculty, University of Indonesia, Depok 16424, West Java, Indonesia

y.masduky@ui.ac.id

Abstract. Carbon steel pipe material API 5L grade B is commonly used for distribution in oil and gas industry. Internal corrosion is one of the biggest corrosion problems in carbon steel material that could impact production and cause economic losses. In this work, corrosion rate, remaining life, fluid composition, and corroded sample of API 5L grade B are analysed and calculated by development of microstructure, mechanical denaturing, and change of chemical composition by comparing with new pipe sample. The results show that the thickness of the pipe API 5L grade B applied as distribution line significantly decreased from 8.4 mm to 7.1 mm in only one year with a high corrosion rate of 1.32 mm/year based on ultrasonic testing inspection during the years of 2000 - 2016. It was found that the deterioration of pipe material was caused by gas composition contain mostly gas hydrocarbon with CO_2 10.1% mole and water composition containing 57.3 mg/L Cl^- ion. This chemical composition decreased the pH of liquid to 4.5, creating a good corrosion environment for this pipe. It caused the active corrosion rate significantly increase in just one year. Based on calculations, the remaining life of this pipe line was found to be just 3.6 years because of internal corrosion. XRF, XRD, and SEM analyses were done to determine the composition of the corrosion product. Based on those analysis, the main causes of the corrosion are found to be CO_2 and O_2. It is concluded that, the mechanical properties of this corroded pipeline still satisfy the standard for API 5L grade B with a tensile strength of 441 MPa and hardness of 74.

1. Introduction

Pipelines and piping are the most economical and efficient means of large scale fluid transportation for natural gas and crude oil. Pipeline and piping are commonly made of carbon steels due to some reasons: carbon steel have good mechanical properties, low cost and wider availability despite their corrosion resistance is relatively low. This corrosion happen due to the presence of high corrosion agents such as CO_2, H_2S and chlorine compound. The corrosion agents are dissolved in the fluid that can accelerate corrosion process inside of the pipelines. Internal corrosion due to dissolved CO_2 is commonly occurred in natural gas pipelines. The content of formation water (containing chloride ions) and dissolved CO_2 can cause uniform corrosion in pipeline. At distribution of natural gas in pipeline, formation water is a distribution medium for natural gas produced from the wells [1]. Therefore, a big concern must be given to formation water in controlling CO_2 corrosion. Corrosion could cause production and economic losses that is followed with pollution to the environment and disaster [2].

Flare Gas Recovery Unit (FGRU) is a process units for recovering waste gas that are usually vented or burned, so they can be used as fuel gas elsewhere in the facility. This process can reduce emissions and production cost. The existence of units in addition to having a positive impact to the environment. After a few years, at the FGRU there are several points that occur rapid depletion of the pipe thickness. The impact when there is leaking in the unit due to corrosion is very dangerous for the environment such as fire and environmental pollution. In addition, the company will suffer huge losses.

By analyzing and investigating the cause of corrosion, we could prevent losses and avoid similar incidents. It is useful to determine the remaining life of the piping system. The remaining life is simulated to the level of risk that the output of forming, planning, inspecting, and making integrated maintenance strategy. In order to support the implementation of the activities, it is necessary to include data on aspects of design, metallurgical, fabrication, measurement and field testing UT in the various operating conditions and operational historical data [3].

2. Experimental method

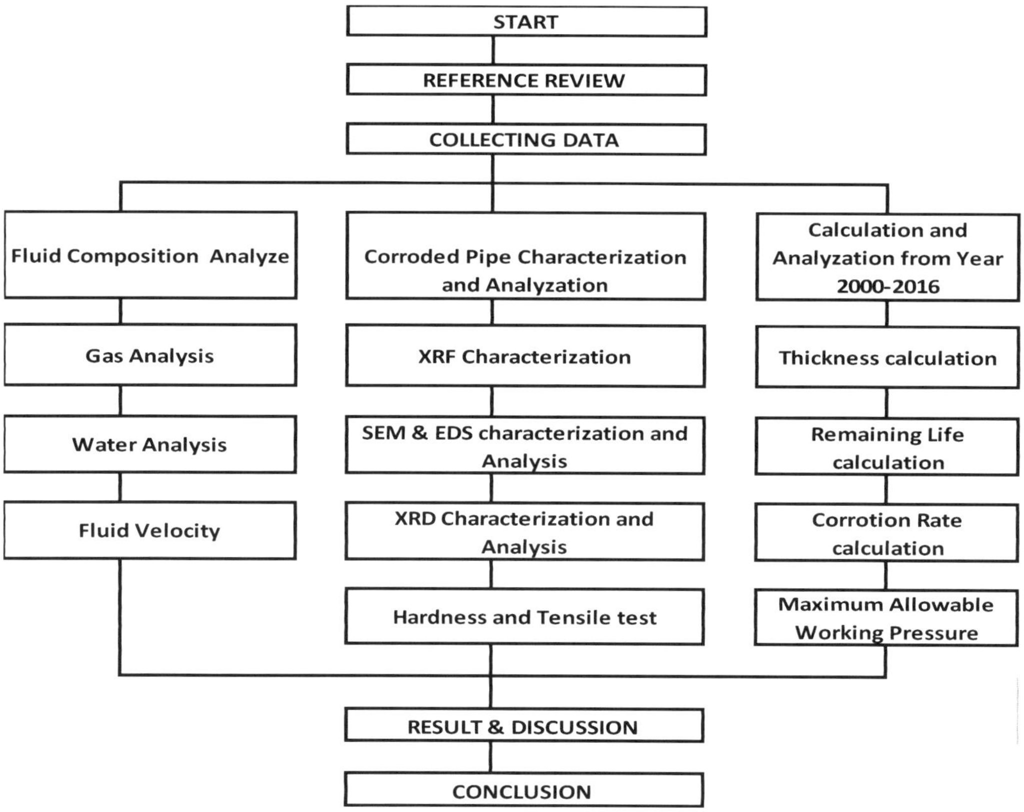

Figure 1. Experimental method.

This experiment is focusing to analyse corrosion which occured on flare gas recovery unit. The corrosion rate, remaining life, fluid composition, and corroded sample of API 5L grade B were analysed and calculated by developing of metallographic, mechanical denaturing and change of chemical composition comparing with standard pipe sample. Corrosion rate, maximum allowable working pressure (MAWP) and remaining life data are taken and calculated from Ultrasonic data testing from year 2000 until 2016.

Four samples were characterized by using XRF, XRD and SEM EDS to know the composition, metallography, and morphology of the material [2, 5, 6]. New pipe and corroded pipe carbon steel samples are used to study mechanical properties of API 5L grade B by using tensile test and hardness test. Figure 1 show the experimental diagram of the research.

3. Experimental method

3.1 Piping integrity

Based on the data Table 1, there is a significant corrosion rate change at the 6 o'clock direction. The measurement of pipe thickness used ultrasonic tesine measurement which is done since 2000 until 2016. This fact is caused by the corrosive nature of the liquid carried to the pipe and the gas is not sufficiently dry. Changing of the operating condition on the piping, as resulting liquid is temporary hold up that will cause an internal corrosion of the pipe. Significantly corrosion rate changes between 2015 and 2016 to 1.32 mm/year with the remaining life 3.61 years. Assessment and examination need to be done to determine the cause of this significant corrosion rates, maintenance and inspection strategies to reduce the likelihood and risk of leakage occurring.

Table 1. FGRU piping integrity by UT measurement.

Line Number		Point	Thickness				Corrotion Rate		Remaing Life	MAWP
			2000	2008	2015	2016	Long	Short		
10-G-B10A-7728	1056	W/12	9.70	8.78	8.62	8.57	0.02	0.05	261.92	33.09
10-G-B10A-7728	1056	W/03	9.50	8.96	8.75	8.64	0.04	0.11	218.28	33.37
10-G-B10A-7728	1056	W/06	9.00	8.52	8.38	7.06	0.17	1.32	3.61	27.12
10-G-B10A-7728	1056	W/09	9.50	9.87	9.68	9.16	0.08	0.52	20.35	35.44

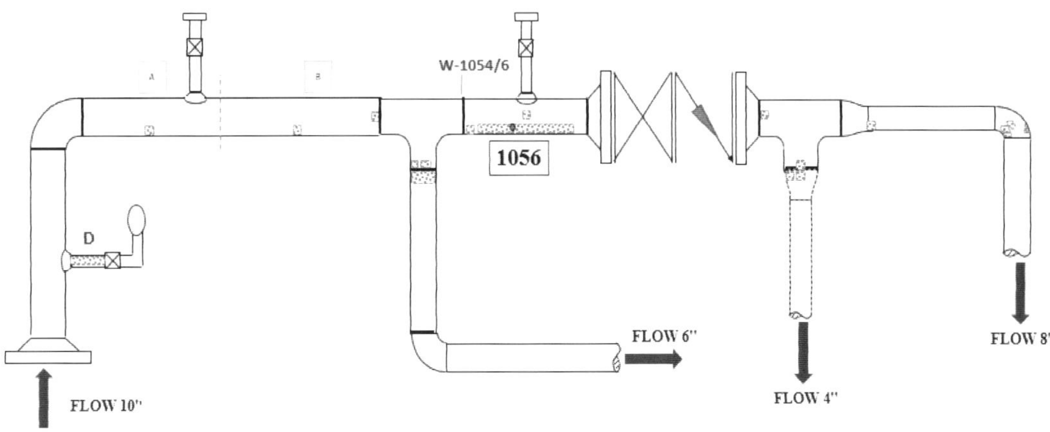

Figure 2. Inspection and corroded pipe point.

At Figure 2 High Corroded point is marked by red marking, most of the corrosion is happened at 6 o'clock direction especially at point 1056 at the upstream of pressure control valve. This control valve is set up at 5 Barg operating pressure. When the pressure is lower than 5 Barg, the valve will be closed and create liquid hold up at the bottom of the pipe. This phenomenon can propagate internal corrosion caused by water.

3.2 Fluid analysis

Operating parameter of FGRU shows in Table 2, where operating pressure is 5.5 barg (low pressure) and the pipe size is 10 inch. This condition make the fluid is easy to condense then trapped at the point 1056. The temperature of the process is 37 °C. When temperature is below 40 °C, it will reduce precipitation of iron carbonate ($FeCO_3$). It means the less protective layer is formed, the less corrosion rate increment [4].

Fluid of this unit contain gas and water, for gas is analysed by gas chromatography and water by ten ions with pH measurement. The gas composition contains most hydrocarbon with C1 to C10 89.9%, N_2 gas 0.067% and CO_2 gas 10.1%. The water composition contains ion Cl^- 57.3 mg/l. The water containment can propagate corrosion and lowering pH. pH is measured 4.5 showing that corrosion rate increase at this point. Solution pH plays an important role in the corrosion of carbon steels by influencing both the electrochemical reactions that lead to iron dissolution and the precipitation of protective scales [3]. Ion Cl^- can increase conductivity of the water. Three major effects of CO_2 will be addressed: The effect on the water chemistry, the effect on the electrochemical reactions, and the impact on the initiation and growth of corrosion product films [5].

Table 2. Operating parameter and piping size.

Specification	Value
Pipe OD	10 in
Wall thickness	7.1 mm
Material	API 5L grade B (Carbon Steel)
Operating Pressure	5.5 barg
Operating Temperature	36 °C
Flowrate	10 mmsfd
Fluid	Gas + Water

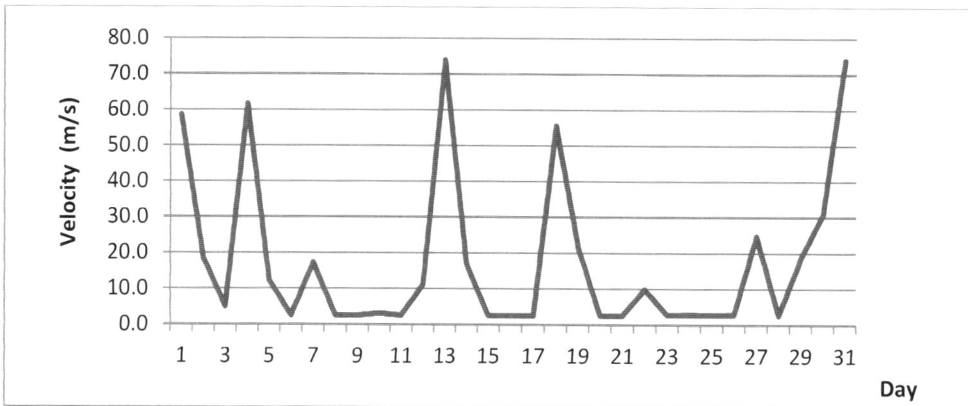

Figure 3. Fluid-velocity at FGRU in one month.

Flow of the FGRU fluid fluctuated between 0 – 13 mmscfd, depending on feed from oil processing. Figure 4 shows fluid velocity of FGRU in one month. Erosion will occur in that condition, based on API 14E RP fluid velocity 4.5 m/s. It can cause the erosion, in 10 inch pipe means 0.730 mmscfd, that was below flow average of this current operation. Erosive fluids can damage the protective film, and remove small pieces of material as well, leading to a significant increase in penetration rate. The damage to the protect film may be the results of the fluid-induced mechanical forces or flowing-enhanced dissolution [8]. Beside the erosion, flow rate of gas containing water and CO_2 has two principal effects on film formation and corrosion rate and form. First, it prevents the formation and slows down its growth by reducing the local supersaturation. Second, flow can damage film locally to cause localized corrosion, especially mesa attack [4]. An increasing flow rate can increase the corrosion rate by enhancing the transport of corrosive species toward, or remove protective product from, the surface.

3.3 Visual examination

Figure 3 shows internal pipe visual examination result is uniform corrosion attack at the bottom of pipeline (6 o'clock). This type of corrosion leads to assume that corrosion is happened by electrochemical process. This process usually caused by water hold up as electrolyte that might contain be ion Cl^- and CO_2. The color of this corrosion product is brown that might be contain hematite. Mesa type attack also formed at the bottom of pipe. Mesa attack is a type of localized corrosion and occurs in low to medium flow conditions where the protective iron carbonate film forms but it is unstable to withstand the operating regime [3]. However, this analysis need to be clarified and verified more detailed by some characterizations as shown later.

Figure 4. Visual examination internal pipe.

3.4 Analyze of corroded FGRU's piping

Characterizations of the pipeline material were conducted using chemical composition, metallographic examination, and mechanical property test including tensile test and hardness test measurement. Table 3 below shows composition of corroded samples result by using XRF found that Fe contain reduce to 70.5 % compared with API 5L grade B standard is 98 %. The composition showed that the corroded pipeline composition still fulfills that API 5L grade B standard composition.

Table 3. Composition of API 5L grade B and corroded pipeline.

Material	Fe	C	Mn	P	S	Ti	V	Ni	Mo	Cr	Cu	Cl
Pipeline	70.5	-	0.636	0.135	3.739	0.09	0.009	0.11	0.02	0.15	0.13	0.153
API 5L grade B	98	0.22	1.2	0.03	0.02	0.04	0.05	0.15				

The Scanning Electron Microscope (SEM) examination provides information about the result of the interaction and process between the corrosive environment and steel pipe, specifically with regard to the morphology of the metal surface, deposits and corrosion products. Deposits may be crystalline or amorphous [9]. Figure 4 shows morphology of corrosion layer at the corroded steel pipe. Corrosion deposits from steel formed in solution are mainly composed of insoluble products, undissolved constituents and trace amounts of alloying elements. They are formed in various oxides and carbides.

Figure 5. SEM micrograph of FGRU corroded pipe.

In Figure 5 the image shows that corrosion products are formed almost all of the surfaces in the pipeline where the corrosion product has the same shape. To find out the content of corrosion layer is also tested by Energy Dispersive Spectroscopy (EDS) at 7 point in 1000x magnification. In corrosion studies EDS is used to provide chemical composition about the material being examined by SEM. The relative amounts of elements detected are categorized as major, minor and trace values.

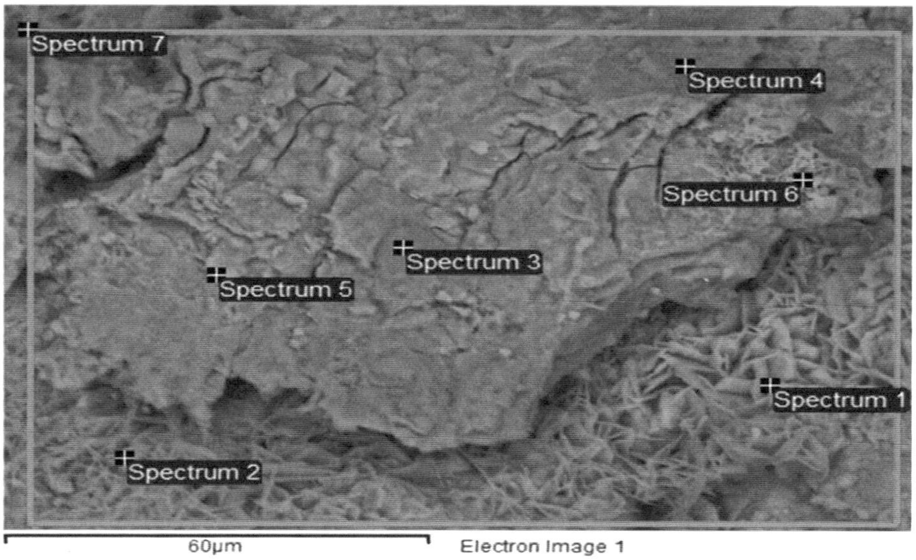

Figure 6. EDS micrograph of FGRU corroded pipe.

Table 4. EDS of FGRU corroded pipe.

Element	Spectrum Location						
wt%	1	2	3	4	5	6	7
C	5.58	6.17	7.1	7.16	5.82	6.23	7.33
O	47.95	49.04	54.88	51.23	45.46	42.29	44.93
S	0.82	1.04	0.9	1.38	1.05	1.58	1.19
Mn	0.37	0.4	0.44	0.2	0.47	0.33	0.36
Fe	45.27	43.35	36.69	39.63	47.2	49.56	46.19
Al				0.17			
Si				0.23			

Major elements in FGRU corroded pipe samples of corrosion products are C, Fe and O. Minor elements are Mn, Al, and Si. At points 1 to 7 it was found that the Fe content ranged from 36 to 49% wt while the oxygen content was 42 - 54% wt. This two elements will produce metal oxide as a form of corrosion oxide process, it shown the corrosion products that stick and form on the wall of the pipe. These high levels of oxygen content comes from the water due to turbulence processes that occur in the fluid flow process. Oxygen enters into a flow system in the form of dissolved. Oxygen comes from the reservoir when the gas is produced from the well. In addition, oxygen also comes from outside air processes when internal inspection work is done. Carbon content (C) is present with a minor amount of 5 - 7 wt%. This element is derived from CO_2 gas that also produced with natural gas. It will form formation of corrosion products such as siderite ($FeCO_3$) and carbides (Fe_3C).

The presence of other elements in corrosion products such as silicon (Si), possibly derived from fine sand (SiO_2) or other compounds that participate in the fluid and flows into the pipe. The sulfur element (S) is derived from a gas fluid containing H_2S, which can cause corrosion. The existence of flow turbulence in the pipeline causes the protective layer detached and dissolved again at the corrosion process. This intermitten and high speed flow will also sweep and form new corrosion products.

Metallographic examination is done by analysing phase formed and surface morphology. Phase formed based on X-Ray Diffraction refinement found corrosion products are Fe_2O_3 (Hematite), $FeCO_3$ (Siderit), FeO_2 (Geothite). Table 5 shows the percentage of corrosion product phase.

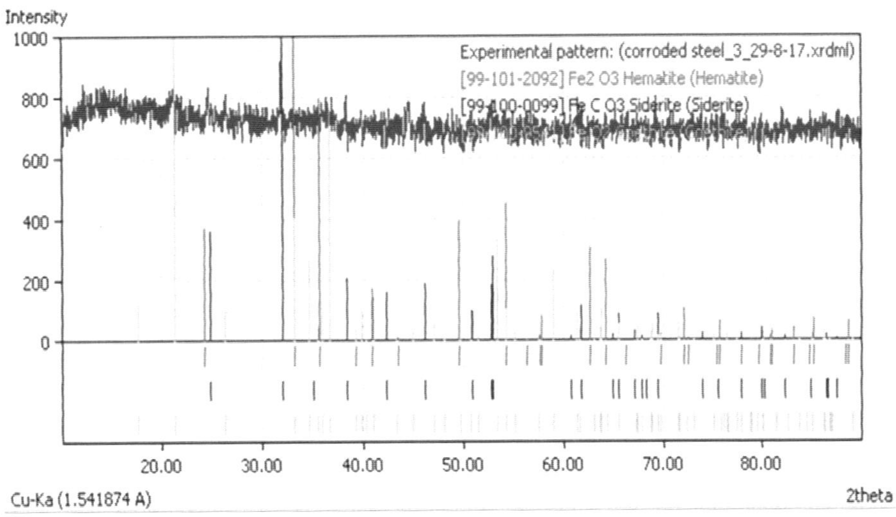

Figure 7. XRD analysis of FGRU corroded pipe.

Table 5. Corrotion product.

No	Name	Phase	Massa (% wt)
1	Hematite	Fe_2O_3	37.8
2	Siderite	$FeCO_3$	37.9
3	Geothite	FeO_2	24.3

Atom C present in corrosion products probably comes from hydrocarbon and/or bicarbonate (HCO_3^-) whereas chloride ions (Cl^-) dissolved in water. CO_2 is one of the acid gas causing internal corrosion in the gas pipeline. This gas is not corrosive if it is dry and not dissolved in water. If dissolved in the water of this gas will form a weak acid H_2CO_3 that is corrosive [5]. Generally, CO_2 dissolved in water will form carbonic acid by reaction because when CO_2 is dissolved in water it is partly hydrated and forms carbonic acid.

$$CO_2 + H_2O \rightarrow H_2CO_3 \quad (1)$$

Carbonic acid is diprotic and dissociates in two steps:

$$H_2CO_3 \leftrightarrow H^+ + HCO_3^- \quad (2)$$

$$HCO_3- \leftrightarrow H^+ + CO_3^{2-} \quad (3)$$

When the steel corrodes, Fe^{2+} and an equivalent amount of alkalinity are released in the corrosion process.

$$Fe + 2H_2CO_3 \rightarrow Fe^{2+} + 2HCO_3^- + H_2 \quad (4)$$

The pH in the solution increases and when the concentrations of Fe^{2+} and CO_3^{2-} ions exceed the solubility limit, precipitation of $FeCO_3$ can occur:

$$Fe^{2+} + CO_3^{2-} \rightarrow FeCO_{3\,(S)} \quad (5)$$

The rate of corrosion depends on the partial pressure of the CO_2 gas (PCO_2) and the operating temperature. Partial pressure gas CO_2 determines the pH of the solution and the concentration of solute based on its temperature. An increase of CO_2 partial pressure (PCO_2) typically leads to an increase in the corrosion rate. PCO_2 concentration in H_2CO_3 will increase and accelerate the cathodic reaction and ultimately the corrosion rate [7]. In this FGRU unit the CO_2 partial pressure is 0.55 Barg, based on API level the corrosion level is unlikely the corrosion type is uniform corrosion with corrosion rate <0.1 mm / year [4]. At 37^0C temperatures the corrosion rate increases, because the higher the temperature the kinetic energy of the reacting particles will increase to exceed the magnitude of the activation energy and consequently the rate of reaction (corrosion) will also accelerate. While at a temperature of 80^0C $FeCO_3$ solubility in the solution decreased and high supersaturation led to precipitation of $FeCO_3$ which became protective scale formation [10].

Result of tensile and hardness test are shown in Table 6. The yield and tensile strength of the pipeline are 382 and 441 Mpa, respectively these data fulfill the minimum stresses specified by API 5L grade B. the hardness value of the pipeline does not exceed the maximum hardness specified by API 5L grade B standard. Based on the result of chemical composition and mechanical property tests. It is conclude that the pipeline material is closely match to standard API 5L grade B carbon steel.

Table 6. Tensile and hardness test.

Sample Code	Yield Strength Mpa	Tensile strength Mpa	Hardness
Standard API 5L Grade B	245-450	415-760	82
Corroded API 5L Grade B	382	441	74

4. Conclusion

Corrosion rate formed and increased at point in the 6 o'clock direction between 2015 and 2016 to 1.32 mm/year with the remaining life 3.61 years. That calculations are based on ultrasonic testing measurement since 2000 – 2016. Localized corrosion of this FGRU piping is caused by not only by water hold up in low point that contain Cl- increase the conductivity and CO_2 gasses which dissolved to the water, that can reduce pH to 4.5 but also the velocity of the fluid sometimes is changed. Alteration of fluid velocity cause erosion at the corrosion product. Examination of corroded pipe found that is corrosion product contain C, Fe, and O as major products. Phase formed are Fe_2O_3 (Hematite), $FeCO_3$ (Siderit), and FeO_2 (Geothite). The FGRU piping characterization and mechanical testing are still in the standar of API 5L grade B.

References

[1] Kermani M B Gonzales J C Turconi G L Perez T Morales C 2005 *Mater. Optimiz. In Hydrocarbon Product.* NACE
[2] Dariva C Galio F 2014 *Corrosion Inhibitors – Principles, Mechanisms and Applications. Developments in Corrosion Protection*
[3] Solihin Yudi M 2004 *Piping System Oil and Gas Production By Using RBI Method 1st ed.* Jakarta
[4] Kermani M B & Morshed A 2003 *Carbon Dioxide Corrosion in Oil and Gas Production—A Compendium*
[5] Ilman M N Kusmono 2014 *Analysis Of Internal Corrosion In Subsea Oil Pipeline*
[6] Rustandi Andi A Fadly M Subekti E Norman 2012 *Makara J. of Tech.*
[7] Nešić S 2007 *Corrosion Science* **49(12)** 4308–38
[8] Baotung L 2013 *Research and Reviews in Mater. Sci. and Chem.* **2** Issue 1 19-60
[9] Martinez S V Groznadic Ivankovic A 2012 SEM/EDS Analysis Of Corrosion Products From The Interior Of a Crude Oil pipeline *Scientific Paper*
[10] Morland B H Dugstad A Svenningsen G 2016 *13th Int. Conf. on Greenhouse Gas Contr. Tech.*

IOP Conf. Series: Materials Science and Engineering **432** (2018) 012017 doi:10.1088/1757-899X/432/1/012017

The effect of silicon as a reactive element for NiCrAl coating performance on hastelloy substrate

S Ramandhany[1,2], E Sugiarti[1], R D Desiati[1], E Martides[3], E Junianto[3], B Prawara[3] and A Sukarto[1]

[1] Laboratory of High Temperature Materials and Coatings, Research Centre for Physics, Indonesian Institute of Sciences (LIPI), Serpong 15314, Indonesia
[2] Physics Division, Faculty of Science and Technology, State Islamic University of Jakarta 15411, Indonesia
[3] Research Centre for Electrical Power and Mechatronics, Indonesian Institute of Sciences (LIPI), Komplek LIPI, Sangkuriang St., Building 20, Bandung, West Java 40135, Indonesia

enis002@lipi.go.id

Abstract. The isothermal oxidation behavior, microstructure, phase formation and microhardness of NiCrAl coatings with and without silicon addition were investigated. Characterization of heat treated and oxidized samples were carried out by X-ray diffractometer (XRD), scanning electron microscope (SEM) and vicker hardness tester. The NiCrAl and NiCrAlSi coatings were deposited with high velocity oxygen fuel (HVOF) on the surface of hastelloy substrate. Oxide scales consists of Al_2O_3 and $NiCr_2O_4$ phases were formed on the surface of coating sample after isothermal oxidation at temperature of 1000 °C for 100 hours. The addition of silicon increases the oxidation resistance of NiCrAl coating through the formation of Al_2O_3 phase, whereas on NiCrAl coating without silicon addition, mass loss was observed, which indicated scale spallation. The addition of silicon at 0.4wt% can prevent outward diffusion of refractory elements such as Mo from the hastelloy substrate. The measured hardness of NiCrAl and NiCrAlSi coatings were 317.53 and 330.9 Hv, respectively. The experimental results showed that NiCrAl with silicon addition improved the resistance of high temperature oxidation and hardness.

1. Introduction

Materials used for turbine blades should be able to be operated on extreme condition, where high temperature and mechanical properties is required [1]. Nickel-based alloy such as hastealloy or inconel offers the mechanical properties required to be used as turbine blades [1]. However, these superalloy metals properties are not sufficient enough to be operated on high temperature [2]. Therefore, coatings, such as thermal barrier coating (TBC), have been applied to the alloy to improve its performance at higher temperature [3]. A bond coat as part of TBC layer must be introduced as an interface to the metals and the coating. The MCrAlY (M: Ni, Fe, Co or combination of them) materials are widely used as a bond coat in many complicated and aggressive service environments [4].

In the MCrAl systems, addition of reactive elements such as silicon is well known to improve the oxidation resistance [2, 5]. Silicon has ability to increase oxidation resistance comes due to its ability to prevent refractory diffusion, such as tungsten or molybdenum [2]. Moreover, silicon addition to the alloy has been proven to increase its hardness, thus increasing its mechanical properties [6]. Similar studies of the addition of silicon as reactive elements have been investigated by J. Ma, *et al.* (2012) and J. Cao, *et al.* (2017) [4,7]. However, both studies have not presented microhardness test result to determine the mechanical properties of coating. Therefore, further investigation on the effect of silicon addition to the MCrAl bond coat system is still important.

Present study will explore the effect of silicon addition as reactive element to the performance of material coatings. Oxidation test at 1000 °C for 100 hours and hardness test will be carried out to investigate the oxidation resistance of coatings and the material strength, respectively. Coatings morphology and phase formation will also be investigated to obtain information regarding the effect of silicon addition.

2. Experimental procedures

Hastealloy C-276 with 1.6 mm thickness was prepared as substrate with a size of 15x15 mm^2. Prior to coating, the substrate was polished with silicon carbide sand paper #100 up to #800. NiCrAl coating was deposited into the substrate using HVOF spraying technique. The parameters of HVOF coating is summarized in Table 1 and the composition of coating was specified in Table 2.Coating powder was milled beforehand with a planetary ball mill (SFM-1 Desktop type) for 36 hours with a rotation speed of 1500 rpm.

Table 1. Composition of coating powder.

N_2 pressure	4.8 bar
O_2 pressure	8.2 bar
Propane pressure	5.2 bar
Air screw compressor	4 bar
Fuel gas	60 lt/min

Table 2. The composition of HVOF coating.

Coating	Composition (wt%)			
Powder	Ni	Cr	Al	Si
NiCrAl	Bal	24	7	-
NiCrAlSi	Bal	24	7	0.4

Samples were grit blasted with silica powder in order to improve adhesion between coating and substrate. After coating process, sample was sintered at 1100 °C for 4 hours under vacuum condition to ensure the uniformity of coating and formation of expected phases. Oxidation test was carried out on the sample after heat treatment with isothermal oxidation method at temperature of 1000 °C for 100 hours.

The samples were characterized with X-ray diffraction (Smart Lab, Rigaku) with Cu-Kα wavelength and scanning electron microscope (JIB2010F, JEOL) equipped with energy dispersive spectroscopy. The influence of silicon addition on hardness was investigated through micro-vickers hardness tested (LM100AT, LECO) under 300 gf load and dwell time of 13 seconds.

3. Results and discussion

3.1 Isothermal oxidation

Figure 1 shows the mass change of NiCrAl and NiCrAlSi coated samples after isothermal oxidation process. Both samples show a rapid increase in mass during the early stage of oxidation test due to the formation of oxides on samples [8]. To investigate the mass change rate, the graph was sectioned into two parts and mass change rate are calculated from the equation:

$$(\Delta m/A)^2 = k_p . T \quad (1)$$

During the early part, from 0 hours until 31 hours of oxidation test, both samples have a positive mass change rate, which means that the formation of oxide scales still occur until 31 hours of oxidation. Samples coated without silicon additives are found to have higher mass change rate than the samples that are coated with silicon additives.

On the later part of oxidation test, from 31 until 100 hours of oxidation, mass change process started to differentiate between both samples. Samples coated without silicon addition was found to have a significant mass loss with a rate of -3.29 x 10^{-11} $mg^2cm^{-4}s^{-2}$, which means that a breakdown condition occurs on the coating. This breakdown was indicated through scale spallation. On the other hand, samples coated with silicon addition did not have a breakdown phenomenon. The NiCrAlSi coated sample have a steady mass increase from 31 hours with a rate of 3.83 x 10^{-13} $mg^2cm^{-4}s^{-2}$, which means that the sample successfully form a protective layer and prevent the coating from breakdown phenomenon.

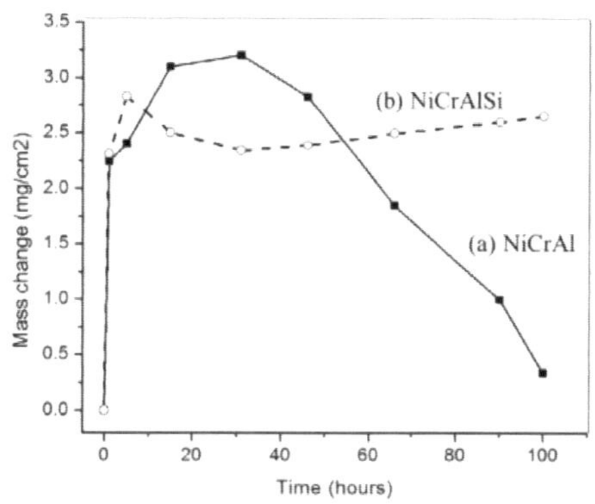

Figure 1. Mass change after isothermal oxidation in 1000 °C of types coated sample (a) NiCrAl, (b) NiCrAlSi.

Table 3. Parabolic constant of curve mass change.

Stage	*Kp* ($mg^2cm^{-4}s^{-2}$)	
	NiCrAl	NiCrAlSi
0h-31h	9.19 x 10^{-11}	4.93 x 10^{-11}
31h-100h	-3.29 x 10^{-11}	3.83 x 10^{-13}

3.2 Microstructure morphology

Morphology of coated samples are able to be observed from Figure 2. It can be identified that the samples are composed of an oxide layer, coating layer and secondary reaction zone (SRZ). The oxide layers were formed when the NiCrAl powder reacts with oxygen to form oxide phases, such as Cr_2O_3 and Al_2O_3. Most of this oxide layer are formed on the top part of coating which are exposed to the air. Inset of each sample coating images show that most oxide is located on top of the coating. These oxide layers plays an important part to increase sample oxidation resistance [5, 11].

Both NiCrAl and NiCrAlSi coating layer are easily identified from the cross-section images, this part of coating is the thickest and composed of γ'-Ni_3Al and alumina (Al_2O_3) phases. Alumina phases were formed when oxygen able to diffuse through the porous coating and reacts with aluminium, thus preventing the oxygen from moving further into the substrate. Therefore, a formation of alumina phases indicated that oxygen diffusion is limited [11].

Precipitates of molybdenum of element which are formed below the coating layer are able to be identified as the secondary reaction zone (SRZ). The SRZ would improve the material creep resistance; however, large amount of SRZ will instead decrease mechanical properties of the coating. The spallation on NiCrAl coated sample while oxidation test was caused by the excessive amount of SRZ.

3.3 Phase identification

Figure 3 shows the XRD characterization result for NiCrAl (a, a.1) and NiCrAlSi (b, b.1) coated sample, before and after oxidation. In addition, spallation NiCrAl coating was also characterized with XRD (a.3). From the characterization result, it can be observed that NiCrAl coating produce different peaks compared to the NiCrAlSi coating. On the NiCrAl coated sample chromium oxide are detected, while on NiCrAlSi coated sample no chromium oxide was present. This could happen due to the addition of silicon as reactive elements. The presence of silicon as additives cause the oxygen to diffuseless and reacts with γ'-Ni_3Al to form protective layer of alumina according to equation (2).

$$2\ \gamma'\text{-}Ni_3Al + 3[O] \rightarrow Al_2O_3 + 6Ni \quad (2)$$

Without the presence of silicon oxygen diffuses freely and reacts with alumina to form θ-alumina and chromium to form chromium oxide according to equation (3).

$$2[Cr] + 3[O] \rightarrow Cr_2O_3 \quad (3)$$

After the oxidation test it is found that metastable θ-alumina change phases into α-alumina which are more stable. NiCrAl coated sample have a less amounts of α-alumina phases after the oxidation test, however α-alumina phases of NiCrAl coating are found to be located on the spallation. This led to the conclusion that large SRZ have lesser adhesion with alumina, and caused the coating to spall. In addition, the presence of chromium oxide led to the formation of spinel structure that is detrimental to the coating performance. The formation of spinel is presented in equation (4).

$$NiO + Cr_2O_3 \rightarrow NiCr_2O_4 \quad (4)$$

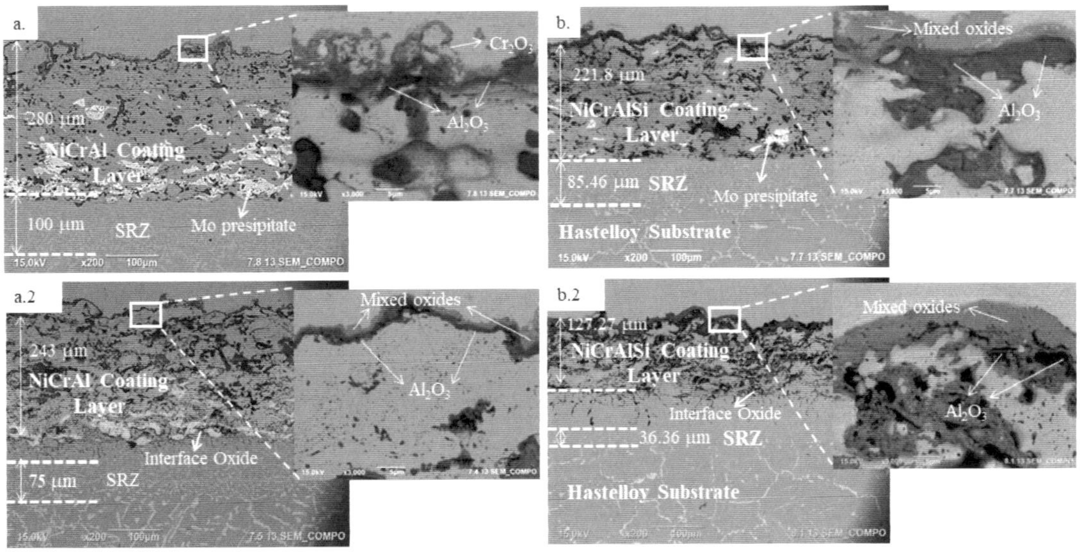

Figure 2. Cross section microstructure of (a) NiCrAl and (b) NiCrAlSi coated sample before oxidation test. Cross section microstructure of (a.2) NiCrAl and (b.2) NiCrAlSi after oxidation test.

3.4 Microhardness

Figure 4 and Table 4 show the result of microhardness characterisation using the micro-vickers hardness tester. Hardness value of NiCrAl and NiCrAlSi coated sample are found to be similar before the oxidation test. However, after the oxidation test the hardness value of substrate and coating changes differently. NiCrAlSi coating shows an increase in hardness value due to the formation of oxide protective layers, while on NiCrAl coating a decrease in hardness value was observed due to the spallation.

The coated substrates show an increase in hardness value due to the oxidation test. After the oxidation test, SRZ can be observed to shrink and an interface oxide layer are formed on the substrate. The interface may form whenever oxygen diffuse through the porous coating and reacts with the substrate. Higher hardness value in NiCrAl coated sample means that it has less protective ability that cause more oxygen to react with the substrate.

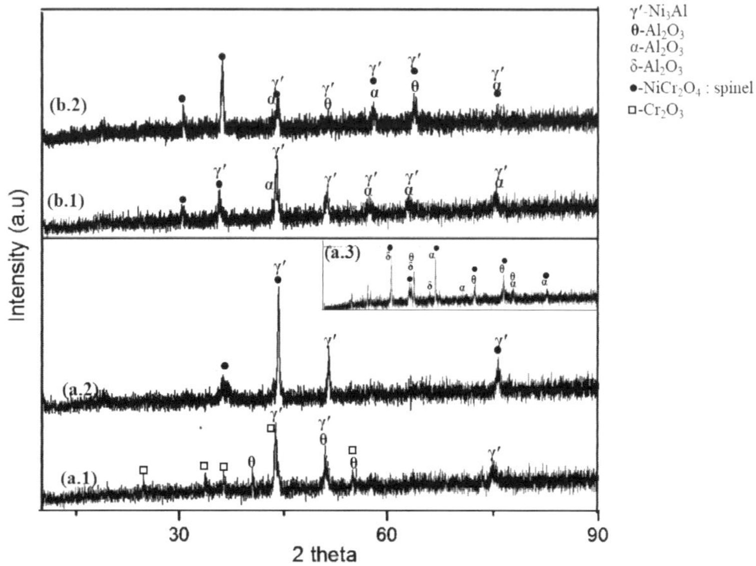

Figure 3. XRD pattern of the coated sample: (a.1) before and (a.2) after oxidation of NiCrAl coated sample with (a.3) spallation of NiCrAl, (b.1) before and (b.2) after oxidation of NiCrAlSi.

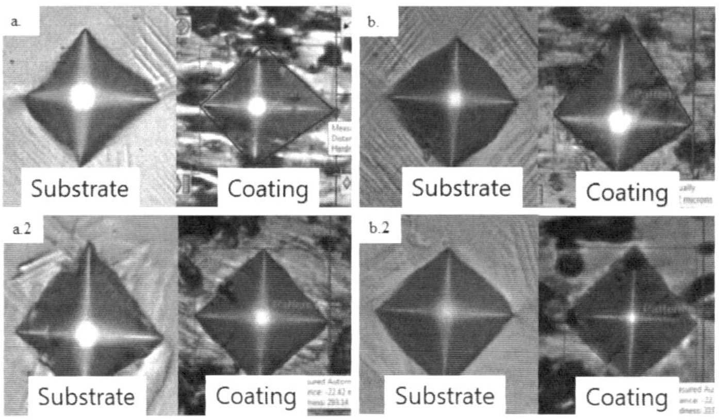

Figure 4. Vickers indents on (a)NiCrAl and (b) NiCrAlSi coated sample before oxidation.Vicker indents on (a.2) NiCrAl and (b.2) NiCrAlSi coated sample after oxidation test.

Table 4. Vickers microhardness at substrate and coating layer samples.

Samples	Before Oxidation (HV)		After Oxidation (HV)	
	Substrate	Coating	Substrate	Coating
NiCrAl	182.84	355.43	243.43	317.53
NiCrAlSi	185.44	288.3	205.84	330.9

4. Conclusions

The characterization of isothermal oxidation NiCrAl and NiCrAlSi coating on hastealloy substrate has been investigated. The conclusions of the investigation can be drawn as follows: (i) The presence of silicon as reactive element prevents breakdowm condition of isothermal oxidation testing; (ii) Silicon addition to the NiCrAl coating are found to increase its hardness and promote the formation of α-Al_2O_3 as protective oxides; (iii) NiCrAlSi coated sample able to withstand 100 oxidation test with mass gain rate of 2.56 mg/cm^2

Acknowledgments

The authors would like to give thanks to Kegiatan Unggulan LIPI, INSINAS IRPI grant No. 04/INS-2/PPK/E/E4/2017 and 2016 Toray Science Technology Research Grant for financial support of this work.

References

[1] Osorio J D Toro A and Ortiz J P H 2012 *Dyna* **79** 149
[2] Cao J Zhang J Hua Y Chen R Li Z and Ye Y 2017 *Surf. Coat. Technol.* **311** 19
[3] Smialek J L and Jacobson N S 2014 *Oxidation of High-Temperature Aerospace Materials in High Temperature Materials and Mechanism* (New York: CRD Press) 50–1
[4] Ma J Jiang S M Li H Q Wang W X Gong J and Sun C 2011 *Corros. Sci.* **53** 1417
[5] Naumenko D Pint B A and Quadakkers W J 2016 *Oxid. Met.* **86** 1–43
[6] Zeren M 2005 *J. Mater. Process. Techl.* **169** 292
[7] Omoniyi F Olubambi P and Sadiku E 2016 *Mater. Sci. Eng.* **5** 1–5
[8] Sugiarti E Zaini K A Wang Y Hashimoto N Ohnuki S and Hayashi S 2015 *Adv. Mater. Res.* **1112** 353
[9] Sugiarti E Zaini K A Sundawa R Wang Y Ohnuki S and Hayashi S 2017 *Phys. Conf. Ser.* **817** 1-6
[10] Mu N 2007 *High temperature oxidation behavior of γ-Ni+γ′-Ni3Al alloys and coatings modified with Pt and reactive elements* (Ames: Iowa State University) 35-6
[11] Wang J Chen M Yang L Liu L Zhu S and Wang F 2016 *Appl. Surf. Sci.* **366** 245
[12] Daroonparvar M Hussain M S and Yajid M A M 2012 *Appl. Surf. Sci.* **261** 287

Synthesis and thermomechanical characterization of peg/zircon composites

N F Muwwaqor[1], N A Fauziyah[1], Musyarofah[1] and S Pratapa[1*]

[1] Department of Physics, Faculty of Sciences, Institute of Technology Sepuluh Nopember, Sukolilo, Surabaya, 60111 East Java, Indonesia

suminar_pratapa@physics.its.ac.id

Abstract. Poly(ethylene) glycol (PEG)/$ZrSiO_4$ composites have been prepared and their thermomechanical propertis have been characterized using a dynamic mechanical analyzer (DMA). PEG 4000 was used as the matrix and a $ZrSiO_4$ powder, extracted from puya sand Kereng Pangi, Central Kalimantan, was used as the filler. $ZrSiO_4$ weight fractions of 0, 5, 10 and 20% were employed. The composites were made by a liquid method, where the matrix and the filler were mixed at a temperature when PEG was in a liquid phase. Characterization of FTIR and XRD was used to confirm the absence of any new phase and the composites were formed. The composites were subjected to DMA measurement in a shear mode at a frequency of 1 Hz. Results showed that, in general, the storage modulus of the composites has been improved by the addition of $ZrSiO_4$. The highest storage modulus was achieved by the composite with 20% $ZrSiO_4$, i.e., nine times of the pure PEG. The 20% $ZrSiO_4$ composite also exhibited the largest damping factor (the widest width of $tan\ \delta$ peak) and the highest melting temperature (T_m), i.e., 57.1 °C, increasing 14.1 °C from that of pure PEG.

1. Introduction

Poly(ethylene) glycol (PEG) is a versatile polymer material [1] that has a great potential in advanced technology. In the medical field, the research had resulted that PEG can be used as a biomaterial for dental implants and orthopedic. In building industry, PEG was applied as a PCM material for temperature control and heat energy storage [2]. The broad applications were due to the characteristics of PEG such as biocompatible, non-toxic, low vapor pressure, and relatively chemically and thermally stable [3]. Unfortunately, PEG has a relatively low tensile modulus of approximately 100 MPa [4]. Therefore, in its applications, blending PEG with reinforcement materials, such as ceramic [5], is required.

Zircon ($ZrSiO_4$) is an interesting filler for PEG to form composites. Zircon offers supporting physical, mechanical, thermal, and optical properties [6] to improve the applicability of PEG. Zircon has a tetragonal structure with lattice parameters a = 6.607(1) Å, c = 5.982(1) Å, and Z = 4 and space group of $I4_1/amd$ [7]. Zircon has a low thermal conductivity (5.1 W/m °C at room temperature and 3.5 W/m °C at 1000 °C), a relatively low thermal expansion coefficient (4.1×10^{-6} °C^{-1} in range 25 and 1400 °C) and a high mechanical resistance to heat (more than 1400 °C) [8], [9]. Moreover, the shear modulus of zircon is much greater than pure PEG, i.e. 109 GPa [10]. Research on the use of zircon as a filler of polymer-based composites had not been found in the literature. Therefore, it would be interesting to study the characteristics of such composites with PEG 4000 as the matrix and zircon powder as the filler.

In addition, the availability of zircon minerals in Indonesia is abundant, especially in Kalimantan and Sumatra. The utilization of zircon as a filler is expected to increase its economy value and support the development of advanced materials.

One of interesting aspects on the application of a polymer-ceramic composite is its thermomechanical behavior, which describes the response of the material under simultaneous mechanical and thermal loadings. Dynamical Mechanical Analyzer (DMA) is an appropriate tool to study the behaviour. If shear mode is used, a sinusoidal shear stress loading is applied on the sample and the sample generates two types of modulus (response), i.e. storage and loss moduli. They are related through the measured complex shear modulus (G^*) according to

$$G^* = G' + iG'' \quad (1)$$

and through the damping factor [11]

$$\tan\delta = {G''}/{G'} \quad (2).$$

In addition to the mechanical loading, DMA also shows a material response to thermal loading. The material response to the thermal loading varies, so that the dependence of the viscoelastic property of the material can be examined. In the shear mode, the storage modulus (G') measures the stored energy, while the loss modulus (G'') measures the energy loss as heat.

This paper reported the shear modulus and *tan* δ characters of PEG/zircon composites. From the measurements, the thermomechanical properties of PEG 4000/$ZrSiO_4$ were revealed.

2. Experimental method

The materials used were zircon and PEG-4000 powders. The zircon powder was synthesized from puya sand of Kereng Pangi, Central Kalimantan. The extraction process of puya sand consists of washing, magnetic separation, milling, and HCl 2M immersion. The immersion was followed by a hydrothermal process using sodium hydroxide (NaOH) 7M to remove quartz and produce pure zircon powder.

The matrix of the composite was poly(ethylene) glycol 4000 (Merck Schuchardt OHG 85662 Hohenbrunn, Germany) or PEG 4000. The mixing of matrix and filler used a liquid method. The PEG was melted at 60 °C. When the PEG melted, a dispersant ($C_{12}H_{25}NaO_4S$) was added to increase the particle distribution of the zircon powder. After stirring for 15 minutes, the zircon powder was poured to and mixed with the melt at various compositions (i.e. 0, 5, 10 and 20% wt zircon), and then stirred for 30 minutes.Then, the composites were molded and pressed. The nomenclature of the samples was denoted as L-PZR0, L-PZR5, L-PZR10 and L-PZR20 following the zircon content. The mixture was cast with a square-shaped sample holder of 5×5 mm^2 and thickness of 1 mm. The composites were confirmed for the absence of new phases during the synthesis process so that Fourier Transform Infrared (FTIR) Spectroscopy and X-Ray Diffraction (XRD) tests were performed.

In the Dynamic Mechanical Analysis (DMA), the experiment was in a shear mode. The DMA was Mettler Toledo SDTA 861e. Measurements of the modulus were carried out in the temperature range of 25 °C to 75 °C at 1 Hz frequency with 1 μm displacement.

3. Results and discussion

Figures 1 and 2 show the XRD and FTIR patterns from the pure PEG, pure zircon and the three composite samples. Figure 1 clearly reveals the formation of the composites. The XRD patterns (Figure 1) exhibited no new peaks than those of PEG4000 and $ZrSiO_4$ (PDF# 6-266). The XRD patterns also impart the differences in the level of crystallinity in the composites, as indicated by background level and peak intensity. Composites with more zircon content exhibited improved peak intensity indicating the increase of crystallinity level. Furthermore, the FTIR patterns (Figure 2) show the absence of the new absorption peaks. The peaks are belong to only two compounds, namely PEG and zircon. From the FTIR pattern, peaks A, B, and C are the absorbance peak of pure PEG. The peaks are still dominant in the composite L-PZR5, L-PZR10 and L-PZR20 due to the percentage of PEG is more than zircon. If more precise, there is a slight change in peak C more pointed on all composites. That's combined a peak of PEG and peak of zircon. For the absorption peak D of zircon is slightly visible on L-PZR20, although in L-PZR5 and L-PZR10 are not observable. A largest percentage of zircon (20 wt%) in LPZR20 causes the zircon absorption peak to be observed. The absence of other phase can then be concluded that the testing materials were PEG/$ZrSiO_4$ composites.

The DMA plots of the PEG/$ZrSiO_4$ composites can be seen in Figure 3. In Figure 3(a), the G' curve of the composites decreases with increasing temperature. The reason for the decreases is that the molecular motion of the free polymer sequences caused the reduction of the stored energy [12]. At temperatures around 60 °C, the curves decreased significantly. The decrease indicated that, at the designated temperature, the polymer melted or had reached the melting temperature (T_m), so that the carbon-bond on the polymer group weakened.

In general, the absolute value of G' composite increased with the addition of the filler. This phenomenon was caused by the shear modulus value (G') of $ZrSiO_4$ which reached 109 GPa [13], a much greater value than that of pure PEG. In theory, when zircon powder mixed with PEG, it fills the space at polymer chains and the surface of zircon powder interacts with them. The presences of zircon and surface interacting in PEG matrix were reducing the freedom of movement of the PEG 4000 polymer chains when the sample was given shear load. This interaction will increase the mechanical strength of the composites well above the polymer strength [14]. The G' values for the 5, 10 and 20 wt%-added samples were 545.5, 642.9, and 837.3 MPa. Fadly *et al.* (2016) studied the storage modulus of PEG/quartz composites and resulted in the storage modulus of PEG/quartz with 40 wt% filler was 730.6 MPa [5]. Therefore, zircon is more effective in increasing the storage modulus of the composite than quartz.

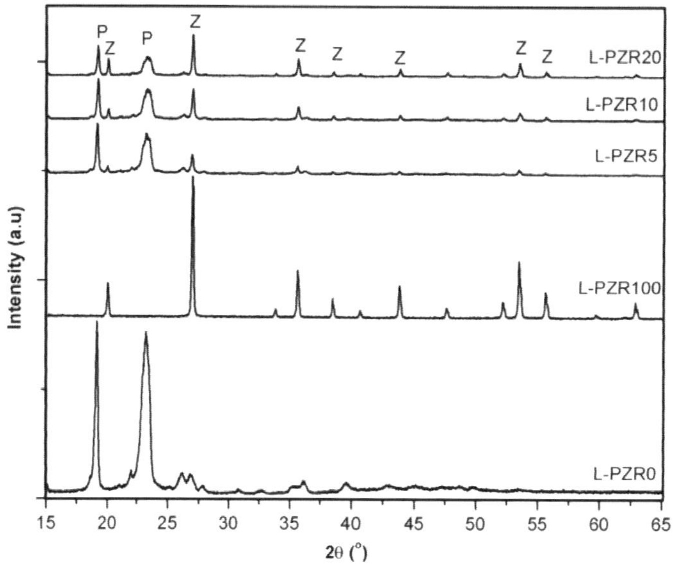

Figure 1. XRD patterns (CuKα radiation) of PEG/$ZrSiO_4$ (L-PZR) composites. P=PEG, Z=$ZrSiO_4$.

The *tan* δ characteristics of the samples can be described from Figure 3(b). The highest *tan* δ value was achieved by the pure PEG sample. The *tan* δ is the damping factor suffered by the composite material due to the irregularities of chain motion in the polymer functional group. When the composites were given a shear load, the movement of polymer chain was damped by the filler. One possible explanation is that this damping results from friction at the interface between filler and matrix [15]. An increase in filler content made fewer polymer chain movements of the figure also shows that the *tan* δ peak decreased with filler content. Meanwhile, the peak position of *tan* δ plot describes the melting temperature (T_m) [16]. From the figure, one can observe that the addition of zircon causes a shift of the *tan δ* peak to the right. It was because the melting temperature of the polymer was lower than the zircon. When a composite was given the heat treatment, the movement of the polymer molecule was inhibited by the zircon molecule. As a result, the polymer in the sample melts at a higher temperature. Further observation shows that the melting temperatures of L-PZR5, L-PZR10, and L-PZR20 are 54.9, 55.8, and 57.1 °C, respectively. The addition of zircon improved the melting temperature of the composite, since T_m for PEG4000 is 43 °C [5].

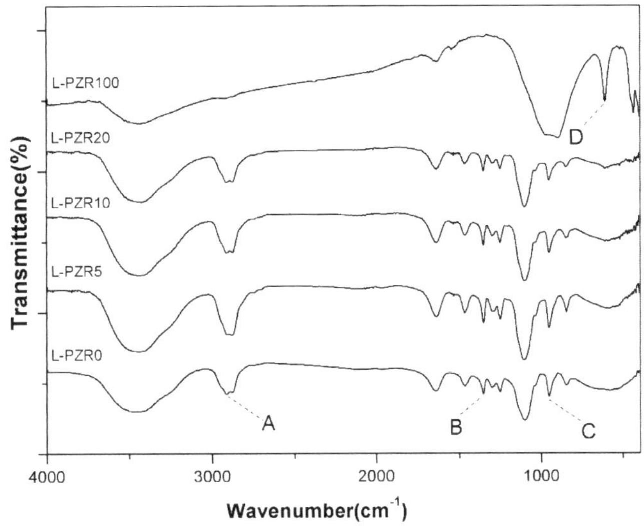

Figure 2. FTIR spectrum of $PEG/ZrSiO_4$ (L-PZR) composites. [A = B = C = the absorbance peak of pure PEG; D = the absorbance peak of zircon].

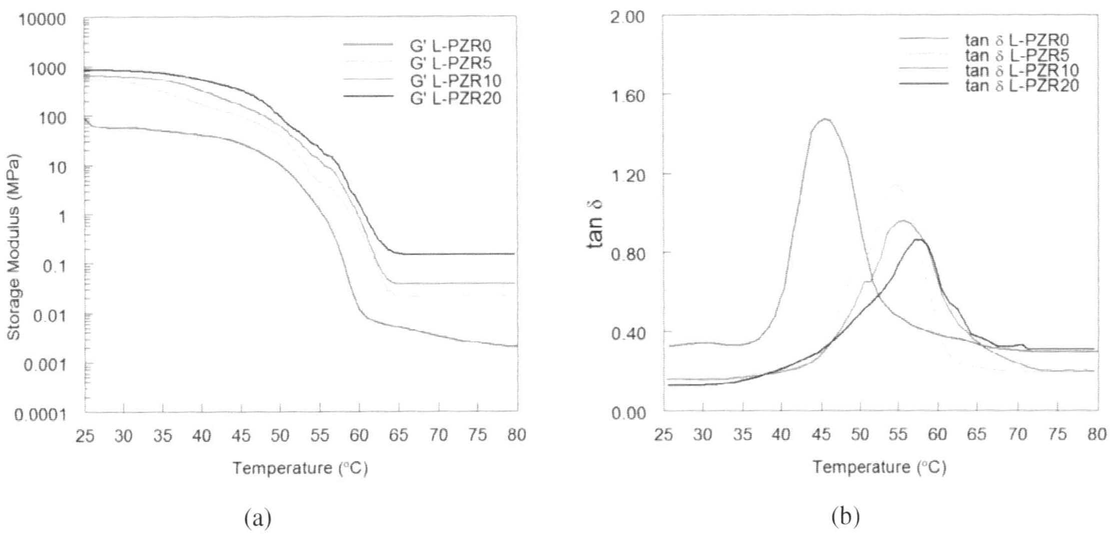

Figure 3. (a) Storage modulus and (b) tan δ plots for the $PEG/ZrSiO_4$ (L-PZR) composites as a function of applied temperature.

4. Conclusion

Zircon-reinforced poly(ethylene)glycol (PEG4000) composites were successfully prepared using a liquid method. The addition of zircon significantly improved the storage modulus and the melting temperature of the composite as shown by the DMA tests. The storage modulus of PEG/zircon composite improved substantially in the addition of zircon filler, i.e., six times and up to nine times of that of pure PEG for 5% and 20 %wt zircon-added respectively. Meanwhile, the presence of zircon also increased the melting temperature in the composites by 11.9, 12.8, and 14.1 °C for the 5, 10, and 20 %wt composites.

5. Acknowledgment

The authors are thankful to Kemenristek-Dikti of the Republic of Indonesia and LPPM ITS for the financial support through Competency Based Research provided to SP under contract scheme of PBK-2017 No. 528/PKS/ITS/2017.

References

[1] Kim S-J *et al.* 2012 *Kaohsiung J. Med. Sci.* **28(3)** 123–9
[2] Karaman S Karaipekli A Sarı A and Biçer A 2011 *Sol. Energy Mater. Sol. Cells* **95(7)** 1647–53
[3] Alkan C Sari A and Uzun O 2006 *AIChE J.* **52(9)** 3310-4
[4] Feng Q Yang J Liu Y Xiao H and Fu S 2014 *J. Mater. Sci. Tech.* **30(1)** 90–6
5] Fadly T A Fauziyah N A Rosyidy A Mashuri and Pratapa S 2017 *ICESNANO 2016* **1788** 030033_1-030033_4.
[6] Yamagata C Andrade J B Ussui V de Lima N B and Paschoal J O A 2008 *Mater. Sci. Forum* **591–3** 771–6
[7] RouNSow K Grsns G V and eNn P H 1971 *Am Miner.* **56** 782–90
[8] Suárez G Acevedo S Rendtorff N M Garrido L B and Aglietti E F 2015 *Ceram. Int.* **41(1)** 1015–21
[9] Veytizou C Quinson J-F and Jorand Y 2002 *J. Eur. Ceram. Soc.* **22(16)** 2901–9
[10] Salje E K H 2006 *Appl. Phys. Lett.* **89(13)** 131902
[11] Saba N Jawaid M Alothman O Y and Paridah M T 2016 *Constr. Build. Mater.* **106** 149–59
[12] Kinloch A J Mohammed R D Taylor A C Eger C Sprenger S and Egan D 2005 *J. Mater. Sci.* **40(18)** 5083–6
[13] Salje E K H 2006 *Appl. Phys. Lett.* **89(13)** 131902
[14] Kinloch A J Mohammed R D Taylor A C Eger C Sprenger S and Egan D 2005 *J. Mater. Sci.* **40(18)** 5083–6
[15] Lewis T B and Nielsen L E 1970 *J. Appl. Polym. Sci.* **14(6)** 1449–71
[16] Ramdani N Wang J Wang H Feng T Derradji M and Liu W 2014 *Compos. Sci. Tech.* **105** 73–9

IOP Conf. Series: Materials Science and Engineering **432** (2018) 012019 doi:10.1088/1757-899X/432/1/012019

Study of temperature effect on carbide layer formation behaviour of dual elements thermal reactive deposition on SUJ2 steel substrate

M Ariati[1], W Narottama P[1], A Cipto[1]

[1] Department of Metallurgical and Materials Engineering, Faculty of Engineering, University of Indonesia, UI Depok Campus, Depok 16424, Indonesia

myrna@metal.ui.ac.id

Abstract. In this research, the formation of carbide layer coating on SUJ2 steel by pack Thermal Reactive Deposition method was studied. Mixture of 35/65 weight percent of ferrochromium/ ferrovanadium (FeCr/FeV) powder was applied as coating elements. The process was carried out at temperatures of 900, 950, and 1000 °C for 6 hours for each temperature. The effects of temperature on layer thickness, microstructures, homogeneity, hardness, and wear resistance were analyzed. The result showed that the higher the temperature, the thicker the layer formed on substrate surface. The higher percentage of vanadium in the coating layer compared to chromium found by EDS Linescan results indicate that vanadium has higher affinity to carbon than chromium. This result also means the possibility of vanadium carbide formation will be higher than chromium carbides, due to the lower Gibbs energy for vanadium carbide formation. XRD results showed that the layer formed in this process consists of vanadium carbide (V8C7 and V6C5), chromium carbide (Cr23C6 and Cr7C3), and complex carbides. The average micro hardness and wear rate of all coatings with three different temperature variations were 2100 HV, and 3 x 10^{-4} mm^3/m respectively. This hardness was close to FeV hardness as single carbide former at approximately 2400 HV.

1. Introduction

The production of thin and hard carbide layer on steel is a widely used treatment to enhance the lifetime of tools, dies, and other mechanical parts. Nowadays, there are three common ways to produce hard carbide coating namely chemical vapor deposition (CVD), physical vapor deposition (PVD), and thermal reactive deposition (TRD) [1]. CVD and PVD have some drawbacks such as high investment cost, and using high vacuum or higly controlled atmosphere [2]. Compared to those process, TRD process has much lower investment cost because it operates at atmospheric pressure and does not need complex equipment.

TRD process can be applied in three different methods i.e. molten salt bath [3, 4], fluidized bed [1, 5], and pack cementation [6, 7]. In these process, vanadium, niobium, tantalum, chromium, molybdenum, or tungsten were used as carbide or nitride forming element. The carbon or nitrogen from steel substrate will diffuse and react with the carbide and nitride forming element, creating a dense deposit, and metallugically bonded carbide or nitride coating [8].

Vanadium and chromium are very frequently used as carbide former for TRD process. Comparing these two elements as carbide forming layers, vanadium carbide has excellent adhesion properties to the substrate, high hardness, and good wear resistance [9-13]. Meanwhile, chromium carbide provides layer which has a good resistant against corrosion and oxidation. Furthermore, chromium carbide former (such as ferrochromium) cost less than vanadium carbide former (such as ferrovanadium) [14]. Using the mixture between FeV and FeCr would combine the advantages of each elements, and simultaneously reduce the production cost. Therefore, the objective of this study is to produce and observe the effect of temperature on mixed vanadium and chromium carbide layer on steel subsrate by pack cementation method. Ferrochromium and ferrovanadium powder were used as carbide formers to produce mixed vanadium and chromium carbide layer on SUJ2 steel.

2. Experimental method

The substrate used in this study was SUJ2 bearing steel. It was a cylindrically-shaped bearing, 2.51 mm in diameter and 6.07 mm in height. Its chemical composition was shown in Table 1. All samples were cleaned by washing with KC99 to produce surface roughness at 0.15 μm. The chemical composition of ferrochromium and ferrovanadium powder were shown in Table 2.

Table 1. Chemical composition of SUJ2 Steel (weight %).

Fe	C	Si	Mn	Ni	Cr
Balance	1.03	0.24	0.28	0.06	1.68

The TRD process performed by the pack cementation method done in this research was using a powder mixture consisting of 195 g of ferrovanadium, 105 g of ferrochromium, 15 g of ammonium chloride, and 185 g of alumina. This powder mixture was then mixed with SUJ2 bearing steel as substrate weighing approximately 500 g. The process was carried out in steel box at 900, 950 and 1000 °C for 6 hours holding duration on each temperature in the furnace, then followed by furnace cooling.

The sample cross section was cut to check the carbide coating and measure its layer thickness. Nital solution (2% HNO_3 dan 98% alcohol) was used to etched the sample, and then observed by optical microscope. Microhardness measurement of the coated layers were carried out with Vickers microhardness tester with 100 gram load and 10 seconds holding time.

Table 2. Chemical composition of : (a) FeCr and (b) FeV (weight%).

(a)

Fe	Cr	C	Si	P
Balance	79%	0.03	1.94	0.024

(b)

Fe	V	C	Si	P	Al
Balance	80	0.15	0.97	0.04	1.2

Wear resistant test were performed using Ogoshi high speed universal wear testing machine with 400 meter sliding distance, and 120 N load at 2.38 m/s sliding speed. The radius and thickness of revolving disk were 14.87 and 3.53 mm respectively.

For phase identification on surface layer, X-ray diffraction (XRD) analysis was conducted for 2θ angles varying from 10° to 100°. The composition of carbide layer was analysed by energy dispersive spectroscopy (EDS) linescan attached within a scanning electron microscope (SEM).

3. Results and discussion

3.1. TRD process temperature effect on the formation of carbide layer

Figure 1 showed the influence of TRD process temperature on carbide layer thickness. It can be seen from the graph that increasing process temperature will produce thicker carbide layer. The mean thickness were 14.5 µm, 24.3 µm, and 30 µm for 900 °C, 950 °C, and 1000 °C respectively.

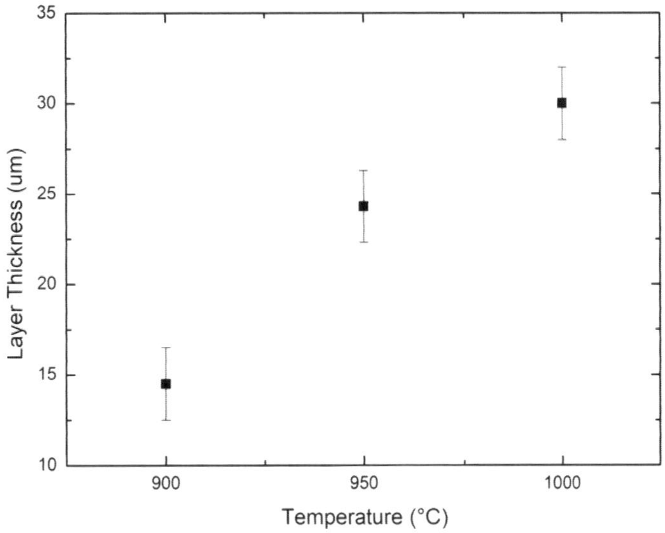

Figure 1. Influence of temperature on thickness.

The relatioship between temperature and layer thickness can be expressed as Arrhenius equation :

$$\frac{d^2}{t} = K = K_0 \exp\left(-\frac{Q}{RT}\right) \tag{1}$$

where d is layer thickness (m), t is time (s), K_0 is pre exponential constant, K is diffusion coefficient (m/s^2), Q is activation energy (joule), and R is the gas constant (8.314 J/mol K). The equation showed that the temperature will affect the diffusion rate of carbon and the carbide former elements [1, 3], whereby increasing the temperature will speed up the forming reaction of carbide layer. Consequently, the coating thickness increases with higher TRD process temperature.

3.2. TRD process temperature effect on carbide layer hardness

The effect of temperature on average layer hardness can be seen in Figure 2. It showed that process temperature did not have a significant impact on average layer hardness. The hardness value tend to be stable at 2100 HV with increasing process temperature from 900 to 1000 °C. By holding for 6 hours at elevated temperature, the carbide compositions formed in the layer were the same for all process temperatures, therefore the hardness values were relatively equal. It indicated that process temperature only affect the layer thickness instead of layer hardness.

This hardness was close to FeV hardness as single carbide former at approximately 2400 HV [2, 15]. The average hardness in this study is lower compared to FeV single carbide former because the chromium carbide phases exist in the layer has lower hardness value than vanadium carbide phases.

The process temperature, in the other hand, has siginificant effect on hardness distribution as shown in Figure 3. The distribution data was taken from 20 samples with 5 measurement points in each samples, totalling in 100 hardness value obtained for each temperature variation. It showed that at 900 °C, hardness is almost distributed evently from 1600 HV to 2800 HV. This might be due to the slower rate of diffusion at lower temperature, so that there is no tendency towards certain phases formed on the layer.

IOP Conf. Series: Materials Science and Engineering **432** (2018) 012019 doi:10.1088/1757-899X/432/1/012019

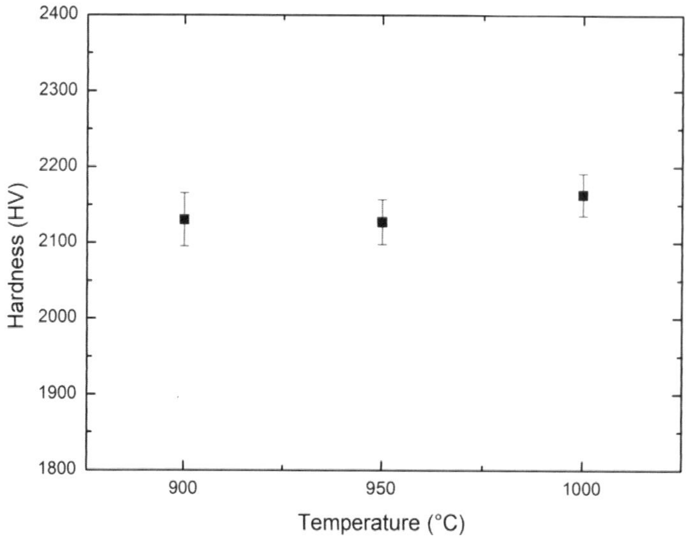

Figure 2. Influence of temperature on layer hardness.

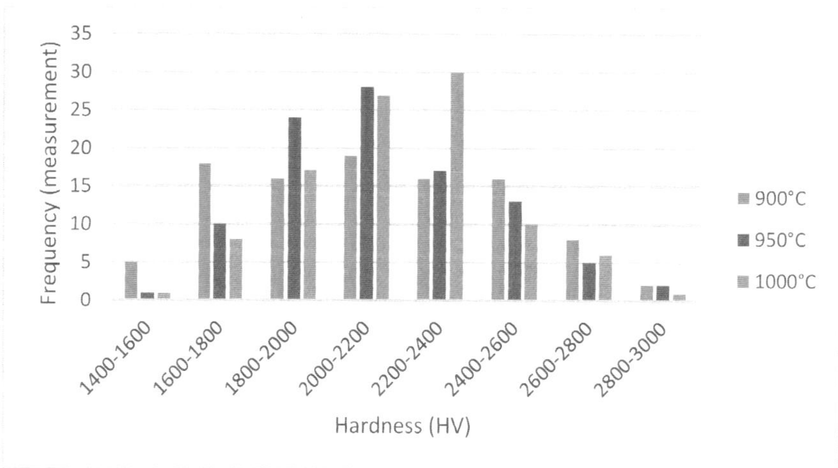

Figure 3. Hardness distribution.

3.3. Characterization of carbide layer

The cross section micrograph of the coated SUJ2 steel which had been processed at 950 °C can be seen in Figure 4. Two distinct regions can be identified on the cross section i.e. vanadium-chromium carbide layer and steel substrate. There is no interdiffusion region observed between steel substrate and carbide layer. This result is similiar with research conducted by X.S Fan [4] and Mattia Biesuz [16], both explaining that the lack of interdiffusion region is due to low solubility product of vanadium and carbon in austenite at TRD temperature.

Different results, however, reported by Aliakbar Ghadi *et al.* when producing mixed vanadium and chromium carbide layer with higher content of chromium in the mixture [10]. The research mentioned ferro-chromium solid solution formed between steel substrate and carbide layer. Therefore, it may be concluded that different result obtained in this research is probably because of low chromium content in the mixture used so that ferro-chromium solid solution was not formed.

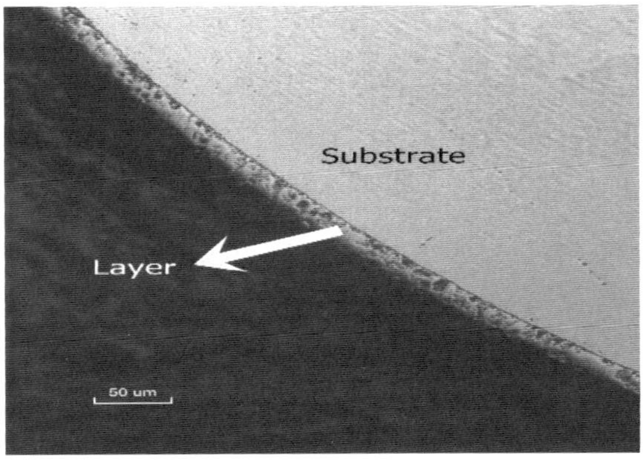

Figure 4. Cross section optical micrograph.

The EDS analysis result proved that interdiffusion ferro-chromium solid solution was not formed between steel substrate and carbide layer. Linescan showed very sharp Fe concentraton profile between substrate and carbide layer as seen in Figure 5. From that figure it can also be observed that vanadium concentration was higher than chromium in the layer. This is because vanadium is stronger carbide former than chromium [17] so that vanadium carbide is more easily formed than chromium carbide.

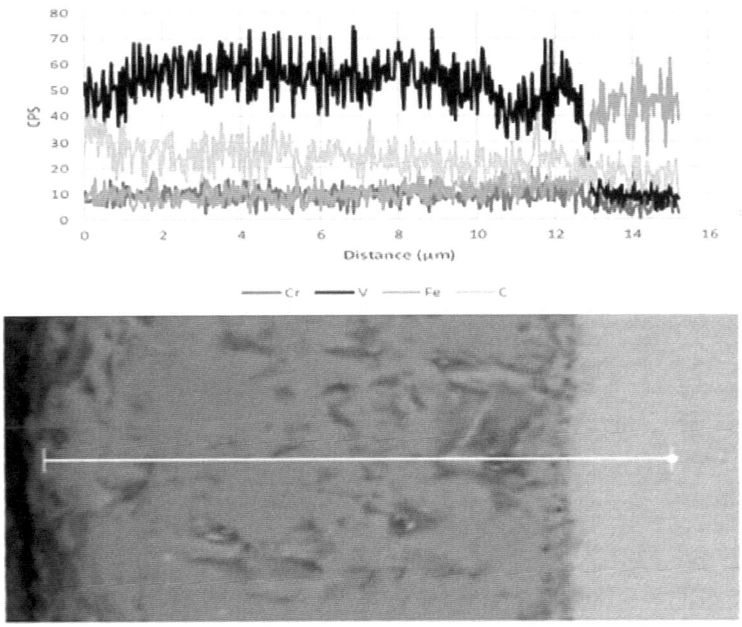

Figure 5. Linescan for TRD process at 900 °C.

Figure 5 also showed that the diffusion of carbide former elements were faster in elevated temperature. During six hours TRD process at 900 °C, the vanadium diffused only 13 μm from the steel surface, while at 1000 °C it could diffuse up to 27 μm. This higher diffusion rate at higher temperature is a well-known thermodynamic phenomenon, expressed by the Arhennious equation in the previous discussion. At elevated temperature, the atoms have more kinetic energy; therefore, they diffuse at a higher rate.

IOP Conf. Series: Materials Science and Engineering **432** (2018) 012019 doi:10.1088/1757-899X/432/1/012019

From Figure 6, regions with brighter colour can be observed. From linescan data on that particular region, the concentration of vanadium drop, but the concentration of iron was going up, while concentration of carbon and chromium were still remain the same. Based on this information, we assume that complex carbide which consists of vanadium, chromium and iron was formed in the layer.

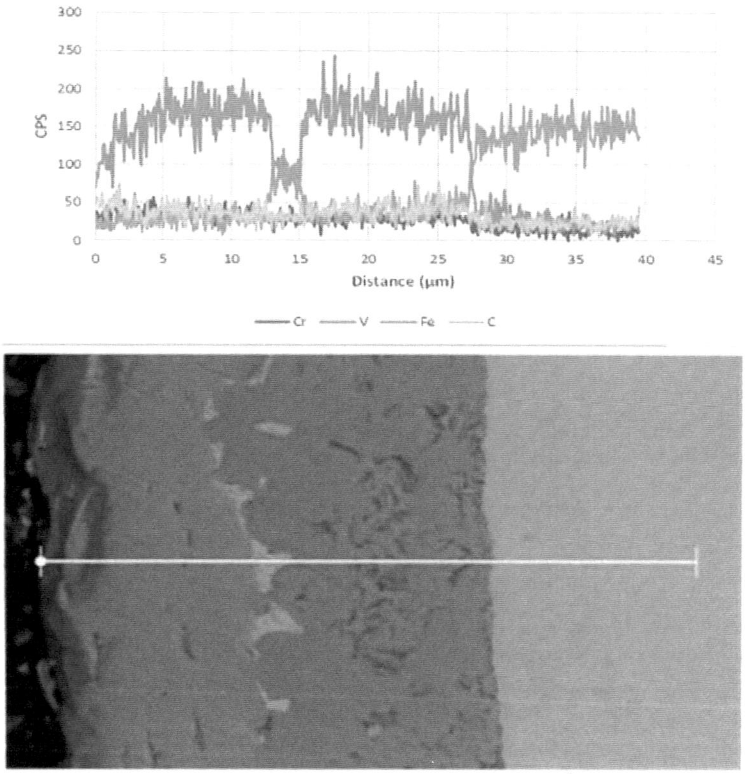

Figure 6. Linescan for TRD process at 1000 °C.

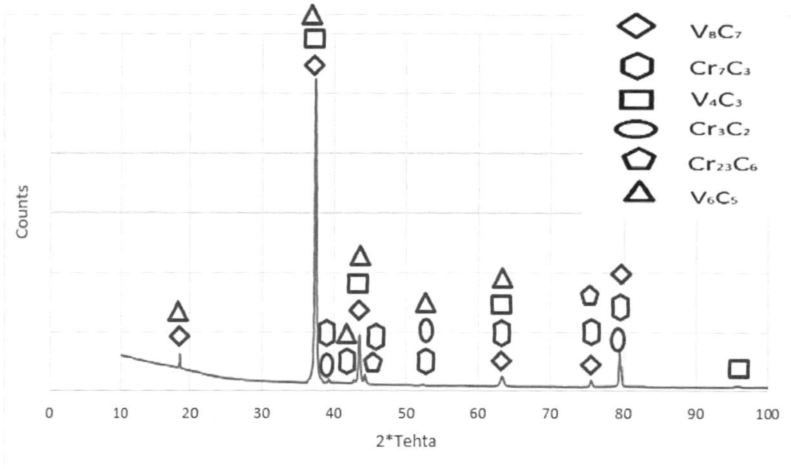

Figure 7. XRD pattern for 900 °C process.

The XRD analysis result of the TRD coating on SUJ2 steel processed at 900 °C is presented in Figure 7. It shows that the layer consists of both vanadium carbide and chromium carbide. Vanadium carbide phases formed in the layer were V_8C_7, V_4C_3 and V_6C_5, while chromium carbide phases formed in the layer were $Cr_{23}C_6$, Cr_3C_2 and Cr_7C_3.

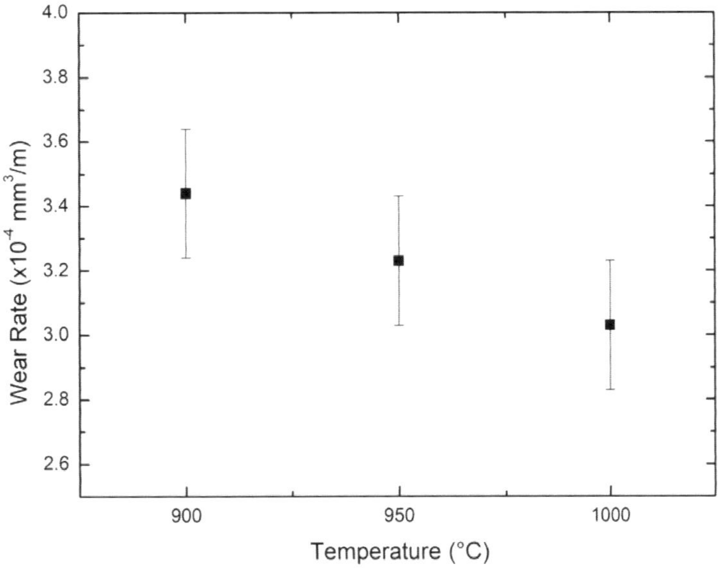

Figure 8. Influence of temperature on wear rate.

3.4. TRD process temperature effect on wear properties

Effect of process temperature on wear rate was shown in Figure 8. It showed that higher process temperature will slightly decrease the wear rate of material from $3.4x10^{-4}$ mm^3/m at 900 °C to $3.03x10^{-4}$ mm^3/m at 1000 °C. It may because of higher process temperature, the layer formed become relatively more thicker, denser, and harder. The specific wear rate of this testing was shown in Table 3.

Table 3. Specific wear rate.

Temperature (°C)	Specific Wear Rate (mm^3/N.M x 10^{-6})
900	2.91
950	2.73
1000	2.55

Research conducted by M.Aghaie Khafrie using FeV as single carbide former, showed a specific wear rate about 6 – 8 x 10^{-6} mm^3/Nm at 140 N load [2]. Compared with mentioned research, the specific wear obtained in this study by using chromium-vanadium as carbide former was somehow have similar values.

4. Conclusion

In this study, a mixed chromium and vanadium carbide coatings layer were successfully formed on SUJ2 steel substrate by pack cementation process. Based on the characterization results, it can be concluded that higher process temperature, the thickness of carbide layer formed is increased. The diffusion occurs faster in higher temperature, where the average thickness obtained at 900, 950 and 1000 °C were 14.5, 24.3 and 30 μm respectively. The resulting thickness was quite homogeneous, showed by only less than 4 μm thickness difference on substrate surface for each process variations. The process temperature, however, has no significant effect on the average coating hardness. The hardness obtained tends to be similar i.e. 2130 HV at 900 °C, 2127 HV at 950 °C and 2163 HV at 1000 °C. This also affects the wear rate which tends to be similar in the three process variations, which is about 3 x 10^{-4} mm^3/m.

EDS Linescan showed higher vanadium content than chromium in the carbide layer, indicating the possibility of higher vanadium carbide formation due to lower Gibbs-forming energy. XRD results showed that the layer formed in this process consists of vanadium carbide (V_8C_7, V_6C_5), chromium carbide ($Cr_{23}C_6$, Cr_7C_3), and complex carbides. The average micro hardness of all coatings were 2100 HV averages. This hardness was close to FeV hardness as single carbide former at approximately 2400 HV.

Finally, mixture of ferrochromium and ferrovanadium powder as carbide former elements can be recommended for Thermo Reactive Deposition process on SUJ2 steel substrate. This mixed carbide former elements could reduce the process cost without sacrificing the mechanical properties, as the hardness value and wear resistance did not decrease significantly compared to the coating using ferrovanadium as single carbide former.

Acknowledgment

Authors would like to thank the Directorate of Research and Public Services, Universitas Indonesia for financial support and project administration respectively to conduct this research under PITTA Project on 2017 fiscal year. Author would also like to thank PT Federal Superior Chain for their support to this research. The assistance from PT Federal Superior Chain for thermal reactive deposition process and equipment is gratefully acknowledged.

References

[1] Chen F Lee P and Yeh M 1998 *Mater.Chem. and Phys.* **53** 19-27
[2] Aghaie-Khafri M and Fazlalipour F 2008 *J. of Phys. and Chem. of Solids* **69** 2465-70
[3] Khalaj G Nazari A Khoie S Javad M K and Pouraliakbar H 2013 *Surf. & Coat. Tech.* **225** 1-10
[4] Fan X Yang Z Zhang C Zhang Y and Che H 2010 *Surf. & Coat. Techn.* **205** 641-6
[5] King P Brownrigg A Long J and Reynoldson R 2004 *J. of Mat. Eng. and Perform.* **13** 431-8.
[6] Sen S 2005 *Vacuum* **79** 63-70
[7] Sen U 2005 *Materials and Design* **26** 167-74
[8] Arai T and Harper S 1992 ASM Handbook Volume 4 Heat Treating, Ohio: ASM International
[9] Zarchi H Jalaly M Sltanieh M and Mehrjoo H 2009 Steel Research Int. 80 859-64
[10] Ghadi A Soltanieh M and Zarchi H 2012 *Defect and Diffusion Forum* **326-328** 377-82
[11] Fan X and Yang Z 2010 *J. of Alloys and Compounds* **505** 15-8
[12] Arai T Fujita H Sugimoto Y and O Y 1987 *J. of Mat. Eng.* **9** 183-9
[13] Liu X Wang H Li D and Wu Y 2006 *Surf. & Coat. Tech.* **201** 2414-8
[14] Ghadi A Soltanieha M Saghafian H and Yang Z 2016 G *Surf. & Coat. Tech.* **289** 1-10
[15] Oliveira C Benassi C and Casteletti L 2006 *Surf. & Coat. Tech.* **201** 1880-5
[16] Biesuz M and Sglavo V M 2016 *Surf. & Coat. Tech.* **286** 319-26
[17] Brooks C R 1996 Principles of the Heat Treatment of Plain Carbon and Low Alloy Steels, Ohio: ASM International

IOP Conf. Series: Materials Science and Engineering **432** (2018) 012020 doi:10.1088/1757-899X/432/1/012020

Effect of synthesis parameters on crystal structure and magnetic properties of cobalt zinc ferrite ($Co_{0.5}Zn_{0.5}Fe_2O_4$) nanoparticles

A T Puspitasari[1], H Kiswanto[1], E Suharyadi[1*], T Kato[2] and S Iwata[3]

[1] Department of Physics, GadjahMadaUniversity, North Sekip BLS Yogyakarta, 55281, Indonesia
[2] Department of Electronics, Nagoya University, Japan
[3] Institute of Materials and Systems for Sustainability, Nagoya University, Japan

*Corresponding author: esuharyadi@ugm.ac.id

Abstract. $Co_{0.5}Zn_{0.5}Fe_2O_4$nanoparticles have been successfully synthesized by coprecipitation method with varying NaOH concentrations and synthesis temperatures. X-ray diffraction (XRD) analysis confirmed that $Co_{0.5}Zn_{0.5}Fe_2O_4$ nanoparticles have spinel cubic crystal structure with crystallite size ranging from 7.9 to 21.4 nm. The crystallite size increase relatively by increasing NaOH concentration and synthesis temperature which related to the nucleation and crystal growth mechanism of nanoparticles during synthesis. The transmission electron microscope (TEM) image showed that the sample was aglomerated. The selected area electron diffraction (SAED) image showed the diffraction ring as representation of miller plane and confirmed that sample was polycrystalline. Magnetic measurement indicated that $Co_{0.5}Zn_{0.5}Fe_2O_4$ nanoparticles shows ferimagnetic behaviour and tend to be superparamagnetic with decreasing crystallite size. The increase in coercivity and maximal magnetization also have been found by increasing NaOH concentration and synthesis temperature. It is attributed to single domain existence and an increasing of crystallinity degree. The Fourier transform infrared spectroscope (FTIR) spectra also confirmed the existence of symmetric stretching vibration bands of metal ions near 563-609 cm^{-1} and 316-493 cm^{-1} in tetrahedral and octahedral sites, respectively.

1. Introduction

Studies of spinel ferrites are highly relevant to modern technologies, so the synthesis and sintering of ferrites have become an important part of modern science research. Nanosized ferrites may have extraordinary electric and magnetic properties that are comparatively different from microstructured materials, tailoring them to modern technologies, as well as providing novel applications such as ferrofluids, magnetic drug delivery, high density information storage, photocatalysis, gas sensors, etc [1]. Nano-crystalline ferrites have been widely studied and has become a field of interest of many researchers because of their unique and improved properties over the bulk crystalline materials [1]. Indrayana*et al.* have studied effects of Zn concentration and sintering temperature on crystal structure, magnetic properties and band gap energy of $Mn_{1-x}Zn_xFe_2O_4$ Nanoparticles [2]. Siregar *et al.* have studied effect of synthesis temperature and NaOH concentration on microstructural and magnetic proerties of $Mn_{0.5}Zn_{0.5}Fe_2O_4$ nanoparticles [3]. Praveena*et al.* have studied ferromagnetic properties of Zn substituted spinel ferrites for high frequency applications [1].

The Co substituted mixed ferrites are among those spinel ferrites which the authorsare interested in. Cobalt ferrites has attracted considerable interest because of theirlarge magnetic multi-axial anisotropy, high saturation magnetization, high Curie temperature and extraordinary chemical stability [4-7]. Cobalt ferrite nanoparticles are suitable for the isolation and refining of genomic DNA, the parting of polymerase chain reaction ready DNA and especially in hyperthermia treatment [1]. Zn ferrites are useful for low and high frequency applications generally for power transformers, power inductors, microwave devices, read and write heads for high speed digital tape, etc. because of their high resistivity, low losses, mechanical hardness, high Curie temperature and chemical stability [8-13]. If cobalt ferrites and zinc ferrites were combined into Co-Zn ferrites, they would present the new characteristics of crystal structure and magnetism properties which then become one of the soft ferrites used in electronic devices such as transformer cores, electric motors and generators [14]. Recently many efforts have been made to develop low power loss Co-Zn ferrites in order to miniaturize electrical devices [15, 16].

Studies on spinel ferrites synthesis methods have led to the development of different chemical synthesis techniques, all of which have a common feature, namely that all reagents are mixed at theatomic or molecular level. The most popular methods of synthesis is coprecipitation. Coprecipitation method is relatively simple, having a short reaction time, a high reaction yield, and yielding small-size particles.The microstructure of nanoparticle directly affects their magnetic, electrical, dielectric, and optical properties [17]. The microstructure of nanoparticle can be tuned by controlling the synthesis parameters,such as metal precursors type, pH, reaction temperature and reaction rate [18]. All of the parameters directly affect the crystallite size of nanoparticle [18]. Treatment temperature sintering for Co-Zn ferrite would be affected by crystal structure and magnetic properties [19]. In the case of coprecipitation synthesis, a reaction temperature of solutes and pH is very sensitive in determining the reaction yield. There has been lot of work done toward Co-Zn ferrite, but still lack of systematic and comprehensive investigations on thepreparation condition, structural and magnetic properties. To date, there has not been a single research done to study the synthesis of cobalt zinc ferrite by varying NaOH concentration and temperature synthesis by coprecipitation method on crystal structure and magnetic properties. In this study the effect of NaOH concentration and synthesis temperature on crystal structures and magnetic properties of Co-Zn ferrite nanoparticles is investigated comprehensively.

2. Experimental method

Solution prepared by dissolving stoichiometric quantities of $CoCl_2.6H_2O$ (Merck, Germany) and $ZnSO_4.7H_2O$ (Merck, Germany) in 25 ml distilled water. Solution of $FeCl_3.6H_2O$ (Merck, Germany), was also prepared in 25 ml distilled water. Both solutions were subsequently mixed into homogenous solutions by stirring them in moderate speed for 5 minutes. Then, it was added with 3.7 ml of HCl (Merck, Germany) and it was stirred again for 5 minutes. NaOH solution was made by mixing NaOH (Merck, Germany) salts with 50 ml distilled water. The precipitation was carried out by dropping the alloys solution into 1.5 M, 3 M, 6 M and 12 M of NaOH solution at temperature 70 °C called sample M1, M2, M3, and M4 respectively for 60 minutes and stirred at 1000 rpm [20]. In the case of the synthesis temperature variations, the samples are synthesized at 30, 50, 70, and 90 ^{o}C in 6 M of NaOH solution labelled sample S1, S2, S3 and S4 respectively for 60 minutes and stirred at 1000 rpm. The synthesis reaction process was maintained until the formation of $Co_{0.5}Zn_{0.5}Fe_2O_4$ nanoparticle product which is indicated by the forming of dark brown precipitates. Subsequently, synthesized nanoparticles were cooled to room temperature. During the cooling process, dark brown slurry of nanoparticles will separate from the solvent. To get free particles from incomplete chemical precursors compounds during synthesis, the precipitate was then washed six times with distilled water to remove the excess surfactant from the solution. Those processes will give rise to thick black precipitate which is ready to be heated in a furnace at a constant temperature of 90 °C for 4 hours. Finnaly, the acquired substance was then grinded into a fine powder of $Co_{0.5}Zn_{0.5}Fe_2O_4$ nanoparticle.

The XRD patterns of $Co_{0.5}Zn_{0.5}Fe_2O_4$ confirmed by Shimadzu XD X-ray diffractometer with CuKα (λ=1.5406Å) radiations. TEM and SAED methods by JEOL Jem-4000 electron microscope were used to study overall morphology of nanoparticles. For the study of functional group, FTIR analysis was performed by Shimadzu Prestige-21 IR Spectrometer. Magnetic measurements at room temperature were conducted using a vibrating sample magnetometer (VSM) of Riken Denshi Co. Ltd . All the hysteresis magnetic measurements were carried out at room temperature in the applied static magnetic fields up to 15 kOe. Particle size (t) was calculated from XRD peak broadening by using Scherrer formula, the values of interplanar distance (d) were estimated by using Bragg's equation and peaks reflection (hkl) was used to estimate the lattice constant (a).

3. Results and discussion

3.1. Crystal structure analysis

Figure 1 and Figure 2 show the XRD pattern of samples for both NaOH concentration and synthesis temperature variations. The diffraction pattern referred to the standard data of Co ferrite (JCPDS no. 22-1086) and Zn ferrite (JCPDS no. 22-1012) [1]. The profile of diffraction pattern could be indexed as (220), (311), (400), (422), (511) and (440), which are characteristics of a cubic spinel structure and some other structures, i.e. ZnO (JCPDS no.36-1451) [3]. Those phases indicate that $Zn(OH)_2$ were exhibited incomplete chemical reaction during synthesis.

The diffraction pattern of sample M3 has the highest degree of crystallinity due to the optimum NaOH concentration. In other hand, samples M1, M2 and M4 have a lower degree of crystallinity. Precipitation process is accelerated by the increase of NaOH concentration which lead to increase nucleation and crystal growth rate. The XRD peaks also become narrower and sharper with the increase of synthesis temperature which indicate the increase of the crystallinity degree. The increasing thermal energy has been attributed to increasing synthesis temperatures, which enables the enhancement of nucleation and crystal growth rate.

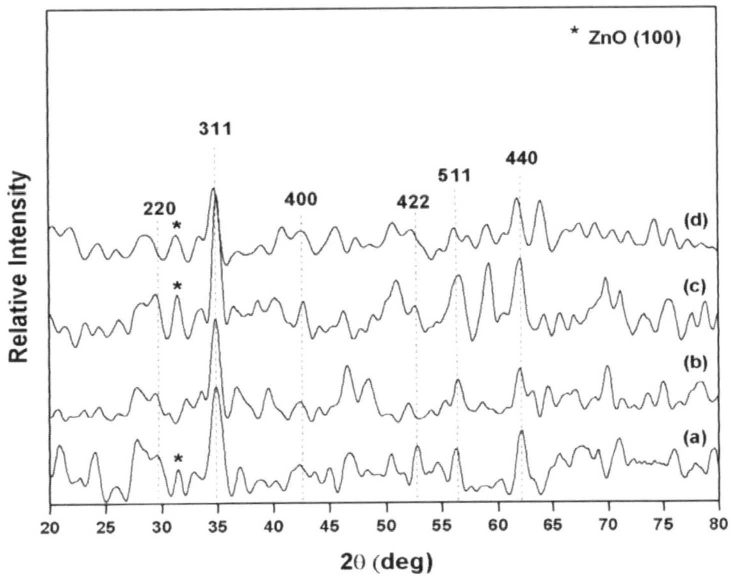

Figure 1. XRD Pattern of $Co_{0.5}Zn_{0.5}Fe_2O_4$ with NaOH concentration (a) 1.5 M, (b) 3 M, (c) 6 M, and (d) 12 M.

The XRD profile peaks were found to be slightly broader in the samples, because of strain effect and crystallite size. The strain is estimated to appear due to a crystal lattice defect. This defect might be caused by synthesis temperature. Increased synthesis temperature contributes to the increase of thermal energy in the process of nucleation and growth of nanoparticles which increases the reaction kinetic energy so that the nucleation becomes more effective. Nucleation process is very sensitive to reaction temperature that agitates thermal energy for the reaction [3]. Low temperature synthesis will limit the nucleation process and crystal growth resulting in a defect of the crystal lattice. The uniform strain occurring in the crystal lattice of nanoparticles brings about some changes in the site of metal cations. The ion distribution in octahedral and tetrahedral sites of crystal lattice will contribute to lattice size and crystallite size of nanoparticles [21]. Lattice parameter and X-ray density are relatively constant with the increase of synthesis temperature because of the composition of Co^{2+},Zn^{2+} and Fe^{3+} are homogeneously in spinel structure of all samples. The crystallites size shows fluctuating results with increasing synthesis temperature. Too high or too low of synthesis temperature caused the process of nucleation and crystal growth is not optimal.The quantitative calculation results of crystallite size (t) [2], lattice constant (a) [2] and X-Ray density [2] are all presented in Table 1.

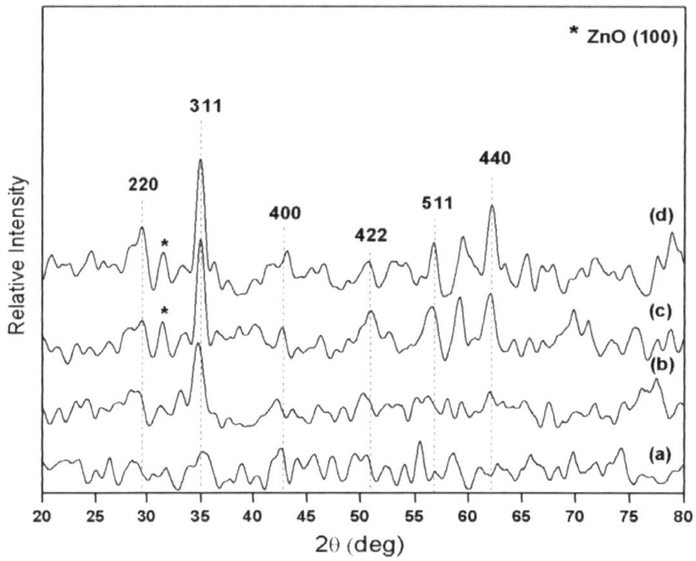

Figure 2. XRD Pattern of $Co_{0.5}Zn_{0.5}Fe_2O_4$ with synthesis temperature (a) 30°C, (b) 50°C, (c) 70°C, and (d) 90°C.

Table 1. Crystallite size, lattice constant and X-Ray density of $Co_{0.5}Zn_{0.5}Fe_2O_4$ nanoparticles synthesized at different NaOH concentration and synthesis temperature.

Sampel	Synthesis parameter variable	Temperature (°C)	NaOH concentration (M)	Lattice constant (Å)	X-Ray density (g/cm^3)	Crystallite size (nm)
M1		70	1.5	8.52	5.1	7.9
M2	NaOH	70	3	8.54	5.0	12.4
M3	concentration	70	6	8.49	5.1	14.1
M4		70	12	8.52	5.1	15.8
S1		30	6	8.24	5.6	21.4
S2	Synthesis	50	6	8.54	5.0	9.3
S3	Temperature	70	6	8.49	5.1	14.1
S4		90	6	8.50	5.1	12.9

The value of the lattice parameter is constant with increasing the NaOH concentration. The increasing of NaOH concentration is not sufficient producing significant deviation of cation distribution. The strain calculation shows a slightly decreased value with increasing NaOH concentration due to reduced defect in the crystal. Density values tend to be constant indicating volume changes followed by significant molar mass changes. It is shown in Table 1 that the crystallite size increases with increasing NaOH concentration. The availability of OH^- ions influences the nucleation process and enhances the crystal growth of $Co_{0.5}Zn_{0.5}Fe_2O_4$ nanoparticles.

3.2. Morphology analysis

The morphology of $Co_{0.5}Zn_{0.5}Fe_2O_4$ samples with NaOH concentration 6 M and synthesis temperature 70°C were investigated using tranmission electron microscopy (TEM) and the results are shown in Figure 3. It is shown that nanoparticle is in the form of aggregate. The observation of the image reveals that the particles are approximately spherical in shape and fine particles are agglomerated. In other words, the decrease in particle size led to agglomeration of magnetic fine particles. Because of the small size, single domain particles experience a magnetic moment proportional to their volume and get permanently magnetized. The SAED image shows that the formation of diffraction rings composes the non-continuous diffraction pattern. The sample can be considered as polycrystalline $Co_{0.5}Zn_{0.5}Fe_2O_4$ nanoparticles which possess low crystallinity compared than bulk material [3].

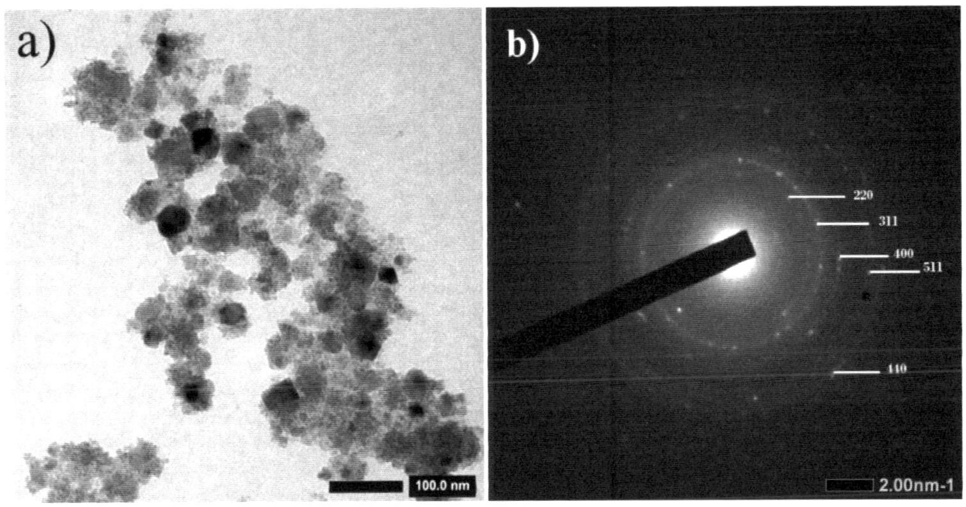

Figure 3. (a) TEM image and (b) SAED image of $Co_{0.5}Zn_{0.5}Fe_2O_4$.

3.3. FTIR spectrum analysis

FTIR spectra of as-synthesized $Co_{0.5}Zn_{0.5}Fe_2O_4$ nanoparticles are presented in Figure 4. Two main broad metal-oxygen (M-O) bands are seen in the FTIR spectra of spinel ferrites. The absorption peaks in wavenumber of 563 - 609 cm^{-1} and 316 - 493 cm^{-1} were observed which corresponded to intrinsic stretching vibration of metal-oxygen at tetrahedral and octahedral sites. The absorption bands of 1049.28 cm^{-1}, 1342.46 cm^{-1}, 1519.91 cm^{-1}, 1627.92 cm^{-1} corresponds to bending vibration of O-H and 2337.72 cm^{-1}, 3356.14 cm^{-1}, 3410.15, 3749.62 cm^{-1}, 3873.06 cm^{-1} can be assigned to hydrogen-bonded O–H stretching vibration arising from surface hydroxyl groups on nanoparticles and adsorbed water.

IOP Conf. Series: Materials Science and Engineering **432** (2018) 012020 doi:10.1088/1757-899X/432/1/012020

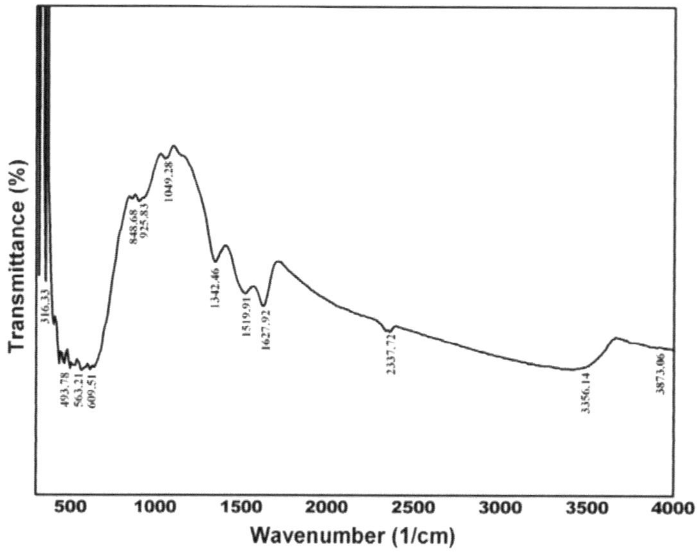

Figure 4. IR spectra of $Co_{0.5}Zn_{0.5}Fe_2O_4$ systems.

3.4. Magnetic measurement

The magnetic properties of samples were measured by using Vibrating Sample Magnetometer (VSM) in room temperature. Figure 5 and Figure 6 show the hysteresis loops of $Co_{0.5}Zn_{0.5}Fe_2O_4$ samples for various NaOH concentrations and synthesis temperature, which indicate that the samples have ferimagneticbehaviour with coercivity and remanent magnetization value larger than zero ($Hc>0$ and $Mr>0$). Itsalso shown that they do not fully saturate at maximum external field of 15 kOe which indicated they tend to have single domain structure. These curves present very low coercivity and low residual magnetization after removal of magnetic field. This behaviour is known as superparamagnetism, which occurs only at nanoscale dimension.

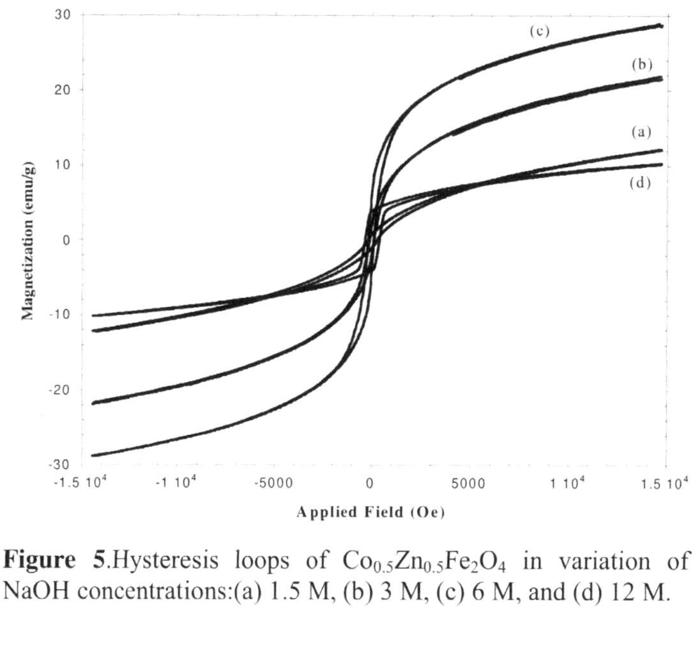

Figure 5.Hysteresis loops of $Co_{0.5}Zn_{0.5}Fe_2O_4$ in variation of NaOH concentrations:(a) 1.5 M, (b) 3 M, (c) 6 M, and (d) 12 M.

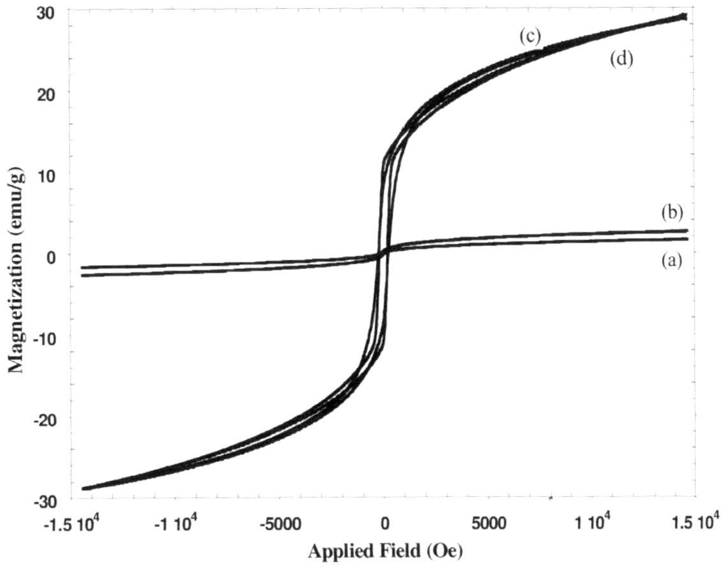

Figure 6. Hysteresis loop of $Co_{0.5}Zn_{0.5}Fe_2O_4$ in variation of synthesis temperatures : (a) 30°C, (b) 50°C, (c) 70°C, and (d) 90°C.

The measurement results of coercivity (H_c), maximum magnetization (M_{max}) and magnetic anisotropy constant (K) is indicated in Table 2. According to Table 2, the value of coercivities (H_c) increases in range 51.1 to 545.56 Oe with increasing crystallite size for the sample with variations of NaOH concentration. This result indicates that the coprecipitation process with increasing NaOH concentration can increase the growth rate and the crystallization progress for the magnetic nanoparticles. This variation of H_c with crytallite size can be explained on the basis of domain structure, critical diameter and the anisotropy of the crystal [22-23].

Table 2. Crystallite size, coercivity, maximum magnetization and magnetic anisotropy constant of $Co_{0.5}Zn_{0.5}Fe_2O_4$ nanoparticles synthesized at different NaOH concentration and synthesis temperature.

Sampel	Synthesis parameter variable	Temperature (°C)	Crystallite size (nm)	H_c (Oe)	M_{max} (emu/g)	K (erg/g)
M1		70	7.9	51.1	12.4	316.9
M2	NaOH	70	12.4	126.9	21.9	1388.0
M3	concentration	70	14.1	257.4	29.0	3732.0
M4		70	15.8	545.6	10.3	2815.1
S1		30	21.4	52.5	1.6	42.8
S2	Synthesis	50	9.3	123.5	2.6	163.6
S3	Temperature	70	14.1	257.4	29.0	3732.0
S4		90	12.8	251.3	29.2	3668.8

In the single domain region one would observe that as the crystallite size decreases the coercivity also decreases because of the thermal effect [24]. Above blocking temperature, the magnetic anisotropy energy barrier of the single domain nanoparticles is overcome by thermal energy and then the spins fluctuate randomly and will easily magnetized by an external magnetic field. The coercive field is associated withthe strength of the magnetic field that is required to overcome the anisotropy barrier. According to the Stoner - Wohlfarth theory, magnetic anisotropy energy (E_A) for non-interacting single domain particles is given by [22]

$$E_A = KV\, sin^2\theta \quad (1)$$

where$K$, V, and θ are magnetic anisotropy energy constant, volume of the particles and the angle between magnetization direction and the easy axis of a nanoparticle, respectively. With the reduction of magnetic anisotropy, the magnetic anisotropy energy barrier (E_A) decreases, which causes lower external magnetic field for spin reversal. When the crystallite size increases, then the volume of the nanoparticles also increases and the magnetic anisotropy energy is enhanced.

For the sample with variations of synthesis temperature the value of coercivities also increases in range 123.5 to 257.4 Oe with increasing crystallite size due to the same reason with samples of NaOH concentration. But for S1 there is some deviation which shows a larger crystallite size with lower *Hc*value. It could be atrributed to either of two reasons. First, it may due to the expected crossover from single domain to multi domain behaviour with increasing size. Second, such phenomena can arise from crystal stucture of nanoparticles. Nanoparticles which aresynthesized atlow synthesis' temperatures may have low crystallinity degree caused by the nucleation and crystal growth process which are not optimal. The maximum magnetization shows an increasing trend with an increasing NaOH concentration and synthesis temperature. The value of maximum magnetization is strongly influenced by the degree of crystallinity of the samples.

4. Conclusion

Powder of $Co_{0.5}Zn_{0.5}Fe_2O_4$ nanoparticles has been successfully synthesized by co-precipitation method. Variations on synthesis parameter i.e. NaOH concentrations and synthesis temperature were found to influence the crystallite size and crytallinity degree due to their contribution in the nucleation and crystal growth process during synthesis. TEM image confirms the reflection peaks shown in the XRD pattern. IR spectral showsvibration in the tetrahedral and octahedral sites that confirm the spinel structures. The VSM magnetic measurement shows the hysteresis loops that indicate ferimagneticbehaviour and tend to be superparamagnetic with decreasing crystallite size. The coercivity and maximum magnetization increases relatively by increasing crystallite size with variations of NaOH concentration and synthesis temperature. It is attributed to the single domain structure existence and crystallinity degree of nanoparticles.

Acknowledgment

The reported work was financially supported by a Grant (*Hibah*)Competency Based Research(PBK) of the Ministry ofResearch and Technology of High Education for the period of 2015 – 2017. Authors would like tothank Institute of Materials and Systems for Sustainability, Nagoya University for providing VSM instrument to magnetic measurement.

References

[1] Praveena K and Sadhana K, 2015 *Int. J. of Sci. and Research Pub.* **5** 2250-3153
[2] Indrayana IPT, Siregar N, Suharyadi E, Kato T, and Iwata S, 2016 *J. Phys: Conf. Series* **776** 012021
[3] Siregar N, Indrayana IPT, Suharyadi E, Kato T, and Iwata S, 2017*IOP Conf. Series: Mater. Sci. and Engin.* **202** 012048
[4] Saafan SA, Assar ST, Moharram BM, and El Nimr M K, 2010 *J. Magn. Magn. Mater.***322** 628-32
[5] Lu X, Liang G, Sun Q, and Yang C, 2011 *Mater. Lett.***65**674-6
[6] Zhou B, Liu Z, Wang X, Sui Y, Huang X, Lu Z, and Su W, 2010 *Physica B: Cond. Matt.***405** 374-8
[7] Rohilla S, Kumar S, Aghamkar P, Sunder S, and Agarwal A, 2011 *J. Magn. Magn.Mater* **323**

897-902
[8] Igarash H and Okazaki K, 1977 *J. Am. Ceram. Soc.* **60** 51-4
[9] Kulikowski K, 1984 *J. Magn. Magn. Mater.* **41** 56-62
[10] Ravindranathan P and Patil KC, 1987 *J. Mater. Sci.* **22** 3261-4
[11] Abraham T, 1994 *Am. Ceram. Soc. Bull.* **73** 62-5
[12] Slick PI, 1980 *Ferromagnetic Materials* (Amsterdam :Wohlforth North-Holland) 196
[13] Bhise BV, Dongare MB, Patil SA, and Sawant SR, 1991 *J. Mater. Sci. Lett.* **10** 922-4
[14] Teixeira AM, Ogasawara T, and Nobrega MC, 2006 *Material Research* **3** 257-62
[15] Hong SH, Park JH, Choa YH, and Kim J, 2005 *J. Magn. Magn. Mater.* **290-291** 1559-62
[16] Ma L, Chen L, and Chen S, 2009 *Mater. Chem. Phys.* **114** 692-6
[17] Gubin SP, 2009 *Magnetic Nanoparticles* (Weinheim :Wiley)
[18] Desai R, Davariya V, Parekh K, and Upadhyay VR, 2009 *J. Phys.* **37** 765-80
[19] Hu P, Yang HB, Pan DA, Wang H, Tian JJ, and Zhang S G 2010 *J. Magn. Magn. Mater.* **322** 173-7
[20] Puspitasari AT, 2018 *Synthesis of Cobalt Zinc Ferrite ($Co_{0.5}Zn_{0.5}Fe_2O_4$) Nanoparticles and Study of Its Crystal Structure and Magnetic Properties*, Thesis, Department of Physics, Universitas Gadjah Mada, Indonesia.
[21] Cullity B D 1956 *Element of X-Ray Diffraction* (New Jersey: Prentice Hall)
[22] Cullity B D 1972 *Introduction to Magnetic Materials* (Massachusetts : Addison-Wesley Pub. Comp. Inc.)
[23] Chikazumi S 1959 *Physics of Magnetism* (New York: Wiley)
[24] Singhal S, Namyal T, Bansal S, and Chandra K 2010 *J. Electromag. Analys. and Applic.* **2** 376-81

IOP Conf. Series: Materials Science and Engineering **432** (2018) 012021 doi:10.1088/1757-899X/432/1/012021

Development of 410S-Y_2O_3 oxide dispersion strengthened steel prepared by a powder metallurgy technique

T Sudiro[1]*, R C Ekaputra[2], K A Z Thosin[1], A Trenggono[2], A Afandi[1], B Hermanto[1]

[1] Research Center for Physics, Indonesian Institute of Sciences, Puspiptek Serpong Area, South Tangerang 15310-Indonesia
[2] Department of Metallurgy Engineering, Faculty of Engineering, Sultan Ageng Tirtayasa Univ., Jend. Sudirman Km. 3 Street Cilegon, Banten 42435-Indonesia

toto011@lipi.go.id

Abstract. In the present study, the Oxide Dispersion Strengthened (ODS) steels were synthesized by adding various concentration of Y_2O_3 (0, 0.25, 0.35 and 1 wt. %) into 410S steel. The sample preparation was carried out using a powder metallurgy technique. The steel-Y_2O_3 powders were wet mechanically alloyed for 3 h, dried and then compacted into green body. The samples were sintered in a vacuum furnace at elevated temperatures of 1000 °C and 1100 °C for 3 h. The phase analysis and structural characterization were carried out by XRD and SEM-EDX, respectively. The bulk density and hardness of the samples were measured using Archimedes principle and Vickers hardness tester, respectively. According to the results of metallographic characterization, ODS steels are composed mainly by (Fe,Cr) phase. The SEM observations indicate the presence of precipitates in the sintered steels. The density and hardness of the sample sintered at 1000 and 1100 °C increase with the increase of Y_2O_3 concentration. At 1100 °C, however, the density and hardness of ODS steel containing 1 wt. % Y_2O_3 have a tendency to decrease due to the formation of a compound consisting of Fe, Cr, Mn, Si and O.

1. Introduction

High temperature applications such as advanced coal fired power plant (A-USC) and nuclear reactor require material that are resistant not only to oxidation and corrosion but also to mechanical loading. Moreover, for nuclear energy systems it must withstand to high radiation damage, swelling and embrittlement [1, 2]. Currently, oxide dispersion strengthened (ODS) steel is the most promising candidate material for the aforesaid applications due to the fact that it exhibits an excellent high temperature properties over the conventional heat resistant steels without oxide particle dispersoids [3]. Structurally, the fine oxide particles are dispersed uniformly in the matrix. Their presence can improve the creep resistance, mechanical strength and irradiation resistance [4, 5].

Mostly, the ODS steel is based on Y_2O_3 fine particles which was added into Fe-Cr [5-6] and Fe-Ni based powder [7]. The particles were acted as a stabilizer of free dislocation and sub-grain structure [8]. Consequently, their structure, size and distribution become important factors that determine the ODS properties. The finer dispersoids have a significant effect on the hardness and strength of ODS material. Previous study reported that the addition of Ti to Y_2O_3 strengthened ODS steel can refine the Y-Ti-O oxide particles [3, 4] into 2-6 nm and increases the number density of oxide particles in the matrix. In comparison to the Y_2O_3-ODS steel without Ti addition, Y-Ti-O strengthened ODS steel exhibits a better creep and radiation resistance, and also higher elevated temperature strength [2].

Several techniques such as spark plasma sintering have been used to synthesize the ODS steel. However, the issues of anisotropic properties and production cost of this technique are often taken into consideration in its use [5]. On the contrary, due to its cost effectiveness and uniqueness, conventional powder metallurgy technique offers distinctive benefit for fabrication of high temperature materials. This method usually consisted of subsequent process as powder blending, addition of pressing aids, compaction, sintering and further thermal treatment to obtain the desired microstructure and properties. Mostly, a mechanical alloying technique was used to homogenize metallic powders and disperse the fine ceramic powders as Y_2O_3 to the matrix [3-5, 8]. The small concentration of sintering aids or binder can be added to obtain the high strength of green body. The powder compaction or consolidation was performed at room temperature by applying compressive stress to the powders. This results a green body with a porosity of about 10-60%. In order to get high end of final density, the compact was sintered at elevated temperature in a carefully controlled atmosphere [9]. The aforementioned points suggest that the material properties of final product are strongly depending on the processing parameter used.

In the present study, the fabrication of ODS steels was investigated by adding the varying concentration of Y_2O_3 into steel 410S. This material was considered as starting material because it contains Ti element which may play a role in the formation of fine dispersed oxide particles. The structures of ODS steels after sintering at 1000 and 1100 °C were studied by means of XRD and SEM-EDX. In this paper, the density and hardness of ODS steels were also presented.

2. Experimental method

2.1. Sample preparation

The ODS steels powder containing varying amount of Y_2O_3 were prepared using a conventional sintering method. First, the steel powder starting material was weighted according to the stainless steel powder 410S composition, as shown in Table 1.

Table 1. Chemical element of steel 410S.

Element	Fe	Cr	Mn	Ti	Si
Weight percentage (wt.%)	Balance	14	1	0.2	1

The steel powder was then added with varying concentration of 0, 0.25, 0.35 and 1 wt. % Y_2O_3 powders with the size of about 5-10 μm. Each composition was mechanically alloyed in steel vial under wet milling condition with balls and powder ratio of 10 to 1 using a high speed shaker mill at oscillation frequency of 700 rpm/min for 3 h. After milling, the mixture powder was placed in desiccators and dried at ambient temperature and environment. The mechanically alloyed powder was then packed in the steel dies and then compacted under a compressive stress of 8 Ton for 3 min. The samples were then sintered in vacuum atmosphere of 10 Pa at elevated temperature of 1000 °C and 1100 °C for 2 h. Finally, it was cooled down in the same chamber until the furnace reached a room temperature.

2.2. Sample characterization

For metallographic characterization and properties measurement, the surface of sintered steels was ground to SiC abrasive papers from #800 to #2000 and DP-polishing cloth for up to mirror finish. The phase composition of ODS powders before and after sintering was determined by X-ray diffractometer (XRD Smartlab Rigaku). The XRD scanning was performed using a Cu Kα irradiation (40kV, 30mA) at diffraction angle of 10-90 deg, step width of 0.01 deg, and scan speed of 0.25. The microstructures of the steels were characterized by a scanning electron microscope (SEM Hitachi SU3500) attached with energy dispersive X-ray spectrometer (EDX Horiba). The sample density was evaluated according to the Archimedes principle. The hardness was measured using Vickers microhardness tester (Leco TM-100 Microhardness) with an applied load of 13 kgf for 28 sec.

3. Results and discussion

3.1. Morphology of the powder after milling

Figure 1 shows the XRD profiles of ODS powders with different amount of Y_2O_3 content after milling for 3 h.

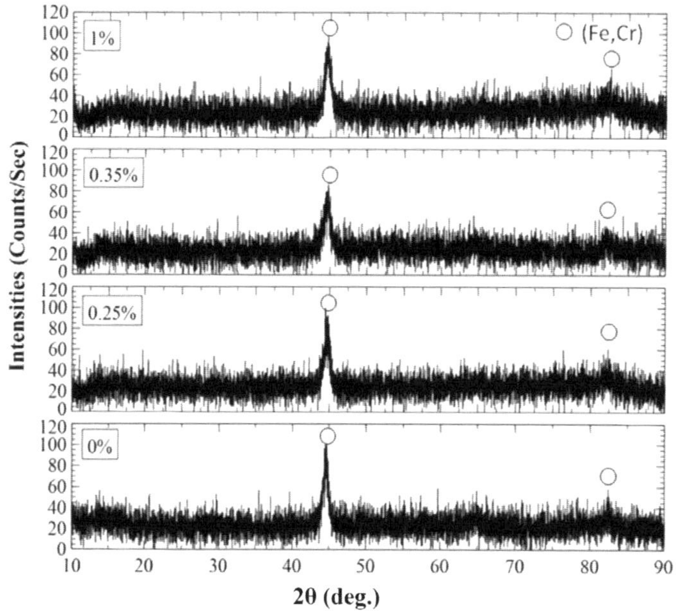

Figure 1. X-ray diffraction patterns of ODS powders containing 0, 0.25, 0.35 and 1 wt.% Y_2O_3 after milling for 3 h.

It can be seen that there is no significant difference in XRD patterns of ODS powders containing varying amount of Y_2O_3. The graph clearly shows that all powders after milling for 3 h have low diffraction intensity and considerable line broadening. This is attributed to the grain refinement or the formation of nanosize structure. Mainly, the diffraction peaks are found at the diffraction angle of about 44° and 82°. According to the results of XRD analysis, the peaks are detected as (Fe,Cr) phase [3]. It was notable that atomic radius of Fe is almost equal to the Cr. Moreover, they also have BCC crystal structure [10]. Accordingly, the diffraction peaks of both elements are overlap between each other and preferentially form (Fe,Cr) solid solution. The results of XRD characterization also reveal that the XRD analysis didn't detect the Y_2O_3 reflection. During mechanical alloying process, the Y_2O_3 and steel powders were subjected to high energy ball impacts. Thus, the brittle Y_2O_3 powders are likely to fragment into fine particles. The particles were deeply incorporated or dissolved in the steel matrix during MA [3-4, 11]. Accordingly, their X-ray diffractions become very weak compared to the (Fe,Cr) diffraction peaks.

3.2. Structure of ODS steel after sintering at 1000 °C

Figure 2 shows XRD profiles of ODS steels with various content of Y_2O_3 after sintering at 1000 °C for 2 h.

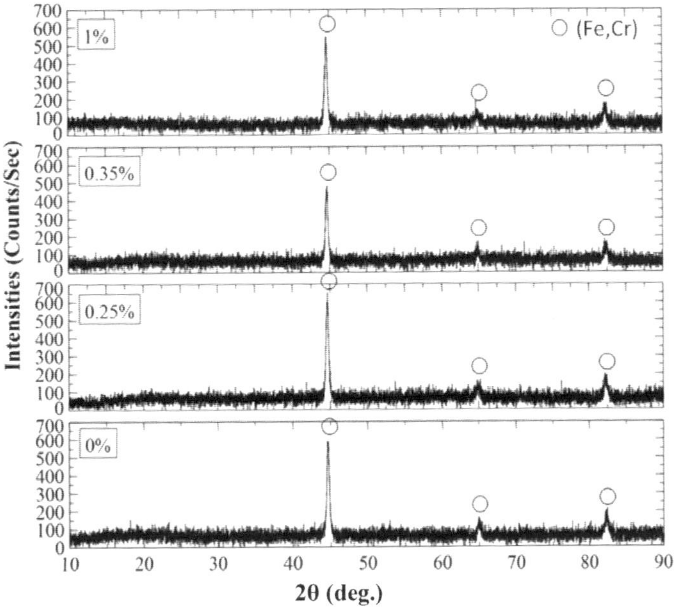

Figure 2. X-ray diffraction patterns of ODS steels containing 0, 0.25, 0.35 and 1wt.% Y_2O_3 after sintering at 1000 °C for 2 h.

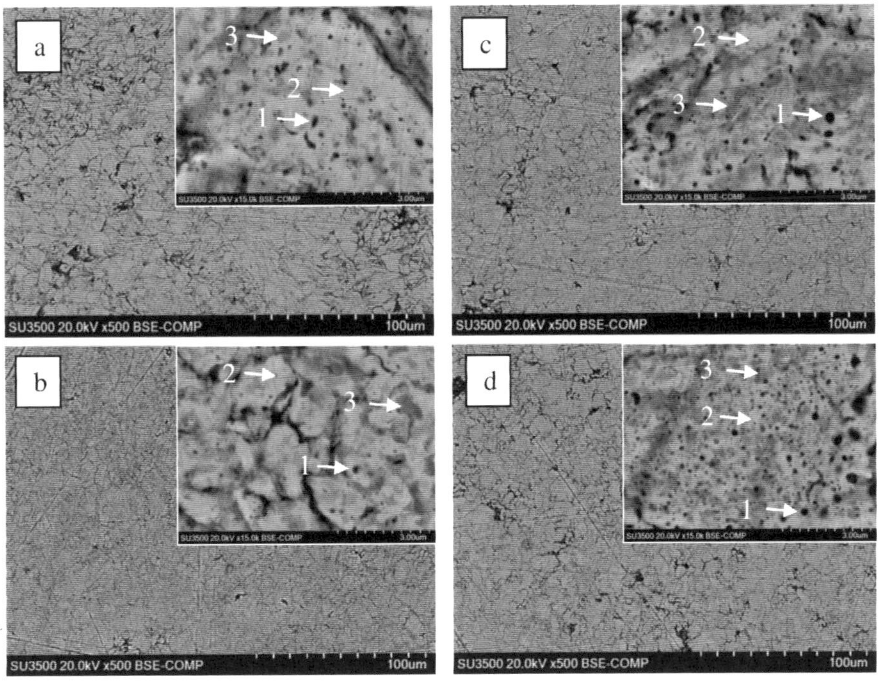

Figure 3. BSE Comp images of ODS steels containing (a) 0, (b) 0.25, (c) 0.35 and (d) 1wt.% Y_2O_3 after sintering at 1000 °C for 2 h.

It is apparent that the diffraction peak of ODS steel after sintering at 1000 °C for 2 h become sharp and stronger compared to the powder after milling. This could be due to the improvement of crystallinity and grain growth of the powder after sintering. It is worth mentioning that due to the high energy of grain boundaries, the nano-crystalline or ultrafine-grained materials are generally susceptible to grain coarsening at elevated temperatures [12]. Accordingly, the high sintering temperature and long sintering time allow the powder to grow. As shown in Figure 2, the third peak is found at the diffraction angle of about 65 deg in the ODS steels containing varying amount of Y_2O_3. According to the results of XRD analysis, all diffraction peaks correspond to (Fe,Cr) phase which is the same as powder after milling. No diffraction peak of Y_2O_3 was also detected in the ODS steel with different Y_2O_3 content.

In order to investigate the effect of Y_2O_3 concentration on the microstructure of ODS steels, the SEM observations were carried out. Figure 3 shows the typical surface morphology of ODS steels after sintering at 1000 °C for 2 h.

As can be seen in Figure 3, observation under low magnification can observe the grain boundaries between particles which suggest that ODS steels with different Y_2O_3 content seem to be not fully sintered at 1000 °C for 2 h. Metallographic characterization also observes the presence of pores. Mostly, it was found in the Y_2O_3-free steel. With the increase of Y_2O_3 concentration, the porosity is likely to decrease. Moreover, it is important to note that further detail characterization at high magnification observes mainly three different areas in the sintered ODS steels: black precipitates (No. 1), bright area as matrix (No. 2) and grey areas (No. 3) as shown in Figure 3. The grey and black areas with different in size were randomly distributed in the steel matrix.

3.3. Structure of ODS steel after sintering at 1100 °C

Figure 4 shows the XRD profiles of ODS steels with different Y_2O_3 content after sintering at 1100 °C for 2 h.

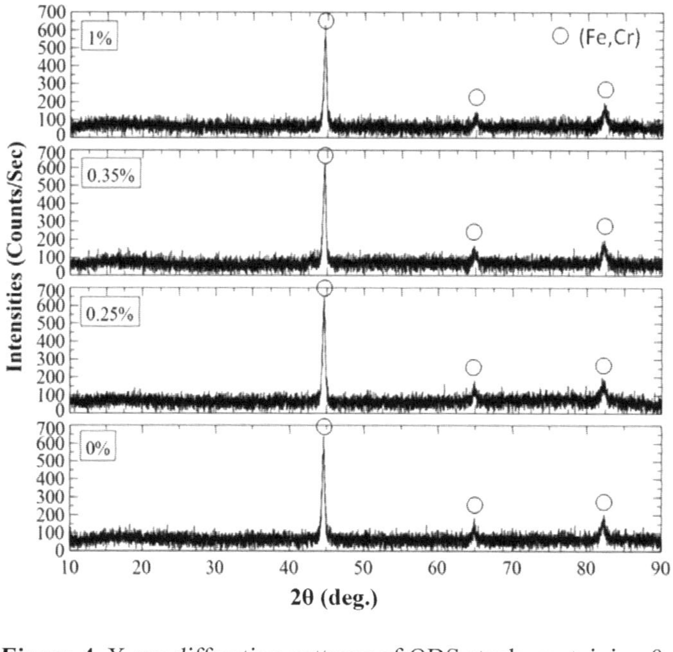

Figure 4. X-ray diffraction patterns of ODS steels containing 0, 0.25, 0.35 and 1wt. % Y_2O_3 after sintering at 1100 °C for 2 h.

Similarly to the results as presented in Figure 2, the sample sintering at 1100 °C for 2 h leads to narrowing and increasing the XRD diffraction peaks. This is attributed to the densification of crystallites and grain growth. The X-ray diffraction analysis also confirm that the phase of ODS steels with different Y_2O_3 content after sintering at 1100 °C is similar to that of at 1000 °C.

Figure 5 shows the surface morphology of ODS steels with different Y_2O_3 content after sintering at 1100 °C for 2 h. It can be seen that the sample sintered at 1100 °C is denser than that of at 1000 °C. The microstructural characterization also reveals that the samples consist of bright area (2) associated with small grey areas (3) and black precipitates (1), similarly to the microstructure of the sample sintered at 1000 °C for 2 h. However, the size of grey areas and black precipitates in the ODS steel sintered at 1100 °C is bigger than the previous one due to the grain growth with the increase of sintering temperature. The EDX analysis at high magnification images needs to be done to determine the composition of aforesaid structure.

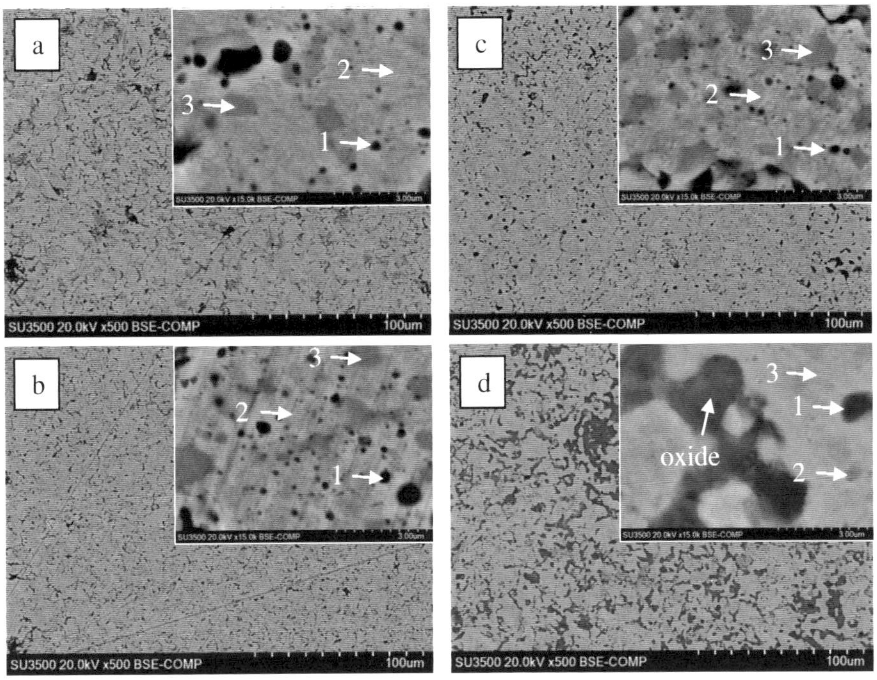

Figure 5. BSE Comp images of ODS steels containing (a) 0, (b) 0.25, (c) 0.35 and (d) 1wt.% Y_2O_3 after sintering at 1100 °C for 2 h.

Moreover, unlike the ODS steel containing 0, 0.25, and 0.35 wt. % Y_2O_3, the significant change in the ODS morphology is found in the ODS steel containing 1 wt. % Y_2O_3. We carried out the EDX elemental mapping to observed the distribution of elements in the ODS steel containing 1 wt. % Y_2O_3 with the results as presented in Figure 6.

According to the results of SEM-EDX analysis, we can clearly observe the distribution of Fe, Cr, Mn, Si and O in the dark area of BSE image. It can be conclude that some elements was somehow oxidized after sintering at 1100 °C for 2 h. This is probably due to that the dissolution of Y and O takes place [4] which leads further formation of a compound consisting of Fe, Cr, Mn, Si and O. This will significantly affects the properties of ODS steels.

3.4. Density and hardness

Figure 7 presents the density and hardness of ODS steels with various concentration of Y_2O_3 after sintering at 1000 °C and 1100 °C for 2 h.

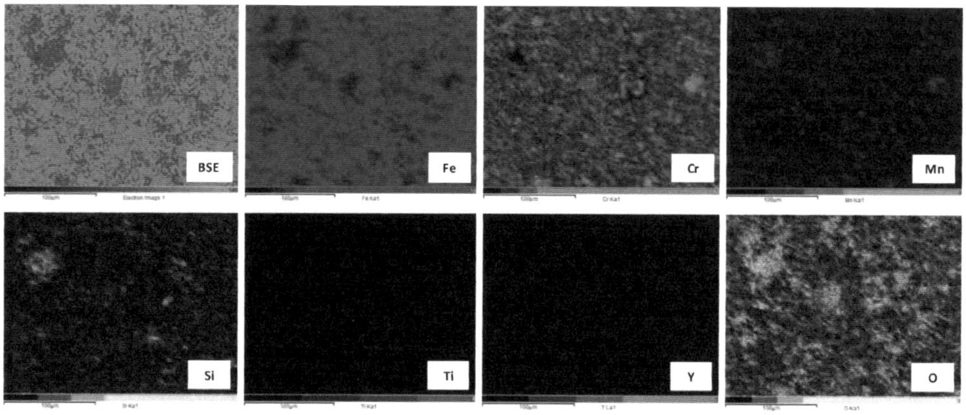

Figure 6. BSE comp image and corresponding EDX elemental maps of ODS steel containing 1 wt. % Y_2O_3 after sintering at 1100 °C for 2 h.

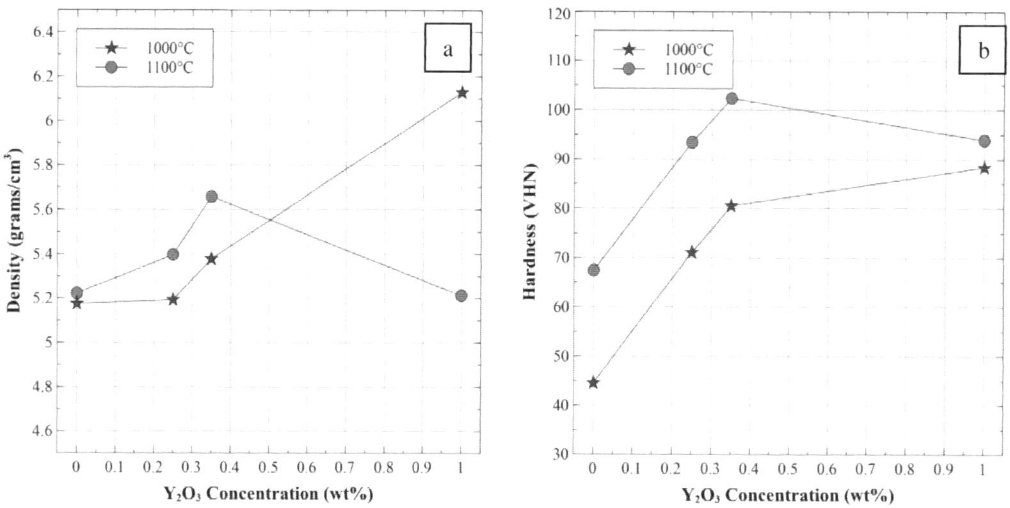

Figure 7. (a) Density and (b) hardness of ODS steel containing varying concentration of Y_2O_3 sintered at 1000 °C and 1100 °C for 2 h.

The curve shows that the Y_2O_3-free steel exhibits low density and hardness compared to ODS steel containing Y_2O_3. The density of the samples after sintering at both temperatures has a tendency to increase with increasing of Y_2O_3 concentration, as shown in Figure 7a. In addition, a higher sintering temperature produces a denser ODS steels. However, a deviation is found in the 1 wt. % Y_2O_3 steel after sintering at 1100 °C. The density calculation clearly indicates that the sample density drops to 5.21 g/cm^3, suggesting that the sample tends to porous. The density value of present study is smaller compared to the density of ODS steels prepared by hot-isostatic press (HIP) [5, 13] and spark plasma sintering (SPS) [5]. This is mainly due to the compressive stress used in this study is lower compared to aforementioned studies. In general, a higher compressive stress results in a higher sample density.

On the other hand, the graph as shown in Figure 7b, reveals a steady increase in hardness of the samples (up to 0.35 wt.% Y_2O_3 concentration) in both sintering temperatures. More than 0.35 wt.% Y_2O_3, there has been a slight increase in ODS hardness after sintering at 1000 °C and a decrease in hardness after sintering at 1100 °C. Since the processing parameter for all samples is same, the different just in the ODS steel composition, it can be assumed that the hardness increase is attributed to the denser microstructure with the increase of Y_2O_3 concentration and dispersion oxide particle in the matrix. The presence of Y_2O_3 in the steel can suppress the growth of the steel grains [2], resulting a high sample density and high sample hardness. Based on the results as presented above, the decrease of density and hardness of 1 wt.% Y_2O_3 steel after sintering at 1100 °C could be linked to the results of microstructure analysis as shown in Figure 6. It is reasonable to assume that that the microstructural change after sintering at 1100 °C for 2 h leads to decrease in density and hardness of ODS steels.

4. Conclusions

The ODS steels containing varying concentration of Y_2O_3 (0, 0.25, 0.35 and 1 wt. %) are successfully synthesized using a powder metallurgy technique at 1000 °C and 1100 °C. After sintering at the aforesaid temperatures, the ODS steels with different Y_2O_3 content are mainly composed by (Fe,Cr) phase. Metallographic characterization at high magnification indicates the distribution of precipitates in the ODS steels. Some elements are somehow preferentially oxidized in the ODS steel containing 1 wt. % Y_2O_3 after sintering at 1100 °C to form a compound consisting of Fe, Cr, Mn, Si and O. The ODS steels sintered at 1100 °C exhibited a higher density and hardness compared to the samples sintered at 1000 °C. In addition, the density and hardness of the sample increase with increasing of Y_2O_3 concentration. However, futher increase in Y_2O_3 concentration (1 wt. %) leads to decrease in density and hardness of ODS steels.

Acknowledgment
This work is financially supported by Program Insentif Riset Sinas (Development of ODS Steel from Commercial Steel with Addition of Rare Earth Metal Oxide Y_2O_3 for High Temperature and Anti Radiation Applications). The authors would also like to acknowledge Research Center for Physics–LIPI for providing the characterization facilities.

References

[1] Chen C-L and Dong Y-M 2011 *Mater. Sci. and Engineer.* **A528** 8374
[2] Sundararajan G Vijay R and Reddy A V 2013 *Current Sci.* **105** 1100
[3] Xu H Lu Z Wang D and Liu C 2016 *Nuclear Mater. and Energy* **7** 1
[4] Zhang H Huang Y Ning H *et al.* 2015 *J. of Nuclear Mater.* **464** 61
[5] Auger M A de-Castro V Leguey T Muñoz A and Pareja R 2013 *J. of Nuclear Mater.* **436** 68
[6] Li N Yadav S K Xu Y Aguiar J A Baldwin J K Wang Y Q Luo H M Misra A and Uberuaga B P 2017 *Scientific Reports* **7(40148)** 1
[7] Moghadasi M A Ahmadabadi M N Forghani F and Kim H S 2016 *Scientific Reports* **6(38621)** 1
[8] Abe F 2016 *J. of Pressure Vessel Tech.* **138** 040804
[9] Kaysser W A and Weise W 2012 *Ullmann's Encyclopedia Industrial Chem.* **30** 1
[10] Singh A and Dahotre N B Laser Induced Carbide Coating on Steel *Proc. of the 22nd Heat Treating Soc. Conf. and the 2nd Int. Surf. Engineer. Congr.* Indiana USA ed Dahotre N B Gaster R J Hill R A Popoola O O ASM Int.et 427-30
[11] Oksiuta Z 2011 *Acta Mechanica et Automatica* **5** 74
[12] Marek I Vojtěch D Michalcová A and Kubatík T F 2016 *Materials* **9(391)** 1-10
[13] Nguyen V T Doan D P Tran T B T Luong V D Nguyen V A and Phan A T 2010 *Adv. Nat. Sci. Nanosci. Nanotech.* **1** 035009

IOP Conf. Series: Materials Science and Engineering **432** (2018) 012022 doi:10.1088/1757-899X/432/1/012022

The performance of the AAA and Acurox XB algorithms in the case of lung cancer

R Zulfani[1], M Fadli[1,2] and S A Pawiro[1*]

[1] Department of Physics, Faculty of Mathematics and Natural Sciences, University of Indonesia, Depok, Indonesia, 16424
[2] Department of Radiotherapy MRCCC Siloam Hospitals Semanggi, Jakarta, Indonesia

supriyanto.p@sci.ui.ac.id

Abstract. The Anisotropic Analytical Algorithm (AAA) and Acuros XB algorithms in advanced technique therapy planning has been implemented for Intensity Modulated Radiotherapy (IMRT) and Volumetric Arc Modulated Radiotherapy (VMAT) techniques. This study aimed to verify the dose simulation of IMRT and VMAT in planning target volume (PTV) of lung cancer cases. All planning were created using TPS Eclipse version 13.0.47. The IMRT planning were created using 7 fields whereas the VMAT technique employed the double arc with target dose prescription of 200 cGy each fraction. The results showed that the small decrepancy between planning and measurements of PTV in left and right lung cancer has occurred in VMAT technique, whereas Acuros XB algorithm had better planning quality than Anisotropic Analytical Algorithm (AAA) with differences about of 0,56%. In the organ at risk of left and right lung cancer, there was no significant difference between Acuros XB and Anisotropic Analytical Algorithm (AAA) with the largest deviation range obtained in simulated heart organ with around of ± 9%. We concluded that the Acuros XB algorithm has better performance than Anisotropic Analytical Algorithm (AAA) by a factor of approximately ± 1%.

1. Introduction

Lung cancer is the leading cause of death in both male and female patients. In 2012, an estimated 1.8 million new cases (12.9% of the total), 15.9 million deaths and 58% occur in less developed countries [1]. Furthermore, based on data from the International Agency for Research on Cancer (IARC) lung cancer was found in males by 34.2%, while lung cancer deaths in males by 30% [2]. In indonesia, cancer slowly begins to shift the position of heart attack as the main cause of death. Rising cancer position as the cause of death due to the high number of new cancer cases that come at an advanced stage. Based on these data required efficient handling to minimize the spread of cancer each year. One of the most commonly used modern radiotherapy techniques is the Intensity Modulated Radiation Therapy (IMRT) and Volumetric Modulated Arc Therapy (VMAT). IMRT (Intensity Modulated Radiation Therapy) technique which is implemented with a number of non-uniform radiation beams optimized to produce high doses of target volume and low doses of normal tissue. In addition, the VMAT is a development of the IMRT technique, in which the dose distribution of 3D is given precisely by providing the photon beam at a rotating gantry at one or more angles [3].

In planning of IMRT and VMAT techniques, Anisotropic Analytical Algorithm (AAA) and Acuros XB algorithm has been implemented in treatment planning system (TPS) for dose simulation in radiotherapy. Furthermore, Acuros XB is the development of the Acuros External Beam (AXB) algorithm and had been implemented in the Eclipse TPS (Varian Medical Systems, Palo Alto, CA) since 2010. Explicitly, thc Linear Boltzmann Transport Equation (LBTE) equation numerical method has been solved in Acuros algorithm [4]. The LBTE equation also assumes that radiation particles (electrons, photons and protons) only interact with matter they are passing through, and not with other particles, an is valid for conditions without external magnetic fields [5].

Previous work on the use of the Acuros XB algorithm and Anisotropic Analytical Algorithm (AAA) on IMRT and VMAT radiotherapy techniques is to assess the dosimetric effects of the Acuros XB (AXB) algorithm compared to the Anisotropic Analytical Algorithm (AAA) in esophageal cancer using a slab of fantom [6]. They did the comparison of the calculated percentage depth dose (PDD) with AXB and AAA algorithms on a phantom slab with a 2 cm air gap thickness. The result of using AXB makes it possible to avoid overly high doses of esophageal cancer planning compared to AAA. The same pattern is also found in the study by Kan *et al.*, who has assessed the dosimetric on IMRT and VMAT for nasopharyngeal carcinoma (NPC) by using acuros XB and AAA algorithm which used 7 field on IMRT planning and triple-arc on VMAT planning for 12 patients NPC using AAA and then recalculated using AXB algorithms [7]. As a result, the use of the AXB algorithm is highly recommended for IMRT and RapidArc in the case of NPCs. In this study, we explored the performance of AAA and AXB algoriths in lung cancer cases.

2. Experimental method

This study was conducted using Linac radiotherapy (Varian Medical Systems, Palo Alto, CA) at Siloam Hospital Semanggi (MRCCC). This Linac produces two X-rays photon energies that are 6 MV and 15 MV which is equipped with 120 MLCs. The TLD LiF 100 rod and TLD reader dosimeters (Thermo Scientific 3500 model Harshaw) were employed in this study. In addition, the gachromic film EBT3 and Flatbed Scanner Epson V700 were used to evaluate optical density measurements on films and then converted into the dose. Dose verification of lung cancer was performed by scanning Rando Alderson thorax section phantom on CT Simulator Philips Brilliance 16 slice. Verification was done by placing the TLD and the Gafchromic EBT3 film for IMRT and VMAT techniques calculated using AAA, AXBDm and AXBDw algorithms. The use of Gafchromic EBT3 film was cut into size of 1.5cm × 0.5cm and positioned at coronal position orientation in rando phantom. The simulated images were sent to the treatment planning system (TPS) for contouring and dose planning.

All planning was created using the TPS Eclipse version 13.0.47. The IMRT planning were created using 7 fields whereas the VMAT technique employed the double arc with target dose prescription of 200 cGy each fraction. For dose verification, the TLD and gafchromic film were took placed inside Rando Alderson phantom. The dose evaluation was done by comparing dosimetry analysis of PTV and organ at risk between the AAA and Acuros XB algorithms for IMRT and VMAT techniques as shown in Figure 1. The point dose difference between planning and measurement can be calculated by following equation:

$$\%diff = \frac{D_{measured} - D_{calculated}}{D_{presciption}} \tag{1}$$

where $D_{measured}$ is the measurement dose, $D_{calculated}$ is the simulated dose of the TPS , and $D_{prescription}$ is the prescribed dose value at the reference point [8]. The performance of algorithms were evaluated using the comparison between the Acuros XB algorithm to AAA. Dose deviations from both algorithms can be calculated by the equation as below [6]:

$$\Delta_{avg}(\%) = \sum\left(\frac{AXB-AAA}{AAA}\right) \tag{2}$$

In equation (2), AXB and AAA represent the calculate dose in TPS as well as the measure dose with TLD dosimeter of 100 LiF and the Gafchomic EBT3 film.

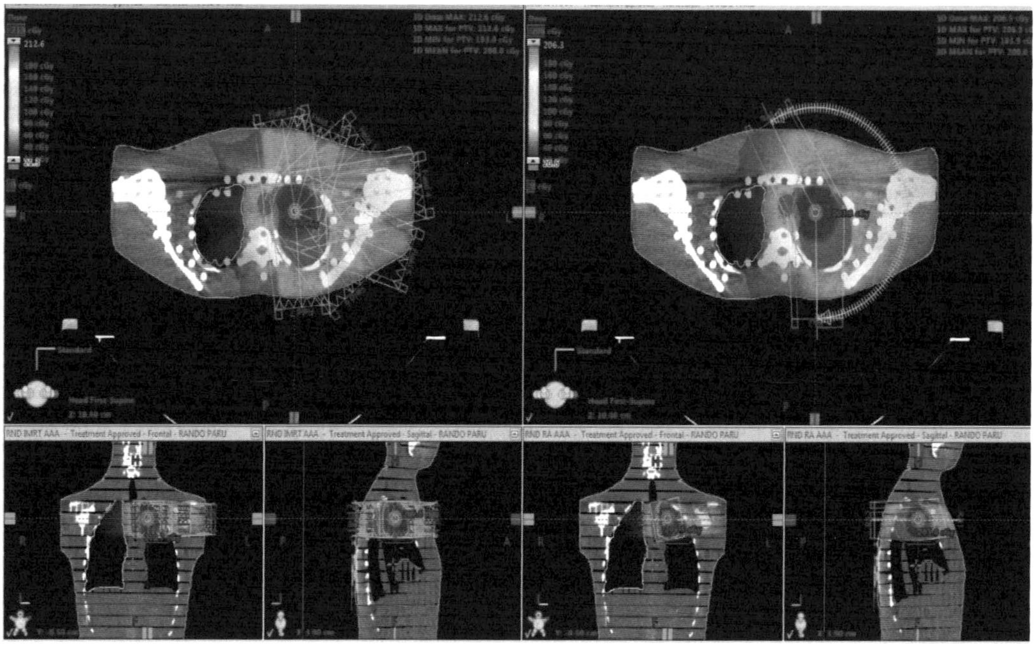

Figure 1. TPS simulation of lung cancer a) VMAT technique b) IMRT technique.

3. Results and discussion

Dose verification of simulated lung cancer was performed by placing TLD 100 LiF rod and Gafchromic EBT3 film in Rando phantom. Measurements of TLD 100 LiF dosimeters and Gafchromic EBT3 films in right lung cases with IMRT and VMAT techniques is shown in Figure 2(a) and 2(b). It indicated the average deviation of the $AXBD_m$ algorithm for PTV tends to be smaller than $AXBD_w$ the same pattern as the left lung case. Furthermore, the difference between the two dosimeters is not very significant but the average deviation of PTV of the IMRT technique and the VMAT algorithm of Acuros XB gives a smaller deviation value than AAA. This result supported Kan et.al., who performed the same dose verification with the IMRT and VMAT techniques in the nasopharyngeal cases and indicated the same pattern [7].

Similar to PTV, the measurement of organ at risk (OAR) using the 100 LiF rod TLD dosimeter and the Gafchromic EBT3 film is shown in Figure 3. The figures explained the percentage of spinal cord and heart deviation on the VMAT technique smaller than the IMRT technique as well as indicated in right lung. Based on the result it is known that the use of AAA and Acuros XB algorithms on VMAT technique is better performance than in IMRT technique. In general, the TLD 100 LiF dosimeter provides a much larger deviation value than the Gafchromic EBT3 film dosimeter. This indicates that the use of TLD dosimeters has considerable uncertainty with poor accuracy of -10.3% on target dosing with uncertainty ± 4.7% as proposed by Fitriandini *et al.* [11]. Meanwhile, the results of measurements obtained in both lung cases with IMRT and VMAT techniques obtained dosage values and standard deviation algorithm $AXBD_w$ tend to be higher than $AXBD_m$ algorithm. This result is also in line with Kan *et al.* and Rana *et al.* research which reported the $AXBD_w$ algorithm computations have high values when compared with $AXBD_m$ algorithm calculations [7, 10]. This is also supported by a study conducted that is to make measurements using films on phantom anthropomorphic alderson (for pulmonary planning on clinical IMRT) [9]. As a results, it is known that AXB is more accurate to predict dose when it comes to air than AAA.

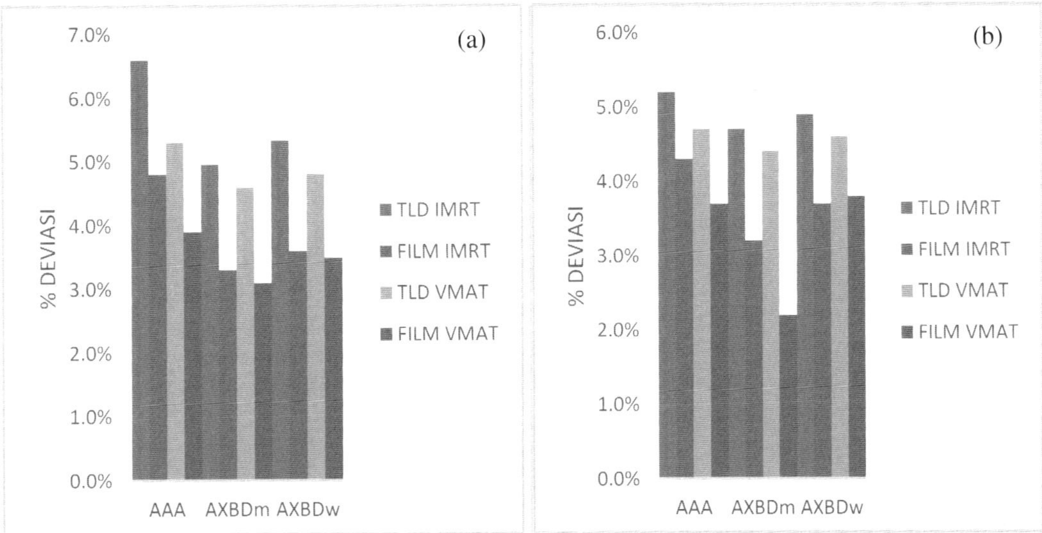

Figure 2. Deviation in PTV: (a) left lung ; (b) right lung anisotropic analytical algorithm (AAA) and Acuros XB IMRT and VMAT techniques.

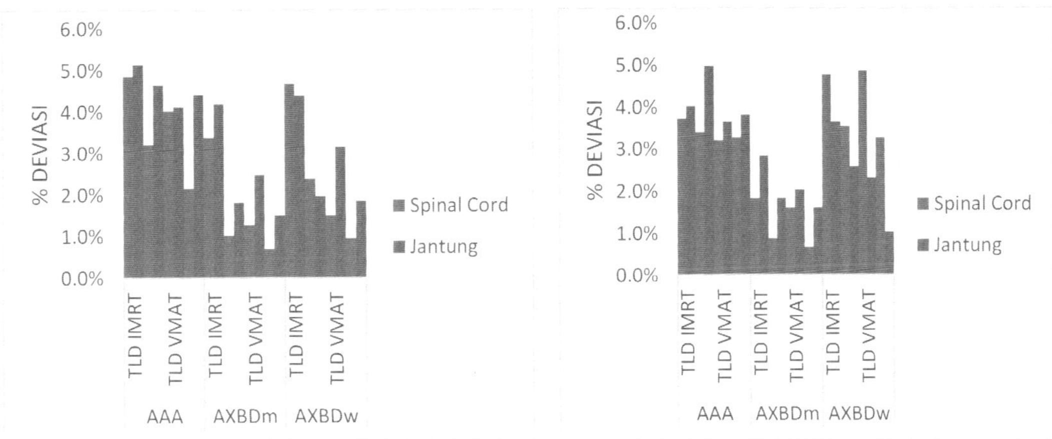

Figure 3. Deviation in OAR: (a) left lung ; (b). right lung Anisotropic Analytical Algorithm (AAA) and Acuros XB IMRT and VMAT techniques.

4. Conclusion

Verification of PTV and OAR doses in both left and right lung cancers indicated that VMAT technique tend had a smaller mean deviation than the IMRT technique using either base on TLD or film dosimeters. Furthermore, the dose of at-risk organs in the left and right lung did not much different either the Acuros XB and the Anisotropic Analytical Algorithm (AAA) algorithms with the largest deviation range obtained by heart ± 9%.

Acknowledgment

We thanks to management of MRCCC for their support in this works. This study was supported by HIBAH PITTA University of Indonesia with contract number 688/UN2.R3.1/HKP.05.00/2017.

References

[1] WHO 2012 *All Cancers (excluding non-melanoma skin cancer) Estimated Incidence* Mortality and Prevalence Worldwide
[2] Bott R 2014 *Igarss* **1** 1–5
[3] Zwan B J Barnes M P Fuangro T Stanton C J O'Connor D J Keall P J and Greer P B 2016 *J. of Appl.Clinical Med. Phys.* **17(5)** 348–65
[4] Lloyd S A M 2011 *Evaluation of a deterministic Boltzmann solver for radiation therapy dose calculations involving high-density hip prostheses*
[5] Failla G A Wareing T Archambault Y and Thompson S (n.d.). *Acuros ® XB advanced dose calculation for the Eclipse ™ treatment planning system*
[6] Rana S Rogers K Pokharel S Lee T Reed D and Biggs C 2013 *J. of Cancer Therapy* **4** 138–44
[7] Kan M W K and Yu P K N 2013 *Experimental verification of the Acuros XB and AAA dose calculation adjacent to heterogeneous media for IMRT and RapidArc of nasopharygeal carcinoma* **40** 1–20
[8] Ezzell G A *et al* 2009 *Med Phys.* **36(11)** 5359-73
[9] Defraene G Crijns W Vanstraelen B and Van D H F 2013 *Radiotherapy and Oncology* **106** S108
[10] Rana S Rogers K Pokharel S and Cheng C 2014 *J. of Appl. Clinical Med. Phys. /American College of Medical Physics* **15(1)** 4474
[11] Fitriandini A Wibowo W E and Pawiro S A *J. of Physics: Conf. Series* 694 012006

Point dose verification in breast cancer radiotherapy: comparison between 3D conformal and IMRT technique

F Manulang[1], F Y Dimitri[1,2], Nelly[2], Y Wahyono[2], R Susworo[2], S A Pawiro[1*]

[1] Department of Physics, Faculty of Mathematics and Natural Sciences, University of Indonesia, Depok, Indonesia
[2] Department of Radiotherapy, Siloam Hospitals Simatupang, Jakarta, Indonesia

supriyanto.p@sci.ui.ac.id

Abstract. The aim of this work was to verify multi point dose of the 3D-CRT and IMRT in the case of breast cancer. The experiment was perfomed at Siloam Hospital using Linac Trilogy and TPS Eclipse ver. 11. For measurements, Gafchromic film EBT3 was employed in this work in which the EBT3 film was distributed at 5 different points (trachea, axilla, and mammari interna, nipple, mamme inferior). Then, the statistical comparison between the dose measurement between 3D-CRT and IMRT techniques was evaluated. Moreover, the comparison of dose measurement between the non mastectomy and mastectomy group was also carried out. The result of differences in dose shows the trachea dose in the 3D-CRT is smaller than the IMRT in both the non mastectomy and mastectomy patients. The dose at the axilla, internal mammary, nipple, and inferior mammae are greater in the 3D-CRT than IMRT techniques in both non mastectomy and mastectomy patients. The dosage difference between TPS and EBT3 film in the range of from 0.66% to 5.67% (3D-CRT) and -0.67% to -7.57% (IMRT). The statistical test showed there was no difference of mean dose (sig 2 tailed > 0,025) between non mastectomy and mastectomy patients group and 3D-CRT and IMRT.

1. Introduction

Breast cancer is the most common cancer found among women worldwide in addition to cervical cancer. Based on Globocan estimation, the International Agency for Research on Cancer (IARC) in 2012 stated that 40 breast cancer incidence per 100,000 women in the world [1]. Radiotherapy is very common technique which is used after surgery for most women diagnosed with early-stage breast cancer [2]. The purpose of radiotherapy is to kill the tumor cell as target volume and minimize the risk of complications in healthy organs. Radiotherapy has the effect of reducing 70% the locoregional region and raising the cure rate by 85-90% [3]. Conventional 3D-CRT techniques have become one of the breast cancer treatment solutions by using tangential beam technique to achieve low doses in the lung and heart areas [4]. A number of studies on 3D homogenous dose distribution for tangential irradiation in the breast was depend on the variation of the output beam, the photon energy and the use of the wedge. Along with technological developments, higher conformal radiotherapy techniques such as Intensity Modulated Radiotherapy (IMRT) has been used for the treatment of breast cancer. IMRT employ a number of radiation field by utilizing a modulated radiation beam with a computer system, Multileaf Collimator

(MLC), and using a fixed gantry angle so that the dose distribution and local tumor can be improved by adjusting the external beam into a specific tumor.

Several study have been conducted to verify the radiation dose distribution in the case of breast cancer radiotherapy. Baird's research explains that the verification of the radiation dose distribution in the midplane phantom of the breast using radiographic film and the results indicates 3% accuracy for dose distribution [5]. Liu also explained that the 3D-CRT technique has a good dose for contralateral lung and liver (P <0.05) and IMRT techniques for both contralateral breast and IMRT have better conformity [6].

In order to elaborate the previous work, this study was to verify multi point dose of the 3D-CRT and IMRT for non mastectomy and mastectomy breast cancer patients.

2. Experimental method

This study has evaluated and measured the dose distribution of 31 breast cancer patients for non mastectomy and mastectomy and has been endorsed by clinical ethics commitee No. 17-04-0396. We have evaluated 25 patients who were treated with 3D-CRT irradiation technique and 6 patients who treated with IMRT irradiation technique. The experiment was perfomed at Siloam Hospital using Linac Trilogy and TPS Eclipse ver. 11. For the measurements, Gafchromic film EBT3 were employed in this work.

Prior to the measurement, EBT3 films with size of size 4 × 4 cm^2 was calibrated using 6 MV photon beams and radiation field size of 10 x 10 cm^2 with dose variation in the range of 0 to - 260 cGy. The irradiated EBT3 film was scanned using an Epson V700 flatbed scanner to evaluate the pixel value. The calculation and plot between netOD and dose was used to obtain the calibration curve which will be used to calculate radiation dose measurement for patients. Figure 1 shows EBT3 film placed horizontally on a phantom slab with an area of 10 cm × 10 cm irradiation and SAD 100 cm.

Before dispensing treatment to the patient, a computerized dose planning has been performed using Eclipse planning system ver.11. The dose planning used a criteria based on ICRU no.50 which is covered at least 95% by the PTV and 100% of GTV which receives 95% of prescription dose, and no more than 5% PTV will receive 107% of the prescription dose. In IMRT technique, the planning has been performed using inverse planning dose calculation mode and 7 field irradiation wheras the 3D-CRT technique used the calculation mode of supraclav and chestwall dose and 4 field irradiation.

Prescribed dose on IMRT and 3D-CRT techniques were 5000 cGy for 25 fractions (200 cGy per fraction) and there were also patients with 4160 cGy prescription doses given for 16 fractions (260 cGy). For dose verification measurements with IMRT and 3D-CRT, EBT3 film (2 cm × 2 cm) were distributed at 5 points (trachea, axilla, and mammari interna, nipple, mamme inferior) as shown in Figure 2(a) and attached to the mask used by the patient during the irradiation as shown in Figure 2(b).

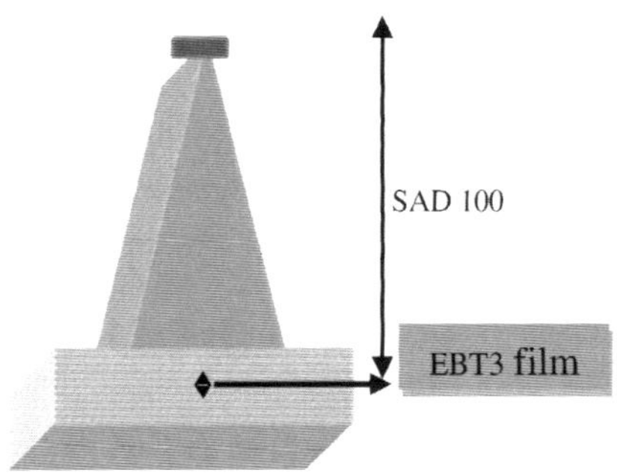

Figure 1. Illustration of calibration of EBT3 film.

An analysis of EBT3 film dosimeters that have been scanned and stored in digital form using imageJ. In order to calculate point dose, calibration curve were plotted between netOD and dose. NetOD were calculated using following the equation [7]:

$$netOD = log_{10}\left(\frac{PV_{un}}{PV_{ex}}\right) \quad (1)$$

where PV_{un} is the average pixel value of film before exposure and PV_{ex} is the average pixel value of the film after exposure.

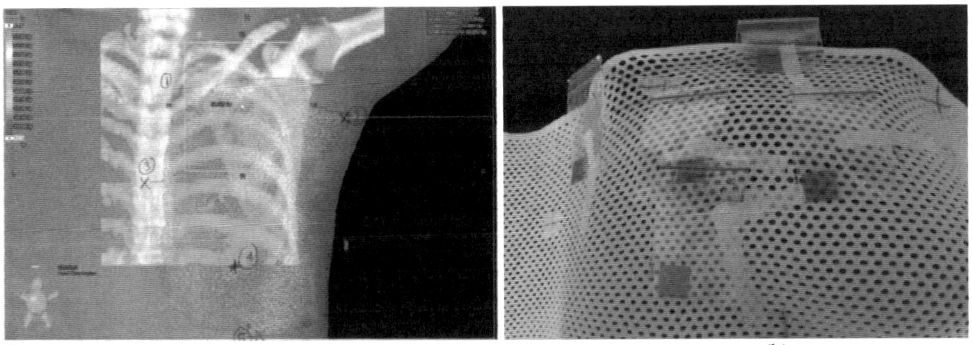

(a) (b)

Figure 2. (a) Illustration of detector placement at 5 points in the patient (b) illustration of detector placement on the mask of the patient.

Evaluation of dosimetric measurement data (D_{meas}) and TPS calculation (D_{Plan}) refers to AAPM recommendation TG 119, that the amount of dose difference (Δ%) between doses of TPS planning results and measurement data is determined by equation:

$$(\Delta\,\%) = 100\,\mathrm{x}\left(\frac{D_{meas} - D_{plan}}{D_{prescription}}\right) \quad (2)$$

where D_{meas} is radiation dose measurement , D_{plan} is the dose value of TPS , and $D_{prescription}$ is the dose prescription . Furthermore, the statistical analysis has been conducted using SPSS 15.00 between the group before and after breast surgery for 3DCRT and IMRT techniques.

3. Results and discussion

The EBT3 films were scanned and saved in .TIFF format and then read on the red channel using ImageJ software to obtain pixel values. The average pixel value was obtained by making the ROI at the center of the image of EBT3 film, it was done to avoid any impact due to cutting on the edge of the film. Furthermore, it is also assumed that the more homogeneous dosage value is at the center of the radiation beam. The dose value and the netOD value are plotted and obtained by linear interpolation technique as shown in Figure 3. From the curve, the authors interpolated between netOD and dose utilizing the following equation,

$$Y = -16493X^{3} + 7957X^{2} + 296.9X + 5.667 \quad (3)$$

where Y is the dose value (cGy) and X is the netOD value.

Point dose measurements between 3D-CRT and IMRT techniques are shown in table 1. Table 1 explaines that the dose difference of PTV50 between EBT3 film measurement and the calculated TPS on 3D-CRT technique was not significant. The smallest dose difference was at the axillary point of 0.30% whereas the largest was athe inferior mammary point of 3.51%. Furthermore, the smallest dose difference was found at the nipple point of 1.12% while the largest was at 8.81% of the internal mammary point for the measurement of EBT3 films via the IMRT techniques.

Comparative analysis point dose results was done by grouping patients based on non mastectomy and mastectomy. In 3D-CRT technique there are 20 mastectomy patients and 5 non mastectomy patients. For the IMRT technique evaluation, four mastectomy patients and two non mastectomy patients were selected. Table 2 and Table 3 shows point dose measurements of film EBT3 using the 3D-CRT and IMRT technique in the patient group non mastectomy and mastectomy. In the group of non mastectomy, the highest percentage difference between the planning dose and the measurement dose was at the trachea point of -2.56% (3D-CRT) and at the inferior mammary point of -11.80% (IMRT) while the lowest was at the nipple point of -0.22% (3D-CRT) and -1.06% (IMRT). Similarly, in the group of mastectomy, the highest percentage difference was at the trachea point of 12.60% (3D-CRT) and at mammaria interna point at -4.85% (IMRT) and the lowest at the nipple point of 1.51% (3D-CRT) and -0.23% (IMRT). Table 4 shows the comparison of mean dose between 3D-CRT and IMRT techniques. In the IMRT procedure, the trachea received a higher dose in comparison with same area with 3D-CRT. In contrast, the axillary, internal mammary, nipple, and inferior mamme point, the 3D-CRT technique received a higher dose than the IMRT technique for both the group of non-mastectomy and mastectomy.

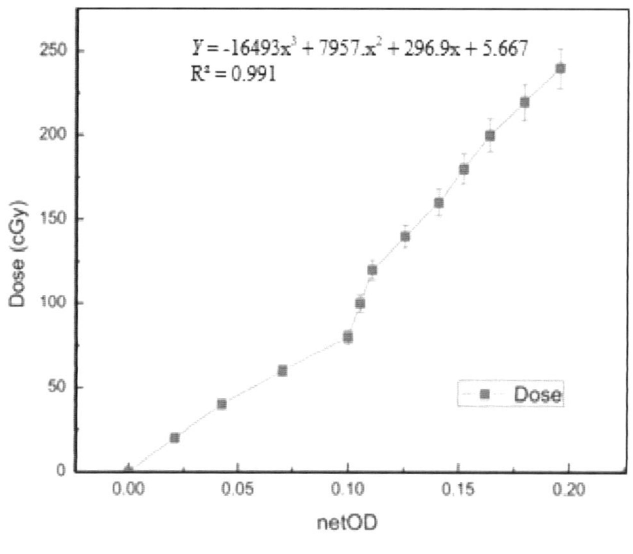

Figure 3. Curve of the relationship between dose values and netOD values of EBT3 film calibration.

The result showed the trachea dose in the 3D-CRT is smaller than the IMRT in both patients non mastectomy and mastectomy. The dose at the axilla, internal mammary, nipple, and inferior mammae are greater in the 3D-CRT than IMRT techniques in both non mastectomy and mastectomy patients. Therefore, using the 3D-CRT is a better treatment option than IMRT to protect trachea as OAR but IMRT is a better treatment option for multi point target tumor of breast cancer than 3D-CRT technique. The percentage of dose difference was relatively higher when compared to the study of Selvaraj [8] and Andrea. In Andrea's study [9], the mean dose of trachea was 48 cGy in both 3D-CRT and IMRT techniques, while the results showed the dose in the range of 40.33 cGy to 68.90 cGy. Although the trachea dose as OAR in this study was higher than Andrea's work, the dose was still within the tolerance limit in accordance with the ICRU report 83 in which the acceptable dose limit of trachea for the average dose was 5 Gy. However, the target is still below tolerance limit which is stated in the ICRU recommendation.

Table 1. Comparison of dose point measurements in 3D-CRT and IMRT techniques of EBT3 film.

	Patient							
	3D-CRT				IMRT			
Points	Measurement (cGy)	TPS (cGy)	Diff*	Error (%)	Measurement (cGy)	TPS (cGy)	Diff*	Error (%)
Trachea	34.21	35.40	1.19	3.36	63.03	66.7	3.67	5.50
Axilla	118.74	119.10	0.36	0.30	155.11	157.3	2.19	1.39
MI	22.48	22.20	-0.28	-1.24	33.01	36.2	3.19	8.81
Nipple	141.91	140.90	-1.01	-0.72	117.70	116.4	-1.30	-1.12
MIn	47.86	49.60	1.74	3.51	27.86	29.4	1.54	5.25

*Diff = different dose between measurements and TPS
MI = *MammariaInterna*
MIn = *Mammae Inferior*

Table 2. Average point dose measurement of EBT3 film using 3D-CRT technique.

	Measurement ± SD (cGy)		TPS (cGy)		Error (%)*	
Points	Non mastectomy	Mastectomy	Mastectomy	Non mastectomy	Non mastectomy	Mastectomy
Trachea	41.90 ± 12.89	57.46 ± 30.15	43.00	51.04	-2.56	12.60
Axilla	111.72 ± 31.44	138.87 ± 30.41	112.08	145.75	-0.32	-4.72
MI	92.00 ± 45.59	95.26 ± 46.37	92.74	97.92	-0.79	-2.71
Nipple	167.20 ± 67.18	176.89 ± 31.60	167.58	174.25	-0.22	1.51
MIn	39.24 ± 17.33	78.72 ± 60.07	40.08	82.81	-2.09	-4.93

*Diff = different dose between measurements and TPS
MI = *MammariaInterna*
MIn = *Mammae Inferior*

Table 3. Average point dose measurement of EBT3 film using IMRT technique.

	Measurement ± SD (cGy)		TPS (cGy)		Error (%)*	
Points	Non mastectomy	Mastectomy	Non mastectomy	Mastectomy	Non mastectomy	Mastectomy
Trachea	78.25 ± 76.93	64.22 ± 14.27	80.15	66.25	-2.37	-3.06
Axilla	89.18 ± 21.86	114.53 ± 30.76	93.65	116.47	-4.77	-1.66
MI	95.40 ± 6.72	68.28 ± 30.43	96.95	71.76	-1.60	-4.85
Nipple	106.11 ± 3.85	96.75 ± 15.28	107.25	96.98	-1.06	-0.23
MIn	53.40 ± 5.75	60.22 ± 30.68	60.55	62.37	-11.80	-3.46

*Diff = different dose between measurements and TPS
MI = *MammariaInterna ;* MIn = *Mammae Inferior*

Table 4. Value of point average dose calculation of EBT3 film on 3D-CRT and IMRT technique.

Technique	Points	Measurement (cGy)		TPS (cGy)		Error (%)*	
		Non mastectomy	Mastectomy	Non mastectomy	Mastectomy	Non mastectomy	Mastectomy
		EBT3 film	EBT3 film			EBT3 film	EBT3 film
3D-CRT	**Trachea**	41.9	57.46	43	51.04	-2.56	12.6
	Axilla	111.72	138.87	112.08	145.75	-0.32	-4.72
	MI	92	95.26	92.74	97.92	-0.79	-2.71
	Nipple	167.2	176.89	167.58	174.25	-0.22	1.51
	MIn	39.24	78.72	40.08	82.81	-2.09	-4.93
IMRT	**Trachea**	78.25	64.22	80.15	66.25	-2.37	-3.06
	Axilla	89.18	114.53	93.65	116.47	-4.77	-1.66
	MI	95.4	68.28	96.95	71.76	-1.6	-4.85
	Nipple	106.11	96.75	107.25	96.98	-1.06	-0.23
	MIn	53.4	60.22	60.55	62.37	-11.8	-3.46

*Error for average value MI = *MammariaInterna* MIn = *Mammae Inferior*

Table 5. Point dose measurement of EBT3 film between groups of patients non mastectomy and mastectomy.

Points	Sig (2-tailed) Non mastectomy vs Mastectomy EBT3 film
Trachea	0.077
Axilla	0.317
MI	0.891
Nipple	0.357
MIn	0.934

(a) 3D-CRT

Points	Sig (2-tailed) Non mastectomy vs Mastectomy EBT3 film
Trachea	0.517
Axilla	0.370
MI	0.719
Nipple	0.551
MIn	0.623

(b) IMRT

Table 6. Point dose measurement of EBT3 film between IMRT and 3D-CRT technique.

Points	Sig (2-tailed) Non mastectomy	Mastectomy
Trachea	0.285	0.431
Axilla	0.407	0.091
MI	0.925	0.336
Nipple	0.279	0.000
MIn	0.331	0.508

MI = *MammariaInterna*
MIn = *Mammae Inferior*

The statistical analysis has been conducted using the independent sample T-test. Results presented in Table 5 (a) and 5 (b) show that there is no difference of mean dose between groups of non mastectomy and mastectomy patients based on the 3D-CRT and IMRT techniques (sig-2 tailed > 0,025). Table 6 shows that there is no difference of mean dose between the 3D-CRT and the IMRT technique using EBT3 film (sig-2 tailed > 0,025) but in the axilla and nipple points of the mastectomy group showed there is a difference of mean dose between the 3D-CRT and IMRT technique (sig 2 tailed < 0,025).

4. Conclusion

The use of EBT3 film on point dose verification of breast cancer patients was succesfully implemented in this study. The measurement dose in the group of mastectomy patient is greater than in the non mastectomy group at the point of trachea, axilla, internal mammary, nipple, and inferior mammary. The trachea dose in the 3D-CRT is smaller than the IMRT in both the non mastectomy and mastectomy patients. The doses received at the axilla, internal mammary, nipple, and inferior mammae are greater in the 3D-CRT technique in both non mastectomy and mastectomy patients.

Acknowledgment

We thanks to Siloam Hospitals Simatupang and Hibah PITTA University of Indonesia with contract number 688/UN2.R3.1/HKP.05.00/2017 to support this reseach.

References

[1] Tutt A and Yarnold J 2006 *Radiobiology of Breast Cancer* 166–178
[2] Poortmans P M *et al.* 2008 *A summ. of the results of the boost vs no boost test EORTC 22881-10882* **12** 565–70
[3] Lin T Chen Y Hossain M Li J and Ma C M 2008 *Medical physics* **35(11)** 5079–87
[4] Hall E J and Brenner D J 1991 *Int. J. of Radiation Oncology Bio. Phys* **21(6)** 1403–14
[5] Baird C T Starkschall G Liu H H Buchholz T A and Hogstrom K R 2001 *J. of Appl. Clinical Med. Phys./American Coll. of Med. Phys.* **2(2)** 73–84
[6] Liu H Chen X He Z and Li J 2016 *Computerized Med. Imaging and Graph.* **54** 1–5
[7] Regulation R P Medicine N and Authority N S 2016 *Gafchromic EBT3 film dosimetry in electron beams-energy dependence and improved film read-out* **17(1)** 360–73
[8] Selvaraj R N *et al.* 2007 *Medical Dosimetry* **32(4)** 299–304
[9] Michalski A Atyeo J Cox J Rinks M Morgia M and Lamoury G 2014 *Medical Dosimetry* **39(2)**163–8

Organ at risk dose verification in nasopharyngeal cancer VMAT planning with collimator comparison between single arc and double arc

F Y Dimitri[1], F Manulang[1], Nelly[2], Y Wahyono[2], R Susworo[2], S A Pawiro[1]
[1] Physics Department, University of Indonesia, Depok, Indonesia
[2] Radiotherapy Department, Siloam Hospitals Simatupang, Jakarta, Indonesia

supriyanto.p@sci.ui.ac.id

Abstract. Volumetric modulated arc therapy (VMAT) is an irradiation technique that enables the improvement of the flexibility of dose delivery from a full range of angles with continuous modulation of beam aperture and dose rate. We compared the homogeneity index (HI), conformity index (CI), dose volume histogram (DVH) and dose differentiation of single arc and double arc VMAT delivered for nasopharyngeal cancer (NPC). Planning for gantry rotation angles of 360° counter clockwise and clockwise (double arc) with collimator (0° and 10°) and 360° clockwise (single arc) were created for ten nasopharyngeal cancer cases applied in RANDO phantom with the Eclipse treatment planning system and delivered using Trilogy linear accelerator. The verification was performed by using gafchromic film EBT3 and placed on the RANDO phantom according to the location of the organs at risk. In this study, we evaluated right parotid, left parotid, right eye, left eye and trachea. The measurement of mean dose using gafchromic film and TPS shows that the resulting values were close to TPS results with the difference being -1.51%. From the result of dose measurement using gafchromic film based on the data from RANDO CT phantom use, it can be concluded that double arc VMAT technique with collimator can reduce the dose in the organ at risk.

1. Introduction

Radiotherapy is one of cancer therapy techniques that uses radiation to kill cancer cell. It is very common combined with surgery and chemotherapy. Radiotherapy normally takes at least 1 week and up to 7 weeks depending on types and stages of the cancer. The goal of radiotherapy is providing the maximum dose in the cancer target volume and the minimal dose in the organs at risk surrounding the cancer target volume. If the dose of organs at risk exceeds the tolerance dose then it can cause diseases. For example the lens has a tolerance dose of 25 Gy during irradiation, if lens received dose exceeds the maximum dose then it can cause cataract disease. Radiotherapy for nasopharyngeal cancer has been technically challenging, requiring delivery of high doses to the target while constraining the dose organs at risk (OARs) around the target. Volumetric modulated arc therapy (VMAT) is a form of intensity-modulated arc radiotherapy which enables increased flexibility of dose delivery from a full range of angles (gantry rotation) with continuous modulation of beam aperture and dose rate [1]. VMAT can improve the dose distribution, reduce the dose to normal tissues and shorten the delivery time [2]. Quick delivery reduces the inconvenience to the patients and also increases the throughput of the linear accelerator. VMAT technique is designed so that optimized plans may be delivered efficiently in a single gantry arc, with high dose conformality using a full 360 deg range of gantry directions and accurately with high resolution sampling of beam directions during planning [3]. The VMAT technique allows for large reductions in the number of MUs and treatment delivery times. Since the VMAT technique is equally effective in producing clinically acceptable plans, but it has the advantage of sparing OARs and normal tissues, the VMAT technique may be more beneficial for NPC treatments than IMRT [4]. Dual arc VMAT for the treatment of NPC produced superior results in terms of planning target volume (PTV) coverage and organs at risk (OARs) sparing, but was slightly less efficient than single arc VMAT [5]. In this study, we would like to evaluate organ at risk verification in nasopharyngeal cancer using VMAT planning single arc and double arc with collimator comparison.

2. Experimental method

2.1. Patient selection
A total of ten simulated treatment planning of nasopharyngeal cancer cases were selected to be registered with treatment planning with RANDO phantom. The clinical stage distribution was five patients each for stage 3 and stage 4, respectively. All simulated planning have employed 3 different of VMAT techniques such as single arc, double arc and double arc with collimator.

2.2. Delineation of target volumes and organ at risk structures
In the Eclipse treatment planning system, the images based on Computerized Tomography of phantom and patient were registered. The target volumes and organ at risk structures were delineated by the radiation oncologists on the transversal slices based on registered images of phantom and patients. According to Radiation Therapy Oncology Group (RTOG) 0225 and 0615A Phase II, the criteria of Locally or Regionally Advanced Nasopharyngeal Cancer, PTV 50 (PTV 60 + Bilateral Nodule) were set up with PTV 60 (Nasopharynx + Nodule), PTV 70 (CTV Primer (CTV-P) + CTV Nodule (CTV-N) with margin 5 mm). For OARs as parotid glands, trachea and eyes, the mean dose of parotid glands < 26 Gy at least 50% of one gland will receive < 30 Gy, the maximum dose of eyes 50 Gy and mean dose of trachea < 40 Gy.

2.3. Treatment planning system
Three technique VMAT plans were optimized on the Eclipse version 11 treatment planning system (Varian Medical Systems, USA), using the 6 MV photon beam of Varian Trilogy Linear accelerators. The optimization software using Progressive Resolution Optimizer (Version 11.0.31) allows users to adjust dose constraints and priorities during optimization to achieve an optimal treatment plan which was equipped with 120 multileaf collimators (MLCs). In the upper jaws and backup jaws covering full 40 × 40 cm^2 and the variable dose rate up to 600 MU/min. CT images from CT simulator machine was transferred to treatment planning system (TPS) via Digital imaging and Communications in Medicine (DICOM) and treatment plan from TPS was transferred to ARIA via DICOM-Radiotherapy (DICOM-RT).

2.4. VMAT planning technique
VMAT plans double arc technique for gantry rotation angles of 360° counter clockwise and clockwise with collimator (0° and 10°) and 360° clockwise or counter clockwise for VMAT plan single arc. All techiques were created for ten nasopharyngeal cancer cases applied in RANDO phantom with the Eclipse treatment planning system and delivered using a Trilogy linear accelerator at Siloam Hospitals Simatupang as shown in Figure 1. Homogeneity index (HI), conformity index (CI), dose volume histogram (DVH) and dose differentiation was compared for double and single arc. The optimization was based on the Progressive Resolution Optimizer (PRO) algorithm. The iterative inverse planning process aimed to optimize simultaneously the instantaneous MLC positions, the dose rates and the gantry rotation speeds to achieve the desired dose distributions.

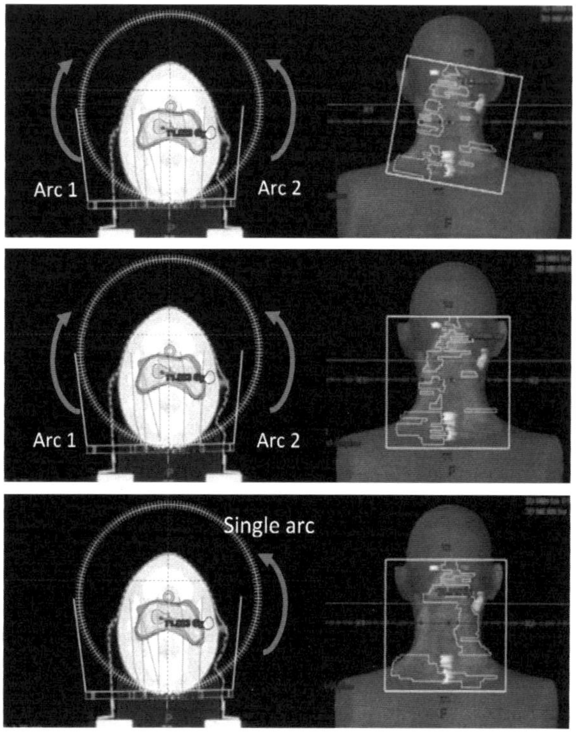

Figure 1. Three Techniques VMAT Plans (a) Double Arc with Collimator, (b) Double Arc, (c) Single Arc.

The dose was calculated using AAA version 11 after optimization. A plan normalization was selected such that 100% of coverage would be delivered to 95% the target volumes.

2.5. Plan comparison

The comparison of three VMAT plans was evaluated using the following terms :

1. Homogeneity Index (HI) : ($D_{2\%}$ - $D_{98\%}$) / $D_{50\%}$, a ratio evaluating the dose homogeneity in PTV, HI of zero indicates the dose distribution is homogeneous
2. Conformity Index (CI) : V_{pres} / PTV_p, a ratio evaluating the coverage of the prescription dose in treatment plans, CI of one indicates the good dose conformity
3. Dose Volume Histogram Organ at Risk : The mean dose of parotid glands < 26 Gy at least 50% of one gland will receive < 30 Gy, the maximum dose of eyes 50 Gy and mean dose of trachea < 40 Gy
4. Dose Differentiation between dose volume histogram and absorbed dose in gafchromic film.

3. Measurement

A Gafchromic film EBT3 was calibrated with size 4 × 4 cm^2 and irradiated using 6 MV photon beams with field size of 10 × 10 cm^2 in dose variation in range of 0 - 200 cGy. The irradiated gafrchromic film EBT3 was scanned using Epson V700 scanner to evaluate the pixel value. Figure 2(a) shows gafchromic film EBT3 inserted horizontally on a slab phantom and irradiated with field size of 10 × 10 cm^2 at SAD 100 cm. The prescribed dose were setup at 70 Gy for 33 fractions (2.121 Gy per fraction). For dose verification measurements with VMAT, the gafchromic film EBT3 3 × 3 cm^2 were distributed at 5 points (right parotid, left parotid, right eye, left eye and trachea) and attached at the mask used by the RANDO phantom during the irradiation as shown in Figure 2(b).

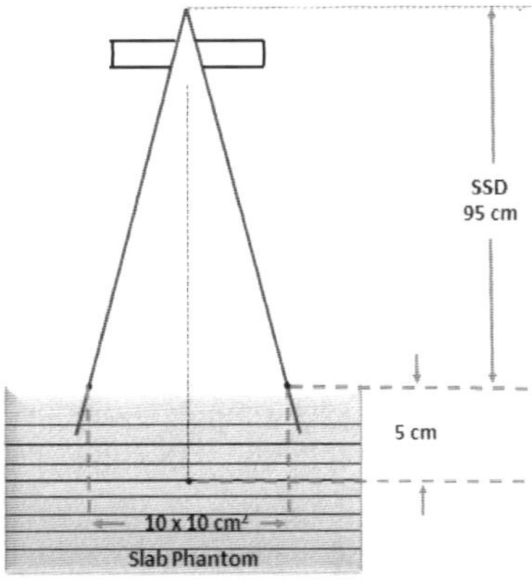

(a)

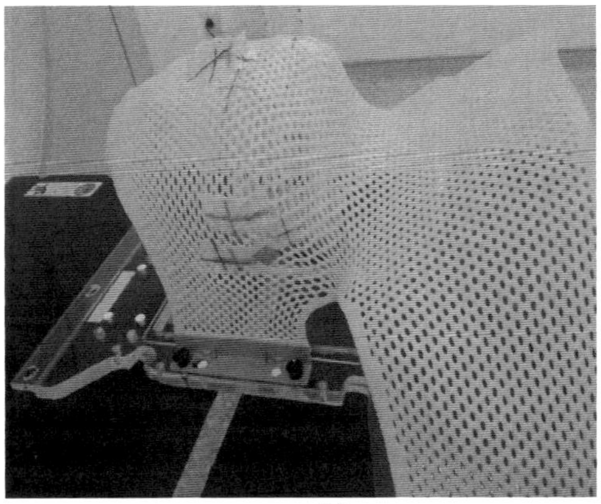
(b)

Figure 2. (a) Gafchromic film EBT3 calibration, (b) gafchromic film EBT3 placement.

4. Result and discussion

The coverage of PTVs of three technique VMAT plans single arc (SA), double arc (DA) and double arc with collimator (DAC) were evaluated by homogeneity index, conformity index, dose volume histogram and dose differentiation. And we irradiated gafchromic film EBT3 and will be compared with dose calculated in TPS. The gafchromic film EBT3 scanning results are obtained in .TIFF format and then read on the red channel using ImageJ software to obtain pixel values. The average pixel value was obtained by making the region of interest (ROI) at the center of the gafchromic film EBT3 images to avoid any impact due to cutting on the edge of the film. In addition, it will get more homogeneous doses value at the center of the radiation beam. The dose value and the netOD value are plotted, a radiation dose calibration curve will be obtained by linear interpolation technique. From the curve, we obtained the relation between netOD and dose as following equation :

$$Y= -16493X^3 + 7957X^2 + 296.9X + 5.667$$

where Y is the dose value (cGy) and X is the netOD value.

4.1. Homogeneity index and conformity index

Homogeneity index and conformity index of three VMAT plans for each PTV are shown in Table 1. Based on the evaluation of PTV54, PTV59,4 and PTV70, VMAT plans double arc with collimator achieved better homogeneity index and conformity index than VMAT plans using single arc and double arc. Homogeneity index for PTV 54 (0.33 ± 0.09), PTV 59.4 (0.23 ± 0.03) and PTV 70 (0.06 ± 0.01) indicates the dose distribution is homogenous with reference value of 0. On the other hand, the conformity index (0.98 ± 0.08) is close to one indicates the good dose conformity. From the data, we found the optimum value of the homogeneity index and conformity index for VMAT plans double arc with collimator.

4.2. Organ at risk dose volume histogram

The Dose Volume Histogram (DVH) comparison of organ at risk for VMAT plans, double arc with collimator, double arc and single arc. The average coverage of 95% of the PTV70 was 100% of the prescribed dose for all techniques VMAT plans. Dose volume histograms for the organ at risk are presented in Table 2. The Table 2 indicated the double arc with collimator VMAT received the lowest dose at all organ at risk. This result supported Ning *et al.*, study that in the single arc technique, all of organ at risk receive a higher dose than another techniques. When single arc VMAT was used in the treatment of NPC, smaller and faster MLCs were sometimes needed to meet the plan optimization requirements [6]. The advantages of using double arc were an increase in the modulation factor during optimization and the provision of more MLC control points than single arc, achieving better dose distribution [7]. The previous results on the maximum dose reported by Verbakel *et al.* [8] showed that parotid dose lower with double arc VMAT (by average 2 Gy) compared with single arc VMAT. Parotid sparing was better with the double arc collimator VMAT plans, with the (average) mean dose of 22.11 Gy (range 20.76-23.65 Gy) for left parotid and 21.49 Gy (range 20.59-22.10 Gy) for right parotid. Trachea doses for double arc and single arc VMAT plans were higher and not acceptable because the trachea received >40 Gy compared with double arc collimator VMAT plan, with a mean dose of (40.61 Gy and 43.87 Gy) versus 39.63 Gy. The left and right eyes received <50 Gy in all techniques VMAT plans. Double arc with collimator VMAT plan led to superior results of dose coverage and spared OARs when compared with single arc and double arc VMAT, without sacrificing the delivery efficiency. The collimator was rotated 10° to cover the entire tumour and to reduce the tongue and groove effect.

Table 1. *Homogeneity index* dan *conformity index* based on CT RANDO.

	Homogeneity Index			Conformity Index		
	DA COL	DA	SA	DA COL	DA	SA
PTV 54	0.33 ± 0.09	0.34 ± 0.10	0.37 ± 0.10			
PTV 59.4	0.23 ± 0.03	0.25 ± 0.02	0.29 ± 00.4			
PTV 70	0.06 ± 0.01	0.07 ± 0.02	0.08 ± 0.02	0.98 ± 0.08	0.96 ± 0.14	0.89 ± 0.10

Table 2. DVH *organ at risk* based on CT RANDO in TPS.

	VMAT Double Arc Collimator (Gy)	VMAT Double Arc (Gy)	VMAT Single Arc (Gy)
Left Eye	28.35 ± 1.06	29.15 ± 0.99	30.55 ± 0.64
Right Eye	38.20 ± 8.49	39.80 ± 8.75	40.16 ± 9.55
Trachea	39.63 ± 0.92	40.61 ± 2.26	43.87 ± 4.05
Left Parotid	20.76 ± 1.77	21.92 ± 5.52	23.65 ± 0.21
Right Parotid	20.59 ± 4.24	21.77 ± 1.41	22.10 ± 2.40

4.3. Organ at risk dose at two clinical stage

The dose differences between stage 3 and stage 4 based on registered CT RANDO phantom to patient are tabulated in Table 3. The classification of stage 3 are $T_{1\text{-}2}$ N_2 M_0 (nasopharynx, oropharynx, nasal cavity, parapharyneal extension, medial and lateral pterygoid muscles ($T_{1\text{-}2}$), bilateral level I_b, II, III and V_a involvement, or the maximum diameter >3 cm, or with extranodal neoplastic spread (N_2) and no distant metastasis (M_0)), T_3 $N_{0\text{-}2}$ M_0 (Bone structures and/or paranasal sinuses (T_3), unilateral or bilateral retropharyngeal node (s), unilateral level Ib, II, III, and Va involvement and the maximum diameter ≤ 3 cm, bilateral level I_b, II, III and V_a involvement, or the maximum diameter >3 cm, or with extranodal neoplastic spread ($N_{0\text{-}2}$) and no distant metastasis (M_0)). For the classification of stage 4 are $T_{1\text{-}3}$ N_3 M_0 (nasopharynx, oropharynx, nasal cavity, parapharyneal extension, medial and lateral pterygoid muscles, bone structures and/or paranasal sinuses ($T_{1\text{-}3}$), level IV and V_b involvement (N_3) and no distant metastasis (M_0)), T_4 $N_{0\text{-}3}$ M_0 (intracranial extension and/or cranial nerves, infratemporal fossa hypopharynx (T_4), unilateral or bilateral retropharyngeal node (s), unilateral level I_b, II, III and V_a involvement and the maximum diameter ≤ 3 cm, bilateral level I_b, II, III and V_a involvement, or the maximum diameter >3 cm, or with extranodal neoplastic spread level IV and V_b involvement ($N_{0\text{-}3}$) and no distant metastasis (M_0)). The doses were within the acceptable limit. Both the parotid glands, eyes and trachea had better sparing in double arc with collimator than double arc and single arc. The volume of normal tissue dose received in double arc with collimator is less than double arc and single arc, which will reduce late effect of low dose radiation (radiation-induced carcinomas). The measurement of dose differentiation using gafchromic film EBT3 and TPS shows that the gafchromic film detector values around of -1.51% in right eye with VMAT plan double arc with collimator.

Table 3. Average dose comparison with 3 VMAT techniques for each stage based on CT RANDO.

		Measurement (cGy) ± SD				Diff (%)	
		Stage 3		Stage 4		Stage 3	Stage 4
		Film	TPS	Film	TPS		
DAC	LE	18.34 ± 1.60	19.90	20.13 ± 0.89	21.40	7.85	5.93
	RE	21.42 ± 1.81	21.10	31.92 ± 0.59	33.10	-1.51	3.56
	T	31.68 ± 0.46	32.50	34.79 ± 1.11	33.80	2.53	-2.91
	LP	26.68 ± 0.70	28.20	32.50 ± 1.41	30.70	5.38	-5.86
	RP	22.54 ± 4.48	20.50	44.67 ± 1.79	46.50	-6.70	3.95
DA	LE	22.21 ± 2.75	23.30	23.52 ± 1.81	24.70	4.70	4.79
	RE	23.01 ± 0.78	22.40	35.90 ± 1.27	37.60	-2.73	4.51
	T	32.86 ± 1.19	34.10	35.53 ± 1.36	37.30	3.62	-6.87
	LP	26.76 ± 1.06	26.10	33.49 ± 0.94	33.90	-2.53	-4.82
	RP	23.71 ± 1.08	24.30	45.52 ± 1.98	50.30	2.42	9.51
SA	LE	31.70 ± 6.52	32.70	29.87 ± 1.26	31.80	3.06	6.07
	RE	27.08 ± 0.91	28.80	40.88 ± 0.76	42.30	5.97	3.36
	T	34.71 ± 2.23	33.30	51.27 ± 2.08	53.10	-4.24	3.45
	LP	33.26 ± 2.60	34.30	36.97 ± 1.09	34.60	3.02	-6.86
	RP	24.95 ± 0.93	26.30	47.76 ± 0.95	49.70	5.13	3.91

5. Conclusion

This study compared VMAT plans double arc with collimator, double arc and single arc for 10 base of nasopharyngeal cases. The measurement ratio of mean dose using gafchromic film EBT3 and TPS shows that the values around of -1.51%. The result of dose measurement using gafchromic film EBT3 based on the registered data of CT RANDO to patient described that double arc VMAT technique with collimator for NPC cases reduced the dose in the organ at risk.

Acknowledgments

We thanks to Siloam Hospitals Simatupang and Hibah PITTA University of Indonesia with contract number 688/UN2.R3.1/HKP.05.00/2017 for support this research.

References

[1] Johnston M Clifford S Bromley R Back M Oliver L Eade T 2011 *Clinical Oncology* **23(8)** 503-11
[2] Teoh M Clark C H Wood K Whitaker S Nisbet A 2011 *Br J Radiol* **84** 967–96
[3] Otto K 2008 *Medical physics* **35(1)** 310-7
[4] White P Chan KC Cheng KW Chan KY Chau MC 2013 *J. of Radiation Res.* **54(3)** 532–45
[5] Lee TF Ting HM Chao PJ Fang FM 2012 *Clin Oncol (R Coll Radiol)* **24** 196–207

IOP Conf. Series: Materials Science and Engineering **432** (2018) 012024 doi:10.1088/1757-899X/432/1/012024

[6] Ning Z-H Mu J-M Jin J-X *et al.* 2013 *Radiation Oncology (London, England)* **8** 237
[7] Zhao L-R Zhou Y-B Sun J-G 2016 *Oncology Letters* **11(5)** 3389-94
[8] Verbakel WF Cuijpers JP Hoffmans D *et al.* 2009 *Int. J. Radiat. Oncol. Biol. Phys.* **74** 252–9

IOP Conf. Series: Materials Science and Engineering **432** (2018) 012025 doi:10.1088/1757-899X/432/1/012025

Moderator material efficiency of neutron energy slowing down on D-T reaction neutron generator for SAMOP

S Santosa [1], A H Anggraini[2]

[1,2] Center for Accelerator Science and Technology, CAST- BATAN, Yogyakarta, Indonesia

anjar.ah@batan.go.id

Abstract. We have studied a collimator design setup for the Subcritical Assembly for Mo-99 Production (SAMOP) at the Center for Accelerator Sciences and Technology (CAST), Nuclear Energy Agency of Indonesia (BATAN). The main purpose of the project is for providing sufficient neutrons to produce Mo-99, to be used to produce Tc-99m for nuclear medical applications. The modelling of the collimator is based on Monte Carlo simulation using MCNPX version 5 software to determine the variation of the suitable materials. The collimator consists of moderator, filter, and reflector, where we expected that the output of neutron will be thermal neutron with a sufficient neutron flux of about 10^8 neutron/cm^2.s. The neutron source used is a D-T reaction neutron generator with a mono-energetic neutron energy of 14.1 MeV. The simulation has been conducted in nine variations, with each experiment uses five types of materials. The best result in terms of neutron flux is seem from the material combination of Al, $C_{25}H_{52}$, LiF, and PbF_2, which produces a thermal neutron flux of 1.24×10^7 neutron/cm^2.s and average energy deposition of 1.88×10^6 MeV/g. This research needs to be developed further for subsequent experiments to optimize the combination of materials and geometry systems to be used.

1. Introduction

The project has the main purpose of producing Mo-99, which thus is used for producing Tc-99m for nuclear medical applications [1]. The Subcritical Assembly for Mo-99 Production (SAMOP) is a new method for radioisotope production. SAMOP utilizes a solvent of uranyl nitrate salt as irradiation target for a neutron beam output. The neutrons are usually produced by nuclear reactor [2], but it also can be produced by using a neutron generator (NG), in which a D-D (deuterium-teuterium) or D-T (deuterium-tritium) reaction occurs [3]. This paper concentrates on the D-T reaction that produces monoenergetic neutrons at 14.1 MeV. In this collimator modelling study, we used an NG as a neutron source, which is considered as good for use on a SAMOP system because of it has a single energy. In a SAMOP system, NG is one of the alternative replacements of neutron sources as a neutron source. Some countries have been using D-T-/D-D-reaction NG as a compact experimental accelerator driven subcritical facility, because of its superiority in, for instance, safety and waste transmutation [3]. We use the D-T reaction NG with an energy of 14.1 MeV in our simulation study.

In Figure 1, shown is the initial design of collimator using a compact neutron generator, which consists of moderator, filter and reflector. The research was conducted to design a better moderator that produces thermal neutrons for SAMOP, for a NG neutron source. The method used is simulation using

IOP Conf. Series: Materials Science and Engineering **432** (2018) 012025 doi:10.1088/1757-899X/432/1/012025

a Monte Carlo N-Particle software package, MCNPX version 5. The simulation was done by optimizing the collimator part to produce the output in accordance with the requirements in Mo-99 production on SAMOP system. Therefore, the variation of material and geometry of the collimator is required to qualify SAMOP system in accordance with the standards of the International Atomic Energy Agency (IAEA) [4]. Part of the materials is varied in order to adjust the degradation of fast neutrons energy into the energy required in SAMOP system using D-T reaction.

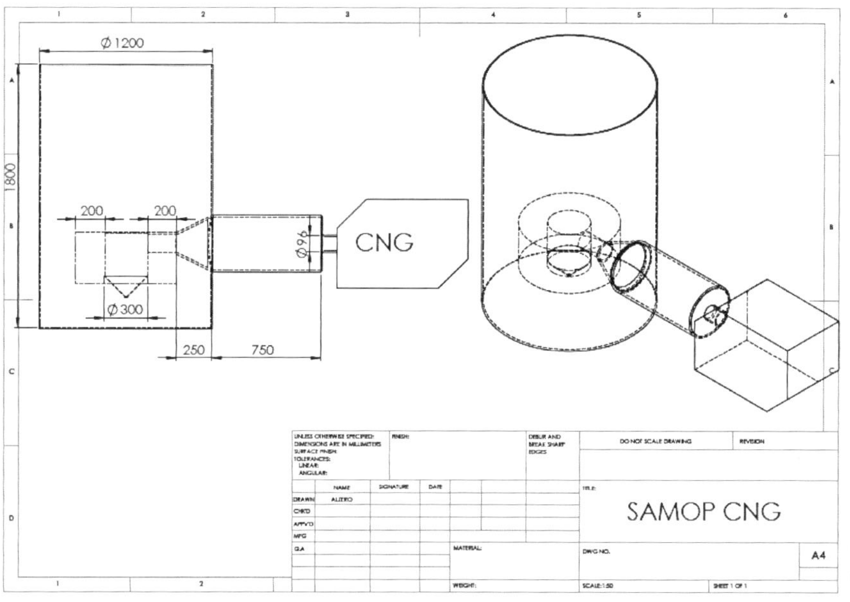

Figure 1. The initial design setup of collimator using CNG.

2. Experimental method

The design of neutron collimator using Monte Carlo N-Particle (MNCPX) method is performed by using cell cards to determine the variation of the materials. It uses a D-T reaction NG with neutron energy of 14.1 MeV. The material on the collimator positioned between the SAMOP system and the NG is divided into five main parts: cell 18, cell 15, cell 12, cell 20, and cell 6, which are moderator, neutron filter, reflector, collimator, and reflector, respectively.

2.1. Moderator

Moderator is a part of collimator that used to reduce the energy of fast neutrons (± 2 MeV) into normal reactor components (± 0.02 to 0.04 eV). As the D-T reaction of NG produces neutron energy of 14.1 MeV, the moderator material must be choosen in accordance with the requirement that it can eliminate most of the fast neutron energy in each collision and small ability to absorb neutrons, as well as a great ability to dissipate neutrons. Materials in this experiment are aluminum (Al), aluminum oxide (Al_2O_3) [4], and paraffin ($C_{25}H_{52}$) [5].

2.2. Filter

Filters are expected to reduce contamination of gamma rays and fast neutron current. The gamma rays could pose high hazard on the human body, so the presence of filters in such a collimator to reduce the radiation of gamma rays is vital.

Fast neutrons can also cause harm to the human body. In general, since the source of neutron has high energy neutron, materials that can reduce the energy is required. The fast neutrons must not be captured because of the intensity of irradiation will be lowered, but the fast neutrons must be moderated in order to keep the high intensity. The filter is normally a material with a small atomic mass number. However, the material must also have a small neutron absorption cross-section, a large scattering cross-section, and a good heat conductivity. Materials used in this experimental setup of simulation are aluminium fluoride (AlF_3), lithium fluoride (LiF), and paraffin ($C_{25}H_{52}$) [6].

2.3. Reflector
The reflector has a function to keep neutrons in the collimator. The reflector material has a scattering cross-sectional requirement of a relatively large atomic mass. The suitable materials that are used in this section of simulation are Pb and PbF_2.

The energy output generated from each material is given in the average flux passing through each cell variation. There are three main parts in the observation of the average flux of neutrons in cell 20 and surface 47, so that the average flux for thermal, epithermal and fast neutrons is known. In addition, the cell 20 is also known deposition energy of the protons and neutrons that have units of MeV/g. In the next section of surface 47, the average flux for protons is obtained.

3. Modelling of collimator geometry
Our collimator design utilized the MCNPX Monte Carlo simulation software. In general, there are two methods: shift and moderator methods. The shift method aims to reduce the flux of neutrons made by moderator materials only, while the moderator method performs neutron flux reduction using moderator component in the collimator. In this simulation, we use both methods for such a simulation.

As we have mentioned, the SAMOP collimator design uses NG as its neutron source, in which the D-T reaction is used, for an output energy of 14.1 MeV. The design requires particular techniques in the process to conform to the requirements of the SAMOP system. Material for such a collimator is varied, so that a large initial energy can be reduced. Figure 2 shows the material variation of the SAMOP collimator design. The design of the collimator is given a variety of materials in such a way on each part of the collimator it is done in order to achieve the purpose of SAMOP system requirement.

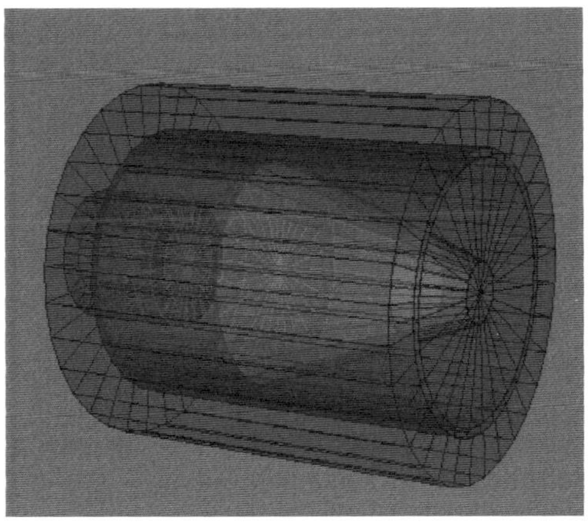

Figure 2. Collimator design geomety on MCNPX software.

4. Results and discussion
The moderator material structure is used to slow down the fast neutron from the DT reactor as the neutron source for the SAMOP. This is actually a neutron generator having a tritium target bombarded by deuterium gas. The process is very important for filtering the resulting gamma rays and slowing down it fast neutrons output. The fast and epithermal neutrons must be filtered, due to the thermal neutron

energy being the only one designed to be used in the SAMOP system. Therefore, the first process of moderation is to moderate neutrons to rapidly reduce neutron energy into thermal neutrons, and then to filter the thermal neutrons in order to support the SAMOP work system and to ensure that its intensity does not pose risk for the human body. In Table 1, the variation of materials used at SAMOP collimator system are compared based on the results of the thermal neutron flux and energy output of each material combination.

Table 1. The variation of material using DT source.

	Material in cell					The average of neutron flux ($neutron/cm^2.s$) at cell 20		The average of photon flux ($particle/cm^2$) at surface 47	The average of deposition energy at cell 20 (MeV/g)
	18	15	20	12	6				
						Thermal	(1.31E+04)±0.0399		
1	Al	AlF_3	Fe	Pb	Pb	Epithermal	(4.15E+06)±0.0072	(6.48E-08)±0.1144	(7.38E-07)±0.0230
						Fast	(7.86E+06)±0.0042		
						Thermal	(2.35E+05)±0.0090		
2	Al_2O_3	AlF_3	Fe	PbF_2	PbF_2	Epithermal	(1.53E+07)±0.0038	(1.07E-07)±0.0884	(3.88E-07)±0.0301
						Fast	(8.53E+06)±0.0038		
						Thermal	(1.89E+06)±0.0464		
3	Al	LiF	Pb	PbF_2	PbF_2	Epithermal	(2.25E+07)±0.0173	(3.75E-08)±0.3569	(4.29E-07)±0.0263
						Fast	(9.77E+06)±0.0148		
						Thermal	(3.05E+06)±0.0341		
4	$C_{25}H_{52}$	LiF	Pb	PbF_2	PbF_2	Epithermal	(2.32E+06)±0.0534	(8.88E-09)±0.7014	(4.19E-07)±0.0250
						Fast	(2.18E+06)±0.0447		
						Thermal	(4.20E+06)±0.0252		
5	Al_2O_3	LiF	Fe	PbF_2	PbF_2	Epithermal	(4.63E+06)±0.0305	(1.33E-06)±0.0747	(3.88E-07)±0.0301
						Fast	(2.66E+06)±0.0311		
						Thermal	(5.60E+06)±0.0065		
6	Al_2O_3	AlF_3	LiF	PbF_2	PbF_2	Epithermal	(2.03E+07)±0.0033	(1.07E-07)±0.0678	(4.19E-07)±0.0250
						Fast	(5.25E+06)±0.0036		
						Thermal	(9.45E+06)±0.0201		
7	Al	$C_{25}H_{52}$	Pb	PbF_2	PbF_2	Epithermal	(9.27E+06)±0.0241	(3.12E-08)±0.3818	(7.38E-07)±0.0230
						Fast	(5.26E+06)±0.0257		
						Thermal	(1.15E+07)±0.0043		
8	Al	$C_{25}H_{52}$	AlF_3	PbF_2	PbF_2	Epithermal	(1.13E+07)±0.0040	(1.62E-07)±0.0643	(1.88E-06)±0.0187
						Fast	(3.10E+06)±0.0046		
						Thermal	(1.24E+07)±0.0183		
9	Al	$C_{25}H_{52}$	LiF	PbF_2	PbF_2	Epithermal	(8.1E+06)±0.0203	(1.87E-07)±0.4427	(1.88E-06)±0.0187
						Fast	(2.63E+06)±0.0148		

Preparation of the moderator materials is required to reduce the fast and epithermal neutrons fluxes, while constantly prevent the epithermal neutrons slightly absorbed. The characteristic of material section is having a high absorption cross section and a high scattering cross section for fast neutrons and epithermal neutrons, as presented in Table 2. In this research, several moderator materials, namely $C_{25}H_{52}$, Al, and Al_2O_3, have been simulated. The effective material value is of Al as it is adequate to produce high thermal neutron flux, as shown in Table 1. Aluminum has a great inelastic scattering cross section to the neutron energy of 1 MeV and fast neutrons, and it has a low absorption at cross section for thermal neutron in particular energies. The suitable material characteristic for moderator is a material with small atomic mass, which thus produces inelastic collisions. In the process, half of the neutron energy can be transferred into atomic nucleus and will be smaller at the collision.

The filter material was selected based several characteristics, namely material's cross section using quantity of neutrons that interact with the atomic nucleus in the surface. The cross section is divided into two parts: microscopic and macroscopic. The microscopic cross-section is sum of two parts, the collision cross section and the absorption cross section. The collision cross section is the sum of of elastic collision and inelastic collision cross sections. The cross section is divided to fission and radiation capture cross sections. Elastic collisions occur due to a collision between fast neutrons and low mass of atomic nucleus. Inelastic collisions occur between neutrons and high atomic number atoms.

Neutrons will provide most of the energy in the atomic nucleus material which the core material will be in the excited state and γ-ray radiation in the inelastic collision. Excessive γ radiation should be avoided as it has a significant effect on the production of Mo-99 on SAMOP systems.

The higher flux of thermal neutron is the combination of collimator materials number nine, it consists of Al, $C_{25}H_{52}$, LiF, and PbF_2 of 1.24×10^7 n/cm^2.s. The selected material on the filter is $C_{25}H_{52}$ (paraffin) which produces thermal neutrons and neutrons faster than epithermal neutrons. Figure 3 shows that paraffin (filter in material number nine) can prevent the reduction of neutrons during the moderation process so that the expected number of neutrons remains stable [8]. In addition to paraffin, AlF_3 and LiF (shown in Table 1) are also good candidates for filling material in the filter section, but the materials produce higher epithermal flux mean in material number three and six (as shown in Figure 3). This causes the precedence of paraffin in filling the filter parts. Paraffin ($C_{25}H_{22}$) also has a low atomic number because it contains more hydrogen atoms than carbon atoms, so this material is more effective for lowering thermal energy compared to AlF_3 and LiF. In Figure 3, it shown that the combination of collimator materials number one, i.e., Al, AlF_3, Fe, and Pb, produces the lowest thermal neutron flux of 1.31×10^4 n/cm^2.s.

Table 2. Material characteristic based on neutron variety [7].

Neutron variety	Energy range	Material characteristics
Thermal	0 – 5 eV	High absorption cross section and high scatter cross section
Epithermal	5 eV – 10 KeV	Low absorption cross section and low scatter cross section
Fast	10 KeV – 20 MeV	Low absorption cross section and high scatter cross section

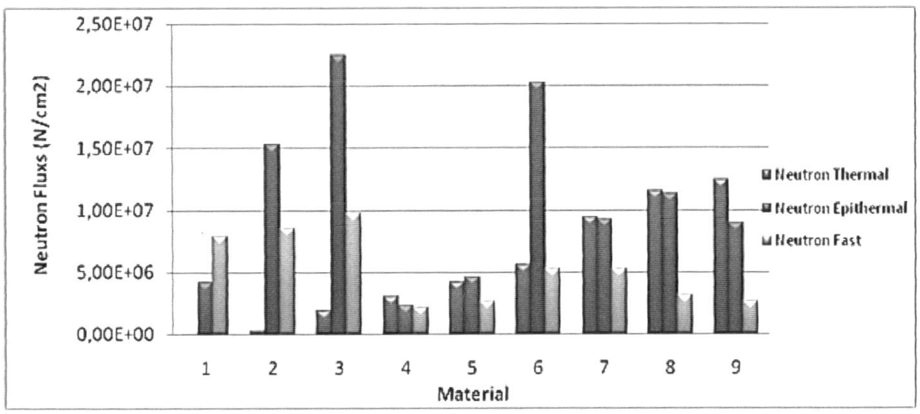

Figure 3. Neutron flux at material sample.

The material characteristics of the reflector section should have a high neutron scattering cross-section. The materials used in this section are Pb and PbF_2. In this research, the DT neutron source is directed to the material so that it can filter fast neutrons and gamma rays (photons) in the composition of the reflector material with PbF_2 material. The reflector part easily resonates where elastic scattering occurs because neutrons react with atomic nucleus of high atomic numbers. This resonance caused largely dissipated energy neutrons.

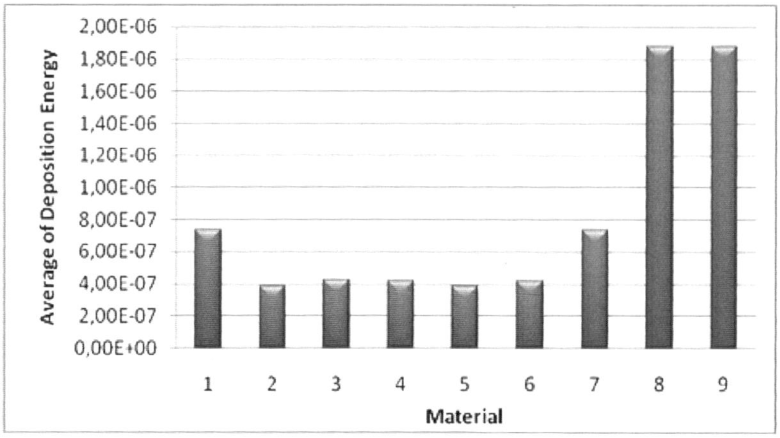

Figure 4. Average of deposition energy at the material sample.

Table 3. Energy range for neutrons and required conditions of neutron fluxes [9].

No.	Neutron	Energy	Neutron flux (n/cm^2s)
1.	Thermal	~0.5 eV	$< 5\times10^7$
2.	Epithermal	0.5 eV-10 keV	$>1.5\times10^9$
3.	Fast	10 keV	$< 10^{12}$

The best result of neutron thermal flux is 1.24×10^7 n/cm^2.s (Figure 3) and average energy deposition of 1.88×10^6 MeV/g (Figure 4). The thermal neutron flux results are still smaller than expected for SAMOP that of 10^8 neutrons/cm^2, but the result is less than the value set by the IAEA of 5×10^7 as shown in Table 3. In the other hand, the minimum acceptable value from others research measurements of thermal and epithermal neutron flux on the neutron extractor is 10^6 to 10^7 (n/cm^2.s) [10, 11].

5. Conclusion

The best result of neutron flux is produced by the material combination of Al, $C_{25}H_{52}$, LiF, dan PbF_2, which resulted in a thermal neutron flux of 1.24×10^7 neutron/cm^2.s and an average energy deposition of 1.88×10^6 MeV/g with D-T neutron source. Thus this research needs to be developed for subsequent experiments to optimize the combination of materials and geometry systems to be used.

Acknoledgment

The authors of the paper would like to thank the Head of PSTA-BATAN and Professor Syarip who have allocated research fund from INSINAS 2016 so that this research can be implemented. They are also grateful to Alzero, Made, and Airlangga University students who have helped the authors in this research. Hopefully they are all given reward by Allah SWT, Aamiin.

References

[1] Setiadipura T and Saragi E 2007 *Int. Conf. on Adv. in Nucl. Sci. and Engin.* **2007** 23–6
[2] Sakai M Fujimoto N Ishii K Murata I and Awazu K 2011 *Nucl. Sci. and Tech.* **1** 513–6
[3] Sinha A Roy T Kashyap Y Ray N Shukla M Patel T Bajpai S Sarkar P S Bishnoi S and Adhikari P S 2015 *Annals of Nucl. Energy* **75** 590–4
[4] Fauziah N 2013 *J. Tech. Reactor Nucl.* **15** 112–9
[5] Bavarnegin E Kasesaz Y and Wagner F M 2017 *J. Instrum.* **12(5)** P05005
[6] Laara S H 2014 Islam State Univ. Sunan Kalijaga (in Indonesia)
[7] de Boer J J 2008 *New Filter Design with Monte Carlo Calculation* TU Delft

[8] Gulo D P Suryasatriya T Santosa S and Sardjono Y 2015 *Indonesian J. of Appl. Phys.* **5(2)** 25–33
[9] Sato A Takizawa Y Hiraga F and Kiyanagi Y 2014 *Phys. Proc.* **60(C)** 15–22
[10] Leal A Campolina D Horizonte B and Revay Z 2011 *Proc. IAEA-CN-188-A19* 1–8
[11] Sutondo T 2015 *Ganendra J. Nucl. Sci. Technol.* **18** 107–13

IOP Conf. Series: Materials Science and Engineering **432** (2018) 012026 doi:10.1088/1757-899X/432/1/012026

Ion beam preliminary testing of DECY-13 cyclotron at the central region using dc extraction voltage

S Silakhuddin[1], R S Darmawan[1], S Suharni[1], F S Permana[1], K Wibowo[1], S Saminto[1], T Atmono[1]

[1] Center for Accelerator Science and Technology, BATAN Babarsari St., Yogyakarta 55281, Indonesia

silakh@batan.go.id

Abstract. The subsystem devices of the DECY-13 cyclotron have been installed except the RF-dee device, so a final ion beam testing has not been tested for cyclically accelerated ion beam or ion beam commissioning. An initial or preliminary test of the ion beam has been performed, for preparing the final beam test. The preliminary test was carried out by measuring ion beam current in the central region of the cyclotron. The extraction of the ion beam from ion source is done by pulling the ion beam using a DC voltage of a puller mounted on a center part of the dee. The test was carried out by observing the effect of the magnitude of the cathode voltage, magnetic field and the puller voltage on the beam current that caught in a beam probe placed at a distance of 3 mm from the puller. The results of the ion beam test show that the degree of influence from the strongest is successively from the components of puller voltage, cathode voltage and magnetic field. The obtained largest ion beam current was 126 μA at the parameters of 0.86 T magnetic field, cathode voltage of 1200 V, 50 mA cathode current, 5 sccm gas flow rate, final vacuum of 8×10^{-6} Torr and puller voltage of 2 kV. These results indicate that the initial ion beam current output qualifies for the steps toward the final test.

1. Introduction

Installation the DECY-13 cyclotron is now up to the stage where the magnet has been installed and integrated into the vacuum chamber, the Penning-type ion source also has been set in the cyclotron vacuum chamber. The cyclotron will accelerate H^- ion beam until 13 MeV energy. The ion beam then extracted to be proton beam directed a target by using a carbon foil stripper. The vacuum chamber has been tested and a pressure is obtained by a vacuum of 7-8 $\times 10^{-6}$ Torr after a gas flow of 5 sccm, and this pressure already qualifies for ion source testing [1]. According to the measurements, the magnetic field in the central region is 1.275 T [2]. Previously, the ion source has been tested in an ion source testing device located out of the cyclotron, and the resulted ion beam current was 35 μA using 8 kV extraction voltage [3]. The ion source also has already been tested the ability of the vacuum, the movement mechanism to adjust the position of the head and is equipped with a hydrogen gas supply device. A center part of dee with a puller on it has also been set in the central region, and the puller is connected to a DC voltage supply.

A final testing of the ion beam *i.e.* testing by measuring ion beam in the target will be carried out when the RF-dee device is already installed in the cyclotron vacuum chamber. In order to prepare the final test an initial testing *i.e.* testing the ion beam in the central region of the cyclotron has been performed. The status of the DECY-13 Cyclotron where the devices have already been available as mentioned above is possible to test the ion beam production in the central region. This testing is called for preliminary testing of ion beam in the cyclotron central region using DC voltage. The testing consists of observing ion beam current related to the influences of the internal performance and the beam extraction efficiency.

The internal performance of the ion source is influenced by the power of ion source cathode *i.e.* voltage and current supply in the cathode, and gas pressure that depends on gas flow rate and cyclotron magnetic field. The extraction efficiency is determined by electrical optics between the ion source head and puller, which in this case is dominantly influenced by the puller voltage. In short, the production of the ion from ion source is influenced by cathode voltage, puller voltage, gas flow rate, and intensity the magnetic field [4]. Minimizing the parameters, in this experiments was limited to observe the influence of the cathode voltage, the magnetic field and the puller voltage. In this case the gas flow rate and the extraction gap (the distance ion source and the puller) were in constant value.

The cathode voltage determines the ion production in the ionization chamber, and the cathode normally. The cathode is normally powered by a -2 kV 2 A pulsed/DC power supply [5], and for the experiment purpose a DC power supply of 2.5 kV 3 A has been prepared. The performance of the Penning ion source in a cyclotron is strongly associated with the cyclotron magnetic field, in this case, the magnetic field used to confine the plasma in the ionization chamber of ion source so that the ionization process occurs be multiplied if compared with no magnetic field. Optimum operation of a Penning ion source of the cyclotron generally occurs in a magnetic field of several kilogauss [5-7]. A 1.25 Tesla magnetic field of the cyclotron is enabled for this purpose, though perhaps that based on the references mentioned above, the magnetic of the DECY-13 cyclotron is in the out of the optimum area. The results of the experiment in this paper will answer it. The puller voltage extracts out the ion beam from the ion source towards an ion collector for measuring ion beam current. The DC voltage could not exceed 2 kV, above the value, the corona on the extractor will occur and the collected negative charge signal increases to several mA quickly [6]. It is prepared a DC power supply that adequate for the puller voltage.

In this paper will be presented the results of an experiment of the characteristics of the ion source under the influence of some operating parameters that are mentioned above. The parameters in the experiments have not been set to the actual value as in the actual operation later. The results of the experiment will be able to predict the ability of ion sources in producing the ion beam current at the operating level during the actual operation of the cyclotron.

2. Experimental method

2.1 Experiment device

The experimental device consists of an ion source, a cyclotron magnet, a central part of dee which installed a puller and beam current probe. The schematic of the experimental device focused in the central region of the cyclotron is shown in Figure 1.

The ion source is also equipped with a hydrogen gas supply component, completed by a smooth flow rate regulator and a monitoring gas flow rate. The magnetic field of the cyclotron magnet in a perpendicular direction to the drawing and can be varied up to 1.8 T. Ion beam extraction from the ion source is ruled by a puller mounted on a central part of the dee. The results of ion beam extraction were measured by using a movable beam probe connected to a microampere. The extractor and the beam should have the same potential, but they are insulated from each other [5]. This arrangement is to protect any ionization of the gas occurring between the puller and beam probe tip. The beam probe tip also emits electrons when the ions impact on it. However, given the strong magnetic field, they will probably not be able to escape from beam probe tip, so the beam intensity measurement will be fine [8].

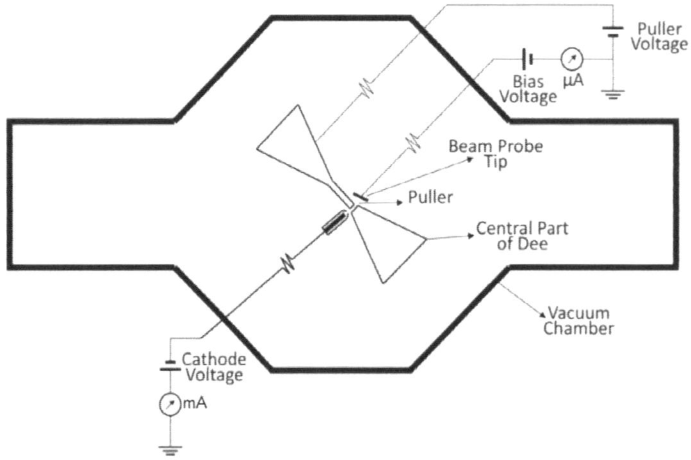

Figure 1. Schematic of the experimental device.

The radial position of extraction point or anode slit of ion source from the central region point is also fixed in advanced and determined by a formula:

$$r = \frac{1}{B}\sqrt{\frac{mV}{q}} \tag{1}$$

where B is the magnetic field of cyclotron in the central region, q is the charge of H- ions, m is ion mass of H^- and V is the effective voltage of puller or dee (*i.e.* peak voltage divided by 1.4). By using equation (1) and by applying B =1.25 T, m = 1.67×10-27 kg, V = 28 kV and q = 1.6 × 10^{-19} C, we obtained the radial position is 0.02 m = 2 cm. This radial position will be applied to determine an optimum value of the distance between anode slit and puller or extraction gap.

The optimization on the extraction gap has been carried out by using a program of Scilab 5.2.1 on the 2 cm radius from the center point of the central region. This program employed a Runge-Kutta numerical algorithm. This method was also performed in Korea for a cyclotron design by using the pwheel program [9]. The distance has varied in between 3 mm and 7 mm, and as a result that the best position of the ion source head in the 2 cm radius is at 4 mm to the puller [10]. This gap is not so much different to other common experiments [11-13].

2.2 Experiment procedure

The first and foremost precondition is a pressure in the vacuum chamber to approximate a pressure of 1-3 × 10-6 Torr and a hydrogen gas flow rate is set at 5 cc/min. The experiment is started with observing the influence of magnetic field to ionization intensity in the ion source (*i.e.* the measured cathode current) in the varying of cathode voltage. From these results, then will be determined the feasible value or value area of the cathode voltage and the magnetic field. By using these feasible values then the influence of the cathode voltage, the magnetic field and the puller voltage on the ion beam current will be observed.

2.2.1. Observation of the influences of the magnetic field and the cathode voltage on the intensity of ionization. The intensity of ionization is indicated by the intensity of the cathode current. In this observation the puller voltage remains 0 V, the cathode voltage as a parameter with the values of 700 V to 1500 V; the value of the magnetic field is varied and cathode currents are measured for each value of the cathode voltage. The magnetic field strengths approximately of 0.2 to 1.3 T are preferred. These experiments will result two things. First, the influence of the cathode voltage and the magnetic field against the ionization intensity will be characterized. Secondly, the best values or feasible values of the cathode voltage and the magnetic field on resulting the intensity of ionization will be determined.

2.2.2. Observation of influence of the cathode voltage on the ion beam current. This experiment and subsequent experiments are carried out by measuring the ion beam caught in the ion beam collector. This observation will explain whether the effect of the cathode voltage on the ionization intensity will has same effect on the extracted ion beam current. By using feasible value range of the cathode voltage and the optimum value of the magnetic field those are resulted in the observation 1 above, that effect will be observed.

2.2.3. Observation of influence of the magnetic field on the ion beam current. The best value of magnetic field with regard to the cathode voltage and the optimum value of extraction gap will be applied to observe the influence of the puller voltage on the ion beam current. Similar to the observation 2 above, this observation will explain whether the effect of the magnetic field on the ionization intensity will has same effect on the extracted ion beam current. The puller voltages of 0 to 3 kV have been applied to pull the ion beam from the ion source.

2.2.4. Observation of influence of the puller voltage on the ion beam current. All results of above experiments above will be referenced in this experiment i.e. the best or feasible values of the cathode voltage and the magnetic field.

3. Results and discussion

3.1. The influence of magnetic field and the cathode voltage on the intensity of ionization

The measurement data of cathode current as an effect of magnetic field on the variation of the cathode voltage are shown in the Figure 2.

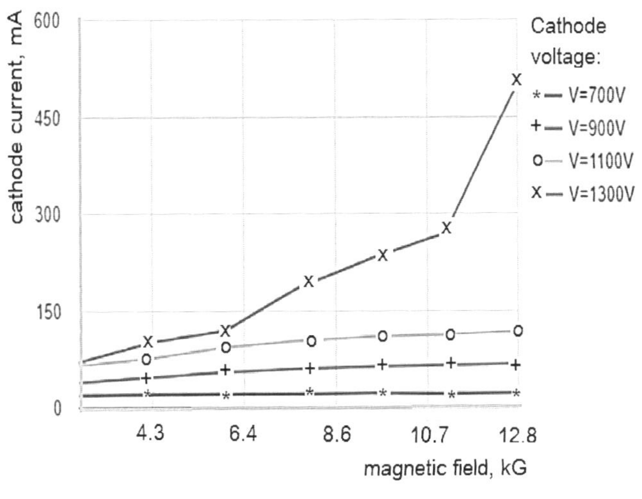

Figure 2. The influence of the magnetic field on the cathode current.

In this experiment, it is obtained that the cathode current increases sharply on the cathode voltage of 1300 V, but in this experiment did not set up a cathode voltage more than 1300 V because the cathode current was not stable and even at 1500 V the current meter was burnt. It is shown clearly that feasible operation of the cathode voltage to get high an ionization is beyond 1300V, and the magnetic field is higher than 4 kG. In this feasible range of the cathode voltage, the influence of the magnetic field is not very sharp.

3.2. The influence of the cathode voltage on the ion beam current
The data in Figure 2 shows that at around 6 kG or 0.6 T, the increase of the cathode voltage results in a significant increase of the cathode current. An expected logical result is that it will also result in a significant increase of ionization intensity in the ionization chamber and then result in a significant increase of the ion beam current. The results that are shown in the following Figure 3 will confirm that expectation. The voltage was limited to 1200 V because on 1300 V the voltage source was not stable caused by instability in plasma.

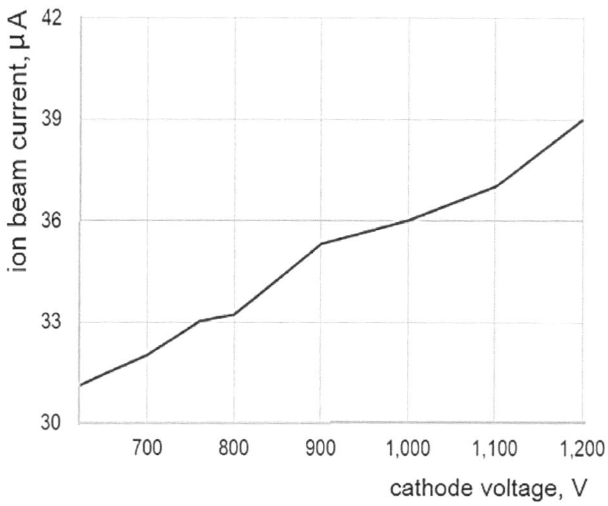

Figure 3. Influence of the cathode voltage on the ion beam current.

In Figure 2, the ratio of the cathode currents in the cathode voltages of 1100 V and 700 V is about 4; but in Figure 3 the comparison of the ion beam currents in the same cathode voltage range is only about 1.2. It means that the increase of ion beam current is not as strong as the increase of the cathode current or it means that increase of the ionization intensity is not effect linearly to the ion beam current. This phenomenon can be explained that due to H^- ions are formed by low energy electron in the plasma [14], while the increase of the cathode voltage will raise the electron energy and then raise the ionization intensity but not so much raise the low energy electron in the plasma.

3.3. Influence of the magnetic field on the ion beam current
Observation of influence of the magnetic field on the beam current was done on the cathode was 1200 V 50 mA, hydrogen gas flow rate was 5 sccm and the final vacuum was 8×10^{-6} Torr. The result is showed in Figure 4.

The ion beam current can start being observed at the magnetic field of 0.4 T and increasing slowly along with the rise of the magnetic field. The magnetic field of DECY-13 Cyclotron will be set at 1.25 T, and from the numerical data obtained, the beam current value can be extrapolated as 135-140. In general, however, the change of the ion beam current is not so high with increasing of the magnetic field. This result also accordance with the experiment by Long *et al.* [6], which no significant influence of the magnetic field on the ion beam current when experiment is done at several kilogauss.

3.4. Influence of the puller voltage on the ion beam current
The experiment was prepared on a relatively flat area of magnetic field in accordance with the curve in Figure 4 and in this case a 0.8 T was applied. The cathode was set at the maximum safe value, which in this case, was 1200 V, and this voltage resulted in a 50 mA cathode current. The puller voltage was varied and the ion beam current was measured. The result of relationship of both as shown in Figure 5.

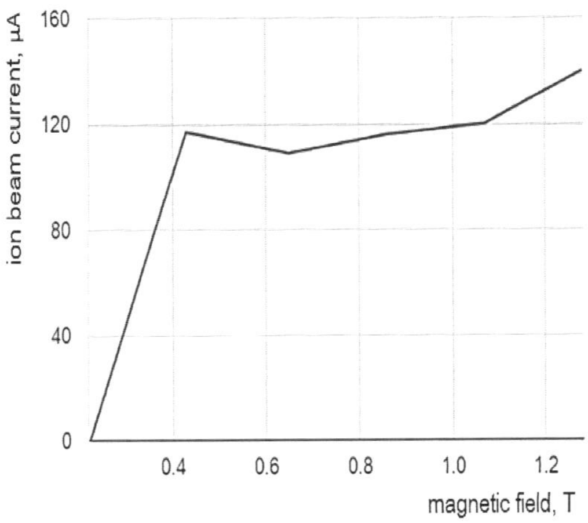

Figure 4. Influence of the magnetic field on the ion beam current.

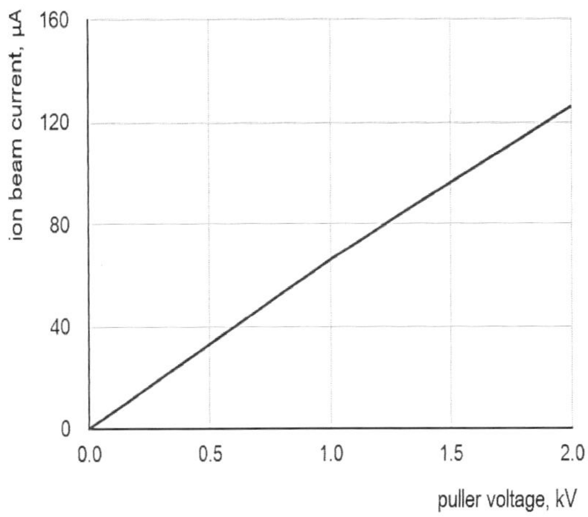

Figure 5. Relationship between the puller voltage and the ion beam current.

The observation was limited to 2 kV puller voltage because at a voltage of 2.5 kV the beam current rose to a very large value. This is in accordance with the experiment carried out by Long *et al.*, that above the 2 kV voltage the ion beam increases to several mA quickly caused by a corona occurring on the puller [6]. Normally, the curve of ion beam current versus puller voltage is proportional to the power of 3/2 [15, 16], but the curve in Figure 5 shows a linear relationship. It can be understood because the voltage is still in low value so it is not yet apparent a curve of power 3/2. A numeric data that is applied to make the curve above show that a maximum ion beam current is 126 µA.

The designation in Figure 3, Figure 4 and Figure 5 show that the puller voltage is most instrumental in influencing the ion beam current, then the cathode voltage and the magnetic field in the last rank. It means that the stability of the voltage source will be most noticed, and in the future when the cyclotron will be tested by the dee voltage, the stability of the RF system must be guaranteed. Although the cathode voltage and the magnetic field have a lower ranking, they must be also be noticed well because both of them have an important role in the plasma formation in the ion source.

These results have not yet figured a final performance of the ion beam current in the cyclotron central region. A stability of operation parameters has not been observed that perhaps can influence to the ion beam performance. In the next experiment will be studied more in depth and more detail regarding the stability problem, and efforts to overcome the stability problem will be done. However, the results of these experiments can indicate an existence of a function of the DECY-13 cyclotron components, especially its function in resulting the internal ion beam current.

As far as the authors know, another testing of ion beam current in the cyclotron central region using DC voltage is not yet carried out. The experiment of Long *et al.* that attained 200 µA using 2 kV puller voltage and the magnetic field of several kilogauss, was carried out in the bench testing out of the cyclotron [6]. The experiment of Jung *et al.* from SKKUY-9 Compact Cyclotron (Korea) obtained an H^- ion beam of 98 µA at a radius position of 1 MeV energy [17], it means that they have been using a RF dee voltage. Compared to those two experiments, the ion beam current of 126 µA on this experiment seems lower. Efforts will be made to increase the beam current, one of them is looking for an optimum value of the gas flow rate but still maintaining the vacuum on feasible value.

4. Conclusion

The preliminary testing of the ion beam for preparing a final test in DECY-13 Cyclotron has been carried out. The testing was carried out in the central region of the cyclotron using DC voltage extraction of the puller. The results of the experiment show an existence of a function of the DECY-13 cyclotron components. The obtained highest value of the H^- ion beam current is 126 μA using 2 kV puller voltage, and efforts are still needed to increase this current by optimizing operation parameters.

Acknowledgement
This experiment is part of R & D on the DECY-13 Cyclotron construction activity at the Center for Accelerator Science and Technology. So, the authors would like to thank the head of the center and the head of the Particle Physics Section and other various parties that have contributions to the financial and material procurement of the R & D relating to the cyclotron construction.

References
[1] Abdessalam F W Abdelrahman M M Soliman B A and Basal N I 2013 *Phys. Rev. & Resear. Int.* **3(4)** 427
[2] Taufik T Hermanto A Anggraita P and Santosa S 2014 *Atom Indonesia* **40** 70
[3] Silakhuddin S Santosa S 2015 *Atom Indonesia* **43** 139
[4] Wang Z Yang P Dong *et al.* 2012 *Rev. Sci. Instrum.* **83** 063302-1
[5] Yang Z Long J D Dong P *et al* 2012. *Chinese Phys. C (HEP & NP)* **36** 100-1003
[6] Long J Yang Z Dong P *et al.* 2013 *Nucl. Sci. Tech.* **24** 040201-3
[7] Yeon J H Kim H S Gad K M *et al.* 2013 *Proc. of Cyclotrons* 197
[8] Obradors D Ahedo M B Arce P *et al.* 2017 *Proc. of IPAC* Copenhagen Denmark
[9] Anggraita P Mulyani E and Kudus I A 2015 *Atom Indonesia* **41** 146
[10] Silakhuddin S Kudus I A 2017 *Atom Indonesia* **432** 85
[11] Oliver C Abramian P Ahedo B Arce P 2013 *et al. Proc. of Cyclotrons* Vancouver BC Canada
[12] Minervini J V 2012 *Technical Report Publishing of Defense Threat Reduction Agency USA*
[13] Mu X J Ghergherehchi M Yeon Y H *et al.* 2014 *Chinees Phys.* C **38** 127004-5
[14] Korenev S 2013 *Proc. of Cyclotrons* Vancouver BC Canada
[15] Eisa M E M Conradie J L and Cellier P J 2013 *World J. of Nuclear Sci. and Tech.* **3** 85
[16] Faircloth D Lawrie S Gabor C Letchford A *et al.* 2012 *Rev. Sci. Instrum.* **83** 02A701-2
[17] Jung S Y Kim H W and Ghergherehchi M 2014 *J. of Instrum.* **9** T04005-11

IOP Conf. Series: Materials Science and Engineering **432** (2018) 012027 doi:10.1088/1757-899X/432/1/012027

Safety assessment of near surface disposal for radioactive waste in Serpong nuclear area using PRESTO software

Sucipta[1*], D Suganda[1]

[1] Center for Radioactive Waste Technology, BATAN, Puspiptek Area, Serpong, Tangerang 15314, Indonesia

scipta@batan.go.id

Abstract. The planning of the construction and operation of Near Surface Disposal (NSD) facilities for Radioactive Waste at Serpong Nuclear Area needs a safety assessment in order to provide assurance that the radioactive waste disposal system functions properly. The purpose of the safety assessment is to provide a reasonable scientific assurance that the disposal system will provide an adequate level of safety and meet the requirements to protect human health and the environment. The NSD facility is vault type with a size of 20.2 × 20.2 m with a depth of 2.6 m from the ground surface (4.43 m above the highest ground water level). The facility was designed to accommodate waste packaged in 200 l drums and 950 l concrete shells. The total capacity of the disposal will be 1350 drums and 144 concrete shells. The waste that was considered in the study contain radionuclides which are dominantly contained in the low level packaged waste and, as reference for the safety assessment, with short half-lifes (less than or equal to about 30 years), namely Co-60 and Cs-137. This study used the PRESTO software version 4.2 to calculate the migration/exposure of radionuclides from the facility into the environment as well as individual doses. The results indicate that the predicted individual doses of the population around the facility at the Serpong site still fulfill the dose limit value of the public based on BAPETEN Chairman Regulation No. 4/2013 on the Protection and Radiation Safety in the Utilization of Nuclear Energy.

1. Introduction

The planning of the construction and operation of near surface disposal (NSD) facility for low level radioactive waste needs a safety assessment. The NSD type that has been chosen was based on the suitability with the waste characteristic and the site. The purpose of the construction and operation of NSD is to dispose the accumulated low level radioactive waste that have been immobilized and packaged since 1988 at the Radioactive Waste Installation (RWI) and stored at the Interim Storage (IS) located at the Serpong Nuclear Area (SNA). The NSD facility in SNA is dedicated also for the Demonstration Plant of Radioactive Waste Disposal (DP-RWD) to demonstrate the safety and durability of the disposal system [1, 2].

In order to provide assurance that radioactive waste disposal system is functioning properly, it is necessary to predict the individual dose as a part of safety assessment. Safety assessment is an iterative procedure to evaluate the performance of the disposal system and its potential impact on human health and the environment [3]. The purpose of safety assessment is to provide a reasonable scientific assurance that disposal system will provide an adequate level of safety and meet the requirements to protect human health and the environment. In the Article 23 of BAPETEN Chairman Regulation No. 4/2013, it is mentioned that dose constraint value for public is 1 mSv/year [4].

The candidate of NSD facility is located in the SNA, PUSPIPTEK Area, Serpong, Banten (Figure 1). The site characterization has been performed since 2010 with engineering geology, hydrogeology, and environmental geology investigations [5].

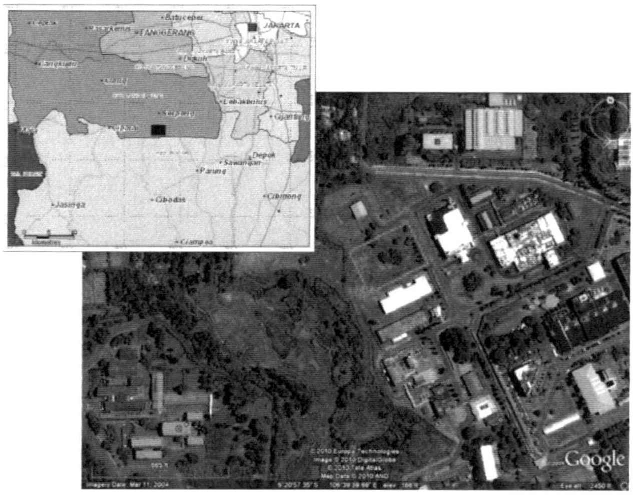

Figure 1. Map of NSD location at Serpong Nuclear Area, PUSPIPTEK, Serpong, Banten [5].

A safety assessment of demonstration plant for radioactive waste disposal (DP-RWD) has been performed in 2012 with different data of waste, conceptual design and site [6], so that the safety assessment must be updated with new data of waste, detail design (ready to be constructed) and some updated site data.

Assessment of radionuclide release scenario to groundwater (well water) and surface water (river) from the demonstration plant of near surface disposal for low level radioactive waste at SNA had been performed [7]. The result of the assessment showed that concentration of radionuclide still fulfilled the BAPETEN Chairman Regulation No. 7/2013 on Boundary Value of Environmental Radioactivity, which set the maximum concentration of radionuclide in water body at 10^2 Bq/m^3 to 10^5 Bq/m^3 [8]. Based on the results of calculation, the maximum concentration of radionuclide in well water was 10^{-10} Bq/m^3 to 10^0 Bq/m^3 and in surface water about 10^{-15} Bq/m^3 to 10^{-1} Bq/ m^3.

As a comparison, in India, a probabilistic safety assessment model for assessing the performance of near surface disposal facilities for low-level radioactive waste had been performed [9]. The annual effective doses derived from the uncertainty analysis fall in a log normal distribution with a geometric mean of $3.6\times10^{-2} \pm 3.9$ mSv and the most probable annual effective dose to a member of the critical group works out to be 2.8×10^{-4} mSv.

Korea also has developed a safety assessment of near surface disposal facilities for LILW using SAGE and VR-KHNP computer code [10]. Through the comparison of radionuclide release rates at near- and far-field, two computer codes show good agreement.

Still in Korea, post-closure safety assessment was carried out for DSRSs generated from 1991 to 2014 in Korea in order to ensure long-term safety of near surface disposal facilities [11]. Two kinds of disposal options were considered, i.e., engineered vault type disposal facility and rock-cavern type disposal facility. Assessment endpoint was individual dose to the member of critical group, which was modeled by GoldSim, which has been widely used as probabilistic risk analysis software based on Monte Carlo simulation in the area of safety assessment of radioactive waste facilities. In normal groundwater scenario, the maximum exposure dose was extremely low, approximately 1×10^{-7} mSv/yr, for both disposal options and satisfied the regulatory limit of 0.1 mSv/yr.

2. Overview of the design of proposed near surface disposal facility

The NSD facility is vault type in 20.2 × 20.2 m size with a depth of 2.6 m from the ground surface (4.43 m above the highest ground water level). The facility was designed to accommodate waste packaged in 200 l drum and 950 l concrete shell. The capacity of the disposal will be 1350 drums and 144 concrete shells [2]. The first emplacement will be started at compartment 1D and 1S for about 675 drums and 72 shells of waste packaged. The other compartments will be developed and operated to accommodate the radioactive waste that would be accumulated in about 25 years later. The general layout of the proposed facility can be seen at Figure 2.

The radionuclides contained in the waste that was considered in the study were radionuclides which dominantly contain in the packaged waste and as reference for the safety assessment with short half-life of less than or equal to about 30 years, namely Co-60 and Cs-137 [12]. The assumption based on the calculation of total activity of Co-60 of 301 032 Bq and Cs-137 of 5 857 920 Bq.

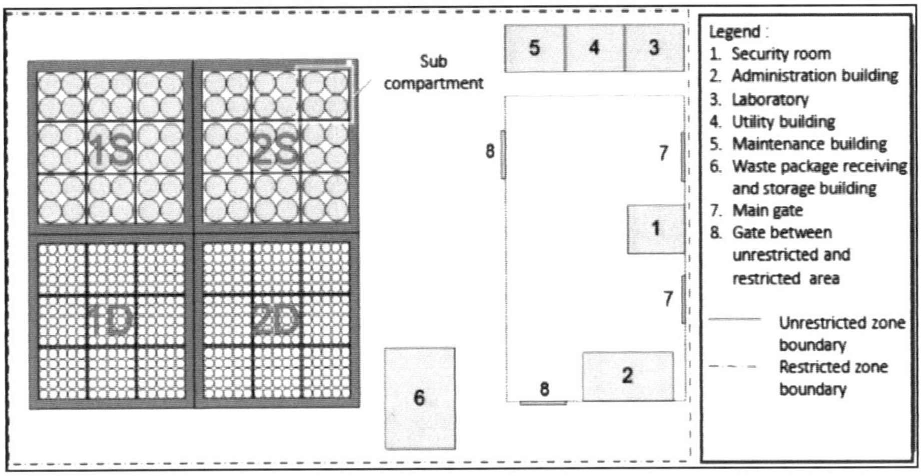

Figure 2. Layout design of proposed NSD [2].

3. Method

The safety assessment of NSD facility in Serpong Nuclear Area uses Prediction of Radiation Effects from Shallow Trench Operation – Environmental Protection Agency – Critical Population Group/ General Population (PRESTO-EPA-CPG/POP) software version 4.2, issued by the U.S. Environmental Protection Agency (US-EPA) [13]. PRESTO software version 4.2 is divided into two PRESTO-EPA-CPG and PRESTO-EPA-POP. PRESTO-EPA-CPG model is designed to calculate the maximum individual dose to the on-site population and critical population group of off-site population. PRESTO-EPA-POP model is designed to calculate the cumulative number of fatal cancer effects and serious genetic effects on the population at the site, off-site residents, and the general population.

Scenarios selected in this safety assessment is the scenario of the migration of radionuclides through the groundwater pathway. In general, hydrologic transport is the main cause of population exposed by radioactivity emanating from the disposal. The main sources of water, which led to the leaching of radionuclides from contaminants matrix, is precipitation. Precipitation water at the site will be infiltrated into the soil, flows over the surface of the ground, or evaporated into the atmosphere.

Transport of radionuclides from the site may occur due to water infiltration or runoff. A dynamic model, which calculates evaporative water loss and water transport based on the dynamic equations, is used to calculate the rate of infiltration [13]. The water that infiltrated into the zone of contamination will leach the radionuclide of the zone. Contaminated water that flows as runoff or infiltrate into the site as a percolation flow, eventually will enter the aquifer.

Radionuclides that eventually reach the aquifer generally will be transported at speeds lower than or equal to the speed of water flow in the aquifer. Retardation as the interaction of radionuclides with the solid media in the aquifer, known as the effect of sorption (uptake). When radionuclides are transported in the aquifer to the wells water, it will be consumed by residents in the site and/or off-site area through the drinking water, irrigation, and livestock feed. Radionuclide remaining in the aquifer is considered transported further and causing a general population health impacts on downstream areas.

Water contaminated at the site would be accumulated if the infiltration rate exceeds the ex-filtration rate out of the contaminated zone. When the volume of water that accumulates in the waste exceeds the cavity / total porosity, contaminated water will overflow to the land surface. Radionuclides in the contaminated water would then mix with surface water flow and subsequently to the flow (river) nearby. Potentially, contaminated water will be consumed by local residents and residents who live downstream through drinking water, irrigation, livestock feed and aquaculture.

3.1. Mathematical models for calculating radiological dose

Many equations are used as mathematical models for calculating the parameters of safety, but due to space limitations, only a simple formula to calculate the annual dose received by the population will be presented. The annual dose of individuals at a site k for the organ l, nuclide i, and an exposure pathway j is expressed as [13]:

$$D_{ijl}\,(k) = (K_j E_{ij}(k).DF_{ijl})/P(k) \quad (1)$$

K_j : numerical factor introduced by units Eij (k),
$E_{ij}(k)$: exposure of radionuclide i in the exposure path j,
DF_{ijl} : dose rate factor of radionuclide i exposure pathways j and organ l, and
P(K) : population on location k.

3.2. Input data

The input data used for the PRESTO software version 4.2 is the primary data obtained from field investigation on engineering geology and hydrogeology of candidate site for NSD facility at Serpong Nuclear Area [4, 14-16], climate data [17], secondary data and some assumption data about site [6], conceptual design of the NSD facility [2], and data of radioactive waste that will be disposed of [12]. The input data can be seen in Table 1, Table 2, Table 3, Table 4, Table 5, and Table 6.

Table 1. Data of NSD basement.

No	Aspect	Parameter	Unit	Data
1	Basement	Fraction of Rn-222 emanation for contaminated soil	--	0.3
		Depth of underground basement	m	3
		Thickness of basement floor concrete	m	0.2
		Porosity of floor concrete	--	0.18
		Length of basement side	m	3
		Negative indoor house pressure	Pa	2.4
		Perimeter shrinkage crack width	m	0.001
		Ventilation rate of basement	change/s	$2{,}78\times10^{-4}$
		Square of basement floor	m^2	9
		Basement occupancy fraction	--	0.6
		Outdoor, onsite occupancy fraction	--	0.2

The sorption capacity of soil at NSD site to certain radionuclide is relatively high. Evidence for this phenomenon was provided by the research that found that the Kd value of the soil to strontium in the range of 1600-2350 ml/g [18] and to cesium in 1400-1900 ml/g [19].

Table 2. Data of NSD facility.

No	Aspect	Parameter	Unit	Data
1	**Cover**	Thickness	m	1.2
		Density	g/cm^3	1.5
		Porosity	--	0.38
		Permeability	m/hr	0.02
		Average slope	m/m	0.02
		Average length slope	m	100
		Pellicular water deficit	m	0.01
		Gravity water deficit	m	3.1
		Pellicular water	--	0.47
		Gravity water	--	0.01
		Diffusivity	m/hr	0.00014
		Hydraulic conductivity	m/hr	1.1×10^{-6}
2	**Waste**	Thickness	m	2.6
		Density	g/cm^3	2.36
		Porosity	--	0.4
		Permeability	m/hr	2.2
		Absorbing waste	--	0.1
		Activated metals	--	0.1
		Solidified waste	--	0.1
		Containerized fraction	--	0
3	**Vertical zone**	Thickness	m	4.43
		Density	g/cm^3	1.48
		Porosity	--	0.3797
		Permeability	m/hr	5.967

Table 3. Atmospheric data.

No	Aspect	Parameter	Unit	Data
1	Atmosphere	Site environment		humid south
		Deposition velocity	m/s	0.01
		Gravitational settling velocity	m/s	0.01
		Onsite dust loading from mechanical disturbance	g/m3	5×10^{-5}
		Fraction of time the wind blows in direction of interest	--	0.31
		Annual average windspeed in direction of interest	m/s	5.14444
		Resuspension equation parameter	--	10×10^{-6}
		Stability category indicator	--	D
		Pasquill-Gifford atmospheric stability class formation	--	0.543
		Height of the inversion layer	m	500
		Hosker's roughness parameter	m	0.01
		Atmospheric transport parameter	--	5.186×10^{-8}

Table 4. Site characteristic data.

No	Aspect	Parameter	Unit	Data
1	**Hydrology**	Annual precipitation	m	1.752
		Annual stream flowrate	m^3/year	7,136,597
		Distance to nearest well	m	50
		Distance to nearest stream	m	175
2	**Aquifer**	Thickness	m	10
		Density	g/cm^3	1.48
		Porosity	--	0.3747
		Permeability	m/hr	1×10^{-7}
		Fraction of water saturation	--	0
		Groundwater velocity	m/yr	0.66
		Angle of dispersion	radian	0.3
		Allow aquifer to stream flow	m	yes
3	**Erosion**	Rain factor	R/year	15.1548
		Soil erodibility factor	ton/acre	0.007
		Plant management factor	--	0.0175
		Erosion control application factor	--	0.6
		Sediment delivery ratio factor	--	1
		Slope steepness-length factor	--	0.56
		Fraction of residual saturation	--	0.17
		Fraction of total annual precipitation for infiltration calculation	--	0.334
		Top soil layer precipitation run-off fraction	--	0.005
		Bottom soil layer precipitation run-off fraction	--	0.1
		Active depth of soil in surface-contaminated region	--	0.1
		Width of the contaminated site measured perpendicular to groundwater flow	--	28.6
		Length of the contaminated site parallel to groundwater flow	--	28.6

Table 5. Human uptake data.

No	Parameter	Unit	Data
1	Human uptake to vegetables	kg/yr	14
2	Human uptake to harvest product	kg/yr	102.4
3	Human uptake to cow milk	L/yr	73
4	Human uptake to goat milk	L/yr	6
5	Human uptake to drinking water	L/yr	73
6	Human uptake to meat	kg/yr	3
7	Human uptake to fish	kg/yr	3
8	Human uptake to soil	kg/yr	1
9	Human inhalation rate	m^3/yr	7300

Table 6. Plants and animals data.

No	Aspect	Parameter	Unit	Data
1	**Agriculture productivity**	Grass	kg/m^2	0.0015
		Other vegetation	kg/m^2	0
2	**Daily water consumption**	Dairy cows	L/d	0
		Dairy goat	L/d	0
		Beef cattle	L/d	0
3	**Time between harvest to consumption**	Grass	hour	12
		Stored feed	hour	0
		Leaf	hour	0
		Fruit	hour	12
4	**Exposure time in the contaminated air**	Grass	hour	1401.6
		Harvest product	hour	1401.6
5	**Food consumption**	Dairy goat	kg/d	0
		Beef cattle	kg/d	0
		Transport time from animal feed to human receptor	hr	48
		Time from animal slaughter to human consumption	hr	480
		Weathering removal decay constant	hr^{-1}	0.0021
		C-14 fractional equilibrium value	--	1
		Absolute saturation of air	gr/m^3	18.172
		Root depth	m	0.75
		Irrigation flow rate	L/m-hr	2880
		Infiltration fraction of precipitation	--	0.5
		Fraction of year that crops are irrigated	--	0.4
		Fraction of year animals graze on the pasture grass	--	0.25
		Fraction of animal's daily feed that is fresh grass	--	0.25

4. Results and discussion

From the calculation results with the PRESTO 4.2 software obtained some output as can be seen in Figure 3, Figure 4, Figure 5, Figure 6, and Figure 7.

At Figure 3 it can be seen that the radionuclide in well water consisted of Co-60 only. The concentration of Co-60 in well water is about 4.0×10^{-10} Bq/m^3 in year 200 after closure and the concentration tend to decrease until about 3.0×10^{-15} Bq/m^3 in year 300 after closure of the NSD. If compared to the BAPETEN Chairman Regulation No. 7/2013 on Boundary Value of Environmental Radioactivity, which the maximum concentration of radionuclide in water body is 10^2 Bq/m^3 to 10^5 Bq/m^3, so the concentration of radionuclide in well water in range between 10^{-15} Bq/m^3 to 10^{-10} Bq/m^3 are still far below the boundary value.

Figures 4 and 5 show the concentration of radionuclide in surface water is consisted of Co-60 and Cs-137. The maximum concentration of Co-60 in surface water is about 1.52×10^{-3} Bq/m^3 in the first year after closure and the concentration tend to decrease until about 9.44×10^{-26} Bq/m^3 in year 401 after closure of the NSD. If compared to the BAPETEN Chairman Regulation No. 7/2013 on Boundary Value of Environmental Radioactivity, which the maximum concentration of radionuclide in water body is 10^2 Bq/m^3 to 10^5 Bq/m^3, so the concentration of radionuclide Co-60 in well water is far below the boundary value.

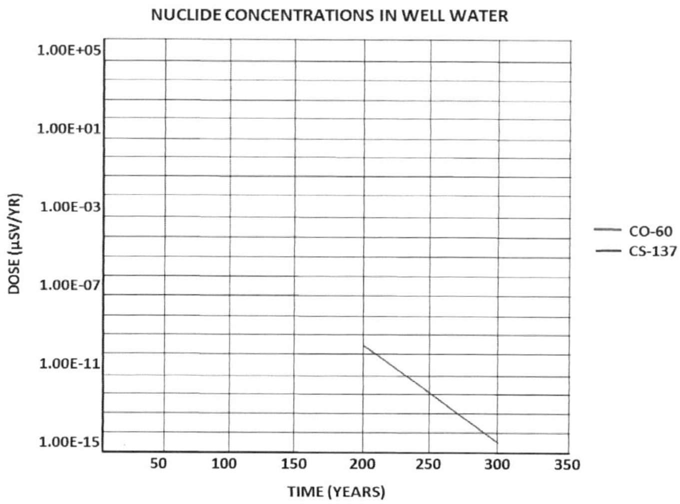

Figure 3. Nuclide concentration in well water.

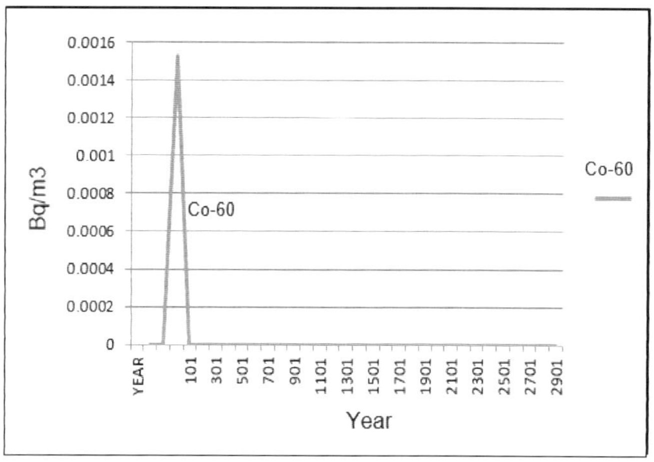

Figure 4. Concentration of Co-60 in surface water

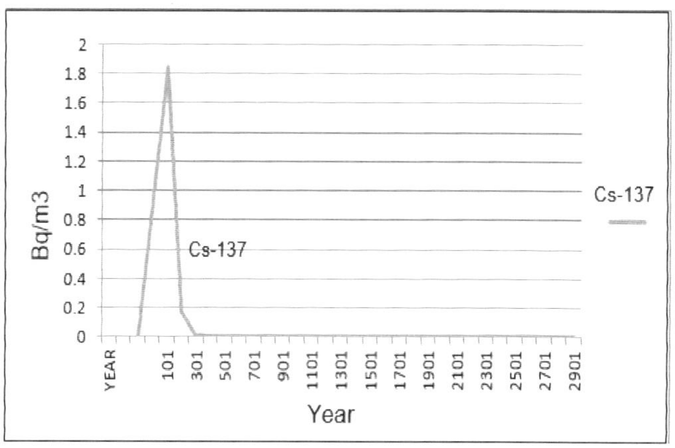

Figure 5. Concentration of Cs-137 in surface water.

The maximum concentration of Cs-137 in surface water is about 1.84 Bq/m^3 in year 101 after closure and the concentration tend to decrease until about 3.31×10^{-29} Bq/m^3 in year 2901 after closure of the NSD. If compared to the BAPETEN Chairman Regulation No. 7/2013 on Boundary Value of Environmental Radioactivity, which the maximum concentration of radionuclide in water body is 10^2 Bq/m^3 to 10^5 Bq/m^3, so the concentration of Cs-137 in well water is still far below the boundary value.

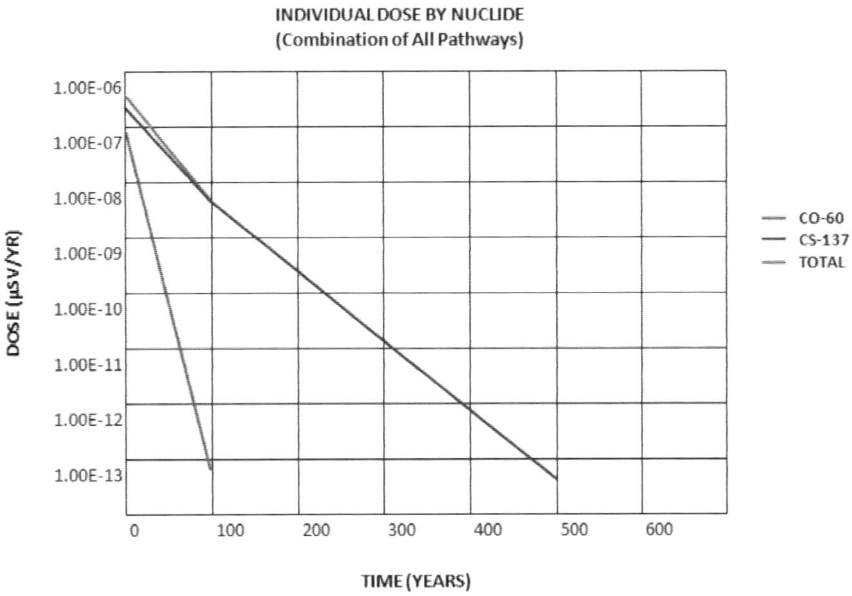

Figure 6. Individual doses by nuclide (a combination of all pathways).

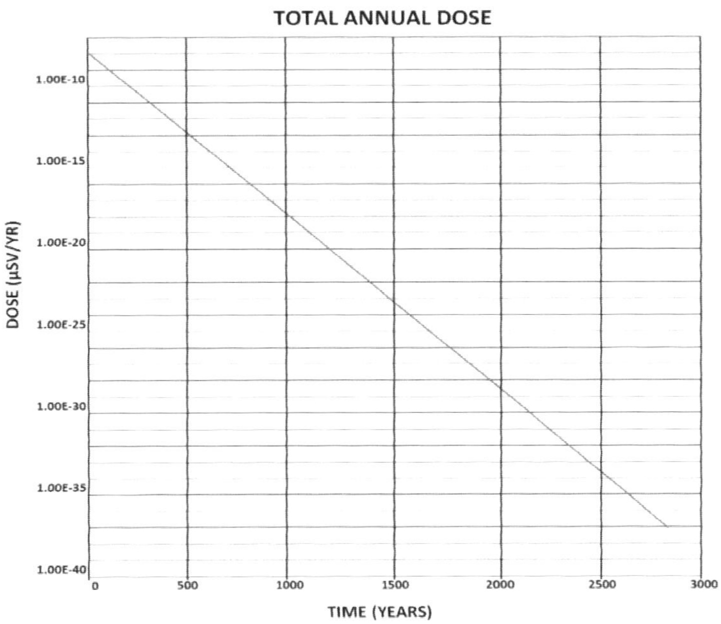

Figure 7. Total annual dose.

From Figure 6 it can be seen that the total contribution of its largest individual dose from Cs-137 radionuclides with a peak value at the beginning of the year amounting to 15.0×10^{-7} μSv/year and the smallest value of 15.0×10^{-13} μSv/year in year 500 after post-closure of NSD.

By looking at Figure 7 seen that the total values of annual dose have a tendency of diminishing from year to year with a peak value of 5.0×10^{-9} μSv/year in the early and the smallest dose is 5.0×10^{-38} μSv/year in year 2600 post-closure of NSD.

The BAPETEN Chairman Regulation No. 4/2013 on the Protection and Radiation Safety in the Utilization of Nuclear Energy defined that dose constraint value for public is 1 mSv/year, so that, the total dose caused from post closure of the NSD is still too far below the dose constraint value.

The result of this safety assessment cannot be compared to the result of the safety assessment that had been conducted in 2012 [6], because the waste data, design of NSD, and the site data are very different. Of course, the result of this safety assessment also cannot be compared to the result of the safety assessment that had been done in India and Korea [9-11], because the waste data, design of NSD and the site data must be different.

If compared to the result of the assessment of radionuclide release scenario to well water and surface water from demonstration plant of NSD that had been performed in 2015 [7], it appears that the difference in result is not so large. The concentrations of radionuclide in well water and surface water in the almost similar ranges of about 10^{-15} Bq/m^3 to 10^{-1} Bq/ m^3, especially for Cs-137.

5. Conclusion

The concentration of Co-60 in well water is 4.0×10^{-10} Bq/m^3 in year 200 after closure and the concentration tend to decrease until 3.0×10^{-15} Bq/m^3 in year 300 after closure of the NSD. The concentration of Co-60 in well water is still far below the boundary value that has been defined in regulation.

The maximum concentration of Co-60 in surface water is 1.52×10^{-3} Bq/m^3 in the first year after closure and tend to decrease until 9.44×10^{-26} Bq/m^3 in year 401 after closure of the NSD, so that, the concentration of radionuclide Co-60 in well water is far below the boundary value.

The maximum concentration of Cs-137 in surface water is 1.84 Bq/m^3 in year 101 after closure and the concentration tend to decrease until 3.31×10^{-29} Bq/m^3 in year 2901 after closure of the NSD, so the concentration of Cs-137 in well water is still far below the boundary value.

The total contribution of its largest individual dose is Cs-137 with a peak value at the beginning of the year amounting to 15.0×10^{-7} μSv/year and the smallest value of 15.0×10^{-13} μSv/year in year 500 after post-closure of NSD.

The total value of annual dose tend to decrease from year to year with a peak value of $5.0\text{x}10^{-9}$ μSv/year in the early and the smallest dose is $5.0\text{x}10^{-38}$ μSv/year in year 2600 post-closure, this value indicate that the total dose caused from post closure of the NSD is still too far below the dose constraint value for public (1 mSv/year).

Acknowledgment

Our gratitude goes to the head and staff of PTLR BATAN who have provided the opportunity to finish the assessment. To the management and staff of Waste Treatment Division we also want to express our thanks for the help in providing the radioactive waste data.

References

[1] Sucipta Heriyanto K Untara and Sastrowardoyo P B 2005 *Proc. of Basic Res. on Sci.and Tech. of Nucl. Energy* PTAPB-BATAN Yogyakarta 60-8 (In Indonesia)
[2] Sucipta 2013 *J. of Waste Manag. Techn.* **16(2)** CRWT BATAN Serpong 47-64
[3] IAEA 1999 *Safety Assessment for Near Surface Disposal of Radioactive Waste* IAEA Safety Standard Series No. WS-G-1.1 IAEA Vienna
[4] BAPETEN Chairman Regulation No. 4/2013 on the Protection and Radiation Safety in the Utilization of Nuclear Energy Jakarta (In Indonesia)

[5] Centre for Nuclear Geology Development 2010 *Final Report of Study on Engineering Geology and Hydrogeology* Radioactive Waste Technology Centre BATAN Serpong (In Indonesia)
[6] Sucipta and Dadang S 2012 *Proc. 2nd Basic Sci. Intern. Conf. Math. Dep. Fac. of Sci.* Brawijaya Univ. Malang Ch 61-65
[7] Dewanto P 2015 *Thesis Civil Eng.* Study Prog. Eng. Fac. Univ. of Indonesia Depok
[8] BAPETEN Chairman Regulation No. 7/2013 on the Boundary Value of Environmental Radioactivity Jakarta
[9] Nair R N and Khrisnamoorthy T M 1999 *Env. Modelling and Software* **14** 447-60
[10] Park J B Park J W Kim C L and Song M J 2005 *5th Waste Manag. Conf.* Tuczon AZ
[11] Lee S and Kim J 2017 Post *J. of Nucl. Eng. and Design* **313** 425-36
[12] Sriwahyuni H 2011 *Rad. Waste Tech. Centre* BATAN 217-24 (In Indonesia)
[13] Hung C Y 2000 *User's Guide For PRESTO-EPA-CPG/POP Operation System Ver. 4.2* U.S. EPA Office of Rad. and Indoor Air Washington DC 20460
[14] Syaeful H and Sucipta 2013 *Eksplorium* **34(1)** Jakarta 35-50 (In Indonesia)
[15] Syaeful H Sucipta and Sadisun I A 2014 *Eksplorium* **35(1)** PUSPIPTEK Serpong Jakarta 13-28 (In Indonesia)
[16] Sucipta and Risdiyana S 2016 *Eksplorium* **37(2)** PTBGN BATAN Jakarta 115-24 (In Indonesia)
[17] BMKG Indonesia 2017 Pondok betung-Tangerang Klimatological Station Monthly Rainfall Data 2003-2016
[18] Setiawan B Mila O and Safni 2014 *AIP Conf. Proc.* **1589** Bandung 325-28
[19] Setiawan B 2014 *Radiocesium Sorption onto SP-4 Site Soil of Serpong Nuclear Centre* Presented on ICICS Pattimura Univ. Ambon

Integration of superior varieties of sidenuk rice production with beef cattle business

Suharyono[1], S N W Hardani[1], T Wahyono[1] and C E Kusumaningrum[1]

[1] Center for the Application Isotopes and Radiation, BATAN, Lebak Bulus Raya 49 St., Jakarta, Indonesia 12440

suharyono@batan.go.id

Abstract. Utilization of nuclear techniques in agriculture, animal husbandry and environment has found technology of superior varieties of rice with the name of SiDenuk, animal feed (feed supplement and concentrate plus) and organic fertilizer worm (kascing). The technologies have been disseminated, but the profitability of livestock farmers only comes from the sale of one commodity, so the income are still not optimal. The purpose of this research activity was to find out how far the profit income from farmers when the three technologies were integrated. The methodology used was the BATAN (INTAAN) integration model between SiDenuk rice with animal feed and kascing fertilizer. The three stages that have been implemented were 1) observation of the production of rice, straw, panicle compared to the national superior varieties, and the economic aspects of SiDenuk rice business unit, 2) the treatment of kascing fertilizer for SiDenuk Rice and 3) the economic aspect of the INTAAN application model. The observed parameters consisted of rice production, rice straw, panicle, rice production and economic aspects.The experimental design used in this study was Completely Randomized Design. The data were analyzed by variance analysis (ANOVA) using SPSS version 20. Step 1, rice production and Si Denuk rice straw tended to be higher than Ciherang, Ir-64, Mikongga and Mentikwangi, while the amount of rice in panicles was significantly different at $P < 0.05$. The value of grain in panicles was 190, 120 and 116 seeds. The economic value of farmers harvests was Rp.1202500 / 1000 m2. Step 2. Treatment by kascing (K), compared with compost (C) and combination of K + C was able to affect rice production, and number of panicles in grain, but the production of paddy was not significantly different, only tended to be higher if given treatment of the K fertilizer. The numbers of rice production and panicle were 720 and 608 vs. 452 kg / 1000 m2 of dry harvest grains respectively, and 170.6 and 162.6 vs. 150.2 seeds/panicles. Step 3. The advantages of using INTAAN model related to Si Denuk rice production integration with beef cattle business were Rp 5594055.56, Rp 5778699.97, Rp 6107833.34, Rp 8383224.35 and Rp 8661321.29. The conclusion shows that rice varieties SiDenuk, grain production, rice straw was higher than national varieties, kascing fertilizer increased rice production, panicle and tended to increase rice straw production compared with fertilization with compost. INTAAN model was able to get the value of sales profit higher than the value of profit from the sale of one commodity.

1. Introduction

Field conditions for agricultural and livestock activities related to the availability of manpower, fertile land and pasture for livestock are limited; therefore a breakthrough is needed to solve the problem.

Human resources for cultivation, planting, maintenance, harvesting and post-harvesting and livestock rising are generally mostly older (over 50 years old) and fewer young people. These conditions resulted in the production of agricultural and livestock activities generated low and the benefits obtained are also less. As a result young workers are not interested in doing business in agriculture and livestock. Another problem faced is the availability of less sustainable livestock feed, and increases production of rice and livestock crops is not achieved.

In order for the production of paddy, livestock and manure products to gain more benefits, then a way that comes from the sale of some commodities of rice, livestock, animal feed, organic fertilizer and worm larvae as seeds. The proceeds from the sale of these commodities are expected to increase the added value of revenues from profits when compared to selling only one commodity, so that young workers will be interested in doing business units in agriculture and livestock. On the basis of that needed a technological breakthrough that able to increase the added value of income more than the results of farming and livestock by farmers. The technology used is the result of research and development (R & D) of BATAN that has been disseminated. Then SiDenuk superior rice varieties are integrated with livestock feed and kascing fertilizer. This integration model is called the BATAN-style integration model (INTAAN).

Rice crop is one of food commodities that plays an important role in the life of the economy in Indonesia. The agricultural sector is the most important field of life to be a source of life for most of Indonesia's population and get the main priority in development that aims to improve the living order of the economy that is able to encourage the improvement of people's standard of living [1].

SiDenuk variety has been released based on the Decree of the Minister of Agriculture no. 2257 / Kpts / SR.120 / 5/2011 on May 2, 2011. SiDenuk came from the Diah Suci variety irradiated by 0.2 kGy gamma ray from Co-60 with a potential yield of 9.1 tons/ha of dry grain milled (DMG) and the number of grains per panicle as much as 175-200 grains [2, 3]. Production capacity of rice and seeds/panicle is high. It is expected that the production of rice straw is also high. This means that the availability of SiDenuk rice straw is more and will be readily available on an ongoing basis, so the need for straw basal feed for cows will be met. Furthermore, it is also reported that rice production from SiDenuk is able to be sold in the form of grain during harvest, or dry milled grain, or dry grain harvested seeds, or conventional milled rice, and modern milled rice.The sales result from each post-harvest product is different, the income will be different and easy to know. To know the extent of SiDenuk varieties, the production of rice, rice straw and grain per panicle will be observed which is compared to the national superior varieties of Ir 64, Ciherang, Mikongga and Menthikwangi. Economic aspect of production of SiDenuk variety will be observed.

Throughout the year, rice harvest produces rice straw in large quantities of about 20 million tons per year, so that rice straw has the potential to become animal feed. Rice straw is part of root without root left after harvest. As it is known that rice is the second important food crop after wheat and more than 90% of Indonesians consume rice as staple food [4].Increased rice production is also accompanied by an increase in rice straw waste. The abundance of rice straw that has not been utilized optimally encourages researchers to develop the potential of rice straw into something that has high economic value [5].

Increased production of livestock, especially ruminant livestock is influenced by feed. In order to live and produce high, livestock requires nutrients that have high quality, balanced, easy to obtain and available throughout the year. However, the availability of feed is the main obstacle faced by farmers, especially during the dry season. Therefore it is necessary alternative feed that is able to meet the needs of livestock continuously. One that can be utilized is rice straw.

Rice straw as a basal feed at this early stage is calculated only from its production, which will then be given to cows which are added to the ConPlus concentrate as reinforcing feed. Cow dung will be processed by cultivation of worms, and will produce organic fertilizer named fertilized worm (kascing). For the next kascing is used for rice fertilization SiDenuk. The two BATAN R & D results covering livestock feed and kascing technology were obtained, respectively through the utilization of nuclear techniques in the form of tracer and radiation which were used for biological evaluation of metabolic results in rumen fluid and microbial stimulation of *Aspergillus niger* [6, 7].

ConPlus is a development of low quality local concentrate and improved quality by the addition of dietary supplements in its composition. This feed supplement is made based on local materials which is a technology that has been produced previously.The development of locally based feed supplement technology is actually the development of a multinutrient urea molasses block (UMMB), or multi nutrient feed supplement (MFS), and mutinutrien feed supplements without molasses (MFSWM).This feed supplement was able to increase daily weight gain of beef cattle, milk production, beef and dairy cattle reproductive performance, also gave positive response to milk nutrition, and feed digestibility of Etawa goat cross breed [8, 9]. Urea Multi nutrient medicated block is also a development of UMMB and able to increase milk production and weight gain after calving of beef cattle [10]. ConPlus has been used to fatten cattle on Idh Adha Day and improve the condition of the female cow body where body condition score (BCS) is only one value.

Based on the above BATAN R & D information, the results of research with the use of nuclear techniques will be integrated between SiDenuk rice crops with livestock, livestock feed and kascing fertilizer. This INTAAN model will be the maximum profit-raising pattern for farmers, since the profits will be greater if compared to the profit value of one commodity sale.

2. Experimental method

Three stages of the activities that have been carried out in the research were 1) observation of the production amount of rice straw and paddy from SiDenuk variety compared with national rice varieties (Mikonga, Ciherang, IR 64 and Menthik Wangi), 2) fertilizer treatment with kascing fertilizer compared with compost, and fertilizer kascings + compost to the production of paddy and rice straw from SiDenuk, and stage 3) economic value calculation of integration model BATAN (INTAAN).

2.1. Stage 1: Observation of paddy and rice straw production of Si Denuk varieties compared with national varieties

The research was conducted in Sompilan, Donoharjo Village, Ngaglik Subdistrict, Sleman District, Yogyakarta Special Region. The study was a case study on farmers. With the initial stage of the research was an approach to farmers to know the socio-cultural and economic conditions through interviews and application of technology. Interviews were conducted with farmers. The application of technology was based on the increase of integrated paddy production , which consists of the use of SiDenuk variety and national superior varieties (Mikongga, Ciherang, IR 64 and Menthik Wangi).

Planting of young seedlings less than 20 days, legowos row planting, organic fertilizer 200kg/1000 m^2, chemical fertilizer 30 kg Phonska and 15 kg Urea/1000 m^2 and utilization of waste from rice straw. The land that will be used as a nursery was processed, then ransacked into the mud. After that, the seed was sown on a plot that has been made. Soil processing was done by hand tractor until the soil forms mud. Planting was done with legowo 2: a 1-row system by planting move 1 stem/clump) [1, 11]. Age of seedlings when planting less than 20 days after seedling. Pest and disease control was done by farmers using observation with integrated disease pest control system. Weeding done using weeding tools and done 3 weeks after planting followed by hand grass removal. Watering was still natural by way of flooding on the surface and added when the water starts to shrink. Harvesting using sickles and threshing was done by banging on a special tool made of wood. Production was taken by tiling 2.5 m^2 x 2.5 m^2 to calculate the production of each hectare. The economic value of farmers' income was also calculated. Parameter measured production of rice and rice straw also panicle. The experimental design used in the study was a Complete Randomized Design. The data were analyzed by analysis of variance (ANOVA) using SPSS 20 version.

2.2. Stage 2: Effect of kascing fertilizer on rice straw and rice production of Denuk varieties

Procedures for planting up to harvest rice as in the activity of Stage 1. The difference was seen in the treatment of fertilizer, and not compared with the national superior varieties. Location of activity was higher than sea level, because it was close to Mount Merapi approximately 10 km, while the stage 1 distance of approximately 25 km. Fertilization with inorganic fertilizer was adjusted to the use of a standard amount of fertilizer which was added with 3 organic fertilizer treatment. Each of kascing fertilizer (K), compost fertilizer (C) and the combination of kascing and compost fertilizer (K+C). The amount of organic fertilizer from 3 treatments was adjusted with the standard. Parameters observed the amount of rice straw, rice production and seeds/panicle of SiDenuk variety. The experimental design used in the study was a Complete Randomized Design. The data were analyzed by analysis of variance (ANOVA) using SPSS 20 version.

2.3. Stage 3: The economic value of the result of BATAN model integration activities

Integration of BATAN model was an integrated activity between BATAN's R and D of rice, animal feed and cattle dung with process of worm cultivation. These results have been disseminated, but integrated has never been implemented. The superior rice variety used was SiDenuk variety. This rice was grown in three farmer groups at planting season I. The farmer group was in the village 1) Wonorejo, Sariharjo, 2) Sompilan, Donoharjo, 3) Suruh, Donoharjo, Ngaglik and Sleman districts. The procedure of planting in accordance with standard operational procedures established by the Department of Agriculture in the area, as well as the harvest of paddy. The Farmers' Group from Suruh cooperates with seed breeders and certification centers for seeds to be sold as seeds. The seed breeder was willing to buy the paddy crop when it was harvested (dried grain of harvest/DGH) or bought in dry harvest (dried milled grain/DMG) condition. A DHG with the price Rp.4500 / kg, this price was higher when compared with the standard price of the DHG price which was Rp.3500 / kg. As for the other two groups were sold at the time of DHG, DMG and sold in the form of ground rice with conventional and modern milling.

The result of the activities of the livestock nutrition group was the plus concentrate (ConPlus) which was a feed provider of energy and protein for beef cattle. This ConPlus was tested on beef cattle to be fattened and sold during Idhul Adha Day. The cattle used for the 3 species of research which were owned by the farmers of livestock, each given concentrate plus diet (ConPlus), and the gift was tailored to the nutritional needs which in line with the desire to increase body weight / head / day. The price of cattle at the beginning of buying was Rp. 14 million/head. Each cattle farmer besides raising cattle also works on rice cultivation. They also exist as owners or landowners. Land used to grow food crops (paddy) area of 1000 m^2.

Cow manure was processed by worm cultivation, the fermentation process was completed 21 days. In this research work together with one of the professors at the Faculty of Agriculture, University of Gadjah Mada (Margino, 2015) which facilitated the house of worm fertilizer (kascing). Manure kascing was packed in sacks 25 kg.

Based on the INTAAN model, the economic analysis of the business unit integration between SiDenuk rice and beef cattle business unit was calculated from the profit of grain or rice sales, which was then added to the profit value of the sale of fattening beef cattle (BC) on Idhul Adha Day, non-productive cow (CRI), animal feed (ConPlus / CP, feed supplement / FS), kascing organic fertilizer (MK) and worm larvae (WL). While SiDenuk rice business unit was divided into two kinds of sales in the form of grain and rice. Sales of dry harvested grains (DHG), or dried unhulled grains (DMG), or dry harvesting grain for seed (GDHS). Sales of rice in the form of traditional milling (TMR) and modern (MMR). So the total value of profits resulting from integration activities of SiDenuk rice business unit with beef cattle business unit can be made with a formula to calculate it. So total profit values are MMR + BC + CRI + FS + CP + MK + WL. Cultivation activities of rice crops, cattle and earthworms (Lumbricus rubillus) were carried out for 3 months, as well as for the production of animal feeds.

3. Result and discussion

3.1. Stage 1: Observation of paddy and rice straw production of Si Denuk varieties compared with national varieties

Paddy and rice straw production among rice varieties in the planting season I showed no significant difference (P>0.05), but the production of panicle showed significant difference (P <0.05). SiDenuk grain production tended to be higher when compared with the Ciherang and Ir 64 varieties, the amount of production were 640 vs 547 and 593 kg/1000 m^2.This was supported by the result of the calculation of the number of grains on the panicle, the number of grain panicle from SIDenuk 190 vs 116 and 120. According to SK Minister of Agriculture no. 2257 / Kpts / SR.120 / 5/2011 on May 2, 2011 the number of grains in panicles was 170-200 [2, 3]. Production of grain, rice straw and number of rice panicles of several varieties during planting season I are shown in Table 1.

Table 1.Paddy, rice straw and seeds/panicle production at the planting season ISuperscript on the same line is significantly different at P <0.05.

Superior varieties	Paddy grain (kg/1000 m2)	Rice straw (kg/1000 m2)	Panicle (kg/1000 m2)
Ciherang	548^{a}	671^{a}	116^{b}
IR64	593^{a}	593^{a}	120^{b}
SiDenuk	640^{a}	640^{a}	190^{a}

Superscript on the same line is significantly different at P <0.05

Observation activities of rice production, rice straw and seeds per panicle have been carried out during the planting season II. The results are presented in Table 2.

Table 2. Paddy, rice straw and seeds/panicle production at the planting season II.

Superior varieties	Paddy grain kg/1000 m2	Rice straw kg/1000 m2	Panicle Seeds/panicle
IR64	805^{a}	853^{a}	119^{b}
SiDenuk	811^{a}	827^{a}	170^{a}
Mikongga	669^{b}	789^{a}	128^{b}
Menthikwangi	696^{b}	800^{a}	160^{a}

Production of rice, and the number of panicle seeds in the dry season II, were statistically significantly different at P <0.05. Production of rice and seeds/panicle of SiDenuk, Ir 64, Mikongga and Menthikwangi were 811 vs 805, 669 and 696 kg/1000 $m^2$and 170, 119, 128 and 160 respectively. However, the production of rice straw was not significantly different. The production of Si Denuk and Ir 64 rice straws were seen to tend to be higher than those of Mikongga and Mentikwangi varieties. The number of panicles is closely related to the ability of the plant to produce the seedlings and the ability to maintain the various physiological functions of the plant. The more saplings are formed, the greater the chances of the formation of saplings that produce panicles. At the time the plant begins to bloom almost all of the photosynthetic results are allocated to the generative parts of the plant (panicles) in the form of flour. In addition also mobilization of carbohydrates, proteins and minerals in the leaves, stems, roots to move to the panicle [12].

Rice in producing the amount of grain is influenced by various factors such as the characteristics of panicle length and nutrient availability. Each rice varieties have different characteristics of panicle length. The longer the panicle, the more grain will be formed. Differences in the number of grains per panicle produced from each variety are caused by genetic factors of each variety. Besides, environmental factors play a role in the high number of per panicle grain, because the sunny weather conditions can increase the rate of photosynthesis, the energy of light used to overhaul the water and the acid gas of charcoal is converted into food, the resulting photosynthate will be stored in the stem and leaf tissues, then will be trans located to grain maturation level [13].

3.2. Stage 2: Effect of kascing fertilizer on rice straw and rice production of Denuk

Based on statistical result, kascing, compost fertilizer, and combination of compost and kascing fertilizer did not give significant difference in rice straw yield (P <0.05). However, based on data obtained, kascing fertilizer tended to give the highest influence on rice straw production from SiDenuk varieties compared with the amount of rice straw production that production from SiDenuk varieties compared with the amount of rice straw production that was fertilized with compost or combination of kascing and compost. The production amount was 1560 kg vs. 1276 and 1348 kg respectively (Figure 1).

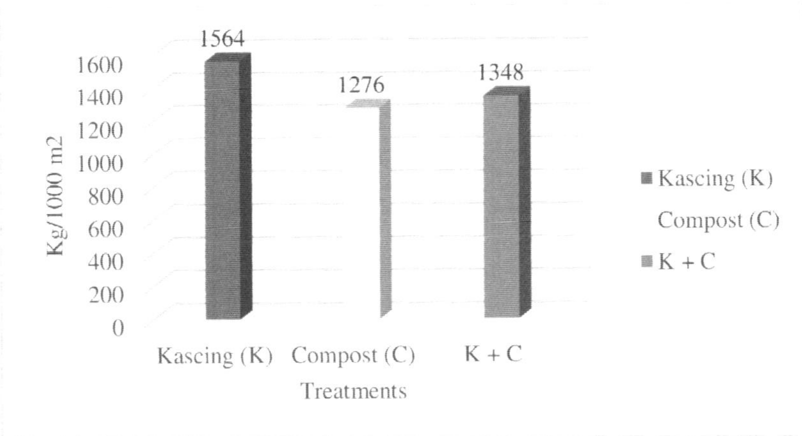

Figure 1. Rice Straw production of SiDenuk.

The statistic shows that compost, kascing and combination of compost and kascing fertilizer gave significant difference to the production of paddy varieties (P <0.05). Treatment of kascing fertilizer gave significant effect (P <0.05) on rice production and higher production compared with composted fertilizer, but with combination of K+C was not significantly different. The amount of production from K fertilized and the combination of K+ C was 720 and 608 kg/1000 m^2 vs 452 kg/1000 m^2 (Figure 2).

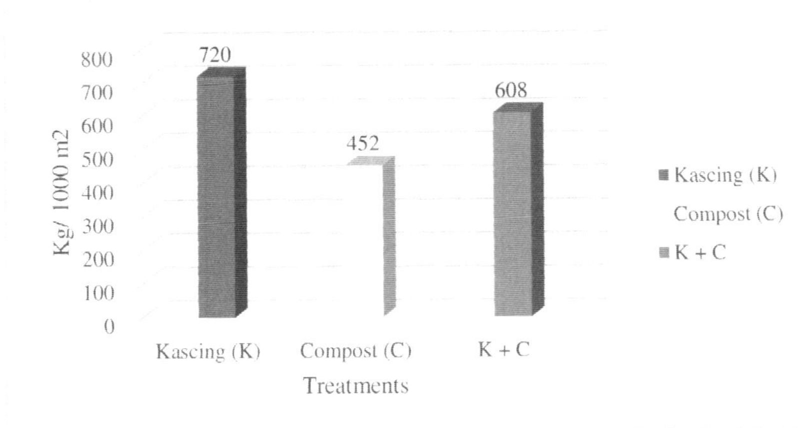

Figure 2. Paddy production of SiDenuk.

Statistical analysis showed that compost, kascing and combination of kascing compost fertilizer gave significant difference to the production of panicle of varieties (P <0,5). Based on the data obtained, kascing fertilizer has the greatest influence of panicle varieties when compared with compost and K+C combination, but the number of grains/panicle was not significantly different P<0.05. The values were 170.6 and 162.6 vs 150.2 grains/panicle (Figure 3).

Based on Figures 1, 2 and 3 it is shown that kascing fertilizer has the greatest influence on grain production and the number of panicle of rice varieties. This shows that kascing fertilizer had sufficient nutrients for productivity of rice crops. This is in line with the opinion of Rohaeni *et al.*, (2012) [14] which states that the availability of nutrients become a very important factor in stimulating the panicle initiation process to be perfect, so that the chances of grain formation become more. If during the process of filling the grain, not balanced with the availability of sufficient nutrients will form a lot of empty grain. Percentage of grain contents is one indicator of crop productivity, the higher the percentage of grain content obtained by a variety indicates that the variety has high productivity [15]. Kascing fertilizer had the greatest influence in the production of rice straw varieties. This was because the nutrientscontained in the fertilizer kascing had a role in cell division, formation of albumin, the formation of flowers, fruits and seeds, accelerate the maturation of seeds and root development. Maximum growing root caused the plant to more easily absorb the nutrients needed for the growth and development of the plant, resulting in average weight of fresh straw obtained in kascing fertilizer treatment was better than with other treatments [16]. Nutrients that have been absorbed by roots, both used in the synthesis of organic compounds and fixed in ionic form in plant tissues, will contribute to the increase of dry weight of plants [17]. Excessive use of inorganic fertilizers will damage soil fertility in paddy-fed rice paddies. Researchers who previously reported that using organic fertilizer had a significant effect on rice growth and productivity. When combined with inorganic rice is able to boost production and while maintaining and improving soil health [18].

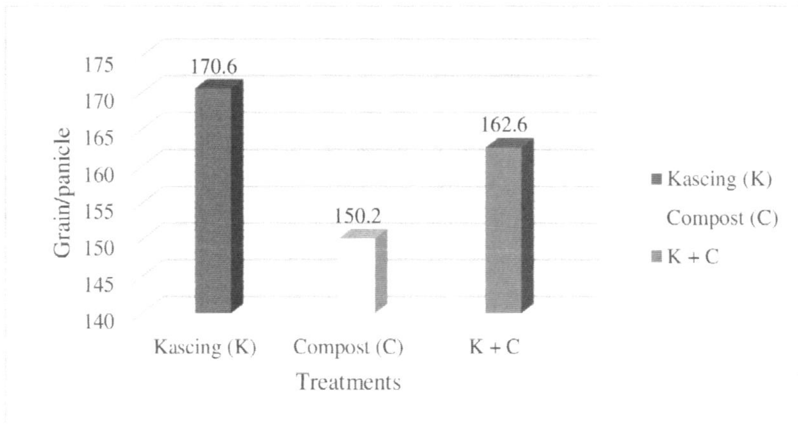

Figure 3. Panicle production of SiDenuk.

3.3. Stage 3: The economic value of the result of BATAN model integration activities

The observation of SiDenuk rice in three locations after 3 months was harvested. The yields were distinguished by three kinds of grain production, the first was dried harvest (DGH), the second, dry milled grains (DMG) and the third, dry seed harvesting grains (GDHS). Observations on traditional milled rice (TMR) and modern milled rice (MMR) were produced from DMG. The profit from each type of post-harvest production was Rp.567388.89; 752033.3; 1081166.67; Rp. 3356557.68 and Rp. 3634654.62. The amount of profit on MMR sales was higher than the profit value of other production sales. The gain disaggregation from MMR sales when compared to DGH, DMG, DGHS and TMR sales was 72.99; 65.71; 54.15 and 3.98%.

Along with the cultivation of SiDenuk rice cultivation, there has also been a business unit of beef cattle which was distinguished from fattening cattle business (BC), unproductive cow (CRI), feed (ConPlus / CP and feed supplement / FS), organic fertilizer from worm cultivation (MK) and larvae of earthworm/WL (*Lumbricus rubellus*) .The respective sale result obtained profit respectively Rp. 2566666.67; Rp. 1000000; Rp. 400000, Rp. 500000, Rp. 500000 and Rp. 60000 (Figure 4a).

In accordance with the INTAAN model integration application, the profits from DGH, or DMG, or DGHS, TMR and MMR sales were added to the benefits of the beef cattle business unit including BC, CRI, CP, FS, MK and WL. So the number of advantages of the INTAAN model integration result was Rp. 5594055.56, Rp. 5778699.97, Rp. 6107833.34, Rp. 838322.35 and Rp. 8661321.29 (Figures 4a and 4b).

Dried grain of harvest (DGH)	**567388.89**
Dried milled grain (DMG)	**752033.3**
Grain of dry harvest seeds (GDHS)	**1081166.67**
Traditional milled rice (TMR)	**3356557.68**
Modern milled rice (MMR)	**3634654.62**
Beef cattle (BC)	**2566666.67**
Cattle reproductive improvement (CRI)	**1000000**
Feed supplements (FS)	**500000**
Concentrate plus (CP)	**400000**
Manure kascing (MK)	**500000**
Worm larvae (WL)	**60000**

(a)

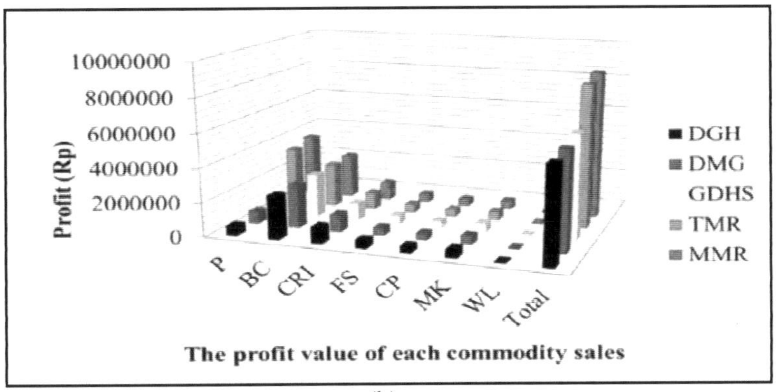

(b)

Figure 4 (a) Detailed information on the summary of words from several commodities and information on the figures from the sale of each commodity (Rp) in the SiDenuk and Beef Cattle business units based on the INTAAN model. (b) The value of profit from sales integration between SiDenuk rice business unit with beef cattle business.

Profit income from the sale of one commodity compared with some commodities from the result of integration between SiDenuk rice and beef cattle business unit was very big difference of profit. This can be seen from the value of profit at the time of the sale of one commodity, the highest profit of MMR, but still lower if compared with the value of MMR sales profit added with the profit from the sale of beef cattle business unit. The value is Rp. 3634654.62 vs Rp. 8661321.29 or the difference in profits up to 40.88%. Especially when compared with the sale of one commodity DGH was Rp.567388.89 vs Rp. 8661321.29, so the difference in profits up to 65.83%.

On the basis of the use of INTAAN model (integration of BATAN research results in the form of SiDenuk rice varieties, feed supplement, Concentrate plus (ConPlus) and fertilizer worm (MK) is very promising to increase farmer income and invite young human resources participate in the development of agriculture and livestock Especially the INTAAN model was developed into a model of integration of organic farming model BATAN, then the value of the profits will be more and more secure, because organic food selling value is higher and will have a positive impact on the welfare, human health and young human resources will be more interested in participating in business units in agriculture and livestock.

The result of integration of SiDenuk rice business unit with beef cattle based on INTAAN model looks bigger profit. Revenue from this integration result has been supported by previous researchers who reported that profit with cattle fattening business alone was only Rp. 611250 / ha/head / year, while the business unit of rice plants only got Rp. 12745000/ha/year. Furthermore, after the integrated activities for beef cattle got profit Rp. 3477380 / head / year, and the cultivation of rice crops as much as Rp. 90517250/ ha/year [19]. Tjeppy (2007) [20] mentioned that the main objective of integrating food crops and livestock was to obtain maximum benefit, further mentioned that one characteristic of integration activities between farmers and livestock.

The above income gains are seen to give hope to the farmers in carrying out the activities of farm or livestock business unit in an integrated way as well as the INTAAN model. Rohaeni *et al.*, 2005 [21] livestock integration system in farming was one of the efforts to achieve the optimization of agricultural production. This effort has been largely undertaken which significantly added value to both farm produce and livestock productivity. Integrated farming was able to reduce production costs, especially on the provision of forage, as a source of labor and can contribute in the purchase of fertilizer.

4. Conclusion

The result of integration activity between rice business unit from SiDenuk and beef cattle varieties shows that SiDenuk's superior rice varieties for paddy and seed production per panicle were higher than National varieties. Production of rice straw tended to be equal to the amount of rice production. Kascing fertilizer contributed to the increased production of rice and grain/panicle from SiDenuk varieties. Application of INTAAN model were able to increase farmer's income more than if without integration (one commodity sold). The highest total profit was obtained by selling milled rice with modern milling (MMR) plus the value of profit of beef cattle business and waste processing with worm cultivation. Total profit value was Rp 8661321.29 (MMR) vs Rp. 5594055.56 (DHG), Rp. 5778699.97(DMG), Rp. 6107833.34 (GDHS), and Rp. 8383224.35 (TMR) respectively.

Acknowledgement

This research was supported by Indonesia National Nuclear Energy Agency. We thank our colleagues from *Center for Application of Isotopes and Radiation (PAIR)* and who provided insight and expertise that greatly assisted the research, although they may not agree with all of the interpretations/conclusions of this paper. We thank people animal nutrition in PAIR for assistance with particular technique, methodology etc for comments that greatly improved of this paper.

Rreferences

[1] Purnomo 1996 *Rice Planting Technology Jajar Legowo System. Agricultural Information Sheet Agricultural Technology Research Institute Karangploso*

[2] Wahyudi B I Rial A and Shiddiq M 2013 *Description of Superior Varieties of Breeding Mutations of Rice, Soybean, Green Beans, Cotton, Sorghum. Nuclear Science and Technology* Dissemination Center National Nuclear Energy Agency

[3] Mugiono Hambali Sutisna S and Yulidar 2011 Inpari Sukiden *Description of Superior Variety of Breeding Mutation Result* Dissemination and Partnership Center National Nuclear Power Agency

[4] Empel Mireille van Makkar H P S Dijkstra J Lund Peter 2016 *Agricul., Veterinary Sci. Nutrition and Natural Resources* **11** 37

[5] Yanuarianto O Amin M Iqbal M and Hasan S D 2015 *J. of Indonesian Animal Sci. and Tech.* **1 (1)** 47-52

[6] Suharyono Shintia N W H and Teguh W 2015 *A Scientific J. for The Applicat. of Isotopes and Radiation* **11** 99-111

[7] Nana M Larasati T R D Nurhasni Meliana N 2015 *A Scientific J. for The Applicat. of Isotopes and Radiation* **11** 13-25
[8] Suharyono 2010 *Development of feed supplements for ruminants and introduction to farmers. Nuclear Sciences, scientific presentations of middle/major researchers* **1(1)** Center for the Dissemination of Nuclear Science and Technology BATAN 1-40
[9] Suharyono Faizal Z Gito W and Asih K 2015 *Proc. of Nat. Sem. on Livestock and Veterinary Techn.* 164-71
[10] Suharyono H Sutanto H Sutanto Purwanti Martanti A Agus and R Utomo 2014 *Atom Indonesia J.* **40(2)** 49-56
[11] Suardana P A Antara M and Nature M N 2013 *J. Agrotekbis* **1(5)** 477-84
[12] Mahmud Y and Purnomo S S 2014 *J. of Sci. Solutions* **1(1)** 1-10
[13] Chozin M Sudjamiko S and Barchia M F 2017 *Australian J. of Crop Sci.* **11** 5
[14] Rohaeni W R Sinaga A and Ishaq M I 2012 *Agricultural Informatics* **21(2)** 107-15
[15] Ikhwani Suhartatik E and Makarim A K 2010 *J. of Food Crops Agriculture Research* **29(2)** 63-71
[16] Syakhril Riyanto and Arsyad H 2014 *AGRIFOR J.* **XIII(1)**
[17] Sulistyono Eko Suwarno Lubis I and Suhendar D 2012 *Agrovigor.* **5(1)** 1-8
[18] Morteza S Alireza N Shankar L L 2011 *J. of Agricultural Sci.* **3(3)** 217-24
[19] Judy M T, Panelewen V V J and Mirah A D P 2014 *Zootek J.* **34(2)** 1-9
[20] Soedjana T D 2007 *J. of Agricultural Research and Development* **26(2)** 82-7
[21] Priyanti Atien 2007 *Impact of the Livestock Integration System Program on Allocation of Working Time, Farmer Income and Expenditure* Dissertation Bogor Agricultural University

Fabrication of gas diffusion layer from carbon ramie fiber by hot press method

D A Gumilar[1,2], N Indayaningsih[1], A Subhan[1], N Prastomo[2] and S Hastuty[2]

[1] Research Center for Physics, Indonesia Institute of Science, Puspiptek Area Build. 440-442, South Tangerang 15314, Indonesia
[2] Departement of Engineering Physics, Faculty of Engineering Surya University, Gading Serpong, Tangerang 15810, Indonesia
[3] Mechanical Engineering, Pertamina University, Jl. Teuku Nyak Arief, Simprung, Jakarta 12220, Indonesia

agum.dimas14@gmail.com

Abstract. Polymer exchange membrane fuel cell (PEMFC) is an electrical energy-generating device that is renewable and environmentally friendly. However, the price of components in PEMFC is still relatively expensive and more studies are needed on making cheaper components in terms of methods and raw materials. In this research, the fabrication of gas diffusion layer (GDL) which is a component of PEMFC made from ramie fiber was investigated. The processes of preparing GDL in this research were divided into two parts: the process of making carbon from ramie fiber through pyrolysis process and fabrication of composite from carbon ramie fiber and urea formaldehyde resin (UFR) by hot press method. The carbon results from the process of pyrolysis of degummed ramie fiber with variations in pyrolysis temperature of 500 °C, 900 °C, 1100 °C and 1300 °C yield electrical conductivity values of 8.7 x 10^{-7} S/cm, 0.84 S/cm, 3.72 S/cm and 4.6 S/cm, respectively. The electrical conductivity of carbon ramie fiber was increased by degumming and increasing pyrolysis temperature. The 1300 °C carbon composite with a hydrophobic coating 10 wt.% FEP has an electrical conductivity of 0.25 S/cm and a water droplet angle of 133° (hydrophobic). Composites of carbon ramie fiber have potential as GDL in PEMFC.

1. Introduction

The Proton Exchange Membrane Fuel Cell (PEMFC) is a device that converts chemical energy directly into electrical energy [1]. One of the important components in PEMFC is the Gas Diffusion Layer (GDL). A GDL main functions are to distribute the fuel gas, to eliminate water produced from the cell and electrical pathway. In order for the GDL to perfom these functions, it has to have electrical conductivity-, porous and hydrophobic characteristics [2]. Research trends in the use of natural fibers as carbon fiber-based GDL are on the rise, some of which involve coconut fiber [3] and oil palm fiber [4]. The electrical conductivity of carbon is affected by the cellulose composition of the natural fibers [5] and the pyrolysis temperature [6].

One of the promising materials for carbon conductive is Ramie fiber (Boehmeria nivea). The Ramie fiber has higher cellulose content of about 68.6 – 76.2%[7–9] than natural fibers in the previous study i.e. coconut [3] and oil palm fiber [4]. With the higher content of cellulose, carbon ramie fiber is predicted to have higher electrical conductivity [5] that makes it a suitable material for GDL in PEMFC.

In this research, the authors' objective is to look for a method to process the ramie fibers to become carbon composite that has good characteristics similar to that of GDL in PEMFC. The hot press method was chosen to produce GDL from carbon. This method is a simple and inexpensive process. Three treatment were selected to achieve the goals; the degumming process to increase the content of cellulose, pyrolysis temperature to produce higher electrical conductivity carbon and the optimum composition of carbon composite to produce a good solid layer.

2. Experimental

This research was divided into two stages. The first stage was the process of producing carbon from ramie fiber through pyrolysis process. Prior to the pyrolysis process, ramie fiber was applied to initial treatment with the degumming process to remove hemicellulose/gum so that cellulose content increases [10][11]. The degumming process was carried out by soaking the ramie fiber in 10%wt NaOH solution for 3 hours at 95 °C. Then the ramie fibers were washed with water and dried in the oven. The pyrolysis temperature used to convert ramie fiber to carbon fiber was based on the results of the thermal analysis (TGA). The resulting carbon fibers were then polycyclicized at the higher pyrolysis temperature (900 °C, 1100 °C, 1300 °C). This re-pyrolysis treatment aims to increase the value of electrical conductivity [12]. The electrical resistance of carbon ramie was measured using LCR-meter.

The second stage was the fabrication of composite using hot press method. The composite is the mixture of carbon ramie and urea formaldehyde resin. The hot press was applied at the temperature of 150 °C, the pressure of 200 kg/cm^2 and for 30 min. In this process, carbon ramie content was varied between 70%, 75%, and 80%. Afterwards, the composite product was coated with hydrophobic coating 10 wt.% FEP solution. For characterization, the properties of samples were evaluated by electrical conductivity, porosity, contact angle and surface morphology. The electrical conductivity of carbon composite was measured using LCR-meter Hioki 3522. The porosity was measured using pycnometer (BS 1902: Part 1A: 1966). The measurement of the contact angle with water droplet using Bashforth and Adams 1883 Table [13]. The surface morphology was observed using scanning electron microscope SEM HITACHI SU-3500.

3. Results and discussion

3.1. *Thermal analysis of ramie fiber*

Figure 1(a) shows the results of Thermal Gravimetric Analysis (TGA) and Differential Scanning Calorimetry (DSC) measurements for ramie fiber without degumming. It shows three intervals of mass loss. The first interval occurred at the temperature of 50 °C – 260 °C, the second interval occurred at the temperature of 260 °C – 370 °C and the third interval at 370 °C – 600 °C. At the first interval, relative mass decreases about 10% due to hemicellulose decomposition of the ramie fibers. This process took place at the temperature between 200 °C – 260 °C [14]. The second interval shows the occurance of cellulose decomposition at temperature of 260 °C – 370 °C with a total mass loss about 47%. The decrease in mass resulting from the release of NO, C_4H_7, CH_4, C_4H_3, C_3H_6, C_3H_3 components of ramie fiber material [14]. At the third interval, the mass decrease of 10% from temperature of 400 °C – 600 °C due to the decomposition of lignin. In the other hand, the DSC curve shows two endothermic peaks at 370 °C – 400 °C indicates cellulosic decomposition. Another endothermic peak temperature range of 770 °C – 780 °C, it was suggested due to the melting of KCl mineral, that occurred at 772 °C [4].

Figure 1(b) shows the results of TGA and DSC for the degummed ramie fiber. It was identified that the results have the same intervals as those of non-degummed fibers. The differences are in terms of the relative magnitude of the mass drop. In the interval two of degummed ramie fibers, the decomposition of cellulose has a relative mass decrease of 58%. There is an increase by 11% of the mass loss. This proves the increased cellulose content in the degummed ramie fibers. Based on these TGA result, the first pyrolysis temperature to convert ramie fibers to carbon is at 500 °C.

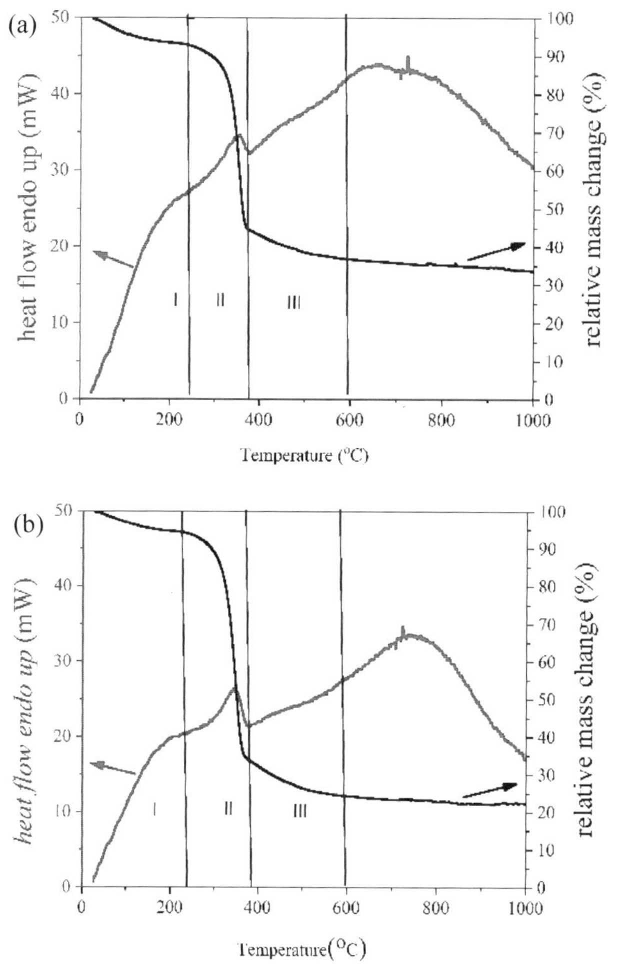

Figure 1. TGA and DSC of ramie fibers (a) without degumming process and (b) with degumming proces.

3.2. Effect of pyrolysis temperature on electrical conductivity of carbon from ramie fiber

The results of carbon pyrolysis temperature at 500 °C, 900 °C, 1100 °C and 1300 °C at 30 kgf.cm on the electrical conductivity are 8.7 x 10^{-7}, 0.84, 3.72 and 4.6 S/cm, respectively. This result is shown in Figure 2. The electrical conductivity of carbon is closely related to the structure of carbon material. The electrical conductivity increases significantly by rising the temperature of pyrolysis, which causes the growth of the carbon structure with turbostratic crystallites structure in the form of irregular graphene sheets [6]. The electrical conductivity of carbon with pyrolysis temperature at 1300 °C is 4.60 S/cm, this result is higher than the electrical conductivity of other natural fibers at the same pyrolysis temperature i.e oil palm fibers of 0,14 – 0,17 S/cm [4] and coconut fibers of 0.12 - 0.18 S/cm[15]. The ramie fiber has a higher cellulose content of about 68.6 - 76.2% [7–9] than the cellulose content of the coconut fiber of 23-43%, and the result of pyrolysis process of cellulose produces a more orderly carbon structure so that the electrical conductivity of carbon ramie is higher [5].

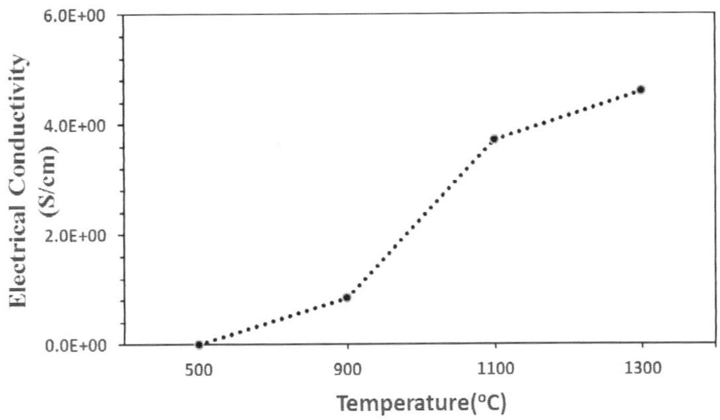

Figure 2. Electrical conductivity of carbon ramie.

3.3. Carbon composite from ramie fiber

3.3.1. Effect of initial treatment (degumming process) and carbon composition on the through-plane electrical conductivity of the composite. Figure 3 shows the electrical conductivity of carbon composites at the pyrolysis temperature of 900 °C with initial treatment (degumming process) of ramie fibers and carbon compositions variations. The electrical conductivity of carbon composites with carbon compositions variations of 70%, 75%, and 80% is 0.03 S/cm, 0.11 S/cm and 0.26 S/ cm respectively. Increasing the percentage of conductive carbon increases the electrical conductivity. The increase in electrical conductivity has proceeded constantly as an increase in the percentage of carbon to a point where the increase in the amount of carbon has increased dramatically, a point called a percolation threshold. The addition of carbon after the percolation threshold does not significantly affect the electrical conductivity. However, it affects the decreasing mechanical properties of the composite [12].

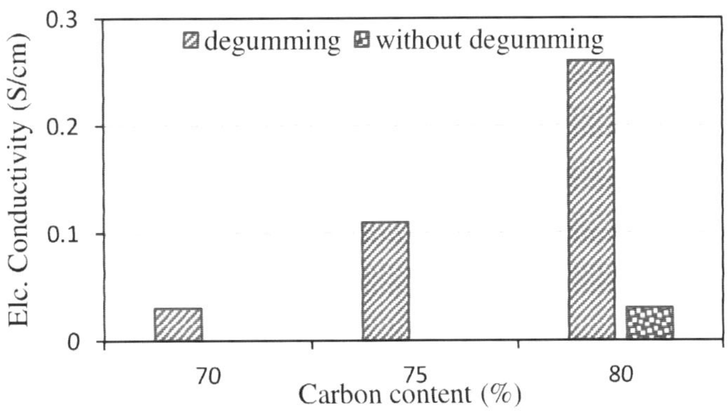

Figure 3. The electrical conductivity of carbon composite.

The through-plane laminar electrical conductivity of the carbon degummed ramie composite is nearly ten times than the electrical conductivity of carbon ramie composite without degumming. The result shows 0.26 S/cm for degummed ramie versus 0.03 S/cm for non-degumming. Higher cellulose content in degummed ramie fiber produces more electrically conductive carbon [5].

3.3.2. Carbon composite with optimum variabeles. After investigations of the effect of degumming treatment, the increase of pyrolysis temperature, and the composition of carbon ramie fibers in the composite on the electrical conductivity, the experimental works are continued with the composing of the composites by using the optimum result of those variables.

Based on the results of the research, the degumming treatment and the increase of pyrolysis temperature increases the electrical conductivity of carbon with a yield of 4.6 S/cm for pyrolysis at 1300°C and the optimum composition of carbon fiber in the composite is 80%. The electrical conductivity of composite of the above three variables has an electrical conductivity of 2.85 S/cm. The result has a higher value than the electrical conductivity of coconut fiber that has an electrical conductivity of 2.22 S/cm [16]. But GDL from coconut fiber uses carbon with an electrical conductivity of 0.18 S/cm. There was an increase in the electrical conductivity by a factor of 12 if one uses the casting process to produce GDL from carbon coconut fiber[16]. Using hot press method with carbon ramie and urea formaldehyde resin produces a lower electrical conductivity value than the original electrical conductivity of carbon ramie fiber itself. There is expected that fabrication GDL from carbon ramie fiber using the same method as the previous study [16] can produce GDL with higher electrical conductivity than GDL using hot press method. But with hot press method result, carbon ramie fiber composite using hot press method has good potential for GDL in PEMFC.

3.3.3. Hydrophobic coating on carbon composite. The hydrophobic coating is used to support GDL capability in avoiding the occurrence of water flooding at the GDL-catalyst layer in the cathode area [2]. The hydrophobic coating used was 10 wt.% FEP. This content was chosen because from the study conducted by Chan Lim and C.Y. Wang on the effect of FEP content used in GDL with the content variation of 10-40%. The GDL with 10 wt.% FEP has higher current density and relatively similar contact angle than higher content [17].

Figure 4 shows water droplets on carbon composite without a hydrophobic coating and Figure 5 water droplets on the composites coated by 10 wt.% FEP over time of 0-60 min observation. The contact angle of the composite without coating is below 90° while the composite after coating of 10 wt.% FEP layer is 133° which is hydrophobic.

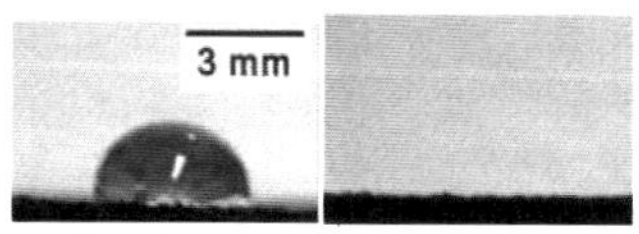

0 minute >0 minute

Figure 4. Water droplet on carbon composite.

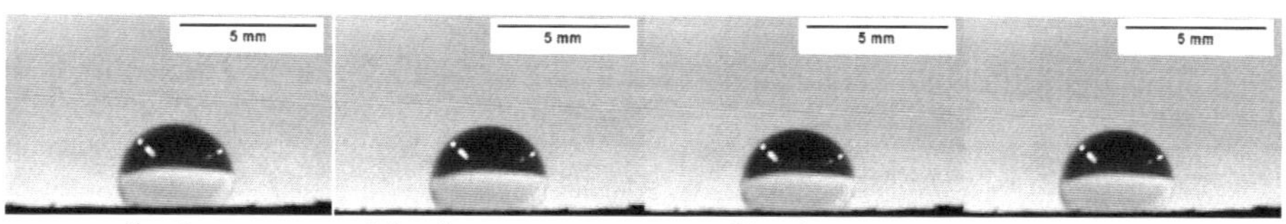

0 minute 15 minutes 30 minutes 60 minutes

Figure 5. Water droplet on carbon composite with hydrophobic coating.

3.3.4. Effect of hydrophobic coating on the porosity of carbon composite. The porosity of the GDL has an effect on PEMFC performance. High porosity has an impact so that GDL is capable of supplying more reactant gases [18]. The porosity of the composite before and after hydrophobic coating decreased from 53.53% to 52.23%. This decrease results from the narrowing of the pores in composite with a hydrophobic coating. However, the porosity of carbon composite after hydrophobic coating is still in the range of porosity characteristics required by GDL in PEMFC that is between 50% - 90% [19].

3.3.5. Effect of hydrophobic coating on the electrical conductivity of carbon composite. The electrical conductivity of the carbon composite with hydrophobic is 0.25 S/cm has a one-order difference with a commercial GDL of 5.10 S/cm under the same measurement conditions. Hydrophobic coating with FEP on carbon ramie composite decreases the carbon composite electrical conductivity from 2.85 S/cm to 0.25 S/cm.The value of this decrease is higher than the effect of PTFE coating which has the same electrical resistance value as FEP> 1018 Ohm.cm. Carbon coconut composite fiber coated with 10 wt.% PTFE decreased the electrical conductivity value from 2.22 S/cm to 2.09 S/cm [16].

Figure 6 shows the surface morphology of the carbon composite from ramie fiber and coconut fiber. The two composites show the size of carbon fiber particles of ramie fiber is smaller than the carbon size in the coconut fiber coconut composite so that the surface area of the carbon ramie fiber composite is greater than carbon coconut composite. The coating of hydrophobic materials on the carbon ramie fiber composite has the higher percentage of hydrophobic content than on carbon coconut fiber composite. So that the coating of hydrophobic materials on carbon ramie fiber composite produces a significant decrease in electrical conductivity.

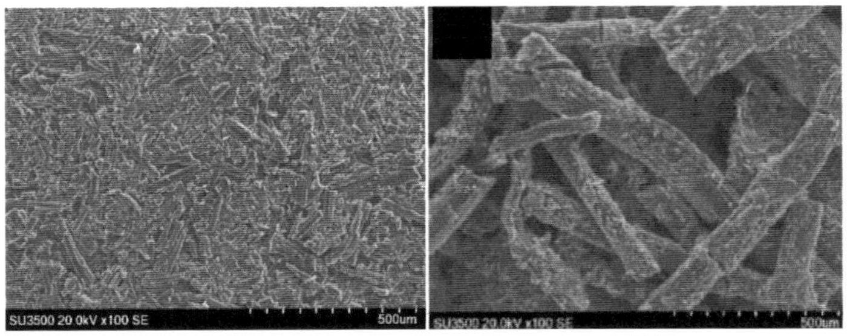

(a) (b)

Figure 6. Surface morphology of carbon composites (a) ramie fiber; (b) coconut fiber.

4. Conclusion

The initial treatment with the degumming process of ramie fiber produces carbon composites with higher electrical conductivity than carbon composites of ramie fiber without degumming treatment. Increased pyrolysis temperature in second stage pyrolysis of carbon ramie fiber leads to increase in electrical conductivity of carbon. In the first-stage pyrolysis with the temperature of 500 °C, the conversion process from ramie fiber to carbon has the electrical conductivity of 8.7×10^{-7} S/cm with measured pressure at 30 kgf.cm (maximum pressure given during measurement). Then in the second stage pyrolysis of carbon at 900 °C, 1100 °C and 1300 °C produce carbon with the increased electrical conductivity of 0.84 S/cm, 3.72 S/cm, and 4.6 S/cm. The optimum composition of carbon content in the composite is 80%, that has good electrical conductivity and compact. The composites with carbon content above 80% yield a fragile composite. Carbon ramie composite has the electrical conductivity of 0.25 S/cm, the porosity of 52.23% and hydrophobicity properties. therefore carbon ramie composite has good potential for GDL in PEMFC.

Acknowledgments

This research was supported by Research center of Physics, Indonesia Institute of Science. The authors would like to thank Indonesia Institute of Science for the financial support.

References

[1] Revankar S T and Majumdar P 2014 *Fuel Cells: Principles, Design, and Analysis*
[2] Cindrella L Kannan A M Lin J F Saminathan K Ho Y Lin C W and Wertz J 2009 *J. Power Sources* **194** 146–60
[3] Indayaningsih N Priadi D Zulfia A and Suprapedi 2011 *Key Eng. Mater.* **462–463** 937–42
[4] Indayaningsih N Zulfia A Priadi D and Hendrana S 2011 *Adv. Mater. Res.* **277** 137–42
[5] Sim I N Han S O Kim H Han I S Kim S Seo D W Seong Y H and Foord J 2014 *Macromol. Res.* **22** 996–1003

[6] Kercher A K and Nagle D C 2002 *Carbon N. Y.* **40** 1321–30
[7] Faruk O Bledzki A K Fink H P and Sain M 2012 *Prog. Polym. Sci.* **37** 1552–96
[8] Kalita B B Gogoi N and Kalita S 2013 *Int. J. Eng. Res. Gen. Sci.* **1**
[9] Pradeep P and Dhas E R J 2015 *Adv. Mater. Sci. Eng. An Int. J.* **2** 1–6
[10] Kundu P K Sarmah R and Sarkar C R 1996 *Indian J. Fibre Text. Res.* **21** 205–9
[11] Jose S Rajna S and Ghosh P 2016 *Asian J. Text.* **7** 1–9
[12] Ali M H F 2014 *Carbon Black Mediated Conductive Polymer Composite* UAE University
[13] Bashforth F and Adams J C 1883 *An Attempt to Test the Theories of Capillary Action by Comparing the Theoretical and Measured Forms of Drops of Fluid* Cambridge: Cambridge Univ. Press
[14] Xia Z Yao C Zhou J Ye W and Xu W 2016 C *Text. Res. J.* **86** 856–67
[15] Destyorini F Suhandi A Subhan A and Indayaningsih N 2010 *J. Phys.* **10** 122–32 (in Indonesia)
[16] Destyorini F Subhan A Indayaningsih N Prihandoko B and Syahrial A Z 2016 *Int. J. Technol.* **7** 1283–90
[17] Lim C and Wang C Y 2004 *Electrochim. Acta* **49** 4149–56
[18] Lee W K Ho C H Van Zee J W and Murthy M 1999 *J. Power Sources* **84** 45–51
[19] Yoon G H and Park Y il 2012 *Int. J. Precis. Eng. Manuf.* **13** 1153–9

IOP Conf. Series: Materials Science and Engineering **432** (2018) 012030 doi:10.1088/1757-899X/432/1/012030

Performance of post-hidrothermally treated xerogel TiO_2 dye-sensitized solar cell (DSSC) and its nanostructure characteristic

B Priyono[1*], A H Yuwono[1], A Z Syahrial[1], M H Mustofa[1], R S Bawono[1]

[1] Department of Metallurgical and Materials Engineering, University of Indonesia, Depok, 16425, Indonesia

bambang.priyono@ui.ac.id

Abstract. Dye-sensitized solar cell (DSSC) is one of photovoltaic solar cells, which is very favorable as alternative renewable energy source to anticipate the diminishing the fossil fuel due to its low-cost, ease of production and eco-friendliness. This work is aimed to investigate the correlation between TiO_2 nanostructure properties i.e.: crystallite size, surface area, band gap energy to obtain the optimum parameter for best performance as DSSC prototype. At this study, TiO_2 nanoparticles were synthesized via sol-gel process using titanium tetra-n-butoxide, ethanol, hydrochloric acid and deionized water with the ratio of (0.4 : 0.83 : 1 and 1.39) respectively upon sol preparation. The followed with multi-step calcinations to obtain anatase phase and post-hydrothermal process with various temperatures (100, 120, and 150 °C). As comparison, TiO_2 of P25 Degussa was used and also assembled to make DSSC. The X-ray diffraction, BET (Brunauer-Emmett-Teller) and UV-Vis DRS (diffuse reflectance spectroscopy) tests were taken to characterize the samples. The performance of DSSC was investigated using open voltage circuit (Voc) measurement. The characterization results of synthesis powder shows indication at temperature 150°C was optimum process of post-hydrothermal with crystallite size 10.55 nm, surface area 95.38 m^2/grams, and band gap energy 3.16 eV. Nevertheless, the open circuit voltage (V_{OC}) measurement reveals that the DSSC fabricated with post-hydrothermal at temperature process 120°C provide the highest voltage (250 mV).

1. Introduction

To overcome the polution problem and world energy crisis, the usage of solar energy would be inevitable due to its cleanliness and abundance resources. Dye-sensitized solar cell (DSSC) is one of the solar cell types which gain more attention today due to economic cost, high photon conversion efficiency and good stability [1]. Many development has been achieved to improve the DSSC performance [2]. One important approach is focussed on the photo-anode field. The DSSC performance improvement in this field relates to base material which is influenced by the structure of oxide layer having characteristic of the high surface area to absorb sensitizing dyes to maximize the performance required. One of semiconductor layer most commonly used in DSSCs is TiO_2 [3]. Many methods have been developed to make the DSSC components, i.e.: photoanodic oxide. Sol-gel method is one of most useful techniques that provide nanoparticles with high purity of precursor, high surface area, larger pore volume and uniform pore-size distribution. In our previous study [4], it has been learned how to synthesize nanoparticle with sol-gel method followed with multi-step calcinations and post-hydrothermal treatment to obtain high crystallinity and well-ordered Ti-O-Ti network. Beside, one important component of dye-sensitized solar cell (DSSC) is the photo-anode, which consists of a monolayer of highly porous semiconductor oxide layer. This study was carried out to develop a preparation technique that should increase the nanostructure properties to be integrated in the DSSC prototype. The method made use of xerogel combined with multi-step calcination followed by post-hydrothermal process as an effort to improve the crystallinity all of samples while maintaining minimum loss in the surface area [5]. The results of post-hydrothermally treated samples would be compared to P-25 Degussa.

The objective of the present study is to investigate the correlation between TiO_2 nanostructure properties such as: crystallite size, surface area, band gap energy, and their performance as a semiconductor material in the DSSC prototype. Hydrothermal process is employed to obtain high temperature and pressure for nucleation and crystallization of nanoparticles. This method is favorable for the formation of a better TiO_2 nano-particles crystallization [6]. On the other hand, super critical extraction technique (SCE) involves CO_2 as the extraction solvent has the potential to augment the surface area of nanoparticles but has a low degree of crystallinity.

This article describes the effects of temperature parameters on the properties of TiO_2 powder prepared by sol-gel method and post-hydrothermal method is used to control the properties and the reaction efficiency of TiO_2 nanoparticles, which are used in dye sensitized solar cells and performance of these cells.

2. Experimental method

TiO_2 sol was prepared by hydrolysis of titanium tetra-n-butoxide (Kanto Chemical, JPN) that was previously mixed with hydrochloric acid in ethanol solution under vigorous stirring and the water was added drop-wise to the precursor until the appropriate ratio was obtained. The molar ratio of Ti-but, ethanol, hydrochloric acid and deionized water were determined as (0.4 : 0.83 : 1 and 1.39), respectively. The mixture was stirred for 3 hours to stabilize the solution, and then sealed in a container for 1-2 days at room temperature. TiO_2 P-25 Degussa nanoparticles powder was obtained using P-25 Degussa Merck™. Next, the gel was processed using a multi-step calcination method in a tube furnace. The multi-steps was carried out at two different conditions, i.e. the first and second temperature were carried out under vacuum condition, while the third one was performed by injecting air for regaining the white powder of the TiO_2 nanoparticle. The first step was conducted at 150 °C in which the solvent would evaporate, whereas the second step was carried out at 300 °C in order to remove the entire remaining organics compound, and subsequently the final step was carried out at 420 °C under an air-injection condition. After calcination process the gel was prepared to undergo a post-hydrothermal treatment in a Teflon-lined stainless steel autoclave which is equipped with a special stand-holder for sample, sealed and placed in an oven at 100°, 120° and 150 °C for 14 hours. Afterwards, it was dried at 70 °C for 24 hours.

Post-hydrothermal samples and P-25 Degussa samples were characterized by XRD method to identify the phase of TiO_2 anatase and to determine the crystallite size via Scherrer's equation analysis. The BET (Brunauer-Emmet-Teller) testing was used to determine the surface area, and UV-Vis with Diffuse Reflectance Spectroscopy (DRS) method is used to determine the band-gap energy calculated via Kubelka-Munk function that transformed to Tauc's equation analysis. After characterization, both samples were prepared to be assembled as DSSC. The performance of dye-sensitized solar cell (DSSC) assembled from the nanoparticles resulted from both of methods tested through the measurement of open circuit voltage (V_{oc}), taken under the light beam source of an overhead projector.

3. Results and discussion

In order to improve the crystallinity of sol-gel, the multi-step calcinations were conducted. After that, samples were post-hydrothermally treated to increase the order of Ti-O-Ti bond. It was confirmed that all of samples which post-hydrothermally treated shows anatase phase. For P-25 Degussa, it shows combination of anatase and rutile phases because of higher manufacturing process above 600 ^{0}C [7].

The XRD patterns of samples of P25 Degussa sample and post-hydrothermal treatment at 100 °C, 120 °C, and 150 °C were shown in the Figure 1.

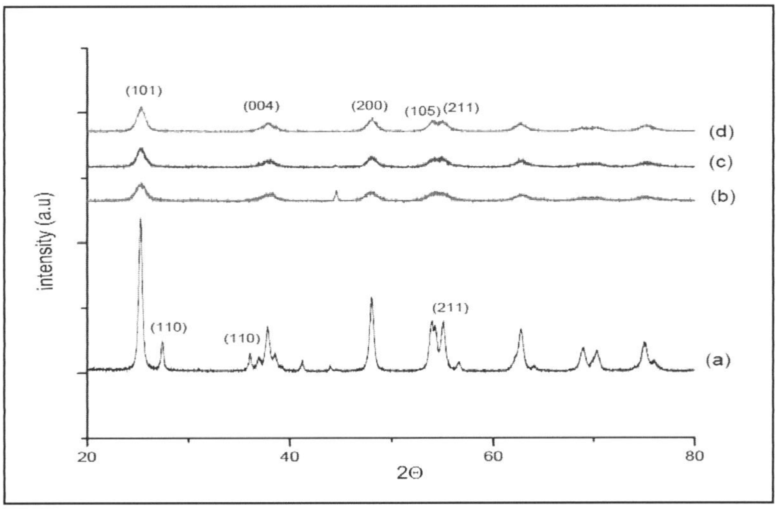

Figure 1. XRD Pattern of (a) P-25 Degussa and Post-hydrothermally treated at temperature, (b) 100 ^{0}C, (c) 120 ^{0}C, (d) 150 ^{0}C.

In Figures 1 (b), (c) and (d), are shown the XRD patterns of as-prepared TiO_2 powders post-hydrothermally treated at different temperatures (100 °C, 120 °C and 150 °C). The as-prepared sample show reflection peaks at 2θ of 25.65°, 37.60°, 38.95°, 48.37°, 54.23°, 55.60°, 63.20°, 68.99°, 70.78° and 75.37°, which are assigned to (101), (004), (112), (200), (105), (211), (204), (116), (220) and (215) lattice planes of anatase TiO2 phase (JCPDS Card no. 21-1272). The well-formed peak shows clearly that the anatase phase was successfully formed by multi-step calcinations process at 150, 300, and 420°C respectively.

Meanwhile, ss comparison, in Figure 1 (a), the P25 Degussa has additional peaks at 27.5°, 36.04°, 41.18°, 44.14°, 61.79°, 68.99°, 70.17° and 76.80° which are assigned to the (110), (101), (111), (210), (002), (301), (112) and (202) lattice planes of rutile phase (JCPDS Card no. 21-1276). The peak intensities of the anatase phase do not diminish even with the appearance of the rutile phase.

Based on those peaks, the broadening on each of peak was measured and furthermore was used to estimate the crystallite size of TiO2 nanoparticles by using the Scherrer equation by measuring the widening of the peak, that is the full width at half of the maximum intensity (FWHM). The results are summarized in Figure 2.

As the temperature of post hydrothermal treatment increases, the crystallite size of the respective samples also increased. The calculated crystallite size shows that the largest size was possessed by the 150°C post-hydrothermally treated sample, i.e. 10.55 nm with a surface area of 95.38 m^2/g. The second largest one is the 120 °C post-hydrothermally treated sample with a value of 8.85 nm, accompanied with a surface area of 92.25 m^2/g. Finally the last sample is a 100 °C post-hydrothermally treated sample, i.e. 5.49 nm, but the surface area was only 117.96 m^2/g. This result is possibly caused as a consequence of an attempt to suppress the fast development of stiff Ti-OH by pre-hydrothermal treatment which led to a largely amorphous particle [8]. The highest crystallite size was obtained by post-hydrothermally treated sample at 150 ^{0}C. Enhancement in crystalline size refers to hydrothermal growth at higher temperature [7]. However, P-25 Degussa samples still has higher crystallite size due to high crystallinity compared to post-hydrothermal samples. The crystallite size of 7.29 nm was obtained by calculation for the noncalcinated TiO_2 powder. The XRD patterns of TiO_2 nano-powders calcinated at different temperatures show that the anatase phase was detected for the calcinated samples till a calcinated temperature below 500 ^oC.

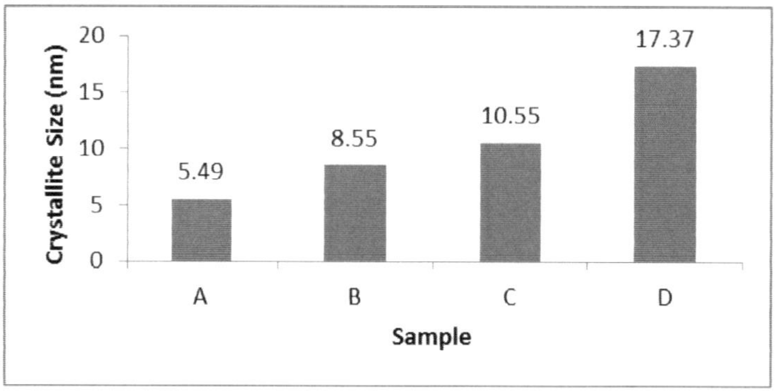

Figure 2. Crystallite Size estimation of TiO_2 nanoparticles Post-hydrothermally treated at temperature of 100 ^0C, (b) 120 ^0C, (c) 150 ^0C and (d) P-25 Degussa.

The result of BET testing is depicted in Figure 3. Based on BET testing, it could be established how extensive the size of the surface area is, which is obtained via those processes. One also would then understand how effective the multi-step calcinations process is, to minimize the drop of surface area while gaining the crystallinity of the nanoparticles.

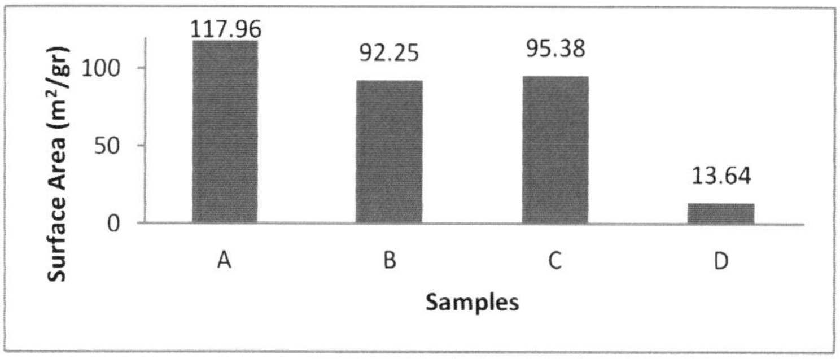

Figure 3. Surface Area of TiO_2 nanoparticles Post-hydrothermally treated at temperature 100 ^0C, (b) 120 ^0C, (c) 150 ^0C and (d) P-25 Degussa.

As the temperature increases, there is an indication that the surface area of the post-hydrothermally treated sample has decreased which causes the growth of the crystal grains. However, a unique result obtained by a post-hydrothermally treated sample at 150 ^oC shows that there is a slight increase compared to the 120 ^oC treated sample. P-25 Degussa sample has the lowest surface area due to high crystallinity and also the lowest value of crystallite size because of the presence of the rutile phase, where the average rutile's diameter size is greater than the diameter size of the anatase [9] .

The optical characterization by UV-Vis with DRS mode was aimed at observing optical response of TiO_2 synthesized nanoparticles. Figure 4 shows the absorbance spectra of all four samples.

From the figure bellow, it can be seen that the post-hydrothermal 150 ^{0}C move toward to red shift compared to 120 ^{0}C sample. DR/UV–vis directly provides some insight into the interactions of the photocatalytic materials with photon energies [10]. The obtained result of UV-vis spectroscopy is the UV spectral intensity versus wavelength of each sample which is then converted into band-gap energy.

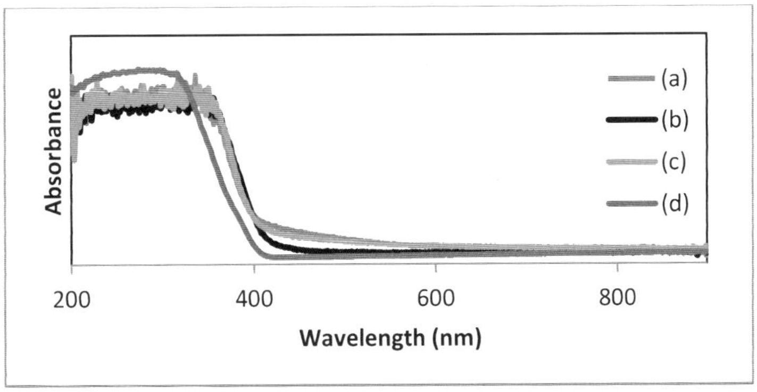

Figure 4. Photo-activity of TiO_2 nanoparticles Post-hydrothermally treated at temperature 100 ^{0}C, (b) 120 ^{0}C, (c) 150 ^{0}C and (d) P-25 Degussa.

This finding is in accordance with the quantum size effect, where the nanocrystals with larger size exhibited lower energy and displayed red shift. Conversely, the nanocrystals with smaller size displayed blue shift because of their higher energy. In our experiments, a red shift was observed for the products consisting of rutile nanorods compared with pure anatase particles, because the size of rutile nanorods was larger than anatase particles [11]. As a comparison, P-25 Degussa has very smooth line which was blue shifted from post-hydrothermal samples. The band gap energy of samples was further determined by using Kubelka-Munk equation and the result is shown in Figure 5.

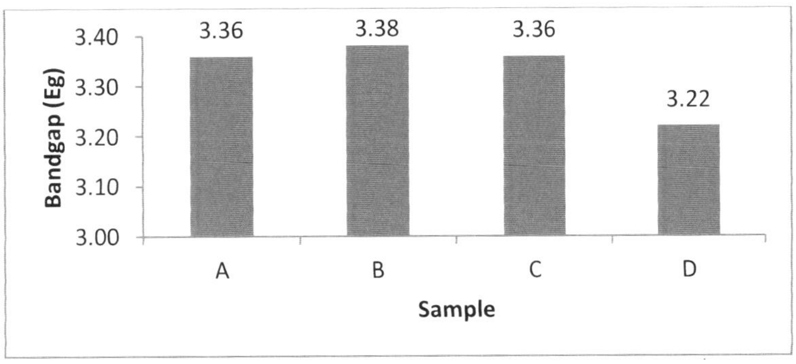

Figure 5. Band gap energy of TiO_2 nanoparticles Post-hydrothermally treated at temperature 100 ^{0}C, (b) 120 ^{0}C, (c) 150 ^{0}C and (d) P-25 Degussa.

Figure 5 shows that post-hydrothermally treated sample at temperature 100 °C and 150 °C has the lowest band gap value among post-hydrothermal samples. Both samples obtained band gap energy i.e. 3.36 eV. The lower band gap energy value indicates that the sample has higher crystallinity. In Figure 5, it is shown that the higher crystallite size belonging to the 150 °C post-hydrothermal sample did not provide the lowest band gap energy value. To clarify this phenomenon further investigation and more accurate calculation must be performed.

Moreover, the DSSC performance of the sample was examined by measuring the open voltage circuit (V_{oc}) of the prototype upon exposed in 50 W projector lamp. The result of open circuit voltage is shown at Figure 6.

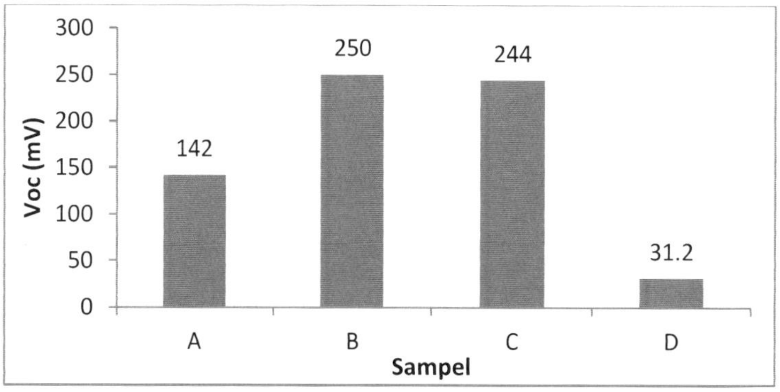

Figure 6. Open Circuit Voltage (V_{OC}) of TiO_2 nanoparticles Post-hydrothermally treated at temperature 100 ^{0}C, (b) 120 ^{0}C, (c) 150 ^{0}C and (d) P-25 Degussa.

DSSC performance of open voltage circuit correlates with crystallinity, surface area, and band gap energy. From Figure 1, the highest crystallite size was achieved by the Post-Hydrothermal-150 treated sample at 10.55 nm, but the V_{OC} value of the sample does not show the maximum value. Post-Hydrothermal-120 shows the highest open voltage circuit at 250 mV. The lowest band gap energy, high crystallite size of up to 8.85 nm and also a high surface area value of 92.25 m^2/gr which also makes this sample to achieve the highest V_{OC}. P-25 Degussa should have achieved highest V_{OC} due to highest crystallite size (17.33 nm) and lowest band gap energy to compared TiO_2 anatase (3,2 Ev) [12], but due to the lowest surface area V_{OC} value of P-25 Degussa obtained the lowest value i.e 31.2 mV.

Based upon these indications, it seems proper to suggest that post-hydrothermal process has made the stiff Ti-OH network to become more flexible and to be arranged as Ti-O-Ti after completion of the hydrolysis process. Ti-O-Ti rearrangement has helped to make the degree of crystallinity of the TiO_2 higher and for the sample to achieve a better performance. The effect of this phenomenon has improved the distribution of Ti-O-Ti network in the hydrothermally treated. The highest V_{OC} of post-hydrothermal at 120 °C sample indicates that in this temperature at 120 °C is the optimum condition of post-hydrothermal treatment in this study.

4. Conclusion

All the TiO_2 xerogel samples in the present study have been successfully synthesized by sol-gel process, and in this work have been successfully assembled to become DSSC and have displayed their activity in converting the light spectrum into electrical energy as shown by their open circuit voltage (V_{OC}) values. The TiO_2 xerogel possesses good nanostructure characteristics in terms of relatively large surface area and a sufficiently high crystallinity and post-hydrothermal treatment with operating parameters and conditions. The increase of the post-hydrothermal temperature gives the effect of increasing the diameter of crystallite size of the samples. However, conversely the surface area and energy band gap values obtained have not shown any downward trend.

The nanostructure TiO_2 properties obtained at temperature 150 °C shows the highest crystalline size (10.55 nm), high surface area (95.28 m2/gr), and bandgap energy (3.36 eV), so it is considered as the optimum value of the present post-hydrothermal process. The performance test of open circuit voltage revealed that the DSSC fabricated with P-25 Degussa provides 3.14 mV, while the highest result belongs to post-hydrothermal at 120 °C with 250 mV followed by post-hydrothermal at 150 °C and 100 °C.

Acknowledgment
The authors would like to thank the financial support from Directorate of Research and Community Services-University of Indonesia (DRPM-UI) through Grants for Middle Research University of Indonesia with contract no. DRPM/RII/175/RM-UI/2013.

References
[1] O'Regan B Gratzel M 1991 *Nature* **353(6346)** 737–40
[2] Khan M Z H Halder P K Aziz M A 2017 *Renew. and Sust. Energy Reviews* **71** October 2016 602–17
[3] Sofyan N Ridhova A Yuwono A H Udhiarto A 2017 *Int. J. of Tech.* 1229–38
[4] Priyono B Yuwono A H Munir B Rahman A Maulana A Abimanyu H 2013 *Adv. Mater. Research* 28–32
[5] Khan M A Akhtar M S Yang O 2010 *Solar Energy* **84(12)** 2195–201
[6] Brodsky C J Ko E I 1994 *J. of Mater. Chem.* **4(4)** 651–2
[7] Kashyout A B Soliman M Fathy M 2010 *Renewable Energy* **35(12)** 2914–20
[8] Yuwono A H Xue J Wang J Elim H I Ji W 2006 *J. of Electroceramics* **16(4)** 431–9
[9] Fujishima A Zhang X Tryk D A 2008 *Surface Science Reports* **63** 515–82
[10] Hussain M Ceccarelli R Marchisio D L Fino D Russo N Geobaldo F 2010 *Chem. Engin. J.* **157(1)** 45–51
[11] Dai S Wu Y Sakai T 2010 *Nanoscale Research Lett.* 1829–35
[12] Langlet M Kim A Audier M Mat L 2002 *J. of Sol-Gel Sci. and Tech.* 223–34

Antioxidant activity and dose enhancement factor of CeO_2 nanoparticles synthesized by precipitation method

I Nurhasanah[1], W Safitri[1], Z Arifin[1], A Subagio[1], and T Windarti[2]

[1] Department of Physics, Faculty of Science and Mathematics, Diponegoro University
[2] Department of Chemistry, Faculty of Science and Mathematics, Diponegoro University, Prof. H. Soedarto St., S.H., Semarang 50275, Indonesia

nurhasanah@fisika.undip.ac.id

Abstract. CeO_2 nanoparticles (CeO_2 NPs) have been considered as promising antioxidant and radioprotectant due to its mixed valence state that can prevent cell damage induced by ionizing radiation. The aim of this study is to investigate the antioxidant activity and dose enhancement factor (DEF) of CeO_2. The NPs were synthesized by using a precipitation method. The structure of CeO_2 NPs was characterized with an X-ray diffractometer (XRD). The antioxidant activity of CeO_2 NPs was analyzed by the reduction of 1,1-diphenyl-2-picrylhydrazil (DPPH) stable free radical using UV-vis spectroscopy. The dose enhancement factor (DEF) of CeO_2 NPs was examined using X-ray radiation with an energy of 6 MV. The radioprotective ability of CeO_2 NPs was evaluated by mixing CeO_2 NP suspension with *E. coli* and exposing it to X-ray radiation with a dose of 2 Gy. The XRD pattern analysis reveals that CeO_2 NPs possess a cubic fluorite structure. CeO_2 NPs show activity with IC50 value of 4.4 mg/ml. CeO_2 NPs also show the ability to absorb X-ray radiation with DEF value < 1 and reduce the *E. coli* damage induced by X-ray radiation. The amount of irradiated *E. coli* in the presence of 0.2 mg/ml CeO_2 NPs was found to be 23 times larger than that irradiated *E. coli* in the absence of CeO_2 NPs.

1. Introduction

The application of nanoparticles in radiotherapy has been a subject of considerable interest. Radiotherapy is the reliable way for cancer treatment by exposing cancer cells to ionizing radiation. During radiotherapy of cancer, the ionizing radiation resulted in an increased rate of not only cancer cell death but also normal cell death. The main challenge in the cancer therapy is to increase the radiation ability to kill cancer cells by increasing the dose effectiveness and minimizing damage to the surrounding normal cells. The dose effectiveness to kill cancer cells could be increased by using radiosensitizer, while radioprotector is used to protect normal cell [1-3]. Many free-radical scavengers or antioxidants have been investigated and used to overcome the effects of radiation on normal cell. Amifostine is the only compound as radioprotector recommended by US FDA (Food and Drug Administration) [4, 5]. However, this compound has limitations due to short life and poor penetration to the location where the free radicals produced.

The current nanotechnology makes it possible to develop a variety of nanoparticles tobe used for biomedical application with superior properties. Nanoantioxidants, including inorganic nanoparticles have shown promise as high-performance therapeutic nanomedicine in attenuating oxdative stresses. However, the small size and large surface area of nanoparticles resulted in particles aggregation and led to difficulty in the handling of nanoparticles in liquid [6]. A few metal oxides nanoparticles were reported as antioxidants, including TiO_2 [7], ZnO [9-11], and CeO_2 [12-17]. Among metal oxide nanoparticles, applications of CeO_2 nanoparticles (CeO_2 NPs) in biomedical have experienced growing attention due to lower toxicity than the other oxide nanoparticles that commonly investigated such as TiO_2 and ZnO. Furthermore, CeO_2 nanoparticles are known as catalysts that have pharmacological potential and promising therapeutic agent. The redox reaction that occurs between Ce^{3+} and Ce^{4+} with superoxide and oxidizing hydrogen peroxide are similar to the behavior of antioxidant enzymes [18, 19]. The antioxidant properties of CeO_2 NPs are determined by the presence Ce^{3+} which depend on shape, size, valence states and synthesis method [13-15].

This paper describes the antioxidant activity and dose enhancement factor of CeO_2 NPs synthesized by precipitation method. The antioxidant activity of nanoparticles CeO_2 was examined using DPPH method. The dose enhancement factor of CeO_2 NPs was determined using the absorbed dose of X-ray radiation measurement. The capability of CeO_2 NPs as radioprotector was evaluated by loading those CeO_2 NPs into *E. coli* culture under X-ray irradiation.

2. Experiment

The CeO_2 NPs were synthesized by precipitation method from cerium nitrate solution as described in previous report [20]. The solution of 0.08 M cerium nitrate (Sigma-Aldrich) was prepared in demineralized water/isopropanol mixed solvent with a volume ratio of 1:6. NH_4OH was then added drop by drop under stirring into the solution until a pH value of 10 was reached. Precipitates were repeatedly washed by isopropanol. The precipitates were then dried at 60 °C for 2 hours and calcined at 500 °C for 2 hours.

The crystallinity and phase of CeO_2 NPs was characterized by X-ray powder diffractometry (XRD) with Cu K_α radiation (λ = 1.54060 A). UV-Vis transmittance of the CeO_2 NPs was examined using UV-Vis spectrophotometer. Antioxidant activity of CeO_2 NPs was evaluated using DPPH (1,1-diphenyl-2-picrylhydrazyl) radical scavenging method. The suspension of CeO_2 NPs with different concentration (1.5, 3.0, 4.5, 6.0, 7.5, and 9.0 mg/mL) were sonicated using sonicator bath at room temperature for 10 minutes to avoid agglomeration. The absorbance of samples (the mixture of CeO_2 NPs and DPPH solution) and control (DPPH solution without CeO_2 NPs) were measured by UV-Vis spectrophotometer. The ability of CeO_2 NPs to scavenge DPPH radical was calculated using equation (1). Furthermore, the required concentration to inhibit 50% of DPPH (IC_{50}) was determined by linear regression.

$$\%\ \text{inhibition} = \frac{A_C - A_S}{A_C} \times 100\% \tag{1}$$

To investigate dose enhancement factor (DEF), the absorbed dose of CeO_2 NPs with various mass of 0,5 g, 1 g, 2 g, and 2,5 g was measured using detector. Measurements were made for photon energy of 6 MV and dose rate of 200 cGy/min, SSD of 100 cm, the field size of (2x2) cm^2, and detector position at 1 cm below phantom surface. The absorbed dose was measured in the absence and in the presence of CeO_2 NPs. The dose enhancement factor (DEF) was defined as ratio of radiation dose in the presence of CeO_2 NPs to radiation dose in the absence of CeO_2 NPs.

The protection effect of CeO_2 NPs was investigated by mixing CeO_2 NPs suspension with *E. coli* culture at various concentrations in the range 0.02 to 0.1 mg/mL. The *E. coli* cultures were irradiated by using a photon energy clinical linear accelerator (LINAC) of 6 MV X-ray with a dose rate of 200 cGy/min, a depth of 10 cm and source-to-surface distance (SSD) of 90 cm. The number of live *E. coli* was determined by using total plate count method.

3. Results and discussion

The X-ray diffraction (XRD) pattern of CeO_2 NPs synthesized using precipitation method is shown in Figure 1. The XRD pattern reveals that the prepared nanoparticles are polycrystalline of cubic fluorite

CeO_2 structure. Four preferred diffraction peaks at the 2θ values of about 28.46°, 32.97°, 47.44° and 56.18° are observed corresponding to (111), (200), (220), and (311) planes, respectively. Those diffraction peaks are in good agreement with the Joint Committee on Powder Diffraction Standard (JCPDS) No. 34-0394. No other peaks related to impurities or other phases are detected in the XRD pattern which confirmed the prepared CeO_2 NPs are single phase crystalline of CeO_2. This result was confirmed by Rietveld refinement as shown in Figure 1(b). The Rietveld refinement shows good agreement between experimental and simulated XRD pattern with the refinement factors being R_{wp} = 9.6 % and R_p = 7.55% with the goodness-of-fit (χ^2) = 1.146. The refined cell parameter of CeO_2 NPs is 5.419 Å which is slightly larger than that of bulk CeO_2 (5.411 Å) indicating the prepared CeO_2 NPs consist of nanocrystallite. The crystallite size was calculated using Scherer formula to be about 9 nm.

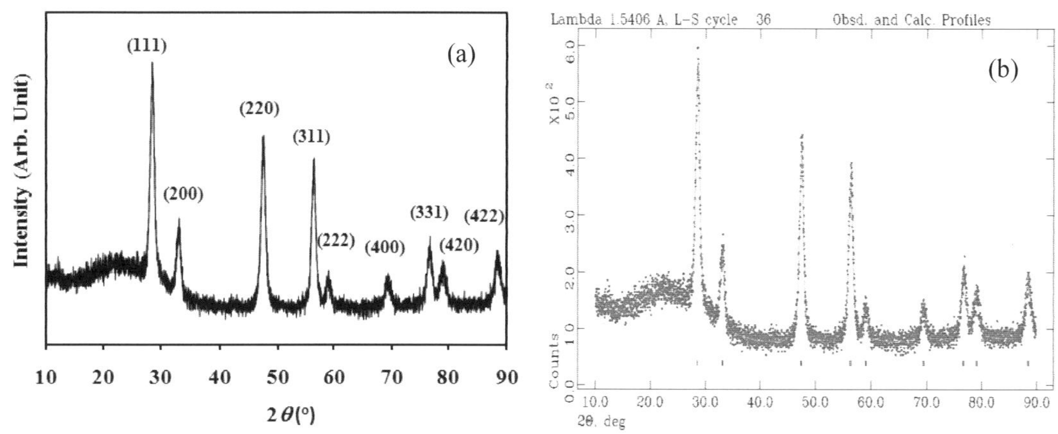

Figure 1. XRD pattern (a) and Rietveld refinement (b) of CeO_2 NPs synthesized by precipitation method.

The UV-Vis absorbance spectrum of the CeO_2 NPs is shown in Figure 2. The peak of absorbance was observed at wavelength of 207 nm and 303 nm corresponding to characteristic absorption peak of Ce^{3+} and Ce^{+4} of CeO_2 NPs. The sprectrum reveals that CeO_2 NPs consist of Ce^{3+} and Ce^{4+} ions. The present of Ce^{3+} ion has an important role on the antioxidant activity of CeO_2 NPs [14].

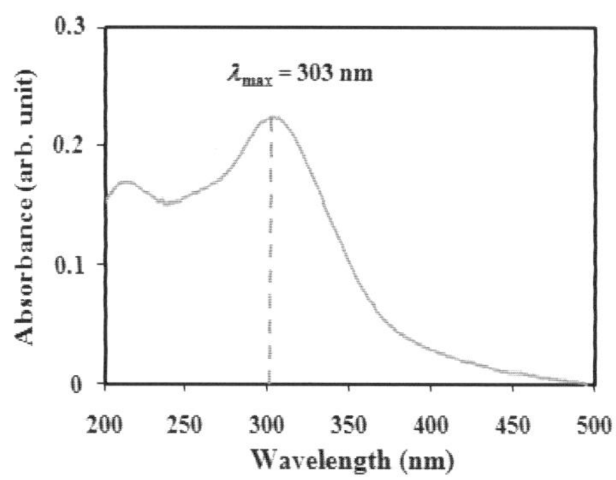

Figure 2. Absorbance spectrum of CeO_2 NPs.

In this study we use DPPH as a stable free radical to determine the radical scavenging activity of CeO_2 NPs. The absorbance of DPPH decreased due to the scavenging of DPPH by donation of electron or hydrogen from antioxidant substance to form the stable DPPH molecule. The radical scavenging activity values were expressed as percentage of inhibition as ratio of percentage of sample absorbance decrease and the absorbance of DPPH solution in the absence of antioxidant substance at absorption peak of DPPH (λ = 520 nm). Figure 3 shows radical scavenging activity of CeO_2 NPs. The inhibition of DPPH increased up to 67% at concentration of CeO_2 NPs of 9 mg/mL. In another study, CeO_2 NPs prepared by co-precipitation method and coated by polysaccharide showed DPPH scavenging activity with % inhibition of 85% [13] whereas, CeO_2 NPs prepared by hydrothermal and solvothermal methods showed DPPH scavenging activity up to 55% and 30% respectively [15]. The DPPH scavenging activity of our CeO_2 NPs is smaller than levan-coated CeO_2 NPs, but higher that that CeO_2 NPs prepared by hydrothermal and solvothermal methods. The CeO_2 NPs posses antioxidant activity with IC_{50} value of 4,38 mg/ml. The CeO_2 NPs exhibited the higher antioxidant activity than ZnO nanoparticles (IC50 ~ 8 – 10 mg/mL) that reported in the literature [9-11]. The antioxidant properties of CeO_2 NPs would be explained as due to transfer of electron from CeO_2 NPs to odd electron located at nitrogen atom in DPPH.

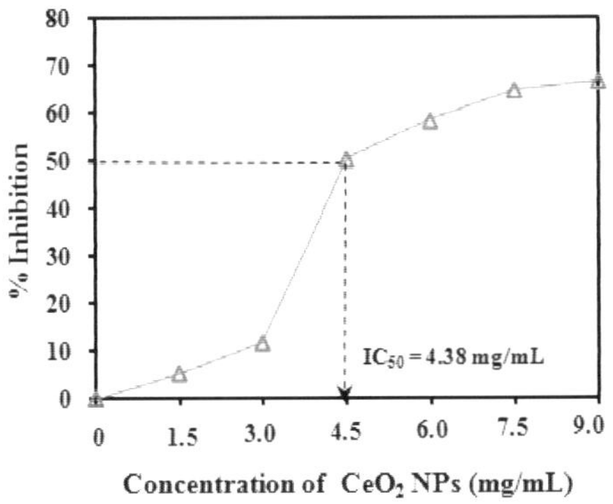

Figure 3. Radical scavenging activity of CeO_2 NPs.

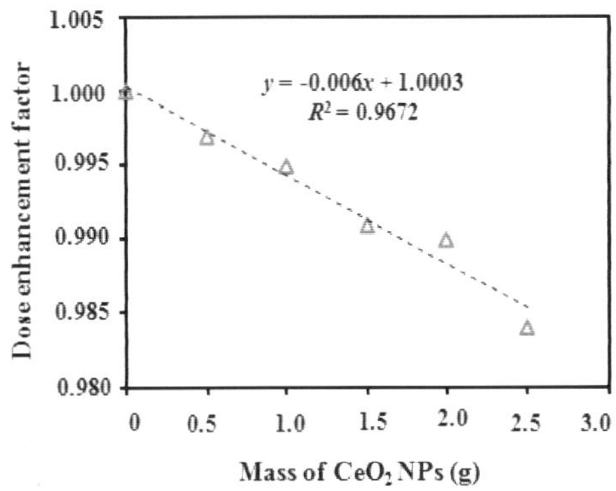

Figure 4. Dose enhancement factor for various mass of CeO_2 NPs.

Figure 4 depicts the dose enhancement factor (DEF) for various mass of CeO_2 NPs. The DEF values of CeO_2 NPs were 0.984 to 0.997 (DEF < 1). The DEF value decreased linearly as increase in mass of CeO_2 NPs. Its show the addition of CeO_2 NPs can decrease absorbed dose of about 1.2 cGy/g. The CeO_2 NPs are potent as radioprotector for preventing normal cell damage induced by X-ray radiation.

The protective effect of CeO_2 NPs was evaluated to protect *E. coli* from damages induced by X-ray radiation. Figure 5 demonstrates the protective effect of CeO_2 NPs. It can be seen that protective effect depends on concentration of CeO_2 NPs. X-ray radiation with dose of 2 Gy killed 98.92% *E. coli*. The increase in the amount of *E. coli* was found in *E. coli* treated with the presence of CeO_2 NPs. At the condition with addition of 0.02 mg/mL CeO_2 NPs, the amount of *E. coli* increased by 23 times larger than that irradiated *E. coli* without addition of CeO_2 NPs. These results suggested that CeO_2 NPs reduced the amount of *E. coli* death and protected *E. coli* damage induced by X-ray radiation. The amount of *E. coli* death increased with the increase in concentration of CeO_2 NPs. The CeO_2 NPs can reduce *E. coli* damage with concentration up to 0.06 mg/mL. However, the addition of CeO_2 NPs with concentration larger than 0.06 mg/ml did not act as radioprotector but as radiosensitizer. The protective effect mechanism of CeO_2 NPs is physical protection due to its ability to absorb radiation as described by DEF value < 1 and chemical protection through neutralization of reactive oxygen species (ROS) produced by radiolysis of water during irradiation [16]. The main protection mechanism that plays a role in our CeO_2 NPs is chemical protection. It is generally attributed to oxygen vacancy corresponding to surface Ce^{3+} fraction and/or Ce^{3+}/Ce^{4+} redox switch.

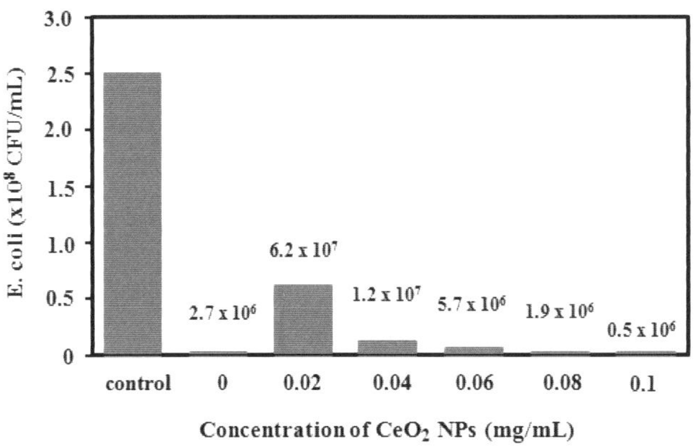

Figure 5. The number of X-ray irradiated *E. coli* with the presence of CeO_2 NPs at various concentration.

4. Conclusion

We have investigated the antioxidant activity and dose enhancement factor of CeO_2 NPs synthesized by precipitation method. The CeO_2 NPs possess good antioxidant activity and can reduce absorbed dose of X-ray radiation. The CeO_2 NPs with concentration of no more than 0.06 mg/ml show a protective effect on *E. coli* damage induced by X-ray radiation. These results indicated that CeO_2 NPs synthesized by precipitation method are a potent radioprotector. However, its effectiveness should be further investigated.

Acknowledgment

This work was financially supported by Ministry of Research, Technology and Higher Education of the Republic of Indonesia, grant No. 007/SP2H/ LT/DRPM/IV/2017.

References

[1] Rancoule C Magne N Vallard A Guy J B Rodriguez-Lafrasse C Deutsch E and Chargari C 2016 *Cancer Lett.* **375** 256-62
[2] Retif P Pinel S Toussaint M Frochot C Chouikrat R Bastogne T and Barberi-Heyob M 2015 *Theranostics* **5(9)**1030-44
[3] Koukourakis M I 2012 *The Britis Journal of Radiology* **85** 313-30
[4] Ormsby R J Lawrence M D Blyth B J Bexis K Bezak E Murley J S Grdina D J and Sykes P J 2014 *Cell Biol. Toxicol.* **30** 55-66
[5] Maier P Wenz F and Herskind C 2014 *Strahlentherapie und onkologie* **190** 745-52
[6] Sandhir R Yadav A Sunkaria A and Singhal N 2015 *Neurochem. Inter.* **89** 209-26
[7] Santhoshkumar T Rahuman A A Jayaseelan C Rajakumar G Marimuthu S Kirthi A V Velayutham K Thomas J Venkatesan J and Kim S K 2014 *Asian Pasific J. Tropical Medicine* **4(12)** 968-76
[8] Nair P M G and Chung I M 2015 *Comparative Biochem. and Physiology* **B 190** 1-7
[9] Nethravathi P C Shruthi G S Suresh D Udayabhanu Nagabhushana H and Sharma S C 2015 *Ceramic Inter.* **41** 8680-7
[10] Suresh D Shobharani R M Nethravathi P C Kumar M A P Nagabhushana H and Sharma S C 2015 *Spectrochim. Acta Part A* **141** 128-34
[11] Madan H R Sharma S C Udayabhanu Suresh D Vidya Y S Nagabhushana H Rajanaik H Anantharaju K S Prashanantha S C and Maiya P S 2016 *Spectrochim. Acta Part A* **152** 404-16
[12] Kim S J and Chung B H 2016 *Carbohydrate Polymers* **150** 400-7

[13] Leung Y H Yung M M N Ng A M C Ma A P Y Wong S W Y Chan C M N Ng Y H Djurisic A B Guo M Wong M T Leung F C C Chan W K Leung K M Y and Lee H K 2015 *J. Photochem. Photobiol. B: Biology* **145** 48-59
[14] Dunnick K M Pillai R Pisane K L Stefaniak A B Sabolsky E M and Leonard S S 2015 *Biol. Trace Elem. Res.* **166** 96-107
[15] Soren S Jena S R Samanta L and Parhi P 2015 *Appl. Biochem. Biotechnol.* **177** 148-61
[16] Zholobak N M Shcherbakov A B Bogorad-Kobelska A S Ivanova O S Ye B A Ye S N and Ivanov V K 2014 *J. Photochem. Photobiol. B: Biology* **130** 102-8
[17] Bandyopadhyay S Peralta-Videa J R Plascencia-Villa G Jose-Yacaman M and Gardea-Torresdey J L 2012 *J. Hazard. Mater.* **241-242** 379-86
[18] Walkey C Das S Seal S Erlichman J Heckman V K K Ghibelli L Traversa E McGinnis J F and Self W T 2015 *Environ. Sci. Nano* **2(1)** 33-53
[19] Karakoti A S Monteiro-Riviera N A Aggarwal R Davis J P Narayan R J Self W T McGinnis J F and Seal S 2008 *JOM* **60(3)** 33-7
[20] Nurhasanah I Sutanto H and Futikhaningtyas R 2014 *Adv. Mater. Res.* **896** 108-11

Fabrication and characteristics of ADC-12 reinforced nano-SiC and nano-Al_2O_3 composites through stir casting route

A Zulfia[1*], E C Putro[1], M Wahyudi[1], D Dhaneswara[1] and B W Utomo[2]

[1] Department of Metallurgy and Materials Engineering, University of Indonesia, Depok, 16436, Indonesia
[2] Astra Honda Motor, Cikarang Barat, 17530, Indonesia

anne@metal.ui.ac.id

Abstract. Aluminum ADC-12 reinforced with nano-SiC and nano-Al_2O_3 have been successfully produced by stir casting method. The effect of addition nano SiC and Al_2O_3 into matrix has been studied with variation of 0.05, 0.10, 0.15, 0.20, 0.30 vf-% for each composites. The addition of Mg was introduced with 10 wt-% to promote wetting between molten Al and ceramic particle reinforcement. The OES result shows decreasing amount of excess Si and Mg in the composite compared to as-cast ADC-12. One of the main intermetallic phases present evenly in aluminum matrix is Mg_2Si which was confirmed by EDS and XRD. The highest tensile strength and hardness of composites were achieved by the addition of 0.15 and 0.30 %vf SiC and Al_2O_3, respectively, with the values of 111 and 115 MPa and 42 and 41 HRB. The increasing hardness generated significant improvement in wear resistance of composites. The higher the reinforcement content, the higher the porosity formed, due to the tendency of dewetting as well as nano particles agglomeration.

1. Introduction

Aluminum and its alloys are metals that are presently widely used in industries. There are around 8.1% aluminum in Earth's crust alone, making it the most abundant metal and third most abundant element after oxygen and silicon. Compared to other metals, aluminum is economically preferred in many industries, due to its light weight, ease of fabrication, good physical and mechanical properties, and good resistance to corrosion. The ADC-12 aluminum alloy, known as A383.0 in ANSI (American National Standards Institute) standard, is commonly used in transportation due to good cast-ability, low density, low chance of shrinkage and high strength [1, 2] due to its main alloying elements which are Si and Cu.

A composite is a system which is composed of reinforcement constituent distributed in a matrix which then makes up combined properties on the composite, taking specific properties from both the reinforcement and the matrix [3]. Good ductility of aluminum combined with good hardness of ceramic reinforcement results in a composite having excellent mechanical properties [4]. Two types of materials that are commonly used as particulate reinforcement in aluminum matrix are Al_2O_3 and SiC. Both of them can be used in many sizes, and nano-sized particulates have been used in recent studies to enhance the hardness properties in aluminum matrix using only small amount of reinforcement particulates. Manu [5] and Mazahery [6] shows that adding nano-sized reinforcement particulates; whether SiC or Al_2O_3, increase the mechanical properties in aluminum based composites.

Ceramics like Al_2O_3 and SiC particles generally have poor wettability on aluminum matrix. Therefore, it is needed to add magnesium as wetting agent to improve the wettability of matrix to generate the formation of $MgAl_2O_4$ phase. This spinel decrease the surface tension of molten aluminium to improve the wettability between particles reinforced and aluminium matrix [7]. In this work, effect of Al_2O_3 and SiC nano particles addition on ADC12 matrix to the mechanical properties of the composite are studied.

2. Experimental method

Variation of reinforcement particles of nano-SiC and nano-Al_2O_3 used were 0.05, 0.1, 0.15, 0.2, and 0.3 vf-%. Aluminum was melted at 800 $^oC^{\pm 20}$ in the electric furnace, then 10 wt-% Mg was added to increase wettability, followed by one minute stirring and 2 minutes degassing process using argon (Ar). Meanwhile, reinforcement particles were pre-heated for 1 h at 800$^oC^{\pm 20}$ to remove moisture, then added to liquid matrix, followed by 1 minute stirring at ±600 rpm, and then by 2 minutes degassing again. Casting process was conducted into a mold which were pre-heated at 500$^oC^{\pm 20}$ for ±8 minutes and coated at the insides using water-based zirconia sand coat to increase mold's thermal resistance and prevent products stick to the mold.

The composites were characterised by both microstructural analysis and mechanical properties analysis. Their chemical compositions were tested using Optical Emission Spectroscopy (OES); then, they were prepared by metallography procedures to analyse their microstructure using optical microscope and Scanning Electron Microscope (SEM) linked to Energy Dispersive Spectroscopy (EDS) to analyse phases present in composites and confirmed by X-Ray Diffraction (XRD). Mechanical tests such as tensile, hardness, wear, and density-porosity test were also performemed, using the ASTM E8 standard for tensile test, ASTM E18 for hardness test with Rockwell B, wear test using Ogoshi refers to ASTM G99, and density-porosity test using Archimedes method.

3. Results and discussion

3.1. Chemical composition of ADC-12/ nano SiC and ADC 12/ nano Al_2O_3 composites

Examination results of chemical composition using OES of unreinforced ADC 12, ADC 12/0.15%Vf SiC and ADC 12/0.30%Vf Al_2O_3 is presented in Table 1. There were differences in chemical composition between ADC-12 and both reinforcing with SiC and Al_2O_3.

Table 1. The chemical composition of the composite using OES.

Materials	Elements (at.%)						
	Al	Si	Fe	Cu	Mn	Mg	Zn
ADC 12	84.80	10.50	0.864	2.33	0.223	0.221	0.637
Composite SiC	77.80	7.16	0.821	1.01	0.196	> 12	0.518
Composite Al_2O_3	77.90	7.17	0.827	1.05	0.191	>12	0.520

Mg content was intentionally added at about 10 wt.% into a matrix to avoid evaporation because operation condition of melting furnace was at elevated temperature (800$^0C^{\pm 20}$). However, no evaporation effect of Mg occurred; even on the final composite sample, Mg content exceeds up to 12 wt.%. The presence of Mg affects the formation of the $MgAl_2O_4$ spinel phase and tends to form Mg_2Si. Mn content of 0.191% by weight is likely to form an intermetallic phase of α-AlMnSi or better known as α-cubic [8]. This phase was considered might lead material into brittle structure especially when the content is high.

The presence of Mg_2Si phase which is clearly seen in the microstructure is confirmed through XRD testing of the as-cast composite. Mg will react with Si to form Mg_2Si phase during solidification process and takes several forms: primary, secondary and tertiary. In addition, the presence of other phases is also confirmed by XRD testing as seen in Figure 1.

The addition of Mg pushes oxygen to create a bonding MgO which subsequently binds with the aluminum oxide coating on the Al_2O_3 and SiC to form a spinel layer ($MgAl_2O4$) due to following reaction [7] :

$$SiO_2 + 2\ Al + Mg \rightarrow MgAl_2O_4 + 2\ Si \quad (1)$$

$$3\ Mg + 4\ Al_2O_3 \rightarrow 2\ Al + 3\ MgAl_2O_4 \quad (2)$$

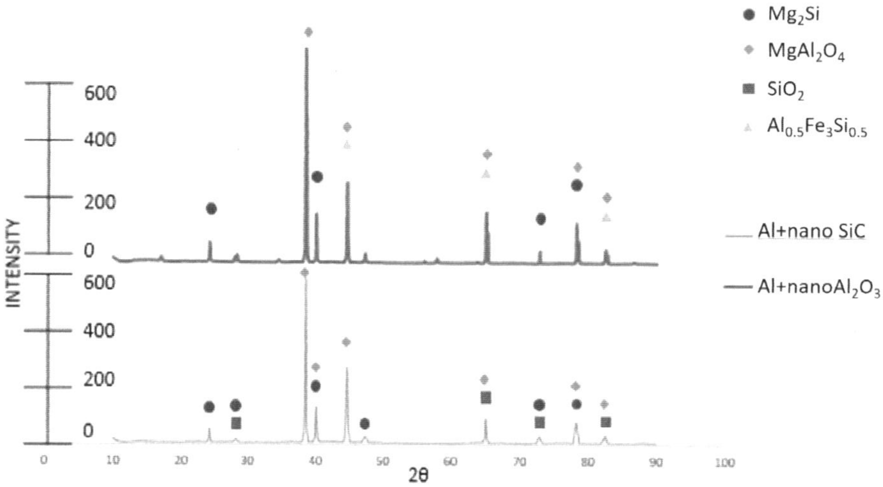

Figure 1. XRD pattern of ADC-12 reinforced nano-SiC and nano-Al_2O_3 composites.

Reaction (1) commonly occurred with SiC and reaction (2) with Al_2O_3 particles. For some cases, reaction (2) may occur on SiC reinforcement, in which the Al_2O_3 comes from oxidized at Al melt surface that sticks to SiC particles when the particles were added. Spinel affluence the infiltration of molten aluminium into reinforcing particle so that effectiveness of wettability can be achieved.

3.2 Microstructure analysis of ADC-12/ nano SiC and ADC-12/ nano Al_2O_3 composites

The microstructure of unreinforced ADC-12 analysed by optical microscope showed needle-like shaped aluminum, which makes it hard to analyse the grain boundaries. The Si element on ADC-12 is eutectic, but during the process, the addition of nano Al_2O_3 and SiC reduce the presence of Si. It is proven by the results of the OES in Table 1, wherein the original Si content of 10.5 wt.% in ADC-12 was reduced to about 7% in the composition of composites with 0.15 vf-% SiC and 0.30 vf-% Al_2O_3 respectively, because Si element binds Mg to form Mg_2Si compound in various forms and is not in a free eutectic state.

Nano SiC and Al_2O_3 will reduce grain size as they act like a grain refiner. As the content of reinforcer increases, the aluminum grain gets finer. Figure 2 shows various appearances in microstructures which are signed by the arrow. The arrow 1 indicates shrinkage porosity, arrow 2 is *a* β-Al_5FeSi phase, arrow 3 is a π-$Al_8Mg_3FeSi_6$ phase, arrow 4 is gas porosity, and arrow 5 is particle porosity.

Mg content of up to 12 wt.% in a chemical composition will lead shifting in a major constituent of a matrix into Al-Mg-Si which results in the formation of Mg_2Si. According to Li *et al.* [9], the Mg_2Si phase can be formed in several morphologies: primary Mg_2Si with polyhedral or dendritic morphology, binary eutectic Mg_2Si in flake-like form surrounded by α-Al, and ternary eutectic Mg2Si which has a small round shape between binary eutectic phases Mg_2Si, thus forming an archipelagic region together with a binary eutectic Mg_2Si phase. The formation of Mg_2Si follows reaction (3) as follows:

$$L \rightarrow L + Mg_2Si_P \rightarrow L + (\alpha\text{-}Al + Mg_2Si)_E \rightarrow (\alpha\text{-}Al + Mg_2Si + Si)_E \quad (3)$$

Primary Mg_2Si can improve the mechanical properties of composite [10, 11] due to its morphology: blocky or snow flake with sharp edges, which can induce localized stress concentration that might increase the tensile strength of the composite. The Fe content which reaches 0.80 wt.% causes Fe to react to form an intermetallic phase by the reaction:

$$L \rightarrow L + Al + \beta\text{-}Al_5FeSi \rightarrow Al + \beta\text{-}Al_5FeSi + Si \quad (4)$$

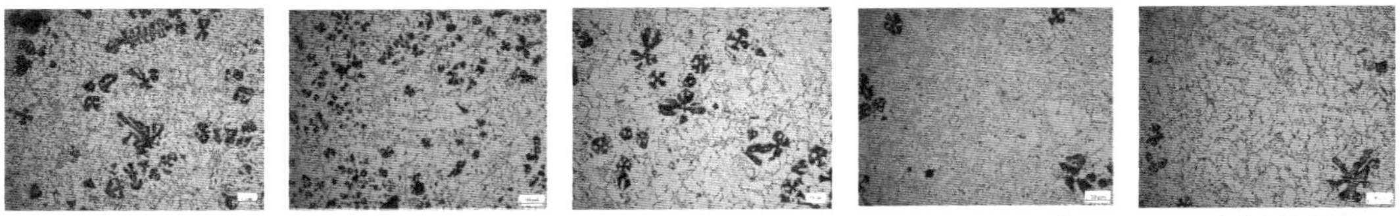

+ 0.05% Vf + 0.10% Vf + 0.15% Vf + 0.20% Vf + 0.30% Vf

(a) nano-Al_2O_3

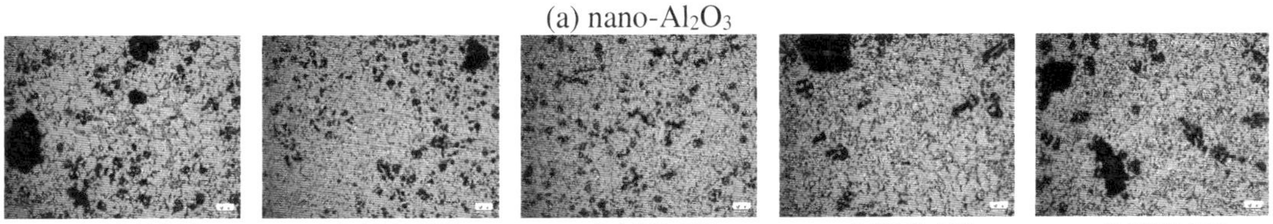

+ 0.05% Vf + 0.10% Vf + 0.15% Vf + 0.20% Vf + 0.30% Vf

(b) nano-SiC

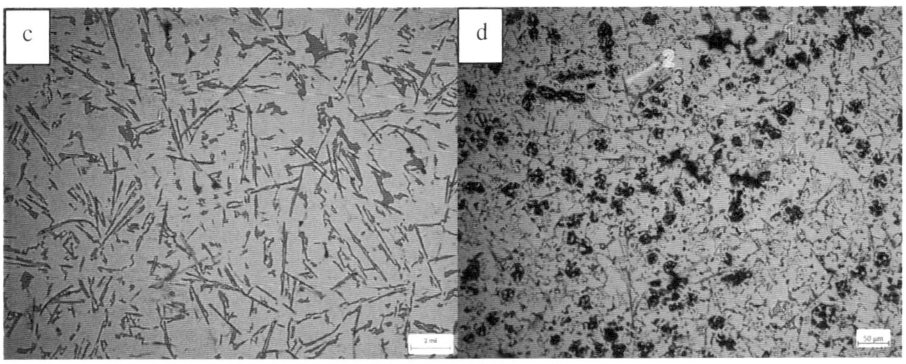

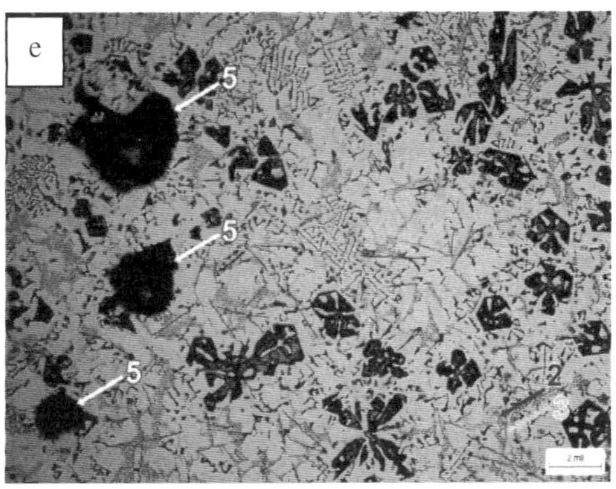

Figure 2. Microstructure of composites; (a) ADC-12/nano-Al_2O_3, (b) ADC 12/nano-SiC, (c) ADC 12 (d) ADC 12 + 0.15% Vf SiC, and (e) ADC-12 + 0.30% Vf Al_2O_3. Keller etchant,10s.

The β-Al_5FeSi phase will have a needle/plate-like morphology. The higher Mg content, the greater the possibility of quartenary phase which is π-$Al_8Mg_3FeSi_6$ formed after Mg_2Si [12]. The further observation of these phases is shown with SEM link to EDS in Figure 3 and indicated points within the Table 2 and Table 3 are also depicted from Figure 3 .

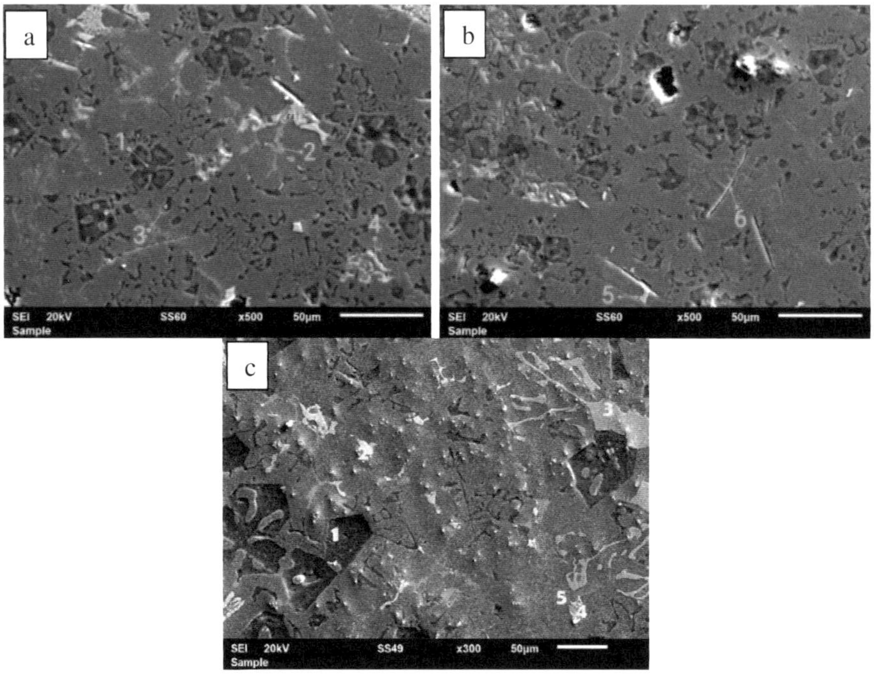

Figure 3. SEM images of ADC-12 with 0.15vf-% SiC composite at magnification of (a) 500x (b) 3000x and (c) ADC-12 with 0.30 vf-% Al_2O_3 composite.

Table 2. EDS result on a composite of ADC-12 with 0.15 vf-% nano-SiC.

Point	Amount of Element (at.%)						Possible phase formed
	Al	Mg	Si	Fe	Cu	O	
1	84.86	1.60	6.08	-	-	7.46	Mg_2Si (primary)
2	86.37	1.92	2.79	3.71	2.44	2.77	π-$Al_8Mg_3FeSi_6$
3	18.32	32.53	26.54	1.38	-	15.11	Al-Mg_2Si (binary)
4	12.80	36.46	27.75	-	-	17.27	Al-Mg_2Si (binary)
5	80.41	-	-	-	8.59	-	Al_2Cu
6	64.44	1.43	9.73	7.00	-	6.14	β-Al_5FeSi

Table 3. EDS result on a composite of ADC-12 with 0.30 vf-% nano-Al_2O_3.

Point	Amount of Element (at.%)						Possible phase formed
	Al	Mg	Si	Cu	Fe	C	
1	-	28.78	28.74	-	-	12.37	Mg_2Si (Primary)
2	52.11	-	-	30.98	-	11.16	Al_2Cu
3	60.48	-	6.96	-	18.98	-	Al-Fe-Si
4	67.10	4.61	4.64	23.64	-	-	Al_2Cu
5	63.10	10.84	19.97	-	6.09	-	Al + Mg_2Si (Binary)

3.3. Mechanical properties of ADC 12/ nano SiC and ADC 12/ nano Al_2O_3 composites

Figure 4 shows the Ultimate Tensile Strength (UTS), for both composites with SiC and Al_2O_3 reinforcement. The UTS of composites greater than that of its matrix. ADC-12 unreinforced has UTS of 101 MPa, while composites with 0.15 vf-% SiC is 111 MPa and 115 MPa for composites with 0.30 vf-% Al_2O_3.

The strengthening mechanism at work in the application of SiC and Al_2O_3 nano-sized reinforcement are load transfer into reinforcer, dislocation density, Hall-Petch and Orowan Strengthening [13, 5]. A decline of UTS is predicted to be the result of agglomeration of nano particles. These particles have a greater surface-to-volume-ratio, leading to agglomeration with each other. Moreover, the increase of porosity level further enhanced by agglomeration of SiC particles leads to the decline of UTS [6]. As the porosity level increases, more crack initiation presents to ease the failure take place.

Generally, Figure 5 shows an increased hardness number with the addition of SiC and Al_2O_3 and decreased wear rate. The nano-sized particles have significantly higher hardness compared to the matrix and make it helpful to harden the composite. In addition, reinforcers act as a barrier during indentation so that higher force is needed to perform plastic deformation. Nonetheless, higher porosity level contributed from the higher content of reinforcing responsible for declining hardness of the material.

The wear rates of SiC- and Al_2O_3-reinforced composites are given in Figure 6. Unreinforced ADC-12 has a wear rate of 2.51 x10^{-5} mm^3/m where the most-reduced wear rate is down to 0.98 x10^{-5} mm^3/m and 1.08 x10^{-5} mm^3/m that are reached by addition of 0.1 vf-% SiC and 0.3 vf-% Al_2O_3, respectively. The decreased wear rate corresponds to the presence of nano particles reinforcement as they created materials harder followed by slight material's cut [14]. To define wear resistance, the hardness is a profound characteristic to be considered. There is a increased of wear rate due to poor microstructure homogeneity leads to undispersed reinforcement, higher level porosity and weak interface bonding between matrix and reinforce [15].

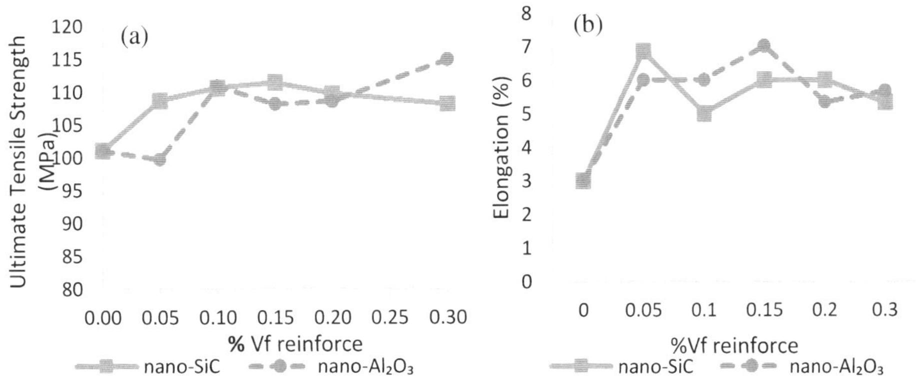

Figure 4. Graphic of (a) UTS and (b) elongation of ADC-12 with an addition of nano-SiC and nano-Al_2O_3 in various compositions.

Generally, the data obtained by impact testing shown in Figure 6 (a), the impact toughness inclined to decrease with the addition of volume fraction of nano-reinforce. Reinforcement hardens matrix material and then decreases impact strength because the material will be broken immediately without absorbing much energy [16]. Current work is in line with the research conducted by Bekir *et al.* [17]. The addition of Al_2O_3 or SiC will lower impact toughness of composite aluminum. However, the increase of impact toughness might result from good distribution of nano particles inside matrix which causes good interface bonding between them [18].

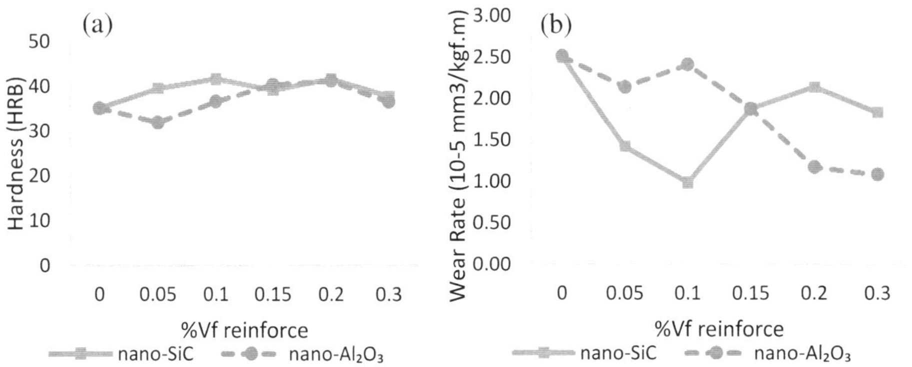

Figure 5. Graphic of (a) hardness and (b) wear rate of ADC-12 with an addition of nano-SiC and nano-Al_2O_3 in various compositions.

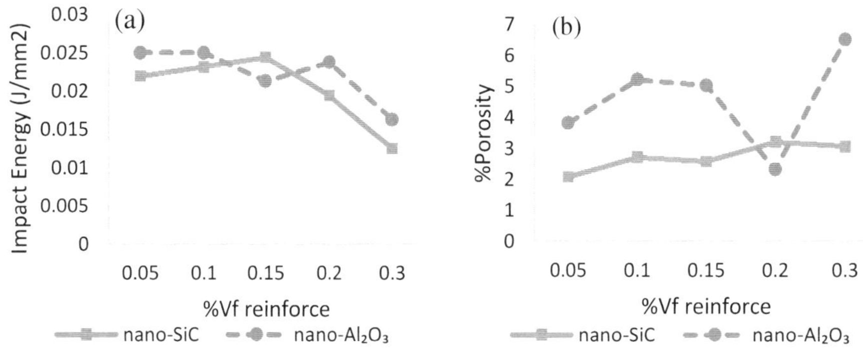

Figure 6. Graphic of (a) impact toughness and (b)% porosity level of ADC-12 with an addition of nano-SiC and nano-Al_2O_3 in various compositions.

3.4. Density and porosity of ADC 12/ nano SiC and ADC 12/ nano Al_2O_3 composites

The density of composite is calculated using Archimedes' test. As-cast composite will have lower density compared to theoretical density due increasing in weight percent of nano particle. In this work, it can be seen in Figure 6 (b). Hydrogen composition arises when metal melting on a temperature fabricated around $800^0C^{\pm 20}$ during stir casting. Even though argon degassing had been applied, hydrogen is remaining in molten aluminum because argon less protective to prevent hydrogen infiltrate into molten aluminum particularly in the open melting furnace [19]. Various kinds of porosity which are present in the microstructure are gas, shrinkage and particle porosity; they significantly reduce the composite's density.

4. Conclusion

Composite ADC-12 with variation of addition 0.05, 0.10, 0.15, 0.20, 0.30% Vf nano-sized SiC and Al_2O_3 were successfully fabricated using stir casting method. The excess of Mg content influences the formation of Mg_2Si which reduces the effectiveness in the formation of spinel layer. In addition, the Fe content is high enough to form an intermetallic phase that affects the mechanical properties of composites. Both of these elements can be seen through SEM-EDS testing. Addition of 0.15vf-% SiC and 0.30 vf-% Al_2O_3 give the optimum UTS of 111 MPa and 115 MPa, respectively. The Orowan mechanism is being the major strengthening mechanism in terms of nano-sized particles addition. The decline of UTS might be a result of agglomeration of reinforcement particles or porosity. The maximum hardness gained by adding of SiC up to 0.1 vf-% is 42 HRB and Al_2O_3 up to 0.2 vf-% is 41 HRB, respectively. However, as the content of reinforcement arise, the possibility of porosity in the matrix increase, thus decreasing the hardness. Nano-sized SiC and Al_2O_3 were significantly reducing wear rate of an unreinforced matrix as materials become harder and reducing volume of material loss. Increased wear rate might be caused by inhomogeneity of microstructures, and impact toughness reduced due to increasing hardness and decreasing ductility of composites.

Acknowledgements
Authors would like to express their gratitude for financial support from the Directorate of Research and Community Services (DRPM), University of Indonesia, through International Publications Index for Final Year Students Projects Grant (PITTA) 2017 with contract number: 737/UN2.R3.1/HRP.05.00/2017.

References

[1] Okayasu M Ohkura Y Takeuchi S Takasu S Ohfuji H and Shiraishi T 2012 *Mater. Sci. and Engin. A* **543** 85–192
[2] Su H Gao W Feng Z and Lu Z 2012 *Mater. Des.* **36** 590–6
[3] Surappa M K 2003 *Sadhana* **28(1–2)** 319–34
[4] Bharath V Nagaral M Auradi V and Kori S A 2014 *Proc. Mater. Sci.* **6** 1658–67
[5] Manu K M S Kumar S A Rajan T P D Mohammed M R and Pai B C 2017 *J. Alloys Compd.* **712** 394–405
[6] Mazahery A and Shabani M O 2012 *Transactions of Nonferrous Metals Soc. of China (English Ed.)* **22(2)** 275–80
[7] Zhou W and Xu Z M 1997 *J. of Mater. Proces. Tech.* **63(1-3)** 358–63
[8] Lacaze J Eleno L and Sundman B O 2010 *Open. Arc. Toulouse Arc. Ouv.* **41** 2208–15
[9] Li C Wu Y Li H and Liu X 2009 *J. of Alloys and Compd.* **477(1)** 212-6
[10] Nasiri N Emamy M and Malekan A 2012 *J. Mater.* **37** 215–22
[11] Zulfia A S J and A H E 2015 *Mater. Sci. Forum* **827** 294-9
[12] Liu Y Kang S and Kim H 1999 *Mater. Lett.* **41(6)** 267-72
[13] Mirza F A and Chen D L 2015 *Materials* **8** 5138-53
[14] Ko M 2006 *Compos. Part A* **37** 457–64
[15] Akbari M K Baharvandi H R and Mirzaee O 2013 *Compos. Part B* **52** 262–8
[16] Lakshmipathy J and Kulendran B 2014 *Int. J. of Refractory Metals and Hard Mater.* **46** 137–44
[17] Ünlü B S 2008 *Mat. Design.* **29** 2002–8
[18] Raturi A Mer K K S and Pant P K 2017 *Mater. Today Proc.* **4(2)** 2645–58
[19] Srivastava N and Chaudhari G P 2016 *Mater. Sci. Eng. A* **651** 241–7

IOP Conf. Series: Materials Science and Engineering **432** (2018) 012033 doi:10.1088/1757-899X/432/1/012033

Annealing temperature effect on structural, vibrational and optical properties of $Co_{0.8}Ni_{0.2}Fe_2O_4$ nanoparticles

J Utomo[1], A K Agustina[1] and E Suharyadi[1*]

[1] Department of Physics, Gadjah Mada University, North Sekip BLS Yogyakarta 55281, Indonesia

*esuharyadi@ugm.ac.id

Abstract. $Co_{0.8}Ni_{0.2}Fe_2O_4$ nanoparticles were synthesized by coprecipitation method and annealed at different temperatures (200, 400, 600, 800 and 1000 °C). The X-ray diffraction (XRD) exhibited the formation of spinel cubic structure. The crystallite size of $Co_{0.8}Ni_{0.2}Fe_2O_4$ nanoparticles at the annealing temperature of 200 °C was 7.0 nm and then increased with the increase of annealing temperature. The reason for this is that the particles grow faster and larger at a higher temperature. The transmission electron microscopy (TEM) images showed that the size distribution of the particles was increased with the increase of annealing temperature. It was also seen from the samples that most of the particles appeared nearly spherical in shape with slight agglomeration, particularly at higher annealing temperature. The Fourier transform infrared spectroscopy (FTIR) spectra of $Co_{0.8}Ni_{0.2}Fe_2O_4$ nanoparticles confirmed the presence of vibration bands of metal ions at 370.30 – 401.9 cm^{-1} and 586.36 – 601.79 cm^{-1} which was in a region of octahedral and tetrahedral sites, respectively. The values of the gap energy of $Co_{0.8}Ni_{0.2}Fe_2O_4$ nanoparticles observed ultraviolet-visible (UV-Vis) spectroscopy were in the range of 4.25 – 4.94 eV. The gap energy was decreased with the increase of annealing temperature and the increase of crystallite size.

1. Introduction

Spinel ferrite nanoparticles having the general chemical formula MFe_2O_4 (M is divalent metal cation) are attractive materials because of their applications in such fields as drug delivery, spintronics, high-density information storage, microwave devices, and optoelectronics [1-3]. They have excellent chemical and physical properties which could not be found in the bulk size due to their large surface area [4]. Among spinel ferrite nanoparticles, recently Co-Ni ferrite are most fascinating materials to be investigated. Cobalt ferrite and nickel ferrite have inverse spinel structure with all of the Co^{2+}/Ni^{2+} ions occupied at octahedral (B) site and Fe^{3+} ions distributed between tetrahedral (A) and octahedral (B) sites [5]. In addition, cobalt ferrite is hard magnetic materials with high chemical stability and mechanical hardness while nickel ferrite is soft magnetic materials with high resistivity more than 10^9 Ωcm [6, 7]. Therefore Co-Ni ferrite is suitable to be applied for high-frequency electronic devices e.g. isolators and circulators due to their high electrical resistivity and good magnetic properties [8, 9].

To achieve these applications, it is necessary to consider the quality of crystal through their properties. Their properties are mainly affected by particles size, shape, crystallinity and cation distribution between tetrahedral (A) and octahedral (B) sites [10, 11]. One of the main factors that determine the quality of crystal is the preparation method. It is important to consider obtaining well ferrite performance and also preparing easily. Coprecipitation method is one of the methods chosen by many researchers because this method is simpler, low cost and used low synthesis temperature. Another reason is that this method could control the production of ferrite particles [12]. However, this method produces tiny particle size causing agglomeration. It mainly affects the weakening of their properties such as structural, vibrational and optical properties. Therefore, the weakening of their properties could be handled by giving annealing temperature after synthesis process [13]. Their properties of spinel ferrite could be improved by the effect of annealing temperature to reach well crystallinity and reduce agglomeration. The objective of this investigation is to explore structural, vibrational and optical properties of $Co_{0.8}Ni_{0.2}Fe_2O_4$ nanoparticles annealed at various annealing temperatures and synthesized by coprecipitation method. Correlation between annealing temperature on structural, vibrational and optical properties will be carried out in detail.

2. Experimental method

$Co_{0.8}Ni_{0.2}Fe_2O_4$ nanoparticles were synthesized by coprecipitation method. Cobalt chloride hexahydrate ($CoCl_2.6H_2O$), nickel chloride hexahydrate ($NiCl_2.6H_2O$) and iron chloride hexahydrate ($FeCl_3.6H_2O$) were used as precursors by dissolving them in distilled water with the concentration of 0.16 M, 0.04 M and 0.4 M, respectively.The mixture was added to HCl as a catalyst. And then, the mixture solution was added drop wise to the NaOH solution (12 M) heated at a synthesis temperature of 90 °C while homogenized at constant stirring (1000 rpm) for 1 hour. The chemical reaction of $Co_{0.8}Ni_{0.2}Fe_2O_4$ nanoparticles could be described as:

$$0.8CoCl_{2(aq)} + 0.2NiCl_{2(aq)} + 2FeCl_{3(aq)} + 8NaOH_{(aq)} + H_2O_{(aq)} \xrightarrow{+\ HCl_{(aq)}} Co_{0.8}Ni_{0.2}Fe_2O_{4(s)} + 8NaCl_{(aq)} + 5H_2O_{(l)} \tag{1}$$

The prepared precipitate was then rinsed five times with distilled water until the rinsing was free from the traces of sodium, chloride and other impurities. The precipitate was subsequently dried at 90 °C in a furnace for 4 hours to remove traces of water. Afterward, the resulting powder was annealed at different temperatures (200, 400, 600, 800 and 1000 °C). The structural properties and morphologies of $Co_{0.8}Ni_{0.2}Fe_2O_4$ nanoparticles at different annealing temperatures were studied by X-ray diffractometer (XRD) using Cu-K_α radiation with wavelength λ=1.5406 Å and transmission electron microscope (TEM) Jeol JEM-1400. The vibrational spectra were recorded by Fourier transform infrared spectroscope (FTIR) Prestige Shimadzu-21 at room temperature. The optical properties of samples were analyzed by UV-Vis spectrophotometer DR 1700 PharmaSpec.

3. Results and discussion

3.1. X-Ray diffraction analysis

Figure 1 shows the XRD patterns of $Co_{0.8}Ni_{0.2}Fe_2O_4$ nanoparticles for as-prepared and annealed at 200, 400, 600, 800 and 1000 °C. There are reflection planes for all samples such as (220), (311), (400), (422), (511) and (440) indicating the formation of cubic spinel structure with space group *Fd3m* (JCPDS card no #22-1086) [14]. $Co_{0.8}Ni_{0.2}Fe_2O_4$ nanoparticles have nearly single phase without any new phases detected. The increase of annealing temperature results in an enhancement of crystallinity degree where the peak intensity (311) increases while the full width at the half maximum (FWHM) decreases for $Co_{0.8}Ni_{0.2}Fe_2O_4$ nanoparticles. The structural parameters such as lattice parameter, crystallite size, microstrain and X-ray density could be obtained from the XRD data. The lattice parameter a is calculated according to the Bragg's formula [7]. The crystallite size D is estimated according to the Scherrer's formula [15].

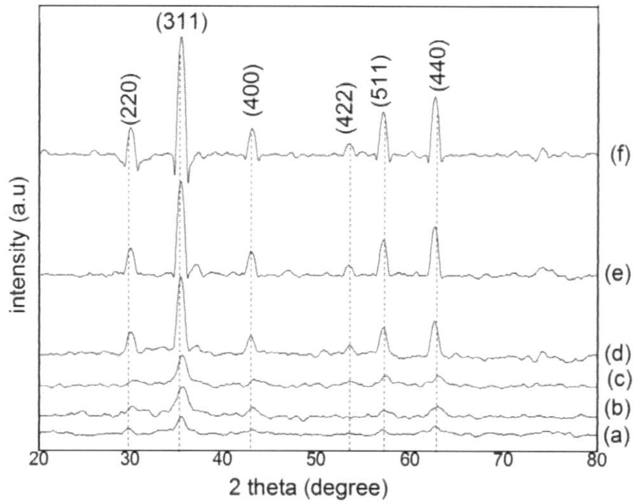

Figure 1. XRD patterns of $Co_{0.8}Ni_{0.2}Fe_2O_4$ nanoparticles at different annealing temperatures (a) as-prepared (AP), (b) 200 °C, (c) 400 °C, (d) 600 °C, (e) 800 °C and (f) 1000 °C.

The structural parameters from the XRD patterns of $Co_{0.8}Ni_{0.2}Fe_2O_4$ nanoparticles for as-prepared and annealed samples are shown in Table 1. The data obtained from XRD analysis reveals an increasing pattern in crystallite size with the increase of annealing temperature as shown in Table 1. This is due to the crystallite size at higher annealing temperature enhances the coalescence process of crystallites into a larger average size of nanoparticles [4].

Table 1. Lattice parameter, crystallite size, X-ray density and strain of $Co_{0.8}Ni_{0.2}Fe_2O_4$ nanoparticles.

Anealling temperature (°C)	Lattice parameter (Å)	Crystallite size (nm)	X-ray density (g/cm^3)	Microstrain (× 10^2/line^2)
AP	8.38	10.4	5.3	4.4
200	8.35	7.0	5.3	6.5
400	8.36	8.3	5.3	5.5
600	8.39	9.6	5.2	4.7
800	8.40	9.8	5.2	4.6
1000	8.39	10.5	5.2	4.3

The lattice parameter has an increasing trend with the increase of annealing temperature which is in the range 8.35 Å – 8.40 Å. This is attributed to faster coarsening process when a higher annealing temperature is applied to the samples. A faster coarsening process causes enhancement in unit cell volume which results in increasing lattice parameter. The X-ray density decreases with the increase of annealing temperature. This is caused by the increase of annealing temperature results in larger unit cell volume of nanoparticles as shown in the following formula [16].

$$\rho_x = \frac{8M}{NV} \tag{2}$$

where ρ_x is X-ray density, V is unit cell volume, N is the Avogadro's number and M is molecular weight of $Co_{0.8}Ni_{0.2}Fe_2O_4$ nanoparticles.

IOP Conf. Series: Materials Science and Engineering **432** (2018) 012033 doi:10.1088/1757-899X/432/1/012033

3.2. Transmission electron microscopic (TEM) images

Figure 2 shows the morphology of $Co_{0.8}Ni_{0.2}Fe_2O_4$ nanoparticles for as-prepared and annealed at 400, 600, 800 and 1000 °C. The particles are initially agglomerated in a spherical shape at an annealing temperature of 400 °C. Subsequently, the particles are nearly in cubic shape with slight agglomeration at annealing temperatures of 600 and 800 °C. And then, the particles are agglomerated at an annealing temperature of 1000 °C. From TEM images, it is confirmed that the particles size increases with the increase of annealing temperature. This result supports XRD analysis that has been studied previously. This phenomenon is caused the particles grow faster and form larger at higher annealing temperature [17]. The selected area electron diffraction (SAED) patterns of the $Co_{0.8}Ni_{0.2}Fe_2O_4$ nanoparticles for as-prepared, annealed at 400 °C and 600 °C display continuous polycrystalline rings while for samples annealed at 800 and 1000 °C display discontinuous polycrystalline. All of the samples reveal polycrystalline nature.

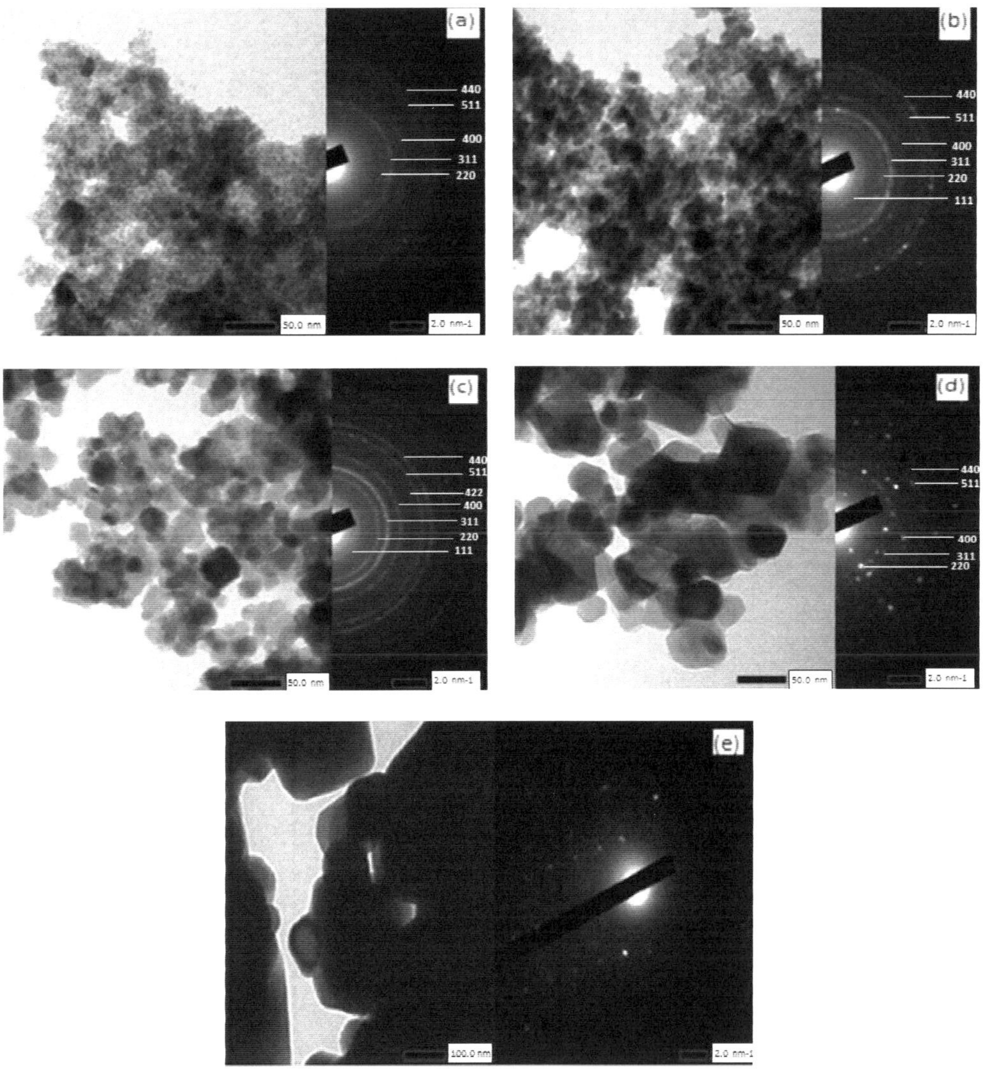

Figure 2. TEM images of of $Co_{0.8}Ni_{0.2}Fe_2O_4$ nanoparticles at different annealing temperatures (a) as-prepared (AP), (b) 400 °C, (c) 600 °C, (d) 800 °C and (e) 1000 °C.

3.3. Fourier transform infrared spectroscopy (FTIR) studies

The vibrational spectra of $Co_{0.8}Ni_{0.2}Fe_2O_4$ nanoparticles for as-prepared and annealed were observed by FTIR spectroscope at room temperature as shown in Figure 4. According to the Figure 3, there are two metal oxygen (M – O) bands which are in wave number 370.3 cm^{-1} – 401.19 cm^{-1} and 586.36 cm^{-1} – 601.79 cm^{-1}. They are respectively at octahedral and tetrahedral sites indicating the formation of spinel ferrite structure.

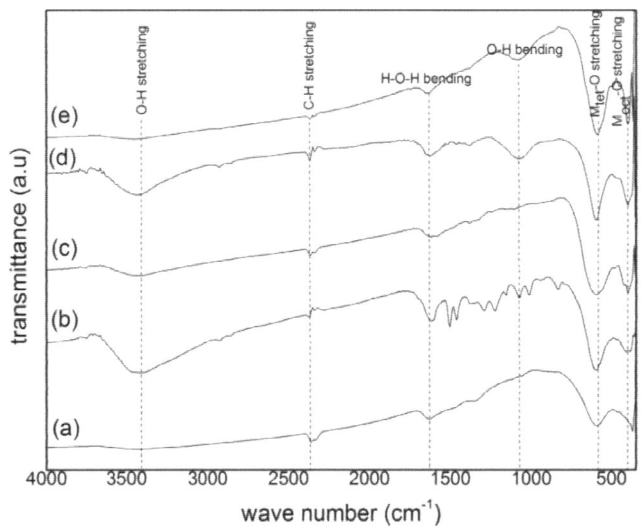

Figure 3. Vibrational spectra of $Co_{0.8}Ni_{0.2}Fe_2O_4$ nanoparticles at different annealing temperatures (a) as-prepared, (b) 400 °C, (c) 600 °C, (d) 800 °C and (e) 1000 °C.

The detailed wave number of M – O at octahedral and tetrahedral sites are shown in Table 2. From Table 2, the wave number of M – O stretching vibration at the octahedral site is lower than the tetrahedral site. This could be attributed to shorter or longer bond length of the octahedral or tetrahedral site. In addition, the wave number of octahedral sites is shifted toward higher wave number with the increase of annealing temperature. This is related to their atomic weight of ions. The absorption band at 3410 – 3448.72 cm^{-1} is attributed to the hydroxyl (O – H) group of stretching vibration. This is also observed that the intensity of this absorption band has a decreasing trend with the increase of annealing temperature. This may be attributed to the loss the residual water in the nanoparticles [18]. The appeared absorption band at 1581 – 1651.7 cm^{-1} for all samples correspond to the existence of H_2O molecules [7, 10].

Table 2. The wave number of metal-oxygen vibration at octahedral and tetrahedral sites of as-prepared and annealed $Co_{0.8}Ni_{0.2}Fe_2O_4$ nanoparticles.

Anealling temperature (°C)	Wave number (cm^{-1})	
	Octahedral site	Tetrahedral site
AP	370.3	594.08
400	385.76	594.08
600	401.19	594.08
800	401.19	601.79
1000	401.19	586.36

3.4. UV-Vis analysis

The optical gap energy of $Co_{0.8}Ni_{0.2}Fe_2O_4$ nanoparticles for as-prepared and annealed was carried out by using UV-visible spectroscope. Figure 4 shows the dependence of the optical gap energy on annealing temperature. According to the Figure 4, the optical gap energy decreases with the increase of annealing temperature.

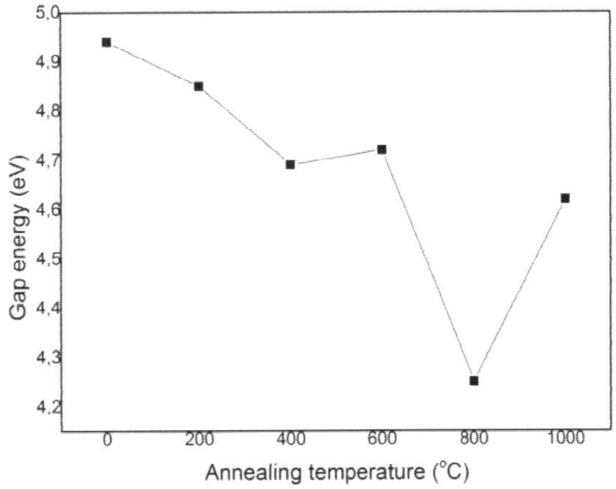

Figure 4. The dependence of the optical gap energy on annealing temperature.

The decrease of gap energy may be attributed to the Burstein-Moss Effect. Also, the reduction of gap energy shows that the photon absorption is shifted toward higher wave length (red shift). This shift could be caused by the strong interaction (s-d and p-d) result into a positive and negative correction to the valence and conduction band edges [19]. In addition, there is a correlation between crystallite size and optical gap energy as shown in Table 3.The optical band energy decrease with the increase of crystallite size. This is related to an enhancement of crystallinity when the higher annealing temperature is applied [4].

Table 3. Gap energy of as-prepared and annealed $Co_{0.8}Ni_{0.2}Fe_2O_4$ nanoparticles.

Anealling temperature (°C)	Crystallite size (nm)	Gap energy (eV)
AP	10.4	4.94
200	7.0	4.85
400	8.3	4.69
600	9.6	4.72
800	9.8	4.25
1000	10.5	4.62

4. Conclusion

We have synthesized $Co_{0.8}Ni_{0.2}Fe_2O_4$ nanoparticles using coprecipitation method and then annealed them at various temperatures. The XRD analysis reveals that $Co_{0.8}Ni_{0.2}Fe_2O_4$ nanoparticles for as-prepared and annealed have the formation of the cubic spinel structure. The crystallite size and lattice parameter display an increasing trend while the X-ray density and microstrain shows a decreasing trend with the increase of annealing temperature. TEM images and FTIR spectra confirm the formation of the cubic spinel structure. The annealing temperature decreases the value of the optical gap energy.

Acknowledgment
The authors are highly thankful to the Indonesia Endowment Fund for Education (LPDP) and the Grant of Competency Based Research of the Ministry of Research, Technology and Higher Education of the Republic of Indonesia (Kementerian Ristek Dikti) for supporting this work.

References

[1] Singhal S Singh J Barthwal S K and Chandra K 2005 *J. Solid State Chem* **178** 3183-9
[2] Kumar A Sharma P and Varshney D 2014 *Ceram. Int* **40** 12855-60
[3] Jiang J 2007 *Mater. Lett* **61** 3239-42
[4] Othman A A Osman M A Ibrahim E M M and Ali M A 2015 *Ceram. Int* **43** 527-33
[5] Mathew D S and Juang R 2007 *Chem. Eng. J* **129** 51-65
[6] Kambale R C Shaikh P A Kambale S S and Kolekar Y D 2009 *J. Alloys Compd* **478** 599-603
[7] Kumar A K Yadav N Rana D S Kumar P Arora M and Pant R P 2015 *J. Magn. Magn. Mater* **394** 379-84
[8] Kafeni K K Msagati T A M and Mamba B B 2017 *Mater. Sci.. Eng B* **215** 37-55
[9] Xiang J Chu Y Shen X Zhou G and Guo Y 2012 *J. Colloid and Interface Sci.* **376** 57-61
[10] Awati V V Rathod S M Mane M L and Mohite K C 2013 *Int. Nano. Lett* **3** 29
[11] Saffari F Kameli P Rahimi M Ahmadvand H and Salamati H 2015 *Ceram. Int* **41** 7352-8
[12] Maaz K Karim S and Kim G 2012 *Chem. Phys. Lett* **549** 67-71
[13] Indrayana I P T Siregar N Suharyadi E Kato T and Iwata S 2016 *J. Phys: Conf. Series* **776** 012021
[14] Kasapoglu N Birsöz B Baykal A Köseoglu Y and Toprak M S 2007 *Cent. Eur. J. Chem* **5** 570-80
[15] Pugazhvadivu K S Ramachandran K and Tamilarasan K 2013 *Phys. Proc* **49** 205-16
[16] Sontu U B Yelasani V and Musugu V R R 2015 *J. Magn. Magn. Mater* **374** 376-80
[17] Abdallah H M I Moyo T and Ngema N 2015 *J. Magn. Magn. Mater* **394** 223-8
[18] Azhagushanmugam S J Suriyanarayanan N and Jayaprakash R 2013 *Phys. Proc* **49** 44-8
[19] Sharma D and Jha R 2017 *J. Alloys Compd* **698** 532-8

Hydrogen storage properties of mechanical milled MgH_2-nano Ni for solid hydrogen storage material

Z Jalil[1], A Rahwanto[1], Malahayati[1], Mursal[1], E Handoko[2], H Akhyar[3]

[1] Department of Physics, Syiah Kuala University, Banda Aceh 23111, Indonesia
[2] Department of Physics, Jakarta State University (UNJ), Jakarta 13220, Indonesia
[3] Department of Mechanical Engineering, Syiah Kuala University, Banda Aceh 23111, Indonesia

zjalil@unsyiah.ac.id

Abstract. Among the metal hydrides, magnesium has the theoretically highest weight capacity for hydrogen storage (7.6 wt.%), lightweight and a reasonably low cost. However, high working temperature (>300°C), slow reaction kinetics (need more than 1 hour to produce 5 wt% of hydrogen) and difficult activation limit the practical application of Mg-based hydrides. In order to improve their performance, MgH_2 was catalyzed with Ni nanoparticles which reactively milled under hydrogen atmosphere. Phase identification and microstructure were characterised by XRD and scanning electron microscope (SEM). Hydrogen sorption properties was studied by gravimetric analysis method. The results showed that, small amount of Ni in nanometer scale proved to be as a suitable catalyst for improvement the kinetics of MgH_2 and at the same time allowed to reduce the milling time process.

1. Introduction

Mg-based metal hydrides are promising candidates for reversible hydrogen storage because of their high capacity (7.6 wt%), lightweight, and low cost [1]. However, relatively high sorption temperatures (>300 °C) and low kinetics are the major difficulties hindering its commercial use. Many efforts have been done to improve the hydriding and dehydriding properties such as element substitution (metal or metal oxides) as catalyst in nanometer scale and modification of ball milling technique have been used to improve sorption and kinetics properties [2-4]. Recently, the reactive ball milling under hydrogen atmosphere was successfully introduced to prepare hydrogen storage materials [5-8]. Aims of this study to observe MgH_2 catalyzed with small amount of Ni in nanoparticles scale and the preparation by reactive ball milling under 10 bar hydrogen.

2. Materials and method

MgH_2 (Goldschmidt, 95+%), Ni (99.9%, Sigma Aldrich, ~90 nm) powders were used to prepare the samples. The powders were filled into a hardened steel vial and sealed together with 13 balls (9.5 mm in diameter) and reactively milled in a planetary mill Fritsch (P6) at a rotational speed of 400 rpm (ball to powder ratio 10:1) for 2 hours under hydrogen atmosphere (10 bar). Phase identification and microstructure were characterised by XRD using a Philips 1050 diffractometer (Co-Kα radiation) and scanning electron microscope (FEGSEM LEO 1530). Hydrogen sorption properties were studied by gravimetric analysis in a wide temperature range for absorption (50-300 °C) and desorption (250-350 °C).

3. Results and discussion

Figure 1 shows the evolution of the XRD diffraction pattern for MgH_2-Ni as a function of milling time. The starting mixture shows the presence of microcrystalline magnesium hydride and Ni nanoparticles. It can be seen that during milling the MgH_2 and Ni diffraction peaks broaden but no changes in the 2θ position at early stage of milling. The as-received sample composed mainly MgH_2 and a small peak of Ni. The same result found after 1 h milling. Then after 2 hour of milling the peaks are broaden. It can be also note that after desorption hydrogen completely react to produce only magnesium. It can be concluded that MgH_2 completely desorp after only two hours milling.

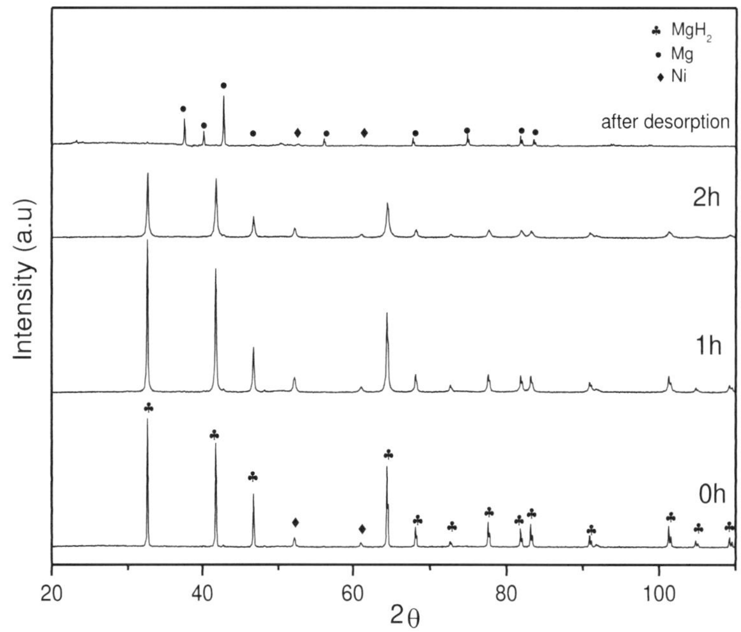

Figure 1. X-ray diffraction patterns of MgH_2-2 mol% nano-Ni reactively milled for 2 hours in 10 bar H_2.

It was shown that after H_2 absorption/desorption cycling, ended with desorption at 350 °C, 50 mbar. It was also observed that with increasing milling time the crystallite sizes decrease further to finally stabilize around 10 nm. Thus, the used of hydrogen pressure during milling promotes the formation of larger amounts of MgH_2 [9-11].

The hydrogen storage and release properties of Ni-catalyzed MgH_2 were measured by gravimetric method. The gravimetric analysis measurements confirm the catalytic effect of nickel nanoparticle on magnesium hydride hydrogenation and dehydrogenation. Figure 2 shows the hydrogen absorption curves of MgH_2-2mol% Ni nanoparticles compare to microparticles and pure MgH_2 samples as well at constant temperatures (300 ^{0}C) and pressures (10 bar) after reactive ball milling under 10 bar H_2 for 2 hours.

The catalytic effect of nickel nanoparticles is magnified especially during absorption: actually, the absorption in the sample catalyzed with Ni nanoparticle took less than 5 minutes for up-taking 5.3 wt% H_2. Compare to the sample catalyzed with Ni microparticles took more than 60 minutes to absorb 5.3 wt% of hydrogen, and similar result for pure MgH_2 as well.

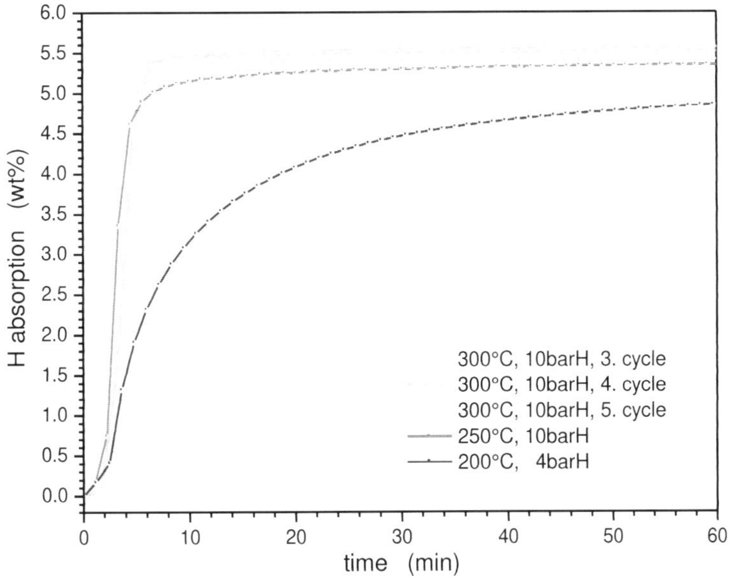

Figure 2. Hydrogen absorption **of** of MgH_2-2 mol% nano-Ni reactively milled for 2 hours in 10 bar H_2 reactive milling under 10 bar H_2.

The results show that the Ni catalyst absorp large amount of hydrogen for short time in the form of nanoparticles. And the large particle catalyst requires a longer time to finish the absorption. It is obvious that Ni has much better as catalyst in nanoparticle size in the absorption and desorption processes. This is also one reason for the great interest in the use of nanoparticles catalyst in order to improve the hydrogen sorption properties of MgH_2. However, Ni nanopaticles has no effect on the operation temperature, which needed high temperature, more than 300 ^{0}C, to complete the absorption.

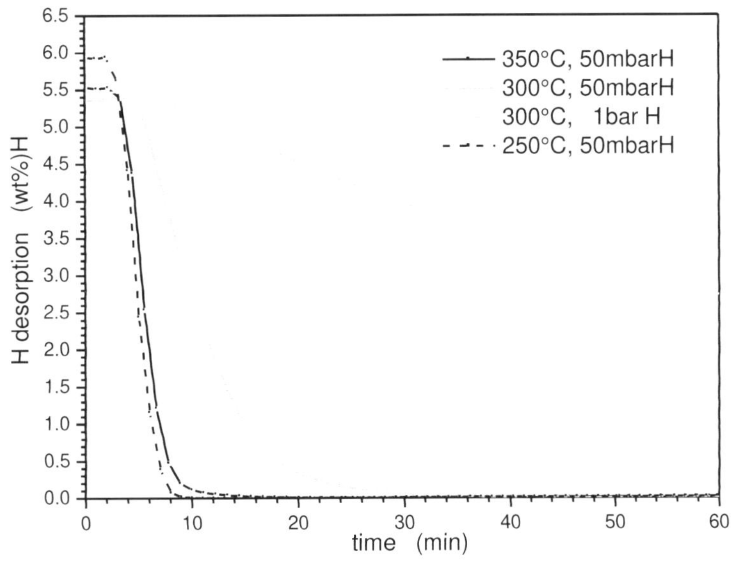

Figure 3. Hydrogen desorption of of MgH_2-2 mol% nano-Ni reactively milled for 2 hours in 10 bar H_2 reactive milling under 10 bar H_2.

IOP Conf. Series: Materials Science and Engineering **432** (2018) 012034 doi:10.1088/1757-899X/432/1/012034

Similar to desorption, its needed 350 ^{0}C to desorp 5.3 wt% of hydrogen as shown in Figure 3. But, the kinetics of absorption/desorption was extremely fast, which work only in 5 minutes. As can be seen clearly, the addition of metal nanoparticles leads to a notable enhancement of both the absorption and desorption kinetics. In this case, it can be concluded that this enhancement in the hydrogen sorption kinetics properties is due to the presence of Ni nanoparticle as a catalyst in MgH_2 and the high pressure of milling as well.

A fundamental study of the role of catalysts in MgH_2 nanoclusters was provided by carrying out based on density functional theory [12]. Interesting to note that the diffusion mechanism may account for the fact that a small amount of catalysts is sufficient in improving the MgH_2 kinetics, which is essential for the use of this material for hydrogen storage application.

The SEM micrographs of Figure 4 show secondary electron image of powders reactively milled in the planetary mill for 2h (a) and 2h (b). The areas with the white contrast (Mg) disappear with increasing milling time leading to improved homogeneity. The surface of the powder is irregular, as a result of the fracturing during the milling process [13, 14].

Figure 4. Secondary electron (SE) image of MgH_2-2 mol% Ni nanopowders reactively milled in the planetary mill for 2h.

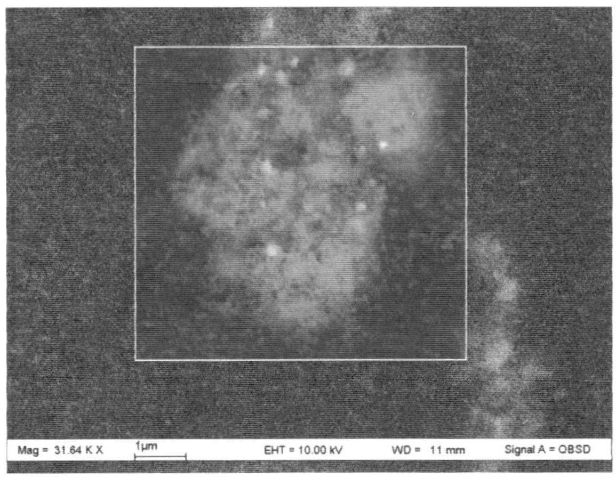

Figure 5. Backscattered electron (BSE) image of MgH_2-2 mol% Ni nanopowders after gravimetric desorption.

Backscattered electron (BSE) image of MgH_2-2 mol% Ni nanoparticles after gravimetric analysis desorption (see Figure 5) shows that the catalyst Ni nanoparticles (white spot) were uniformly distributed on the magnesium surface. This indicates that when small particles of catalyst is uniformly distributed on the metal surface, a small amount of catalysts is sufficient for improving the reaction kinetics.

4. Conclusions

MgH_2 catalyzed with small amount of nano-Ni as catalyst has been prepared via reactive milling. This high energy planetary ball mil (HEPBM) produced a high surface area powders and showed the nano-Ni particles dispersed on the surface of Mg uniformly. The MgH_2-2mol% Ni nanopowders exhibited the sorption kinetics properties, absorp/desorp 5.3 wt% of hydrogen within 5 minutes. It was also clear that, the used of nanopowder Ni and reactive milling method were very effective to reduce the milling process in short time.

Acknowledgement

Authors want to express their gratitude to Syarifah Fathmiyah responsible for x-rays diffractometer and Karmila Sari responsible for particle size distribution calculation.

References

[1] Schlapbach and Zuettel 2001 *Nature* **414**
[2] Sandrock G and Thomas G 2001 *Appl. Phys. A* **72** 153–5
[3] Imamura H Maruta Y Tsuchiya S 1987 *J. Less-Common Met.* **135** 277
[4] Song M Y Kwon S N Bae J-S and Hong S-H 2009 *J. of Alloys and Comp.* **478** 501-6
[5] Yuan H An Y Xu G Chen C 2004 *Mater. Chem. Phys.* **83** 340
[6] Huot J Pelletier J F Liang G Sutton M Schulz R 2002 *J. of Alloys and Comp.* **330-332** 727-31
[7] El-Eskandarany M S Shaban E Ali N Aldakheel F Alkandary A 2016 *Sci. Rep.* **6** 37335
[8] Bobet J-L Akiba E Nakamura Y Darriet B 2000 *Int. J. of Hydrogen Energy* **25** 987-96
[9] Jalil Z Rahwanto A Mulana F Mustanir 2016 *Int. J. of Tech.* **7(8)** 1301-6
[10] Jalil Z Rahwanto A Handoko E Mustanir 2017 *AIP Conf. Proc.* **1826(1)** 020002
[11] Jalil Z Rahwanto A Mustanir Akhyar Handoko E 2017 *AIP Conf. Proc.* **1862** 030023
[12] Larsson P Araujo C M Larsson J A Jena P and Ahuja R 2008 *PNAS* **105(24)** 8227–31
[13] Jalil Z Rahwanto A Sofyan H Usman M and Handoko E 2018 *IOP Conf. Series: Earth and Environ. Sci.* **105** 012093
[14] Jalil Z Rahwanto A Akhyar H Razali R and Handoko E 2018 *IOP Conf. Series: Earth and Environ. Sci.* **105** 012098

IOP Conf. Series: Materials Science and Engineering **432** (2018) 012035 doi:10.1088/1757-899X/432/1/012035

The cycles of mankind: begin from and ended back to nature

R Harjanto

Universitas Prof. Dr. Moestopo (Beragama) Jl. Hang Lekir I/8, Jakarta 10270, Indonesia

abrudyharjanto@gmail.com

Abstract. A classical problem of mankind that has transcended the boundaries of space and time is how to find the essence of life, and the search to find the form of life after death in the universe. Essentially, Taoism suggests that each person should appreciate and put into practice harmony and balancce in their life, which will affect their afterlife. The main problem in this research, is to find what is the meaning of life after death from the perspective of Taoism? Methodologically, the approach used is a phenomenological one. Taoism explains that a person's physical body will go through the process of re-uniting with nature, and then nature will facilitate humans to again go through life, and so forth. Mankind will go through the cycles of coming from nature and reverting back to nature, this process has been recurring throughout the history of mankind, since the beginning when the universe was formed.

1. Introduction

Two of the classical problems that mankind has been grappling with, have transcended the boundaries of space and time, namely: the search to find the meaning of life, and the search for the most fitting form of relationship with the meaning of death. Both are always present and always give rise to sceptical attitudes that reinforces, and at the same time it may become the proverbial glue due to a shared identity and shared frames of reference with regards to leading our lives and taking a stand toward various life events that we live with. These social norms provide a shared guideline in dealing with life as well as death.

Belief systems provide the rules and regulations of how to lead a good (or a bad) life, and is very much influenced by our faith. Faith itself may be a matter of feeling or desires, not common sense. Moral knowledge comes from an association of specific ideas, without necessarily having a rational basis, but rather based on subjective choices that we believe in.

True believers would gladly experience great suffering and hardship, endure physical or mental anguish for prolonged periods, as long as the pain and suffering remains in line with their beliefs and they understand that it will guide them towards eternal life after death. According to Jacques Derrida [1], psychoanalysis has taught us that the dead—for example, dead person—may be more alive, stronger, and more fearsome compared with when they were still alive. The issue here is about spirits or ghosts. Death brings a person to enter life in another dimension. To live means to go on a journey towards eternity.

According to certain belief systems, a person's death is very much determined by their actions or beliefs while they were alive. The survival or sustainability of certain systems of belief is determined directly by the actions of the individuals who belief in said systems, and not by the decision of others.

A number of different religions have a shared belief, that life in another dimension awaits us when we die. Human life becomes only a part of the ritual of universal life that journeys towards death in order to fulfil new life. To become part of the universe, and then the universe creates a new life. The universe creates harmony and balance. The concept of harmony and balance are in line with the ideas put forth by the ancient Taoist philosophers. Taoism is one of the great philosophical and religious traditions, began at around the sixth century b.c.e. or between 500 and 400 b.c.e. [2].

Taoism is essentially a practical philosophy. Taoism suggests that everyone should appreciate and put into practice the principles of harmony and balance in their lives. Practicing balance and harmonization in life, does not always mean being well-planned or deliberate about it, but rather it could be practiced spontaneously while we are naturally performing our normal roles and functions. Without deliberately striving for it, balance and harmony may already be in line with certain local and traditional cultures and for the most part it is influenced by the practice of local values, ethics and morals.

Harmonization is an interesting phenomenon when there is a tendency towards differing perceptions. It becomes even more interesting as we enter the realm of perspectives. At a certain point a dominant group will appear, whose faith will become mainstream. At other times, domination always creates alternatives. It is our differences that enable us to exchange ideas, and to influence each other's ideas about life after death. On the other hand, openness may be interwoven with restrictions (limitations, or "closed"ness), even at the individual level.

Thus the main issue of this research is, what is the meaning of life after death from the perspective of Taoism?

What are the sources of life after death from the perspective of Taoism? On what basis can life after death be verified, according to Taoism? What is basically the relation of verifying life after death from the perspective of Taoism? What is the relevance of the principle of verifying life after death from the perspective of Taoism?

2. Theoretical foundation

The material object of this research is a philosophical reflection about one of the phenomena or actual situations that remains a controversial issue, namely, about life after death. Most of the population of the world today still live with religion. But what's considered as most important for academicians, especially those who are Western oriented, is the pattern of thinking. They seem to think that the human environment consists of all agnostic people who follow a secular philosophy that are cultivated at prominent campuses [3]. They endeavour to develop a rationalization, an environmental code of ethics based on agnosticism, as if this effort will have a major effect on the environmental crisis. As if it were important to realistically consider the world in which mankind lives. To understand the world we need to observe the world. To understand the human self we need to reflect on the human self [4].

The formal object is used to assess the world of the living in order to understand and give meaning to life after death using the perspective of Taoism.

The word "Tao"(or Dao) means 'way' or 'path', or people walking, or walking on a journey. The abstract meaning, which is based on this basic meaning, also covers the concepts of direction, rules, ideals, and operational principles of the universe. Taoism views life as a primordial cosmic symbol of unity and harmony and a manifestation of the phenomenon of duality, or the two main powers that govern the cosmic order of the phenomenal world.

Tao views the human body as a miniature of the universe that is manifested through the spirit (energy) of life. The important life energy is known as Qi. In Daoism, however, Qi is simply the natural operation of the universe, the basic pattern of expansion and contraction. In this more expanded cosmic view of Qi, the energy that gives life to our bodies is nothing less than the basic energy of the universe, which condenses in the form of matter to produce the tangible 'stuff' out of which all things are made. Things like rocks are composed of slow-moving dense Qi. Our bodies are a complex of moderately condensed Qi that forms the material parts (skin, organs, bones) and a relatively light and ethereal Qi that gives us our minds and our spirits. When the Qi or breath is flowing in and out then we have life; when the Qi or breath stops moving then we have death [5].

Although energy can be converted from one form to another, the amount of it is unchanged as time goes on. In other words, energy can neither be created nor destroyed, but is always "conserved," which means that the amount of it stays the same as time goes on [6].

According to the Tao point of view, life should be lived in balance and harmony, mankind should live together in accordance with the harmony of the universe, that is, free from mere personal interest and personal wants. This harmony is reflected in the symbol of Yin Yang, a circle that consists of equal parts of black and white, in which each part is flowing into the other part. This circle describes the principle of nature, as it describes the pattern of interdependence. The Yin-Yang diagram shows two major magnetic poles of the universe, that is, darkness and light, life and death. Negative and positive, female and male, which exist in complete equilibrium and in which each force is equally powerful; together these two forces control everything that exists in nature.

3. Research methodology

Conducting research is a process of gathering and analyzing data, and information about life after death is gathered through an investigative process in order to search for and decide who will be the sources of information. For said purpose, the researcher selects the locations to be studied; then enters the world of the informant and engages in intensive interactions; then separates descriptive notes—portraits of the informant, dialogue reconstruction, explanation about the physical background, reports of specific incidents, and activities—as well as reflections—which provide opportunities to the researcher to make notes about personal thoughts, such as speculation, emotions, problems, ideas, estimates, impressions, and prejudices [7].

An empirical phenomenological approach involves the process of retelling an experience in order to get a fundamentally comprehensive description as a reflective structural analysis that encapsulates the essence of said experience [8].

This method allows the researcher to delve directly in order to understand what is happening, to understand pattern and interactions. In this case, basically the researcher has two roles to play: as a participant, and as an observer/researcher. To quote Lindhof and Taylor, the researcher's interviews were conducted with the objective of developing a view about something that happens between individuals. The interviews that the researcher has conducted were dialogues that had a goal [9].

To analyze data in said research, the researcher will use inductive methods and a qualitative approach. Using this method enables us to bring to reality a theoretical construction, utilizing intuitive knowledge, in order to clarify a logical construct of a philosophical problem, so that the researcher will be able to describe the philosophical structure of the research object. The researcher will completely hold the position of final decision maker regarding any interpretation that is based on an inductive thinking pattern using phenomenological paradigms. The Research was done through February-July 2017 in Jakarta.

4. Results and discussion

An orientation towards earthly life can be either good, or bad. It is good, when daily life is based on behaviour that is in harmony with a valid value-system: a system of social values, law, and religion. It is bad when the reverse happens. The parameters of good and bad are subject to relativity. Social actions actually do not become a problem as long as it remains in the same community or society. Social actions only become a problem when the social interaction 'cross-cuts' between one group and a different group. The parameters of good and bad may be contradictory. Thus, a particular parameter may not be applicable to judge the behaviour of another group in another society which has a different value-system.

What about sacred behaviour? A human being's sacred behaviour is a concrete manifestation of the acknowledgement of a bigger, more powerful and determinant force. Sacred behaviour refers to religious teachings, texts written in Holy Scriptures, sermons given by religious authorities, as well as the written and unwritten rules and regulations with regards to spirituality. Sacred behaviour that can be easily observed are, for example, the devout adherence towards rites that derive from religious decrees. But it is by no means easy to measure a person's spirituality. At this point, spirituality becomes the most private domain of human beings. There are no parameters, whether universal or particular, that are able to measure a human being's spirituality. The process of rationalization is inherent to various social structures, and in an individual's thinking and actions. The individual's actions, are a part of a system of rationality which hold a causal priority position in relation to other religions.

The search to find the essence of divinity has been an age-old problem. This search has never been fully completed. New questions always come up each time an answer is found, and the answers are many. Humans have always been contemplating, reflecting on his thoughts, pondering, thinking, and putting forward the classical existential question regarding life. This never ending effort becomes an important part of the search to find the essence of divinity and at a certain point is a part of the strategy of how to deal with life. Spiritual experiences are unique and very personal. Spirituality does not grow in a vacuum. Rather, it is forged, formed, and requires tenacity. There are a number of influential factors relevant to the search to find the essence of divinity.

Truth is highly correlated with evidence or with facts that have been verified, and merely having the right attitude, faith, or comprehension is not sufficient. Nevertheless, comprehension may be attained as the result of interpretation. Interpretation opens the door to multi-interpretation, even if the text or observable object used as a reference is the very same one. Multi-interpretations may arise due to the diversity of educational backgrounds, value-systems that we have lived by, and interests. As long as the social stake-holders remain consistent in the actualization of social norms, then social life will be harmonious. According to D. Hume [10], based on holy scriptures, and no man shall want to waste their lives, as long as his (her) life still has value to be lived. To lead a life is by itself a goal of life. Kierkegaard [11] has stated that an understanding of life can be attained from the past, however life itself should refer to the future, and that life is not a mere problem to be solved, rather it is a reality that we have to experience within social norms.

To Nietzsche [12] the world is an empty place. The world has no creator, it is able to exist and grow by its own strengths. In such a world, objective knowledge does not exist. What is needed to gain knowledge is just subjectivity, and the ability to make an interpretation. Those two factors, according to Nietzsche, come from the Will to Power itself. By having subjectivity and the ability to interpret, mankind is able to see causal relations in the world. With those two abilities, mankind is able to position itself, and also to position everything else in the world in its place.

The Will to Power drives humans to become an active subject on the journey of life, and at the same time humans are also the ones who interpret the world and who give the world meaning. Having the Will to Power, humans are able to create and organize the world. In this sense, the world is an inhuman place. The world acquires its meaning because of humans; with his subjectivity and his ability to interpret things, humans give it meaning, and thus "humanize" them. From this point we can see how Nietzsche was asking us to look at the world from a positive point of view, by comprehending life and by celebrating earthly life, because after all the world has been created and organized by none other than humans themselves.

The search to find the essence of divinity has never been fully completed. New questions always come up as humans are always contemplating, reflecting, pondering, thinking, and even asking the classical existential question regarding life. As a truth within the realm of religion, catharsis may be given the meaning of a transcendent experience that frees or cleanses the soul. At its apex, transcendental experiences may be in the form of a reverence towards humanity, a merciful attitude, treating human beings in a humane manner, taking an anti-violence stand, having an awareness of the wants and needs of human beings as a reality that has to be experienced within the social norms in going through the journey of life on earth.

On the other hand, the world is an empty place. The world is able to exist and grow by its own strength. In such a world, there is no objective knowledge. Life on earth has been engineered and is governed by mankind by itself, in accordance with their own respective beliefs. Belief systems provide a set of rules and regulations regarding life, in which the concepts of good and bad are very much influenced by subjective choices that one believes in. What common sense provides is merely information. Those who truly believe would gladly endure profound and prolonged pain, both physical as well as mental agony as long as their suffering will bring them towards a life that leads them on a journey to eternal life.

Interpreting the journey of life towards death becomes interesting when we encounter different tendencies. It becomes even more interesting, as we enter the realm of perspectives. At a certain point, a dominant group will appear and become mainstream. At other points, domination always gives birth to alternatives. Having those differences enables us to exchange ideas and to influence each other's ideas. On the other hand, openness can be interwoven with restrictions (or 'closeness), also at the individual level.

One can attain harmony in life by becoming mindfully aware of one's personal ability in comprehending the whole nature of the universe. Mankind's comprehension is indeed limited, while the mysterious complexity of the whole universe is infinite. Therefore, the effort required to change the sequence of spontaneity, to change an attitude towards life's changes through deliberate planned action on the part of humans, may result in unwanted consequences which may be terrifying. It can be dangerous when something that is full of limitations attempts to reach for something else that is infinite. The death of human mentality and an awareness of reality that is like molten silver that is in the process of changing its vitality to become metal, like an idea that once was a bright light.

In order to overcome this obstacle, Taoism—like other philosophies-- involves the search for truth that in some versions even contains elements of mysticism. The main objective, nevertheless, is to help humankind in going through life. Taoism provides guidance and strategies for individuals to go through their daily lives, although its greater objective is to provide insights about how to create and maintain harmony in society as a whole.

According to Taoism, human life depends on harmony, and harmony is a means to make a person become accustomed in his thinking, to accept and allow himself to experience paradoxes without having to eliminate either side: the black and the white, emptiness and fulfilment, personal interest or the greater interest of society in general. Society in general consists of individuals, ideally all that these individuals need to do is to strive to reach what is important for them only in their daily lives, in order to attain what is good. In their striving to reach for what is important, as a matter of course not everyone will be successful, even to the point of failing to get anything at all, except for emptiness.

From the Taoist's perspective, this condition is ideal, that is, the initial condition in which a person has not achieved or attained anything, which will later benefit him (or her), to attain what he (or she) dreams about. Thus, being consistent in accepting such failure, means that basically he (she) is already in the process to reach emptiness, which later on will enable him (her) to attain fulfilment with regards to their final objective. Whoever he is, he will never obtain fulfilment, if he has not yet experienced emptiness.

It is the paradox of emptiness and fulfilment, of fullness and emptiness, just like the harmony we find in the earthly cycle: originating from state of non-existence, then developing to existence, and ending up in the state of non-existence once again.

Having the paradox of emptiness and fulfilment tends to benefit society in life. Psychologically, a person will become more alert and will have more awareness to prepare themselves for what is happening and for what will be happening. A political condition which dominates businesses so that it enables business and industry to conspire against the consumers, will one day reach its end and be replaced by another political condition. When faced with political changes, having an attitude of full awareness will affect producers and merchants in their own interests, and in the long run also affect the lives of mankind.

According to Taoism, human beings must lead their lives by accepting things just the way they are, and each person should endeavour to play his or her own role and function. When humans are playing their respective roles and functions, each element and each component must be maintained. Maintaining the role and function of each element becomes important, because it will guarantee our ability to perform accordingly. Thus the condition of harmony will be attained.

Harmonious conditions will have an indirect and beneficial impact on the human body itself, as well as on society. Life on this earth will run its course by itself, without having human beings who want to govern the world trying to establish their rules and regulations. Thus, the established harmony will no longer need to be ruled in order for life to run harmoniously the natural way.

Mankind's abilities are finite, whereas the complexity of the whole universe is infinite. All of life's elements have to go through the earthly cycles of harmony: originating from non-existence, growing into existence, and in the end reverting to non-existence, and so forth. In evolutive stillness, each element of the body changes; and thereby changing human lives; but for those who truly believe, these changes may even strengthen mankind's attitude of resolve and readiness to enter life in another dimension.

Living life in another dimension according to the Taoists' view is to return back to the natural universe. Taoists believe strongly that all things on earth possess an energy and an interconnected "unifying life force". Energy can be transformed—from mechanical to electrical, or from electrical to heat, for example—but it cannot be created from "nothing" or destroyed [13].

A human being's life does indeed end in physical death, however, in a very broad sense, according to Taoism each and every human being actually never dies. Human beings will be reborn in another form and activity. Death is but the last step taken in life in order to reach unity with the universe, in order to return to our origins before life and death, and in order to return into a new life, probably as a human being leading a new human life. Death is inevitable, and likewise, so is new life. Before life begins we have lost nothing; after death we have lost everything [14].

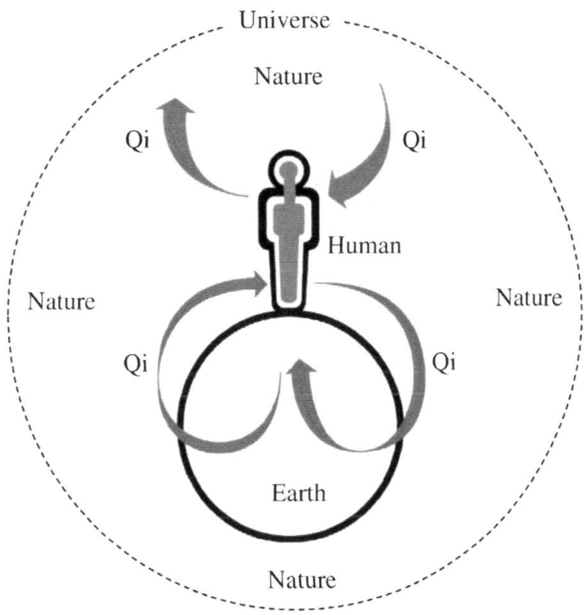

Figure 1. Human body, earth, and qi, united as the part of nature and the universe.

The human body and the earth are two microcosms as the nature, both interlinked not just to each other, but they also form a unity with the macrocosms: the universe. According to Tao philosophy, the human body is living on the earth, which is part of nature; nature provides what is needed, and what is regulated by humans. Mankind appreciates having air to breathe, the body will become whole through breathing, as breathing provides energy.

Energy or Qi regulates the process and movements of the elements of the body, each element has its own function to make the body interact with the cosmic universe. Energy that exits and enters the human body through the process of breathing [exhaling and inhaling] enables human beings to be at one with the universe. Nature provides energy to humans, both to new-born babies as well to those who already exists, and so forth, as shown on Figure 1.

When a person is no longer able to breath, that person is said to be experiencing death. The physical body of humans then goes through the process of uniting with nature, and in turn nature will facilitate humans to have a life, and so it goes on and on. This would involve the most severe fine-tuning of any physical quantity known in nature, without the slightest idea how to adjust it [15].

Human beings go through cycles of coming from nature and reverting back to nature. As the human body transforms as it goes from birth (existence) to death (non-existence) and then again transforms as it goes through rebirth, and so on.

This process has been repeated throughout the history of mankind, since the formation of the universe. Life and death both have structured patterns within the universe.

Earth is in the middle of it, as earth facilitates, earth produces, and earth takes care of the life and deaths of all beings. Earth is both the source as well as the place where everything ends. Earth thus takes it place as the important controller; supporting the essence and the lives of mankind.

According to Taoism, the death of human beings is only a part of the process of universal life. Mankind is always in the process of becoming or reaching eternal life, so long as the universe exists. The universe has a power, a universal power that connects all that exists, to be in harmony with the life of mankind. The more human beings can accept this harmony, the more harmonious it may become.

Tao regards the human body as a miniature of the universe, as part of the physical features of the universe, a multi-faceted manifestation of the spirit or energy of life. According to Tao, this important life energy is known as Qi, and having Qi enables objects and creatures to become individual entities. Qi is a vital cosmic energy that enables beings to survive, and it is this energy that connects them with the universe as a whole. The life of a person is the accumulation of Qi; when we are born we receive the primordial essence of Qi potential which is embedded in our physique.

Within the body, Qi regulates and organizes the circulation of air and food; the Qi energy can flow from one person to the next through sexual, emotional and social interactions.

However, humans can also lose their Qi by inhaling bad air, by burdening their bodies with food and drink, and by engaging in negative emotions, and by engaging in excessive sexual or social interaction. Therefore, Tao teachings introduce the concept of Wu Wei. Wu Wei is sometimes translated as an activity without action, which implies that there is no activity at all. The true meaning of Wu Wei is: an activity that is congruent with the natural pattern of the universe, and therefore completely independent from, and not in any way controlled by personal interest. According to Tao beliefs, mankind should live in balance and harmony, live together in harmony with the universe, free from mere personal interest. For example, this harmony is reflected in the symbol of Yin Yang, a circle consisting of a black side and a white side that flow towards each other. This circle represents the principles of nature, an interdependent pattern. Even though each element is the opposite of the other element, but both form a single and complete unity. The opposing elements could be day and night, happiness and sorrow, or action and inaction. According to Taoism, those opposing conditions can become harmonious parts that work together—whether during ebb and tide-- and together they form a perfect harmony. The two opposing forces are intertwined and inextricably tied together in one unity.

5. Conclusion

The search to find the essence of divinity has never been completed fully. Instead, what remains unanswered is the classical existential question about the meaning of life, after death. True believers would gladly undergo suffering, as long as their suffering will lead them to eternal life after death. The journey towards eternal life is a mystery, and according to Tao, the mystery of this journey to reach eternal life has an infinite universal complexity.

According to Tao beliefs, the meaning of life is about change that occurs in an evolutive stillness, in which each element of the body changes, and this will change the life of human beings; and this change will most probably strengthen and fortify a person's attitude of readiness to lead a new life in another dimension.

According to Taoism, life in another dimension means a return to the universe. All of life's elements always go through cycles in the earthly process of harmonization; originating from non-existence, then growing into existence, and again reverting back to non-existence, and so on.

The source of life after death as seen from the Tao perspective is describing mankind and the earth party part of nature; nature provides what is needed and governed by mankind, using the energy that humans have received from nature. This energy is what regulates the flow and movement of the body's elements so that each of them will function to enable the body to interact with the cosmic universe. Nature provides this energy to mankind, beginning with newborn babies as well as the rest of humanity. When humans are no longer able to breathe, it is said that they are experiencing death.

The verification of life after death according to the Tao perspective is based on the view that humans will be reborn in another form and activity. Death will revert humans to enter a new life. Death is inevitable, as is new life. Tao regards the human body as a miniature of the universe, as a part that has the physical features of the universe, a manifestation of the spirit and energy of life. This spirit and energy enables creatures to continue with life after death, and connects them with the universe as a whole.

This basic relation that is a verification of life after death from the perspective of Taoism explains that a person's physical body will go through the process of re-uniting with nature, and then nature will facilitate humans to again go through life, and so forth. Mankind will go through the cycles of coming from nature and reverting back to nature, the human body transforms from existence to non-existence and then again to existence, and so forth. This process has been recurring throughout the history of mankind, since the beginning when the universe was formed.

References

[1] O'Connor Patrick 2010 *Derrida: Profanations* London Continuum Int. Pub. Group
[2] Hartz Paula R 2009 *World Religions Daoism 3rd edition* New York NY: Chelsea House
[3] Nasr Hossein S 1993 *The Need for a Sacred Science* Surrey Richmond: Curzon Press Ltd.
[4] Bo Mou (Ed.) 2009 *History of Chinese Philosophy, Routledge History of World Philosophies 3* Abingdon Oxon: Routledge
[5] Miller James 2008 *Daoism: A Beginners Guide* Oxford: Oneworld Pub.
[6] Lichtenberg Don 2007 *The Universe and The: Atom* Singapore: World Sci. Pub. Co. Pte. Ltd.
[7] Bogdan Robert C Biklen S K 1982 Boston MA: Allyn and Bacon Inc.
[8] Moustakas Clark E 1994 *Phenomenologic. Resear. Meth.* Thousand Oaks CA: Sage Pub. Inc.
[9] Lindlof Thomas R Taylor B C 2002 *Qualitative Commu. Resear. Meth. 2nd ed.* Thousand Oaks CA: Sage Pub. Inc.
[10] Millican Peter (Ed.) 2007 *David Hume: An Enquiry concerning, Human Understanding* Oxford: Oxford Univ. Press
[11] Moore Charles E (Ed.) 2007 *Provocations, Spiritual Writings of Kierkegaard* Farmington PA: Plough Pub. House
[12] Welshon Rex 2004 *The Philosophy of Nietzsche* Bucks: Acumen Pub. Ltd.
[13] Silverman Mark P 2002 *A Universe of Atoms, An Atom in the Universe* New York: Springer-Verlag New York Inc.
[14] Scarre Geoffrey 2007 *Death*. Stocksfield: Acumen Pub. Ltd.
[15] Krauss Lawrence M 2012 New York: Free Press A Division of Simon & Schuster Inc.

IOP Conf. Series: Materials Science and Engineering **432** (2018) 012036 doi:10.1088/1757-899X/432/1/012036

The effect of molecular weight on the surface wettability of polystyrene treated with nitrogen plasma

Masruroh[1*], T N Zahirah[1], Aulanni'am[2], S P Sakti[1] and D J D H Santjojo [1]

[1] Department of Physics, FMIPA, faculty of Science, Brawijaya University, Malang, Indonesia
[2] Laboratory of biochemistry, Department of chemistry, Brawijaya University, Malang, Indonesia

ruroh@ub.ac.id

Abstract. The wettability of polystyrene thin film surface in the development of biosensor based on quartz crystal microbalance (QCM) is an important aspect in biomolecule immobilization. The polystyrene thin film was produced by a spin coating method. The wettability was achieved by modification of the polystyrene surface using a nitrogen plasma treatment. The aim of this work is to investigate the effect of polystyrene raw material molecular weight to the wettability of the polystyrene surface treated with the nitrogen plasma. A TMS-1200 micro profilometer was utilized to determine the surface roughness. The contact angle of the polystyrene surface was measured by a contact angle instruments to observe the surface wettability. The results showed a decrease in the surface roughness and the contact angle of polystyrene layer after the nitrogen plasma treatment, indicating the polystyrene surface changed to a higher wettability. Furthermore, the higher molecular weight of the polystyrene led to the decrease of the surface roughness and the increase of the wettability. The wettability depends on functional group in the polystyrene. The changes of polystyrene the functional group were determined by an FTIR spectrometry. The infrared spectroscopic measurement has confirmed the presence of polystyrene nitrile (C-N) group in the form of C≡N representing a polar group structure. Moreover, the intensity of the C≡N peak increased in the treated polystyrene with the higher molecular weight. The increase of the absorbance is related to the increase of the surface polarity and hence the wettability.

1. Introduction

On the application of a biosensor, the quartz crystal microbalance (QCM) is used as a biomolecule detector. The QCM is a chemical sensor device which is highly sensitive towards mass changes on its surface. A QCM consists of two electrodes sandwiching a disc of AT-cut quartz crystal. In the biosensor application of the QCM, a coating process is needed to protect the electrodes from oxidizing. Furthermore, the layer is also designed to immobilize biomolecules. A polystyrene layer is generally used as the coating layer as well as the immobilization matrix layer on the QCM [1, 2]. The polystyrene surface is generally hydrophobic and binds biomolecules through physical adsorption. The adsorption of the biomolecules is affected by wettability of the surfaces. The wettability of the polystyrene layer can be improved by modifying the surface morphologies [3-5]. In the previous research, the modification of the polystyrene layer with various molecular weight was carried out by means of UV radiation treatments for 1 and 2 hours [5]. The longer period of the UV radiation on the polystyrene layer resulted in the lower contact angle indicating the surface is becoming more hydrophilic than the result of the short treatment.

It is desirable to establish a shorter treatment period. The surface modification can also be carried out by a plasma treatment. In particular, a plasma treatment of polymeric materials has become one of the most important methods to enhance the wettability and adhesion of biomolecules without affecting the bulk material features. The plasma treatment can change the microstructure and character of the polymer surface which results in the functionalization of materials. There were a significant number of publications related to the modified polymer surfaces using plasma treatment [6-11]. However, the modification of the polystyrene using a plasma treatment on the QCM sensor is still limited.

The surface of the polystyrene layer is modified using a nitrogen plasma to achieve better wettability of the surface. The molecular weight of the polystyrene raw material used in the film production is one of the factors which needs to be examined in the surface modification. In this paper, the effect of the molecular weight on the surface wettability of the polystyrene treated by nitrogen plasma was studied by means of a Fourier transform infrared (FTIR) spectrometry. The FTIR data was used to investigate the functional group of the polystyrene surface before and after plasma treatment.

2. Experimental method

The samples for this study were produced by depositing a layer of polystyrene thin film on the top of a commercial quartz crystal resonator. The resonator was purchased from PT Great Microtama, Surabaya, Indonesia. The polystyrene with various molecular weights (Mw) of 35,000 g/mol, 192,000 g/mol and 280,000 g/mol was purchased from Sigma-Aldrich. The polystyrene was dissolved in a chloroform solvent at a concentration of 3 %. Then, the solution was coated on the quartz substrate by using a spin coating technique. The spin coater was operated at 3000 rpm for 60 seconds. After the deposition, the sample was heated in the oven at a temperature of 150 °C for 60 minutes to evaporate the moisture on top of polystyrene layer. The modified polystyrene surface was produced by nitrogen plasma treatment generated in a mini-reactor using a low power (40 watts) plasma generator for 2 minutes. The process was maintained at a constant pressure by controlling gas flowrate at 20 ml/min. The nitrogen plasma system is shown in Figure 1. The surface roughness of the untreated and treated polystyrene surface was observed by a micro profilometer TMS-1200. The surface wettability of the polystyrene surface was measured by contact angle instruments. The procedure of the contact angle (CA) measurements was reported in the previous paper [5]. The functional group of polystyrene was studied by FTIR (8400S Shimadzu) spectroscopy.

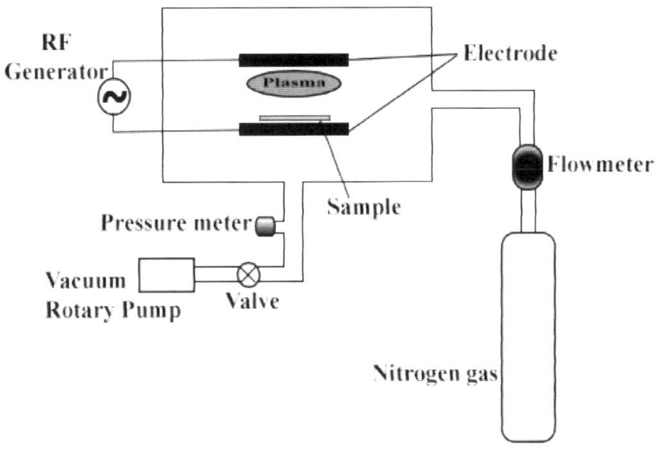

Figure 1. Design of nitrogen plasma system used in this experiments.

3. Results and discussion

Nitrogen plasma treatment has evidently modified the polystyrene surface. Figure 2 shows the surface roughness of the various molecular weight polystyrene before and after plasma treatment. The surface roughness increased with the increase of the molecular weight of the polystyrene. On the other hand, the nitrogen plasma treatment of the surface decreased the surface roughness. The higher of the molecular weights results in the reduction of the surface roughness. The reduction of the surface roughness can be a result of photochemical process, physical knocking by ion bombardment, and/or surface chemical reaction. The photochemical process breaks several polymer chains and opens opportunities to the surface chemical reactions. The process leads to the alteration of the surface architecture. Furthermore, the alteration was enhanced by the ion bombardment onto the surface.

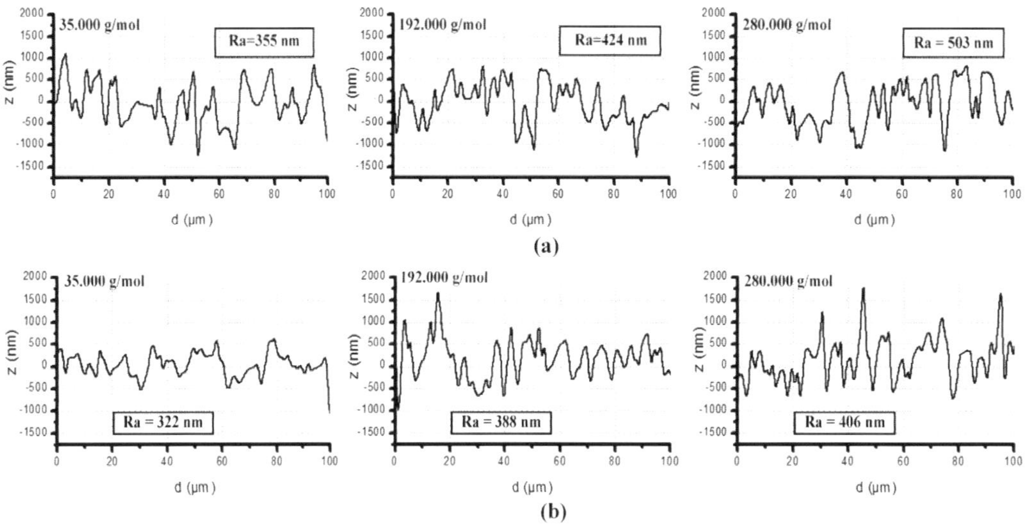

Figure 2. The surface roughness profile of polystyrene with various molecular weight of 35,000 g/mol, 192,000 g/mol and 280,000 g/mol: (a) untreated (b) plasma treated.

The surface roughness of the polystyrene has been known to relate with the wettability [12]. In this study, the wettability was examined using a contact angle measurements. The contact angle (CA) of the untreated polystyrene surface with the molecular weight of 35,000 g/mol was 82.639°. This result indicated that the surface exhibits weak hydrophilicity. After the plasma treatment, the contact angle was reduced dramatically to 42.808°. The reduction trend was also observed with the specimens having higher molecular weights as shown in Figure 3. The summary of the treatment results is shown in Table 1.

Table 1. Summary of the contact angle and roughness of the polystyrene with different molecular weight.

Molecular Weight (g/mol)	Untreated		Plasma Treated	
	Contact Angle (°)	Roughness (nm)	Contact Angle (°)	Roughness (nm)
35,000	82.690	355.42	42.808	329.28
192,000	84.122	427.68	46.966	388.61
280,000	85.836	519.69	52.535	456.56

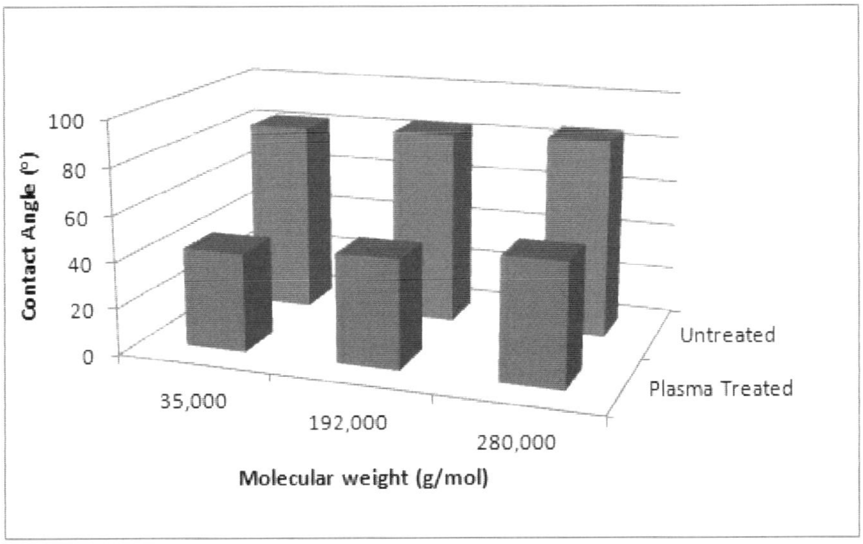

Figure 3. Contact angle vs molecular weight of Polystyrene layers: untreated (blue) and plasma treated (red).

Generally, the contact angle is the measure of the surface energy that controls the wettability of the surface [10]. A lower contact angle exhibits a higher wetting of the surface. The wettability is related to the polarity of molecules or polystyrene's functional group on the surface. The polarity of the functional group can be observed indirectly using an infrared measurement. The IR spectra of the untreated and the treated surface are shown in Figure 4.

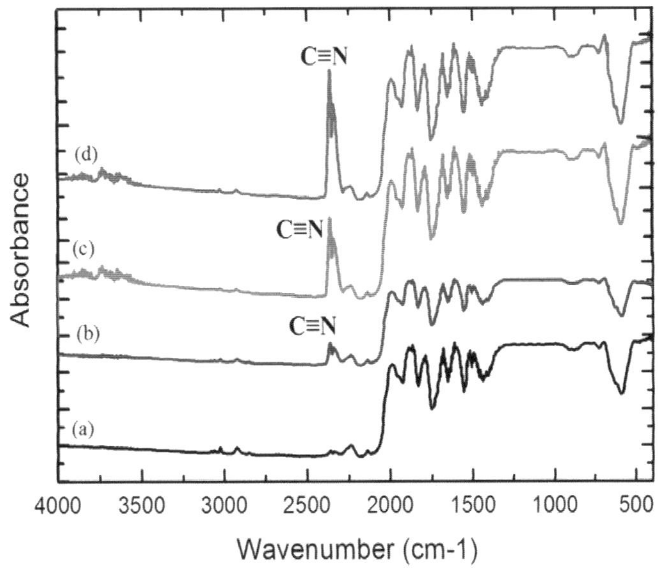

Figure 4. FTIR spectra of Polystyrene layers (a) untreated (b) plasma treated with the Mw of 35,000 g/mol (c) plasma treated with the Mw of 192,000 g/mol (d) plasma treated with the Mw of 280,000 g/mol.

The FTIR spectra show the appearance of a new vibration peak at 2300-2400 cm^{-1}, corresponding to C≡ N bond. The formation of C≡N was initially triggered by the stoichiometry bond breaking of the polystyrene bonds due to the photochemical reaction enhanced by the ion bombardment. The breaking of the bonds was related to the molecular weight of the polystyrene. The broken bonds were passivated by the nitrogen atoms via the surface chemical reactions. The passivation was indicated by the incorporation of the nitrogen atom forming the $C \equiv N$ bonds. Furthermore, the IR absorption intensity of the C≡N peak became stronger with the increase of the molecular weight. This indicates that the polymer with a larger molecular weight breaks more than the smaller molecular weight one. The more breaking has also been observed as the more reduction of the surface roughness.

The C≡N functional group has been known as a polar group [13]. The IR spectra show a doublet indicating two vibrations of the polystyrene's functional group. Both of the two peaks simultaneously increased which confirm that the C≡N has multiple bands of vibration with similar character or polarity. The polar C≡N functional group interact strongly with the polar water molecules results in higher wettability.

4. Conclusion

The nitrogen plasma treatment on the polystyrene was evidently affected by the polystyrene molecular weight. The surface character of the polystyrene changed into hydrophilic after the treatment which is shown with the reduction of the surface contact angle. The contact angle was influenced by the decrease of the surface roughness. Furthermore, the presence of the C≡N was also correlated with the decrease of the contact angle and hence the increase of the wettability. This is due to the C≡N being a polar functional group which was produced by the bond breaking of the polystyrene and surface reaction incorporating N into the molecular structure.

Acknowledgment

This work is supported by Ministry of Research, Technology and Higher Education of the Republic of Indonesia.

References

[1] Masruroh Djoko D J D H Didik L A Rahmawati E Padaga M Abdurrouf and Sakti S P 2014 *Appl. Mechan. and Mater.* **530-531** 54-7
[2] Sakti S P Santjojo D J D H Saputri S N and Aulanni'am 2012 *J. Biosens Bioelectron* **3** 1-5
[3] Yusilawati A N Maizirwan M Hamzah M S Ng K H and Wong C S 2010 *American J. of Appl. Sci.* **7(6)** 724-31
[4] Fahem R and Yousif E 2017 *J. Thin Films Res*. **1** Issue 1 17–23
[5] Sakti S P Amaliya L Khusnah N F and Masruroh 2017 *J. Tech.* **79** 61-7
[6] Choudhury A J Chutia J Barve S A Kakati H Pal A R Mithal N Kishore R Pandey M Patil D S Jagannath 2011 *Progress in organics coatings* **70** 75-82
[7] Truica-Marasescu F and Wertheimer M R 2008 *Plasma Process Polym 5* 44-57 WILEY-VCH Verlag GmbH & Co. KGaA Wemhem
[8] Kong J Yung K L Xu Y Tian W 2010 *eXPRESS Polymer Lett.* **4(12)** 753–62
[9] van Vroonhoven L F C B M and Brunnix E and Benninghoven A 1989 *Polymer* **30**
[10] Gerenser L J 1993 *J. Adhesion Sci. Techn*. **7(10)** 1019-40
[11] Anuradha P and Kalianib A Anu 2009 *J. of Pure and Appl. Phys.* **5(2)** 157–64
[12] Wenzel R N 1936 *Indistrial and Engin. Chemist.* **28** 988
[13] Berstein M P Sanford S A and Allamandola L J 1997 *The Austropical J.* **479** 932-42

IOP Conf. Series: Materials Science and Engineering **432** (2018) 012037 doi:10.1088/1757-899X/432/1/012037

Fabrication of magnesium ecap based maxillary miniplate-typed implant through the method of micro forming

S Supriadi[1]*, H Kristianto[1], Y Whulanza[1], A S Saragih[1], R Dhelika[1], B S Latief[2], Taufiq[3]

[1] Department of Mechanical Engineering, University of Indonesia, Depok, West Java, Indonesia 16424
[2] Faculty of Dentistry, University of Indonesia, Depok, West Java, Indonesia 16424
[3] Metal Industries Development Center, Coblong, Bandung, West Java Indonesia 40135

sugeng@eng.ui.ac.id

Abstract. Research and study on a degradable implant for bone fracture fixation has received great attention. This had to do with a problem, which in some cases, require doctors to do re-surgery to remove the implant from the patient after the fractured bone was healed. The removal surgery could cause several problems, such as the need for another recovery period and the high cost of the removal surgery itself. Several methods are developed to control degradation rate to match with healing rate. One of such method is the grain refinement technique using severe plastic deformation. In this paper, a study of fabrication of a miniplate implant for maxillofacial region using magnesium which passed severe plastic deformation using Equal Channel Angular Pressing (ECAP) was performed. The micro-forming fabrication method is selected to form the miniplate. It is shown that the accuracy precision of the process was achieved with a reasonable result compare to the design. The most crucial part form miniplate design is the hole which shows low variation in diameter, 1.43 %. It shows that the method of micro-forming has great potential to fabricate miniplate using magnesium ECAP. Further work is necessary to conduct finishing process on the miniplate.

1. Introduction

Research on implants has led to the development of biomaterials, both metal and non -metal. While previous researches are focused on the biocompatibility itself, nowadays the standard material of implants has been established. Modern implants are proven enough to be biocompatible, with titanium as the leader of the structural metal implants, and other alloys or ceramics as wear resistant parts. The advance of biomaterial research has brought another inspiration, on how the implants are focused as a temporary solution, not a permanent one, as the previously mentioned titanium and ceramic implants will be permanently resided inside the body, unless removed surgically, which is quite common in the maxillofacial region (face and jaw area) implants. The second surgery required to remove the implant, will be somehow uncomfortable for the patient themselves and meticulous for the doctors. With that in mind, most patient will opt for leaving the implants inside the patient's body, as there is no direct medical side effect of the implants itself. Although, maxillofacial implants tend to affect patient's face proportions slightly, which for some, could be bothersome and not aesthetically pleasing. To refine the fixation method, there has been a proposal on how implants, especially for maxillofacial region, could be made using materials found naturally in human body, so the implant itself could be degraded after bone healing has been done. Magnesium has been a leading material in this regard. With higher strength than most non-metal biodegradable materials, magnesium could provide more structural support and could be used for wider spectrum of implants.

However, magnesium has one weakness, which is a high degradation rate, which is much faster than the healing rate of the bone itself. Therefore, a method of lowering magnesium's degradation rate shall be implemented. There are 3 known method to lower magnesium's degradation rate. The First one is coating. Coating is done using other materials to coat magnesium, so that oxidation will not happen. Another method is alloying, which is combining two or more metals so that the strength and corrosion resistance is increased. The 3rd method is by using grain refinement, which is decreasing the grain size of the material itself, in its pure or alloyed state [1]. The first two methods are not too effective in this application as an implant. The first method, which is coating, will only protect and decrease degradation rate temporarily, as long as the coating present. When the coating is gone, the degradation rate will increase to the initial uncoated rate [2]. The second method, which is alloying, had a different set of problems. Metals used in the alloy, such as aluminum, which when degrade, could be released in a dangerous, toxic amount for the human body. The third method is the most feasible one, as the structure is uniform enough, and there are no alloying metals that will degrade with the magnesium itself [3], and smaller grain size means that degradation rate will drop. There are several methods of grain refining, such as hot rolling, cold rolling, forging, and ECAP (Equal Channel Angular Pressing), and based on the experiment performed by Wang in 2008, it is proven that ECAP method resulted in a more uniform, and overall better mode of degradation, as well as lower rate of degradation [4].

The ECAP method, developed by V.M. Segal in the 1970s, is a method to refine the grain through strain hardening [5]. The heated billet is passed several times through an angular channel, which will bend and straighten the billet itself. 1 pass is equal to one cycle.

This experiment aims to evaluate the method of micro forming in the manufacturing process of magnesium ECAP based miniplate for maxillofacial region, in terms of manufacturing accuracy using important dimensional parameters within the plate.

2. Experiment method

The experiment is done using magnesium that has been through six times ECAP procedure. The magnesium ECAP blanks then cut into rectangular plates of 24 x 3.5 x 0.85 mm dimension. The plates are cut into the outer profile, and then stamped using dies at 250°C with the force of 21046 N. The heating process is done with a muffle furnace for 10 minutes, and the pressing itself is done using hydraulic press. This experiment parameters of the Mg ECAP itself are shown in Table 1.

Table 1. The parameters and values of ECAP process in this experiment [6].

Parameters	Values
Force (N)	25000
Temp (°C)	350
Channel angle (°)	120
Passes	6 with 90° turn every pass

The influence of this process on the magnesium itself, is increase in yield and tensile strength, decrease in grain size, and decrease in degradation rate, as seen in Table 2.

The dies used in this experiment is presented in Figure 1, with SKD-11 as the material, consists of 4 parts, stacked on each other as shown in Figure 1 and sealed with 4 nuts and bolts. The plate blank was put inside the cavity from the oval hole, and closed using the punch (the top part). Both the dies and the blank inside are heated in a furnace to 250°C. Finally, using a hydraulic press, the heated dies and blank were pressed with 21046 N of force. The plate is then extracted by removing the upper dies and the insert.

Table 2. The physical differences between unprocessed Mg and ECAP Mg [6].

Parameters	Pure Mg Without ECAP	Pure Mg ECAP
UTS(MPa)	147.09	186.33
Hardness (Vickers)	51.81	73
Corrosion Density(μA)	172	5
Grain Size(μm)	700	10

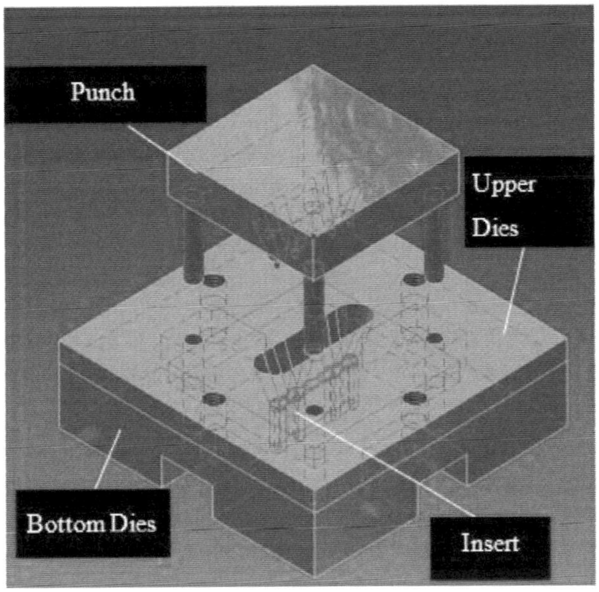

Figure 1. Dies used in this experiment.

The measurement of the result was done using Dino Lite camera, and measured using Dino Capture 2.0 software. The manufacturing process in this experiment is done in 5 samples. 19 data were taken from 5 samples, and each measurement was taken 3 times in each measured part.

3. Result and discussion

Figure 2 shows the plate resulted from the manufacturing process, with a standard pen for comparison.

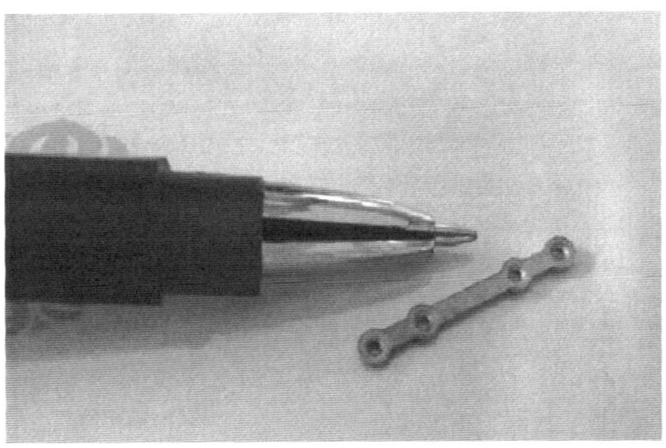

Figure 2. The finished product, pen for scale.

The detailed measurement result of the plate's important dimensional parameters is shown in Figure 3. The important dimensional parameters are: Neck width, hole radius, chamfer radius, and outer circle radius. The measurement method could be seen in Figure 3, with each measurement is done 3 times and averaged.

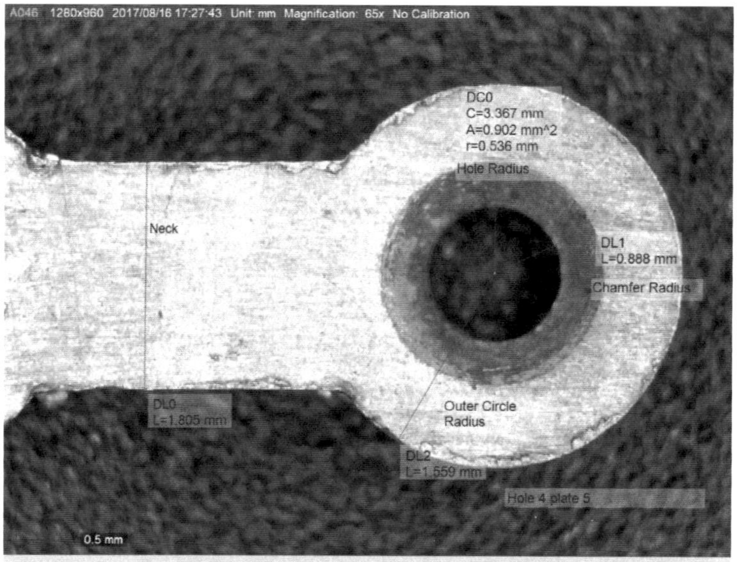

Figure 3. The measured example of the detailed parts of the plate. The part's names are written along the measurement line.

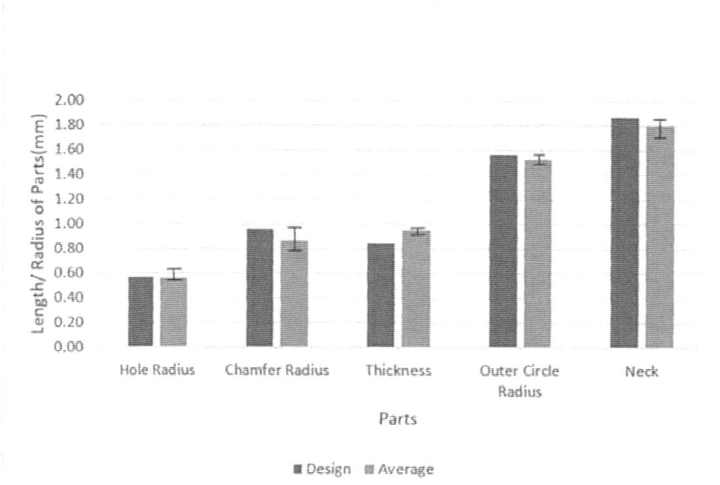

Figure 4. Comparison of Design, Average, Highest Value, and the Lowest Value of each parts measured.

The average result and range (maximum and minimum value) of all the measurement done in this experiment could be seen in Figure 4, and the average error percentage is shown in Figure 5. The formula of specific part's error percentage in this calculation is shown in equation (1),

$$\left|\frac{Designed\ part}{Average\ of\ result}\right| x\ 100\% \tag{1}$$

where: Designed part = length/radius of part according to design; Average of result = average length / radius of each parts resulted from the manufacturing process.

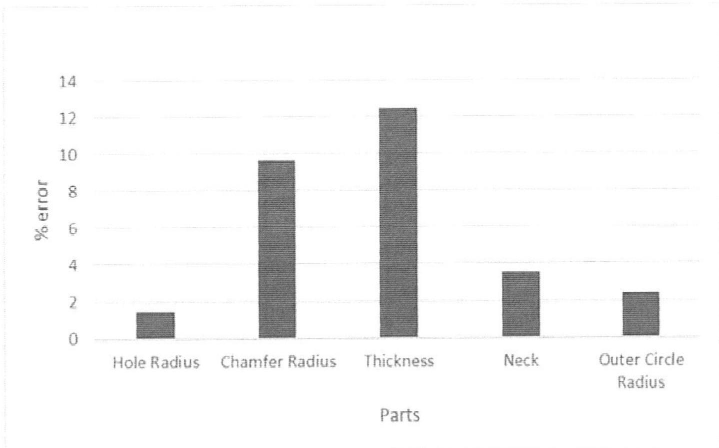

Figure 5. Average Error of each samples' detailed parts.

Table 3. Design, average, highest value, lowest value, average error measurement of the samples' detailed parts.

	Design (mm)	Average (mm)	Highest Value (mm)	Lowest Value (mm)	Average Error (%)
Hole radius	0.58	0.57	0.62	0.55	1.43
Chamfer radius	0.97	0.87	0.97	0.78	9.65
Thickness	0.85	0.96	0.96	0.94	12.47
Outer Circle	1.57	1.53	1.55	1.50	2.42
Neck	1.87	1.80	1.87	1.71	3.55

Looking at the detailed measurement of the samples, we could conclude, that the manufacturing result is quite accurate, based on the International Tolerance governed by ISO 286, in which, using the IT rating of 14, the most important part, the hole, is manufactured within the tolerance, which is 0.15 mm, from the International Tolerance Formula:

$$\mathrm{T} = 10^{0.2\mathrm{x}(\mathrm{ITG}-1)}.\left(0.45\mathrm{x}\sqrt[3]{\mathrm{D}} + 0.001\mathrm{xD}\right) \tag{2}$$

where: T = Tolerance in micrometers
D = Geometric mean dimension in milimeters
ITG = IT Grade, a positive integer [7].

We could see from the result in Figure 5 and Table 3, that the greatest error present is in the thickness, which is quite precise with the average of 0.96 mm. This is caused by the presence of flash that occur when the plate is stamped, and manufacturing limitation that happened in the manufacturing process of the plate itself. The targeted thickness of 0.85 mm is proven to be too thin for the machine too handle, and increase in thickness is the solution to the problem. This thickness problem could be solved by sanding the plate itself until it reaches 0.85 mm. The thickness itself is not the main concern, as the increased thickness will also increase the tensile strength of the plate itself. The increased thickness also will not pose any problems regarding the space occupied either, because the average thickness of such plates is around 1 mm. The most crucial part, however, is the hole size itself. The hole size should fit the screw tightly so the force could be transferred optimally. Average error occurred in the hole section is very small, at 1.43%, making it the smallest error in the measurement, and still within the International Tolerance Standard. Another part's error, which is neck parts, came from the inconsistency of the machining process, which caused burr, and should be sanded down manually with hand. However, sometimes the sanding process just cut too much inside that it caused the neck width itself to decrease. This, could pose some problem for the plate's tensile strength itself. What should be done, is that to leave around one-millimeter excess in the sanding process, to ensure that the neck width will not be below the original design that will result in a decrease in strength.

Figure 6. Flash formed during forming process.

As for the mode of failure, it is mostly ductile, as seen in Figure 6, where the ductile part resulted in flashes, that also contributes to the increase in thickness. In the actual manufacturing process, the flashes should be removed using sandpaper for optimum result as the bottom part of the plate is the part that will be in contact with the broken bones. The top and side part, should be sanded too, as both parts will be in contact with face's tissue, in order not to let the tissue scratched and wounded. The next experiment will be focused on the strength of the plate itself, so that the plate could be used in practical level, as clinical trials on the magnesium ECAP *in vivo* has shown promising results [8].

4. Conclusion

The miniplate made by magnesium for maxillofacial implant was successfully manufactured using microforming process. The forming process was focused on punching and tapering the magnesium miniplate. Microforming is proved to be an efficient method in shaping magnesium with high production rate.

The results of the magnesium miniplate shows an acceptable accuracy based on standard established by ISO 286 (International Tolerance). The most important part, which is the hole, is accurate enough and thus, will not interfere with the plate's functionality. The calculation based on physical properties of magnesium ECAP give an accurate guideline in designing parameters needed for manufacturing process. Further works is needed to improve surface finish of the magnesium miniplate.

References

[1] Staiger M Pietak A Huadmai J and Dias G 2006 *Biomaterials* **27** 1728-34
[2] Uddin M Hall C and Murphy P 2015 *Sci. and Tech. of Adv. Mater.* **16**
[3] Rahayu E F L 2015 *Thesis* University of Indonesia Depok (In Indonesia)
[4] Wang H Estrin Y and Zúberová Z 2008 *Mater. Letters* 2476-79

[5] Valiez R Z and Langdon T G 2006 *Progress in Mater. Sci.* 881-91
[6] Rahmi S 2012 *Thesis* University of Indonesia Depok (In Indonesia)
[7] ISO/TC 213 ISO 286-1:2010 ISO Geneva 2010
[8] Wiwanto S Sulistyani L D Latief F D E Supriadi S Priosoeryanto B P and Latief B S 2017 *2nd Biomed.l Engineer. Recent Progress in Biomater. Drugs Develop. and Medic. Devices* 020013-1 – 020013-7

Thermal properties of thermoplastic natural rubber reinforced by microfibrillar cellulose

I N Indrajati[1], I R Dewi[1], D W Nurhajati[1]

[1] Center for Leather, Rubber and Plastics, Ministry of Industry Republic of Indonesia Sokonandi St. number 9, Yogyakarta, Indonesia 55166

i-novia@kemenperin.go.id

Abstract. Thermoplastic natural rubbers (TPNR) based on natural rubber (NR) and polypropylene (PP) blends had been successfully prepared using dynamic vulcanization. The blends were reinforced with microfibrillar cellulose (MFC). The study aimed to evaluate the thermal properties of TPNR using thermogravimetric (TG) and differential scanning calorimetry (DSC) analyses. MFC was derived from pineapple leaf fiber (PALF) through several treatments. Dynamic vulcanization was carried out using an internal mixer. PP was previously melt-mixed with MFC, then blended with NR compound. The ratio of NR/PP blend was maintained at 60/40, while MFC concentrations were varied ranging from 0, 1, 5, 10, 15 and 20 phr. The results showed that the degradation of TPNR samples occurred in two simultaneous stages. The onset of the degradation of TPNR started ranging from 346 °C to 361 °C. MFC concentration affected the thermal stability of the samples. The highest thermal stability was achieved by incorporating 1 phr MFC. DSC studies revealed that melting temperature (T_m) and heat of fusion (ΔH_f) were independent of MFC concentration. T_m was in the 161-165 °C range, while ΔH_f was around 24-31 J/mol. Keywords: thermoplastic elastomer, natural rubber, polypropylene, thermal properties, cellulose.

1. Introduction

Elastomer often blends with thermoplastics, resulting in a new material with outstanding properties. This blend is known as thermoplastic elastomer (TPE). The unique properties of TPEs are centered in the combination of the processing characteristic of thermoplastics with the wide range mechanical properties of elastomer [1]. The demand for TPEs is presently growing because they provide higher output rates than conventional elastomers. Conventional processing machines of thermoplastics, such as injection molding and extruder, can be used in product manufacturing line. Thus, the production cost will automatically be reduced. Moreover, TPEs offer better material utilization since they are recyclable [2]. These advantages make the TPEs find its market substituting the traditionally used vulcanized elastomer [3]. Thermoplastic vulcanizates (TPV) belongs to the family of TPE, but has different morphology. TPV contains finely dispersed crosslinked elastomer in the continuous thermoplastic matrix and typically produced by dynamic vulcanization [4]. Commercial TPVs are mostly produced based on synthetic rubber rather than natural rubber (NR). When NR is used as elastomer phase, then the generic term of TPV turns into thermoplastic natural rubber (TPNR). Research on the TPNR development is attracting much attention recently. NR is widely known to have outstanding mechanical properties, but when blended with thermoplastics, it results in low technical specification [5]. One possible way for achieving improvement on the technical specification is by addition of fillers. The level of the improvement is affected by many factors, i.e. the nature of the filler (natural or synthetic), particle size, surface area, shape, filler loading, and active functional groups present on the filler surface. These factors determine the interaction between filler and polymer matrix [6], and in turn the final product properties.

Fibers are widely accepted to impart reinforcement of polymer matrix. Recently, natural fiber-based fillers are paid much attention because of environmental issues and the sustainability of feedstock. Natural fibers are lignocellulosic material that are mainly composed of cellulose. Cellulose has excellent mechanical behaviors that are closely comparable to those of synthetic fiber [7]. Moreover, natural fibers are lightweight and renewable resources that are guaranteed the sustainability of the stock. The cellulosic fiber source is abundant in tropical countries. Most of them could be wastes form agricultural activities that are not well-utilized yet. Depending on the preparation method and sources, many types of cellulose-based fiber can be derived from raw cellulose. They include nanocrystalline cellulose (NCC), microfibrillated cellulose (MFC) or nanofibrillated cellulose (NFC) [8]. Nevertheless, the hydrophilic nature of natural fiber could be the major drawback of their application in hydrophobic polymer matrices [9]. This will result in poor interfacial bonding between the fiber and matrix. Surface treatment of the natural fiber should be performed to improve the fiber-matrix adhesion. Pineapple leaf fiber (PALF) is an attractive natural fiber source because of its tremendous mechanical properties due to higher cellulose content and lower microfibrillar angle [10]. Several studies have been conducted on utilization of cellulosic fiber derived from PALF in polyurethane [9], polyamide 6 [11], polypropylene [12] ,and nitrile butadiene rubber matrix [13]. Kalapakdee and Amornsakchai [14] and Yuakkul *et al.* [15] have successfully applied PALF in TPE matrix. The results showed that the mechanical properties are improved by fiber addition. However, those studies employed untreated or alkali-treated PALF. Application of MFC in the polymer matrix is limited. In addition, the studies conducted by Yuanita *et al.* [16] and Thanomchat and Srikulkit [17] have revealed that cellulose possibly act as nucleating agent for PP.

Natural fibers are organic matter that may have lower thermal stability than other components in the polymer matrix. One should consider their thermal properties for the application at elevated temperature. Kaewpirom and Worrarat [18] studied the thermal properties of PALF-reinforced poly lactic acid (PLA). They concluded that the thermal stability was governed by the amount of PALF, in which lower amount of PALF showed higher thermal stability. Other study reported that incorporating rattan nanoparticle into polypropylene matrix lead to an increase in the thermal stability of composites [7]. In this present study, the thermal characteristics of MFC-reinforced TPNR were studied. The TPNR is prepared by blending the natural rubber, polypropylene and MFC in an internal mixer. A fixed ratio of NR/PP blends of 60/40 was taken with various loading of MFC. The thermal properties were analyzed using thermogravimetric analysis and differential scanning calorimetry. Both analyses were conducted in oxygen atmosphere in order to approach real application condition.

2. Experimental method

2.1. Materials

Natural rubber grade SIR 10 was supplied by local estate. Polypropylene Trilene grade HI10HO (injection molding grade, MFI (230 °C/2.16 kg) 10 g/min). The NR compound additives included ZnO (Bratachem), Aflux 42M processing promotor (Rhein Chemie), carbon black N-330 (OCI Korea) and GPF black N-660 (OCI Korea), 2,2,4-Trimethyl-1,2-Dihydroquinoline (TMQ) (Kemai), N-(1,3-Dimethylbutyl)-N'-phenyl-p-phenylenediamine (6PPD) (Northeast), N-Cyclohexylbenzothiazole-2-Sulfenamide (CBS) (Northeast), Sulfur SP-325 (Miwon), Pilgard PVI Pre-Vulcanization Inhibitor. Maleated castor oil (MACO) was used as plasticizer which synthesized in our previous work [19]. Additives of TPNR preparation were antioxidant Irganox 1010 (BASF), stearic acid (purchased from CV. Indrasari Semarang) and compatibilizer PP-g-MA (Sigma Aldrich). The raw PALF was supplied in the form of 0.5 cm-cut by local producer (Subang, West Java Province). Chemical reagents needed for MFC preparation included sodium hydroxide (NaOH), sodium hypochlorite (NaOCl), sulfuric acid (H_2SO_4) 98% and acetic acid glacial (CH_3COOH) 96%. All reagents were industrial grade.

2.2. *Microfibrillar cellulose (MFC) preparation*

Microfibrillar cellulose (MFC) derived from Pineapple leaves Fiber (PALF) were prepared through several processing steps including alkali treatment, bleaching, hydrolysis and mechanical treatment. Before treatments, the raw PALF was grinded using crusher. Alkali treatment was performed by mercerization using 10% (w/w) aqueous sodium hydroxide (NaOH) at ambient temperature for 18 h. The fiber to liquor (FL) ratio was maintained to 1:30. The treated-fiber (TF) were filtered, and washed with water several times until the alkali was completely removed (pH reach normal), and air-dried overnight. Bleaching process was conducted using 1% (v/v) sodium hypochlorite (NaOCl) at 70 °C for 4 h under mechanical stirring 300 rpm. The FL ratio was kept at 1:70. After finished, the fibers were filtered and washed with water for 3 times to remove excess bleaching agent. The TF is then left to dry overnight in room temperature. Hydrolysis was performed using 20% (v/v) sulfuric acid (H_2SO_4) at 80 °C for 2 h under mechanical stirring 500 rpm. The FL ratio was maintained at 1:50. Immediately following the hydrolysis, the suspension was filtered to remove the liquid. Solid fraction was washed with cold water to stop the reaction, and followed by air-dried overnight. After this treatment, the TF was undergo mechanical treatment which performed by ball-milling in acidic solution (1% acetic acid glacial). The suspension was ball-milled at room temperature for 2 h. The surface speed of the ball mill was maintained to 60 rpm. Ceramic balls with various diameter were used as grinding media. After milling, the suspension was then centrifuged for 10 minutes at 3000 rpm. Solid fraction (MFC) was removed from suspension and then washed three times to remove excess acetic acid. Finally, the solid was let to dry overnight at ambient temperature.

2.3. *Thermoplastic natural rubber (TPNR) preparation*

TPNR samples were prepared in two processes. First, NR and the compound ingredients were compounded in a two roll mill according to the formulation on Table 1. The resulted compound was stored in a conditioned room at 25 °C minimum for 24 h. Second, the NR compound was blended with PP and MFC by dynamic vulcanization in Rheomix Polylab 3000 OS with 300 cm^3 mixing chamber (fill factor 80%). The blends were made in a fixed NR/PP ratio of 60/40 with various MFC concentration as detailed on Table 2. The mixing was carried out at 160 °C with rotor running at 90 rpm. The mixing process was run in three stages. First, PP was melted the torque remained stable. Second, MFC was added into the mix and was processed until the torque stable. The last stage was NR compound incorporation. One cycle of mixing process took at least 17 minutes. The resulted blends were pelleted and store in a sealed container.

Table 1. Composition of NR compound.

Ingredients	Concentration (phr)
SIR	100
ZnO	5
Aflux 42	1.5
N-330	20
N-660	15
MACO	5
TMQ	2
6PPD	1
CBS	1.2
Sulfur	1
PVI	0.1

Table 2. Blends NR/PP composition.

Ingredients	Composition (PHR)					
	PN0	PN1	PN2	PN3	PN4	PN5
NR compound	60	60	60	60	60	60
PP	40	40	40	40	40	40
MFC	0	1	5	10	15	20
Stearic acid	1	1	1	1	1	1
PP-g-MA	1	1	1	1	1	1
Irganox 1010	0.5	0.5	0.5	0.5	0.5	0.5

2.4. Thermal analysis

Thermogravimetric (TG) and differential scanning calorimetry (DSC) were used to analyze the thermal properties of TPNR samples. Thermogravimetric analyses were carried out using simultaneous DTA-TG Shimadzu DTG-60 H equipped with TA 60 WS software. Sample was placed in alumina crucible and subjected to a temperature in the range 30 °C to 600 °C with a heating rate of 10 °C/min in oxygen atmosphere. DSC studies were performed using Perkin Elmer DSC 4000. The sample was heated from ambient temperature to 445 °C with a heating rate of 5 °C/min in oxygen atmosphere. The peak of maximum endotherm peak was considered to be the melting point of the sample, while the melting enthalpy was determined from the area of the melting peak.

3. Results and discussion

3.1. Thermogravimetric analaysis

Thermogravimetric (TG) data describe a number of thermal degradation stages, material weight loss in each stage, threshold temperature etc. The TG and its derivative (DTG) tell about the information of the nature and extent of material degradation. While in DSC, the heat flow rate associated with the thermal event that can be measured as a function of time and temperature. This will provide valuable information related to melting and phase transitions of the samples [20].

Figure 1(a, b) depicts the thermogravimetric (TG) and derivative thermogravimetric (DTG) curves of MFC, neat PP and NR compound. The onset degradation temperature (T_{onset}), maximum decomposition temperature (T_{max}), maximum rate of decomposition (R_{max}) and weight loss are tabulated in Table 3. The DTG curve (Figure 1(b)) clearly indicates that the degradation of NR compound is occurred in single stage since one single peak is found. The main degradation of NR compound occurs between 356 and 410 °C. The onset degradation starts at 356 °C, and the maximum degradation is occurred at 379 °C. Similar to NR compound, neat PP shows one major degradation peak with the main degradation takes place in the range 278-311 °C. The decomposition reaches the maximum mass loss at 292 °C, and showing 7% solid residual at the end of the measurement (400 °C).

Table 3. Degradation and thermal data from TG and DSC analyses.

Sample ID	TG								DSC	
	Stage-1				Stage-2					
	T_{Onset} (°C)	T_{max} (°C)	R_{max} (mg/sec)	Weight Loss (%)	T_{Onset} (°C)	T_{max} (°C)	R_{max} (mg/sec)	Weight Loss (%)	T_m (°C)	ΔH_f (J/mol)
MFC	199	236	0.0086	43.16	285	305	0.0057	36.63	-	-
NR-comp.	356	380	0.0113	71.88	-	-	-	-	-	-
Neat PP	278	293	0.0306	92.90	-	-	-	-	165.16	
N0	346	378	0.0495	53.88	420	519	0.0002	41.62	162.39	30.28
N1	361	371	0.0062	58.42	410	414	0.0231	22.40	162.71	27.45
N2	353	380	0.0115	43.25	447	464	0.0148	38.97	162.71	31.65
N3	347	378	0.0065	41.53	413	450	0.0066	33.84	162.58	29.28
N4	354	383	0.0078	40.20	449	467	0.0105	36.76	161.54	27.90
N5	355	381	0.0063	41.51	449	467	0.0089	32.97	161.25	24.75

MFC degrades in two stages indicated by two major peaks on DTG curve. The first degradation occurs at the range of 199-250 °C with T_{max} at 236 °C. This is associated with hemicellulose and cellulose degradation. According to Yuakkul *et al.* [15], hemicellulose starts to decompose at lower temperature (150-350 °C), while decomposition of cellulose takes place at 270-350 °C. The second stage degradation is associated to lignin decomposition. The 2-stage degradation of MFC indicates impure MFC. Therefore, the isolation procedures should be improved so that it provides pure cellulose.

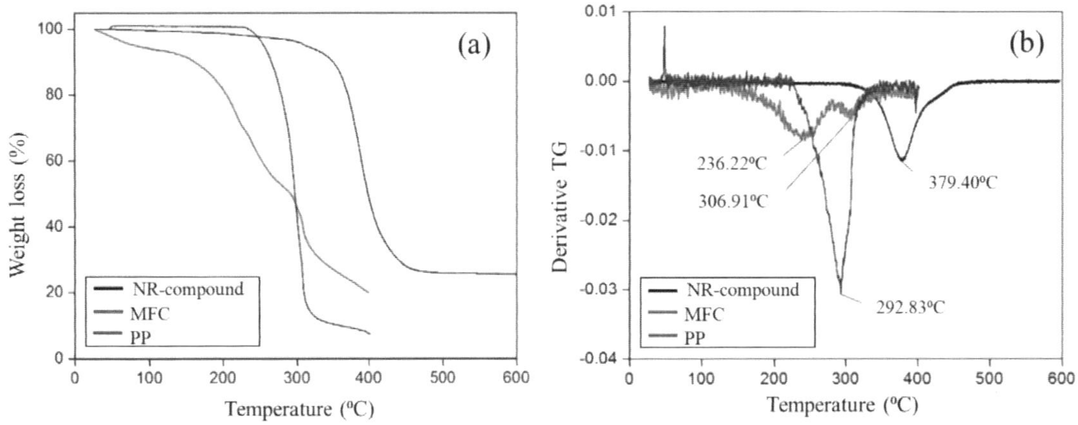

Figure 1. TG (a) and DTG (b) curves of the NR compound, MFC and PP.

Degradation of TPNR samples take place in two stages simultaneously as shown at Figure 2(a). Two major endothermic peaks on DTG curve (Figure 2(b)) clearly indicates this degradation mode. This result is agreed to previous studies [4, 20, 21]. The first degradation corresponds to PP degradation. NR enhances the thermal stability of PP since the onset degradation temperature of TPNR is higher than neat PP. Crosslinked-NR that finely dispersed within continuous PP matrix as a result of dynamic vulcanization acts like a great barrier (Figure 3) thus prevents the volatile PP degradation products [22]. At the second stage, the degradation starts at higher temperature corresponding to the existence of crosslinks formed during dynamic vulcanization [20]. Thus, more energy is needed to break the bonds. It is interesting here that no degradation occurred below 300 °C which is associated to MFC degradation. Perhaps, the decomposition of MFC substituents are prevented by the crosslinked rubber phase and they start to decompose at higher temperature.

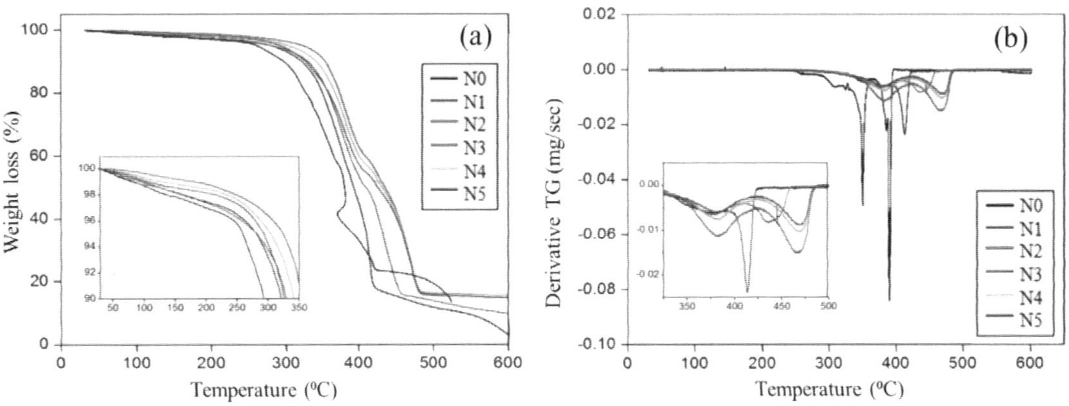

Figure 2. TG (a) and DTG (b) curves of the TPNR samples.

At the first stage degradation, the addition of 1 phr MFC causes an increase in onset degradation temperature to some degree. However, further addition tends to decrease it. Also, the degradation rate significantly reduced to a lower rate by the addition of MFC. Increasing MFC concentration does not give a big influence to the degradation rate. This findings are not in agreement with Yuakkul *et al.* [15] that reported the more rapid degradation rate upon increasing fiber loading caused by lower thermal stability of PALF. Further studies are needed to evaluate the role of MFC in this case. Perhaps, MFC plays a role as nucleating agent for PP that increase the crystallinity of TPNR samples. Therefore, the thermal stability is increased. Overall, the TPNR samples are thermally stable in the range of 345-360 °C, means that they may be safely processed or applied without any degradation.

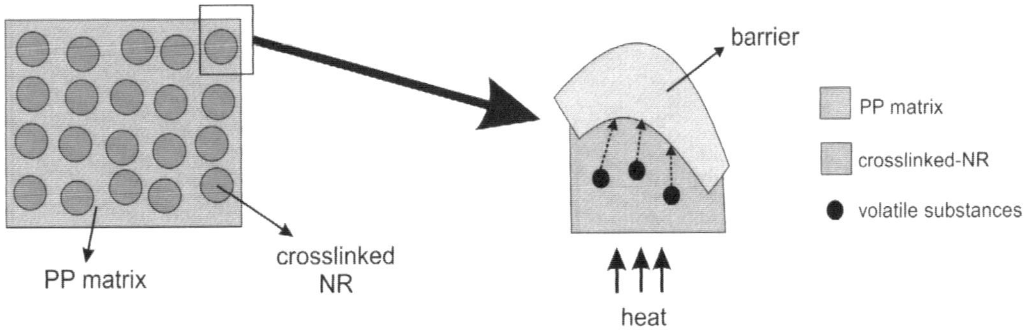

Figure 3. Barrier mechanism of volatile substances in PP matrix by crosslinked NR.

3.2. Differential scanning calorimetry (DSC) analysis

Differential scanning calorimetry (DSC) study is carried out to understand the phase inversion behavior (crystallization and melting) of a material when subjected to heat [23]. This study only discusses the melting behavior of TPNR samples. The DSC thermograms of neat PP and TPNR sample are depicted on Figure 4(a, b), while the melting temperature (T_m) and heat of fusion (ΔH_f) are tabulated on Table 3.

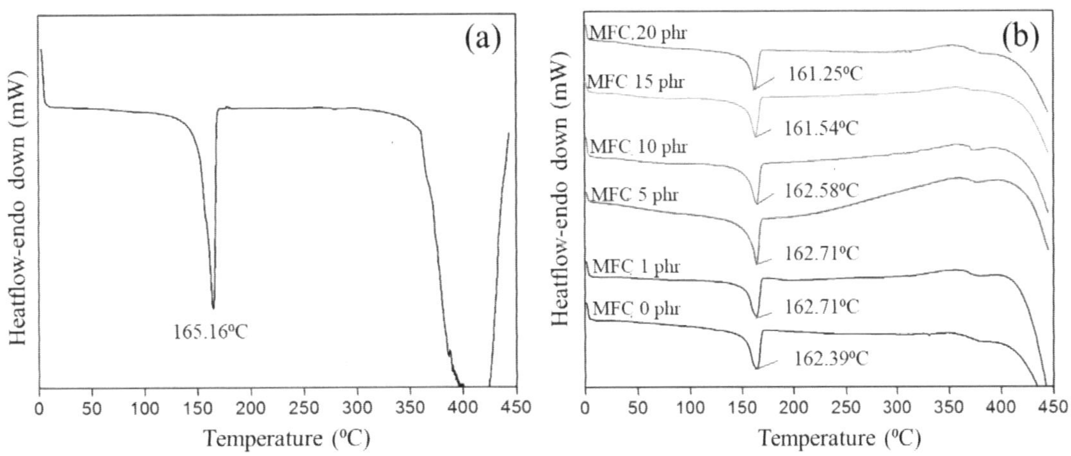

Figure 4. DSC thermogram of a) Neat PP, b) TPNR samples.

Both, neat PP and TPNR, samples reveal single endothermic peak that is corresponded to melting temperature (T_m). It is clearly seen that MFC loading is independent to T_m since no significant increase is found after MFC incorporation (Figure 4b). Melting point of TPNR samples were also similar to neat PP. From Table 3 it is evidence that MFC concentration does not alter ΔH_f. It is a positive indication that the energy required to process the TPNR is quite similar to neat PP.

4. Conclusion

Thermoplastic natural rubber (TPNR) made of NR/PP blend reinforced with MFC had been successfully prepared. The MFC was derived from local pineapple leaf fiber (PALF) through several chemical and mechanical treatments. TG analysis exhibited that the degradation modes of NR compound and neat PP were occurred in single step showed by single peak in DTG curve, while MFC degraded in two steps indicating impurity of the resultant MFC. TPNR samples showed two stages degradation. The onset degradation of TPNR started at 346-361 °C which is higher than its single component. Incorporating rubber phase into PP matrix could enhanced its thermal stability to some extent degree. Melting temperature of TPNR samples did not significantly differ from neat PP, and the heat of fusion was remaining unchanged upon increasing MFC concentration. Overall, the TPNR samples were thermally stable in the range of 345-360 °C, which mean that the samples would be safely processed without any degradation caused by heat. The similarity of melting temperature to those of neat PP gives advantages not only to the processing means but also reducing cost.

Acknowledgement
The authors are grateful to the Center for Leather, Rubber and Plastics, Ministry of Industry of financial support through DIPA 2016. The authors also want to thankful to Dr. Shinta Virdhian and the colleagues at Metal Industries Development Center (MIDC) for the facilitating the Rheomix for the preparation of TPNR samples.

References
[1] Mondal M Gohs U Wagenknecht U and Heinrich G 2013 *Mater. Chem. Phys.* **143(1)** 360–6
[2] Pechurai W Nakason C and Sahakaro K T 2008 *Polym. Test.* **27(5)** 621–31
[3] Varghese S Alex R and Kuriakose B 2004 *J. Appl. Polym. Sci.* **92(4)** 2063–8
[4] Lopattananon N Tanglakwaraskul S Kaesaman A Seadan M and akai T 2013 *Adv. Mater. Res.* **747** 230–3
[5] Bahruddin Ahmad A Prayitno A and Satoto R 2012 *Int. Conf. Innov. Polym. Sci. Tech.* **4** 146–53
[6] Bendjaouahdou C and Bensaad S 2013 *Energy Procedia.* **36** 574–90
[7] Nikmatin S Syafiuddin A Hong Kueh A B and Maddu A 2017 *J. Appl. Res.Tech.* **15(4)** 386–95
[8] Jia X Chen Y Shi C Ye Y Wang P Zeng X and Wu T 2013 *J. Agric. Food Chem.* **61(50)** 12405–14
[9] Cherian B M Leão A L de Souza S F Costa L M M de Olyveira G M Kottaisamy M Nagarajan E R and Thomas S 2011 *Carbohydr. Polym.* **86(4)** 1790–8
[10] Asim M Abdan K Jawaid M Nasir M Dashtizadeh Z Ishak M R and Hoque M E 2015 *Int. J. Polym. Sci.* **2015** 1–17
[11] Panyasart K Chaiyut N Amornsakchai T and Santawitee O 2014 *Energy Proc.* **56(C)** 406–13
[12] Ummartyotin S and Pechyen C 2016 *Carbohydr. Polym.* **142** 133–40
[13] Wisittanawat U Thanawan S and Amornsakchai T 2014 *Polym. Test.* **35** 20–7
[14] Kalapakdee A and Amornsakchai T 2014 *Polym. Test.* **37** 36–44
[15] Yuakkul D Amornsakchai T and Saikrasun S 2015 *Int. J. Plast. Tech.* **9(2)** 388–411
[16] Yuanita E Pratama J N and Chalid M 2017 *Macromol. Symp.* **371(1)** 61–8
[17] Thanomchat S Srikulkit K 2015 *Adv. Mater. Sci. Eng.* **2015** 1–9
[18] Kaewpirom S Worrarat C 2014 *Fibers Polym.* **15(7)** 1469–77
[19] Indrajati I N Dewi I R 2017 *IOP Conf. Ser. Mater. Sci. Eng.* **223**
[20] Joseph P V Joseph K Thomas S Pillai C K S Prasad V S Groeninckx G Sarkissova M 2003 *Compos. Part A Appl. Sci. Manuf.* **34(3)** 253–66
[21] Hernández M Albano C González J Ichazo M N 2006 Polym. Bull. **56(2–3)** 285–91
[22] Roy Choudhury N Chaki T K Bhowmick A K 1991 *Thermochim. Acta* **176(C)** 149–61
[23] Benmesli S Riahi F 2014 *Polym. Test.* **36** 54–61

IOP Conf. Series: Materials Science and Engineering **432** (2018) 012039 doi:10.1088/1757-899X/432/1/012039

Effect of compatibilizers on the mechanical and morphological properties of polycarbonate/poly (acrylonitrile-butadiene-styrene) blends

I R Dewi[1, 2], I N Indrajati[1] and D W Nurhajati[1]

[1] Center for Leather, Rubber, and Plastics. Ministry of Industry. Sokonandi 9 St., Yogyakarta 55166, Indonesia
[2] Departement of Sciences, Faculty of Science and Technology, Prince of Songkla University, Pattani 94000, Thailand

indiah.rd@kemenperin.go.id

Abstract. Blending the existing polymers have an added advantage over the conventional polymers. However, compatibility of the materials is a problem. In this study, metacrylate butadiene styrene (MBS) and polypropylene-graft-maleic anhydride (PP-g-MA) were added as a single compatibilizer on polycarbonate/poly(acrylonitrile-butadiene-styrene) blends (PC/ABS). The effects of the compatibilizer addition on the mechanical and morphological properties were investigated. The PC/ABS blends were prepared in various compositions of 70/30, 60/40, 50/50 with fixed 2 phr MBS or PP-g-MA. PC/ABS blends were prepared by using twin-screw extruder, and test samples by injection molding. Results indicated that MBS exhibited higher tensile strength, hardness, and density, but lower impact strength and elongation compared with PP-g-MA. The tensile strength, elongation, and hardness tended to increase with increasing PC content; however, the impact strength was almost the same. Scanning electron microscopic examination showed that PC/ABS blend compatibilized with PP-g-MA was more porous than the one with MBS. Based on the results, it was found that PP-g-MA was a more effective compatibilizer for the PC/ABS 70/30 blend than MBS. The values of tensile strength, hardness, elongation, density, and impact strength were 602 kg/cm^2, 79.1 Shore A, 15.3 %, 1.05 g/cm^3, and 5.5 kgf/cm respectively.

1. Introduction

Polymer blending technology is one of the major areas of research and development over the past few decades. Blends of the existing polymers have an added advantage over the conventional polymers. Blends can be concluded as compatible or incompatible based on the melt viscosity and mutual solubility of the individual materials [1]. However, compatibility of the materials is a problem. This problem has been solved in many cases by adding a small amount of additives known as compatibilizers. These are generally block and graft copolymers or chemically reactive species which are concentrated at the interface and which act as emulsifiers. The function of the in situ (or in vivo)-formed compatibilizers have been reported to reduce interfacial tension and to improve the size of dispersed phases and the adhesion between two immiscible polymers [2].

IOP Conf. Series: Materials Science and Engineering **432** (2018) 012039 doi:10.1088/1757-899X/432/1/012039

Polycarbonate (PC) is a kind of ductile thermoplastic compared to other polymers such as nylon and polyesters. Acrylonitrile butadiene-styrene (ABS) is the largest-volume engineering thermoplastic resin and is considered a bridge between commodity plastics and higher-performance engineering thermoplastics. ABS resins are commonly used in appliance parts (including electrical/electronic) and automotive/transportation applications [3]. Many researchers struggle to trade off the poor part of PC with ABS, and vice versa, with blending these two polymers.

PC/ABS blends are generally considered as an immiscible pair. As a compatibilized system, they can offer improved mechanical and thermal properties due to the PC matrix, whereas the contribution of ABS focuses on the ease of processing, economy, and more reliable impact strength [4]. Chiang *et al.* [2] reported that three kinds of compatibilizers, methacrylate–butadiene–styrene (MBS), ethylene–vinyl acetate (EVA), and styrene–maleic anhydride (SMA) could be used to improve the flame retardancy phenomenon on the PC/ABS alloy. Tjong and Meng [5] implied that both maleic anhydride (MA)-grafted polypropylene and solid epoxy bisphenol type-A compatibilizers were ineffective to improve the impact toughness of PC/ABS blends containing ABS content ≥40wt%. Farzadfar *et al.*, reported that ABS-g-MA had no significant effect on the tensile strength of the PC/ABS blends while 10 phr of EVA-g-MA decreased the tensile strength. Otherwise, EVA-g-MA could increase the impact strength better than ABS-g-MA in the amount of 5 phr [6]. Triantou and Tarantili [4] succesfully improved the intercalation process in organoclay/ABS/PC nanocomposites with ABS-g-MA, while Han *et al.* used the poly(styrene-co-acrylonitrile)-g-maleic anhydride (SAN-g-MAH) to decrease the droplet size of the ABS on the PC/ABS blends [7]. Polydimethylsiloxane-g-styrene-g-methyl methacrylate copolymer (PSM) in PC/ABS possessed high impact [8] and BS-g-MA performed good compatibilization in various amount of NPCC filler in PC/ABS blends [9]. However, less study were conducted on MBS and PP-g-MA as PC/ABS compatibilizers.

In this paper, two different compatibilizers, MBS and polypropylene graft maleic anhydride (PP-g-MA) were used as PC/ABS compatibilizer. The effect of MBS and PP-g-MA on the mechanical and morphological properties of PC/ABS blends. The crystallinity property was also investigated as supporting data.

2. Experimental method

2.1. Materials

Materials used in this research were PC injection grade resin under the trade name Wonderlite PC-110, ABS injection grade resin under the trade name Toyolac 700 314, methacrylate-butadiene-styrene (MBS) compatibilizer Huaxing ZB-521 ZB-32, PP-g-MA purchased from Sigma Aldrich, and antioxidant under the trade named Irganox 1010 and Irgafos P168.

2.2. PC/ABS blends preparation

The PC and ABS resins were dried at 100 °C and 80 °C for 4 h, respectively, prior to blending. After calculating the necessary quantity accordingly to Table 1, all of the ingredients were mixed in a tumbler mixer for 3 min to form a uniform composition throughout the batch size. The uniformly mixed feed then was prepared using twin-screw extruder and the pellets were made in line with the extruder. The processing temperatures at five zones of the extruder were maintained at 230–240–250–250–250 °C and

Table 1. Formulation of PC/ABS blends.

Materials	Blend 1	Blend 2	Blend 3	Blend 4
PC/ABS	50/50	60/40	70/30	70/30
MBS	2	2	2	-
PP-g-MA	-	-	-	2
Irganox	0.05	0.05	0.05	0.05
Irgafos	0.1	0.1	0.1	0.1

the screw speed was kept at 220 rpm, and the feeding speed was 110 rpm. Extrudes from the die were passed through the water bath at 20–25 °C and then ground into pellets. The pelletizer speed was synchronized with the extruder output and screw speed. Extruded pellets were kept for conditioning (in hot air oven at 80 °C for 4 h) to release the thermal stresses and were subsequently dried.

The conditioned pellets were molded to ASTM standard samples using injection molding. During the sample preparation, temperatures in the front, middle and rear positions of the barrel were kept at 260, 260 and 260 °C, respectively, and followed the setting as mentioned in Table 2.

Table 2. Formulation of PC/ABS blends.

Barrel/Inject				Holding Time		
End Pos (mm)	Time (s)	Flow (%)	Press (%)	Time (s)	Flow (%)	Press (%)
95	2	-	-	3	30	30
85	2	45	45			
50	2	45	45			

2.3. Melt flow index
The Melt Flow Index (MFI) measurements were carried out according to ASTM D 1238 specification (procedure A), in a Tinius Olsen XNR-400B. For each sample 5 measurements were run and the average was obtain. The load and temperature for each sample were given at Table 3.

2.4. Mechanical properties
The density testings were determined with Mirage EW-200SG electro densimeter. The Izod impact tests were carried according to ASTM D256 using Toyoseiki Izod impact tester. The tensile behaviors of the blend specimens were determined using Tinius Olsen H25K universal testing machine, according to ASTM D638.

2.5. Crystallinity properties
X-ray diffraction (XRD) was performed using Shimadzu X-ray diffractometer equipped with XR tubed with the target Cu at the operating voltage and current of 40 kV and 30 mA, respectively and the divergence, scatter, and receiving slits were adjusted at 1 deg, 1 deg, and 0.3 deg respectively. The samples processed as described above were put in the XRD pan where the room temperature was maintained.

2.6. Morphological properties
SEM micrographs were obtained by using Shimadzu SEM with 5K magnification and 10kV. The PC/ABS blends pellets were coated with a thin layer of gold prior to SEM examination.

3. Result and discussion

3.1. Melt flow index
MFI tests have been done to measure the melt viscosity of flow resistance of PC when blended with ABS at a particular shear related to the applied load and temperature. It is also valuable indicator to observe the variations in PC and ABS content in the blends. Table 3 shows that PC/ABS 70/30M blend shows the highest MFI value which means it is the easiest blend to be processed, followed by 70/30P blend. The MFI of the blends is increased gradually with increasing in PC content in the blend. The reason for the increase in MFI value of PC with the addition of ABS may be solvation of the highly entangled structure of molecules of PC due to the presence of ABS molecules. Thus, with increase of shear rate, solvated layers may be sheared away, resulting in an increase in intermolecular interaction and consequent improvement in processability [3].

Table 3. MFI data for PC, ABS, and PC/ABS blends.

Sample	Test Condition (°C, kg)	MFI (g/10min)
PC	270, 2.16	7.34
ABS	200, 5	2.67
PC/ABS 50/50M	230, 3.8	2.41
PC/ABS 60/40M	230, 3.8	3.39
PC/ABS 70/30M	230, 3.8	6.59
PC/ABS 70/30P	230, 3.8	5.56

3.2. Mechanical properties

Blends with fixed concentrations of 2 phr compatibilizer for the various PC/ABS blends (Table 1) were prepared to investigate the effect of using two different compatibilizers on mechanical properties such as tensile (Figure 1), elongation (Figure 2), and impact strength (Figure 3), as well as on physical properties such as hardness (Figure 4) and density (Figure 5). From the Figure 1, it can be seen that the higher PC content, the higher tensile strength of PC/ABS blends. Each additional 10% PC provides increased tensile strength by about 45%, as expected. Also, the tensile strength decreases as the concentration of ABS increases in the polymer blend, accordingly to the law mixture. Similar results were reported by Triantou & Tarantili [4], who compounded at various compositions of PC/ABS. PC has higher tensile strength than ABS. The virgin PC has a tensile strength about 65 MPa, while ABS only 49 MPa [10]. The optimum blend is obtained at 70/30 PC/ABS blend. It has been reported that PC exhibits high yield strength and ductile behavior. In contrast, ABS shows a brittle characteristic and has low yield strength. Addition of ABS lowers the tensile strength, reduces the cold-down ability of the PC matrix and the consequent elongation at break of the blends [5]. The tensile strength of the blends increases almost linearly with the increase of PC contents.

The tensile strength value of this blend compatibilized with MBS is 62.52 MPa, ± 10% higher than the previous work reported by Liu *et al.* [11]. MBS give higher value compared with other compatibilizers, such as PP-g-MA (58.99 MPa), EVA-g-MA (± 55 MPa) [6], ABS-g-MA (± 58 MPa) [4], PSM (± 57 MPa), and ABS-g-MA (± 53 MPa) [9]. Incorporation of various compatibilizers phr acts as a bridge between PC and ABS thereby resulting in reducing the interfacial tension between the both polar and non-polar phases [1] and leads to a further increase in tensile ductility [5]. The tensile strength of PP-g-MA compatibilized blend in this study is 58.99 MPa, slightly higher than which had been reported by Tjong & Meng [5]. As reported by Farzadfar *et al.*, there is a chemical reaction between MA groups of the compatibilizers and hydroxyl end groups in PC as can be seen in Figure 4 [6].

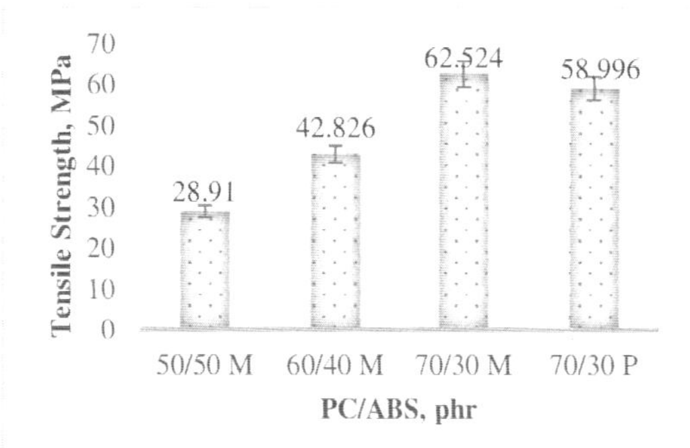

Figure 1. Effect of MBS and PP-g-MA on the tensile strength of PC/ABS blends.

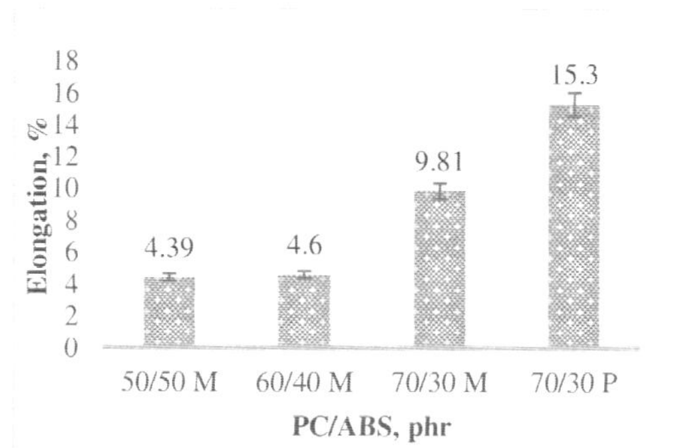

Figure 2. Effect of MBS and PP-g-MA on the elongation of PC/ABS blends.

From Figure 2, it can be seen that the addition of ABS reduces the elongation of the blends, as reported before, ABS exhibits very low tensile ductility. This also implies that the stiffness of PC improves considerably with increasing ABS content. The compatibilizers were added at fix amount of 2 phr. The elongation at break of PC/ABS blends presents a minimum value at compounds containing 50% ABS. These results suggest limited interfacial interactions taking place in PC/ABS blends, which acquires a minimum when PC and ABS are mixed in equal amounts [4].

The effect of compatibilizers on the elongation of the blends is shown in Figure 2. This result shows that the elongation of PC/ABS blends modified with PP-g-MA is 35% higher than that with MBS, as the tensile strength is lower. And the increasing ABS content, increase the elongation. As expected, blending PP with rubbery polymers like ABS can enhance the elongation properties. However, this value is lower when compared with other compatibilizers, such as EVA-g-MA (around 50%) [6], SMA (around 110%) [11], and MAP (around 90%) [5].

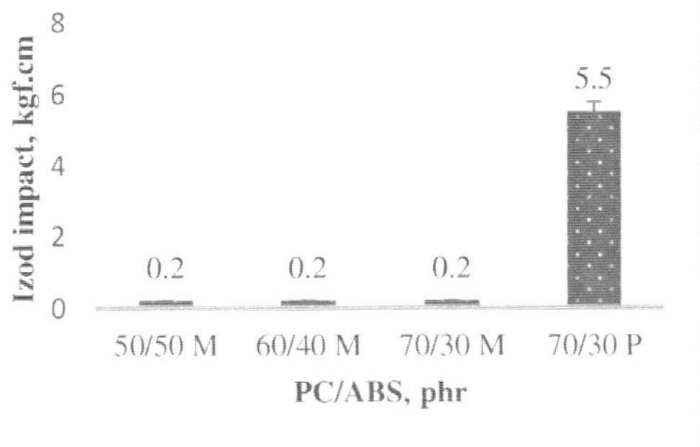

Figure 3. Effect of MBS and PP-g-MA on the izod impact of PC/ABS blends.

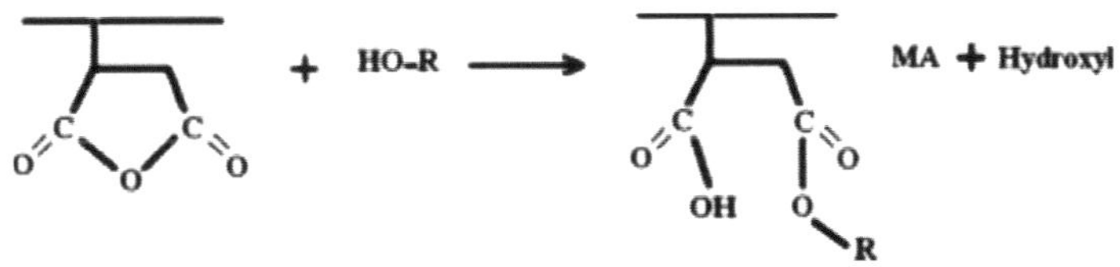

Figure 4. In this case simply justify the caption so that it is as the same width as the graphic.

Figure 3 shows that the best impact strength of the blends occurs when using PP-g-MA as a compatibilizer, with values of 5.5 kgf.cm. This could be due to the chemical reaction between MA groups of the PP-g-MA and hydroxyl and groups in PC as shown in figure 4. It is worth mentioning that the functional groups of maleic-anhydride is very effective to improve impact properties of PC/ABS 70/30 blend [5]. PP-g-MA acts as a suitable impact modifier for PC/ABS blends by modification of the interface, hence the interface strength between PC and ABS was increased. In contrary, 2 phr of MBS compatibilizer gives very low impact strength in each blend. It indicates that the MBS content must be carefully controlled owing to its brittle nature. Too much compatibilizer added to the blend will lead to a deleterious effect on the impact strength [5]. This result has also been observed by Liu *et al.*, the addition of MBS decreased the tensile strength and elongation of blends a lot, and had no significant effect on the impact strength [11].

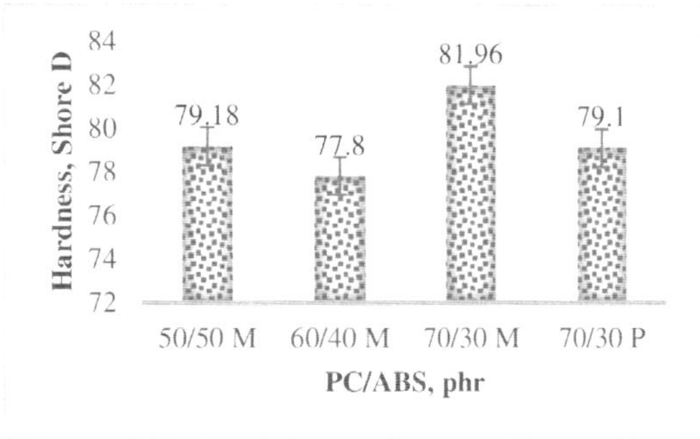

Figure 5. Effect of MBS and PP-g-MA on the hardness of PC/ABS blends.

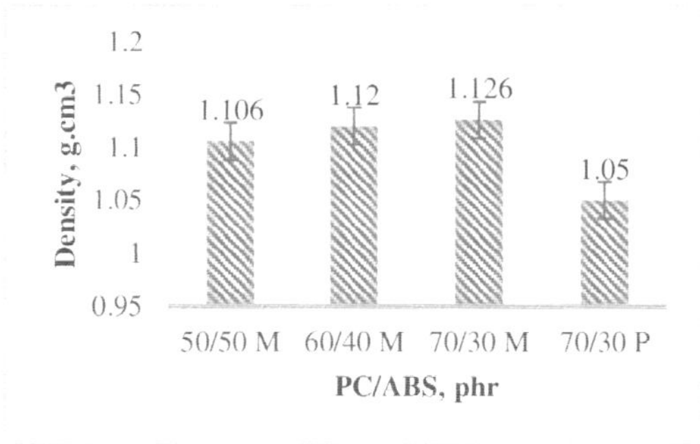

Figure 6. Effect of MBS and PP-g-MA on the density of PC/ABS blends.

Figure 5 shows the hardness variations with the blend composition on PC/ABS blends. The PC/ABS 70/30M blend shows the highest hardness value of 81.96 shore D, but the lowest elongation comes from PC/ABS 50/50M. It shows that the hardness has no correlation with the elongation behavior, and also with the PC/ABS proportion. Krache *et al.* reported that the hardness value increased with the increasing of PC content between 30 to 70%, due to the decrease in flexibility of the ABS chains and by the phase inversion of PC from dispersed to continuous phase when its concentration in the blend was increased from 40% to 70% [12]. MBS give higher hardness value than PP-g-MA by about 3.6%.

The density values of PC/ABS blends (Figure 6) present an increase of density with an increase of the PC content, it is due to the density of virgin PC (1.2 g/cm^3) which is naturally higher than virgin ABS (1.04 g/cm^3) [10]. MBS compatibilizer gives higher density than PP-g-MA about 7.2%. These results exhibit similar phenomenon with the tensile strength.

3.3. Crystallinity properties

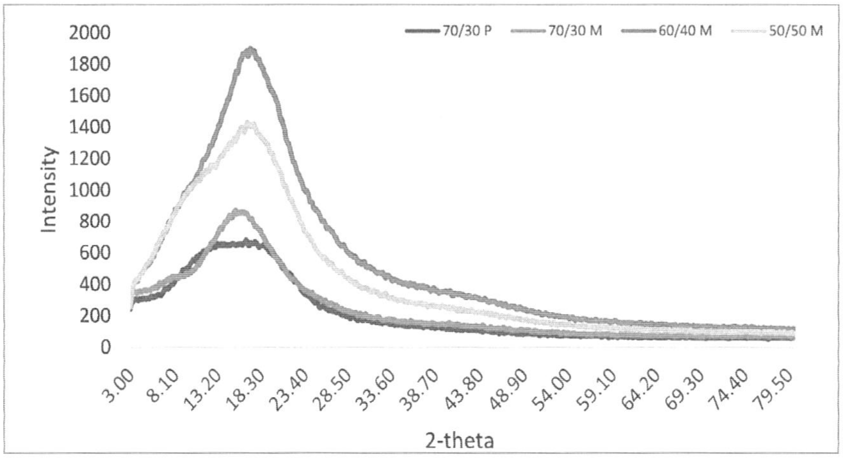

Figure 7. Difractogram of (a) PC/ABS 70/30 with ABS, (b) PC/ABS 60/40 with ABS, (c) PC/ABS 50/50 with ABS, (d) PC/ABS 70/30 with PP-g-MA.

Diffractogram of PC/ABS blends were performed to observe the crystallinity properties of those blends (Figure 7). Nurhajati *et al.* reported the diffraction pattern for origin PC displayed typical peak of A-type semi crystalline and broad peak of amorphous ABS. The single strong peaks were found at 2-theta = 16.26° and 2-theta=18.94° for PC and ABS, respectively [9]. The diffractogram of PC/ABS blends show single broad peak of A-type amorphous polymers with MBS compatibilizer. The strongest peaks are located around 2-theta=17.46° with intensity of 1423, 2-theta=17.68° with intensity of 1861, and 2-theta=17.02° with intensity of 824 for PC/ABS blends 50/50M, 60/40M and 70/30M respectively. The sharpest shape and highest intensity peak is given by 60/40M PC/ABS blend. It can be seen that the peaks are shifted between the characteristics peaks of origin PC and ABS. PP-g-MA compatibilized blend is found to be more amorphous than MBS, with the broad peak located around 2-theta=17.64° with intensity of 672. Even this blend has amorphous phase, it gives the highest impact strength and comparable tensile strength with MBS blends.

3.4. Morphological properties

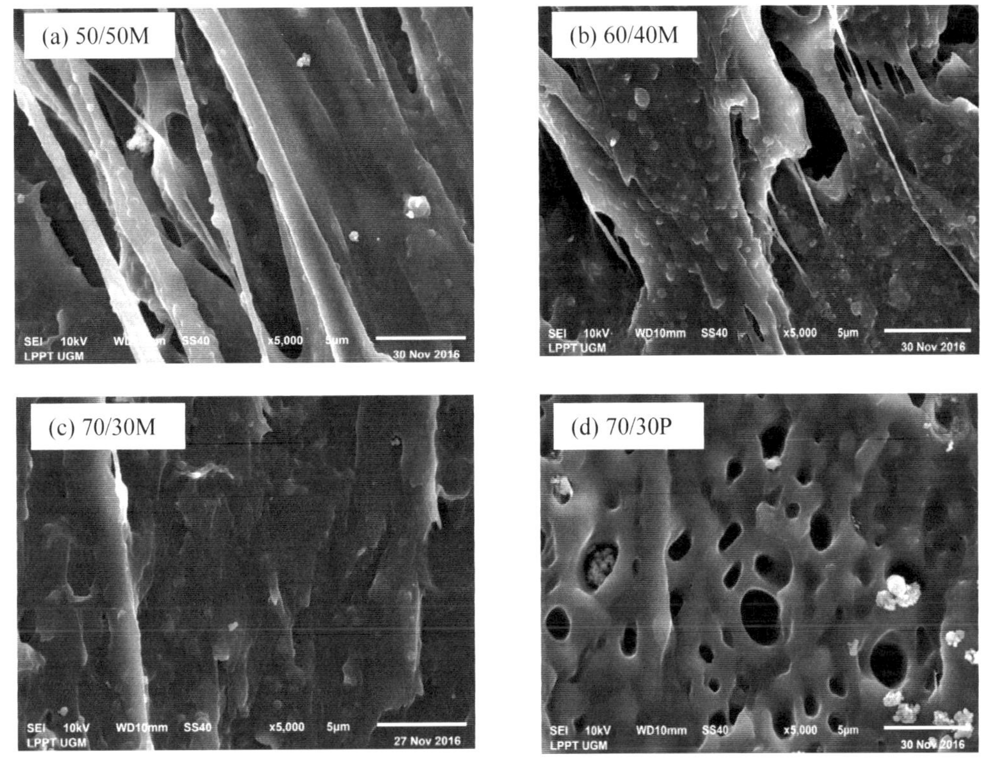

Figure 8. SEM photomicrogrphs (a) PC/ABS 50/50 with ABS, (b) PC/ABS 60/40 with ABS, (c) PC/ABS 70/30 with ABS, (d) PC/ABS 70/30 with PP-g-MA.

The study of the morphology of surfaces by SEM as an interesting tool for identification of different phases. At 50/50 PC/ABS blends, an even distribution of two polymeric phases is relatively clear (Figure 8(a)). In this blend, the continuous phase is ABS, whereas the dispersed one is PC. But when the amount of PC is greater than ABS, PC becomes the continuous phase, and ABS is the dispersed phase. Figure 8(c) exhibit a relatively smooth surface in comparison to 8(b). This result confirms the tensile behavior of the blends. PC/ABS 70/30 has higher tensile strength than the others. Figure 8(d) shows porous surface of PS/ABS with PP-g-MA as a compatibilizer. Even so porous, this blend has the higher impact strength, it is due to PP-g-MA can enhance the ductile properties of the blend. Based on the results, it was found that PP-g-MA played more effective compatibilization role in the PC/ABS 70/30 blend than MBS compatibilizer.

4. Conclusion

Blends of PC/ABS were prepared with two different compatibilizers and some properties such as MFI, density, hardness, tensile strength, elongation, morphology, and crystallinity were investigated. The results show that the addition of MBS at 2 phr level on the PC/ABS blends led to a significant increase of tensile strength, hardness, and density compared with PP-g-MA, but the impact strength and elongation were lower. The tensile strength, elongation, and hardness tended to increase almost linearly with increasing PC content, however, the impact strength was almost the same. Scanning electron microscopic examination showed that PC/ABS blend compatibilized with PP-g-MA was more porous than the one with MBS. Based on the results, it was found that PP-g-MA played more effective compatibilization role in the PC/ABS 70/30 blend than MBS compatibilizer. Those value of tensile strength, hardness, elongation, density, and impact strength were 602 kg/cm^2, 79.1 Shore A, 15.3%, 1.05 g/cm^3, and 5.5 kgf/cm, respectively.

Acknowledgment

The authors gratefully acknowledge to the Head of Center for Leather, Rubber, and Plastics for financial support of this research.

References

[1] Bonda S Mohanty S and Nayak S K 2014 *Iran. Polym. J.* **23** 415–25
[2] Chiang W and Tzeng G 1996 *J Appl. Polym. Sci.* 795–805
[3] Asyadi F Jawaid M Hassan A and Wahit M U 2013 *Polym. Plast. Technol. Eng.* **52** 727–36
[4] Triantou M I and Tarantili P A 2013 *Polym. Compos.* **35** 1–13
[5] Tjong S C and Meng Y Z 2000 *Eur. Polym. J.* **36** 123–29
[6] Farzadfar A Khorasani S N and Khalili S 2013 *Polym. Int.* **63** 145–50
[7] Yun I H, Lee K and Sang H 2014 *J Matter. Sci.* **49** 4522–29
[8] Li J Chen F Yang L Jiang L and Dan Y 2017 *Spectrochim. Acta Part A Mol. Biomol. Spectrosc.* **184** 361–67
[9] Nurhajati D W Sholeh M Indrajati I N and Setyorini I 2017 *Leather, Rubber and Plast. Mag.* **33(1)** 43–48 (In Indonesia)
[10] Nurhajati D W Setyorini I and Sugihartono 2014 *Leather, Rubber and Plast. Mag.* **30(2)** 1–6 (In Indonesia)
[11] Liu Y Li H Ding X Zhu J Zhang L Pan W and Cai R 2014 *American J of Mat. Research* **1(3)** 48–52
[12] Krache R and Debbah I 2011 *Mater. Sci. Appl.* **2** 404–10

Thermal stability of shoe upper leather: Comparison of chestnut and quebracho as vegetable tanning agent

G Griyanitasari[1], I F Pahlawan[1] and E Kasmudjiastuti[1]

[1] Center for Leather, Rubber, and Plastics, Ministry of Industry, Sokonandi St. number 9, Yogyakarta 55166, Indonesia

gresygriyanitasari@gmail.com

Abstract. Thermal stability is one of the most important factors in the leather tanning process and mainly affected by the tannin type used to tan the hides. Vegetable-tanned leather for shoe upper is a material that could be used and preserved for a long time. However, it makes leather easily degraded because of temperature and time. Additionally, one of the stages in shoe making process is the heating process to stick the upper leather into midsole material. Therefore, this study aims to compare the effect of different tanning agents, chestnut and quebracho, on the thermal stability of shoe upper leather. Measured were degree of tannage and thermal stability, using shrinkage temperature, thermogravimetric analysis and differential scanning microscopy. The morphology of leathers tanned with chestnut and quebracho in different concentration were also observed. The shrinkage temperature of leather tanned with chestnut was lower than that of quebracho-tanned leathers. Leathers tanned with 20 and 25% of chestnut showed higher degree of tannage than the ones tanned with quebracho. Thermogravimetric analysis (heating rate of 10 °C min^{-1}; gas flow of 60 ml min^{-1}) showed that the maximum weight loss rate decreased with the increase in the concentration of chestnut or quebracho. The differential scanning calorimetry results showed that quebracho-tanned leather had a higher heat resistance than chestnut-tanned one. The presence of vegetable tanning agent improved the thermal stability of leather. Furthermore, it was shown that leather tanned with 20% and 25% of chestnut resulted in almost equal thermal stability. Quebracho calf leather was more resistant against heat than chestnut calf leather.

1. Introduction

Vegetable tanning is an ancient method of leather tanning. It is still in use, even as its popularity has declined and it has been largely replaced by chrome tanning, and more than 90% leather production in the world are using chrome as tanning agent [1] because of its simple technology application and the reliability of the tanning procedures [2]. Vegetable tanning agent is still used in the leather tanning industry because of its advantages, such as the fullness [3], compatibility with human skin, and its cleaner tanning process waste. Leather-making process consists of three parts: beamhouse operation, tanning, and finishing. Vegetable tanning agents are used in the tanning process [4] to stabilize the raw material into stable materials.

Upper shoes leather usually made of leather tanned with chromium. Chromium-tanned leather has higher collagen stability but has negative effect for the environment. Vegetable tanning is an alternative method for cleaner production of shoe upper leather. Even though vegetable tanned leather is more difficult to biodegrade than chrome tanned leather [3], it is material that would be used and preserved for a long time, which makes leather being easily degraded because temperature and time [5]. Besides, one of the stages in shoe making process is the heating process to stick the upper leather into midsole material. Thus, it is necessary to measure the thermal stability of vegetable tanned shoe upper leather.

Thermal analysis is a useful method to evaluate the thermal stability and behavior, degradation temperature, and thermal decomposition [6]. When the leather is degraded, its physical and chemical properties will decrease. In the leather tanning process, the increased of thermal stability is one of the most important changes [7]. Shrinkage temperature is the most common way to measure hydrothermal stability of leather and as a cook proof at once. Shrinkage temperature of leather is the temperature at which a leather shrinks when heated in water at 2 °C min^{-1} and it is commonly related to degree of tannage [7, 8]. Another way to measure thermal stability of leather is using thermogravimetric analysis [1, 6, 7, 9, 10]. This technique is possible to calculate the thermal stabilities and kinetic decomposition of vegetable tanned leather. Kinetic analysis is a new interesting topic of research because of its importance in thermal analysis and its role in decomposition reaction mechanism [5]. Differential Scanning calorimetry is also one of the technique to measure leather thermal stability [11].

Thermal stability mainly depends on the tannin type used to tan the skin [7]. The most common vegetable tanning agents are mimosa, quebracho [4], chestnut, tara, haraz, and genipin. Beside mimosa, quebracho and chestnut are vegetable tanning agents that easy to obtained and have high tannin content. Vegetable tannins are classified into condensed tannins and hydrolyzable tannin based on its structure and chemical behavior [12]. Tannin in chestnut belongs to hydrolyzable tannin, while tannin in quebracho is condensed tannin [13].

Thermal stability of leather is a new research concern, so that there are still few studies about it. A study by Carsote *et al.* [7] concluded that using micro-DSC, it was known that quebracho tanned calf leather displayed the highest resistance against deterioration than chestnut. There are also still a few study concerning thermal stability of leather, such as study conducted by Yang *et al.* [6] that used thermogravimetric analysis to measure thermal stability of fire retardant leather; Bañón *et al.* [1] that conducted study about thermal stability of leather treated with NaOH, using thermogravimetric analysis; Sanchez-Olivares *et al.* [10] who studied the effect of sodium montmorillonite on the thermal properties of leather. Based on those previous research, there were yet study concerning the thermal properties of quebracho-and-chestnut-tanned calf leather using thermogravimetric analysis and shrinkage temperature. Therefore, this study aims to compare the effect of different tanning agents, chestnut and quebracho, on the thermal stability of calf leather.

2. Experimental

2.1. Materials

Quebracho and chestnut were obtained from chemical suppliers in Yogyakarta, Indonesia. This experiment used cow pickled-hides from Magetan, East Java, Indonesia. Tanning chemicals used in this experiment are alum, NaCl, wetting agent, bating agent, sodium formate, sodium bicarbonate, relugan GT50, basyntan DLX-N, vegetable fatliquor, white syntan, oxalic acid, novaltan PF, neutralising syntan, ipertan 502, retingan R7, levelling agent, dyestuff, synthetic fatliquor, anti-mold, and formic acid.

2.2. Leather tanning procedure

Table 1. Sample identification.

Vegetable tanning agent	Quantity (%)	Sample identification
Chestnut	20	Cx
	25	Cy
Quebracho	20	Qx
	25	Qy

In this study, the quantity of quebracho and chestnut were used to tan calf hides to make upholstery leather and the quantity was varied (Table 1). At the tanning step of the leather tanning procedure, the tanning agent was varied into quebracho and chestnut. The quantity of each was differentiated for 20% and 25%. Other steps but tanning in the leather tanning procedure were controlled.

Table 2. Leather tanning procedure.

Process	%	Materials	Time (min)
Wetting back	200	Water	30
	20	NaCl	
	0,5	Wetting agent	60
	2	Bating agent	
Depickling	100	Water	
	10	NaCl	10
	1	Sodium formate	
	0,5	Sodium bicarbonate	30
Pre-tanning	2	Relugan GT 50	
	1	Basyntan DLX-N	30
	2	Vegetable fatliquor	
Tanning	20;25	Quebracho/chestnut	
	9	Alum	180
Basification	0,75	Sodium bicarbonate	3x15 + 30
	1	White syntan	30
Washing	400	Water	2x10
Bleaching	200	Water	
	1	Oxalic acid	30
	0,5	Wetting agent	30
Neutralising	100	Water	
	2	Novaltan PF	45
	2	Neutralising syntan	30
	1	Sodium formate	30
	0,50	Sodium bicarbonate	3 x 20 + 30
Retanning	100	Water (40 °C)	
	2	Basyntan DLX-N	45
	3	Ipertan 502	30
	2	Dysiamin (RR7)	60
Dyeing	100	Water (50 °C)	
	1	Levelling agent	10
	2	Havana dyestuff	60
Fatliquoring	50	Water (50 °C)	60
	3	Vegetable fatliquor	
	3	Synthetic fatliquor	
	2	Taural ICN	
	1	Tannit LSW	30
	0,05	Anti-mold	
	1,5	Formic acid	3 x 10+ 30
Finishing			

This research used calf pickled pelts as raw material. First, the pelts were soaked in a rotating drum with water, NaCl, soaking agent and run for 30 min, then bated for 60 min. In order to raise the pH of the pelts, depickling process was conducted to achieve the pelts' pH at 5-5.5. Depickling process were done by adding sodium formate and sodium bicarbonate for 30 min. Prior to tanning, the pelts were pre-tanned with Relugan GT50, Basyntan DLX-N and Leathernol BLM for 30 min. Tanning process were based on Table 1 in using vegetable tanning agent.

The experiment used Quebracho and Chestnut, as the vegetable tanning agent with different weight concentration. The concentration of the vegetable tanning agent were 20% (m/m) and 25% (m/m), while alum used in the experiment were 9% (m/m). The vegetable tanning agent were combined with alum and run for 180 min. Following to that sodium bicarbonate were added to reach the pH of pelts at 6-6.5. The wetblue leather, as a result, were then aging and shaving. The shaved-wetblue leather were ready to be processed further for post-tanning as in Table 2.

3. Results and discussion

3.1. Shrinkage temperature

The shrinkage temperature of leathers was determined using Shrinkage Temperature Tester. The Ts is the temperature when the sample shrinks 150 μm. Shrinkage temperature (Ts) of leather tanned with chestnut and quebracho are given in Figure 1 to observe the different hydrothermal stability. The average Ts of finished leathers were higher than semi crust leather for both leather with chestnut and quebracho. Ts of semi crust leathers were obtained after tanning process, while Ts of semi crust leathers were after the finishing process. It was clearly observed that after retanning process, there were some chemicals added to the leather, such as leveling agent, dyestuff, fat liquors, anti-mold, acid, and finishing materials. In the fatliquoring process, fatliquors were added to prevent the collagen fiber resticking when it is dry [13]. During drying, the water was removed and the diameter collagen fibrile shrunk.

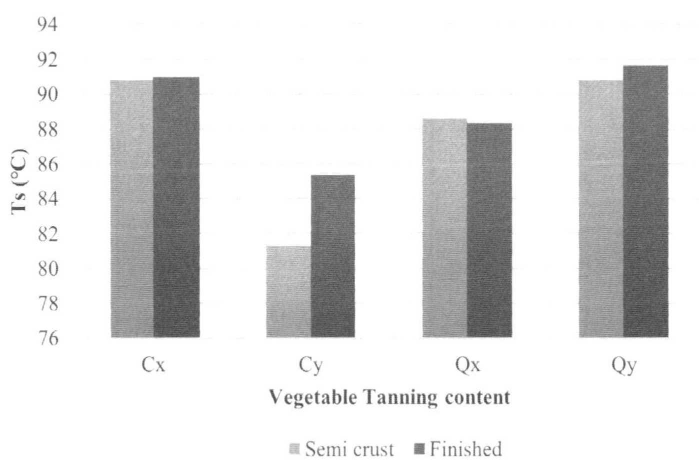

Figure 1. Shrinkage temperature of leather tanned with different vegetable tanning agent.

Shrinkage temperature is used as an indicator to measure the effectiveness of tanning process, while thermal stability gives information about the stability of collagen to heat but it does not reveal other information related to the performance of material [14]. Compared with quebracho, in general, chestnut delivered lower Ts of leather, both for semi crust and finished (Figure 1).

3.2. Degree of tannage

From the Figure 1, it can be inferred that leathers tanned with chestnut have higher tannin bound than leather tanned with quebracho. Leathers tanned with 25% of chestnut (22.4%) have higher tannin bound than 20% of chestnut (21.41%), while leathers tanned with 25% of quebracho (16.97%) have lower tannin bound than 20% of chestnut (21.19%).

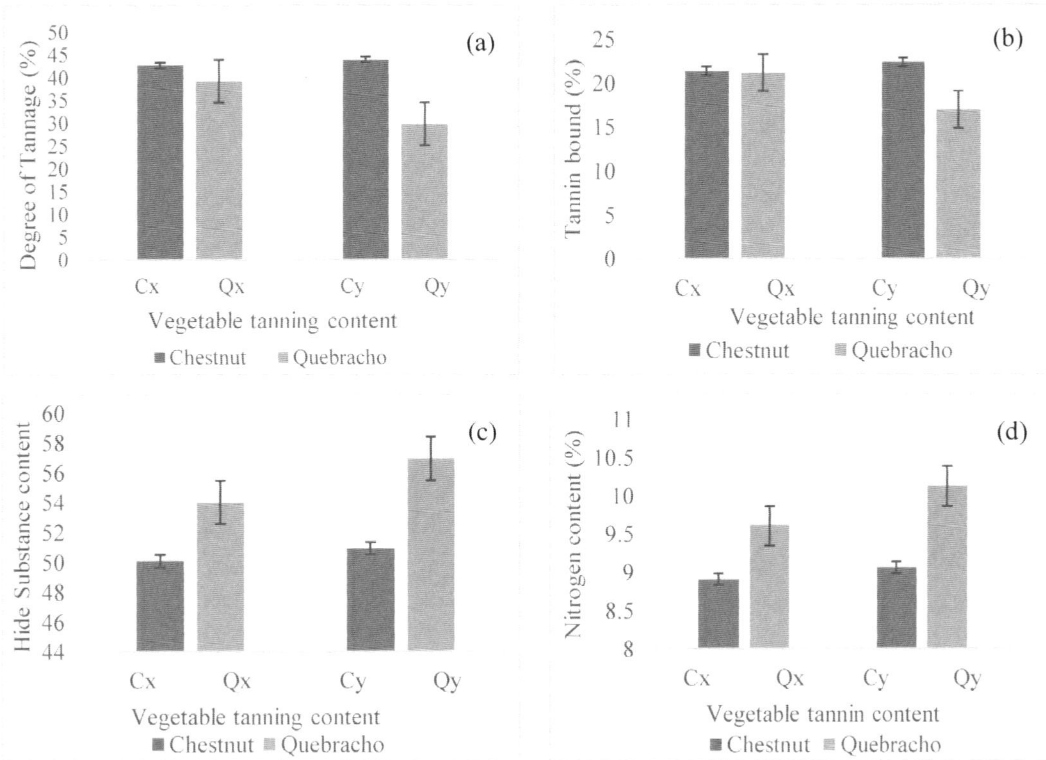

Figure 2. Degree of tannage (a); tannin bound (b); hide substance content (c); and nitrogen content (d) of leather tanned with different vegetable tanning agent.

The degree of tannage of chestnut tanned leathers was higher than quebracho tanned leather, both for 20% and 25%. The use of 20% of chestnut to tan the leathers (42.75%) showed lower degree of tannage than 25% of chestnut (43.99%), while the use of 20% of quebracho to tan the leathers (39.22%) showed higher degree of tannage than 25% of quebracho (29.8%) (Figure 2). These results were in line with the result of the tannin bound test. The higher content of vegetable tanning agent, the higher the percentage of tannin bound. Chestnut-tanned leather showed lower degree of tannage and tannin bound than quebracho-tanned leather.

Different with the Figure 2(a) and 2(b), Figure 2(c) and 2(d) shows contrary results. Hide substance content of leather tanned with 20% of chestnut (50,07) was slightly lower than leathers tanned with 25% of chestnut (50.92). Hide substance of leather tanned with 20% of quebracho (54.01) was lower than that of 25% of quebracho (56.93). Nitrogen content for each 20% of chestnut (8.91%) and 20% of quebracho (9.61%) was lower than that of 25% of each chestnut (50.92%) and quebracho (56.93%). A previous research conducted by Kasmudjiastuti *et al.* [15] reported that high content of nitrogen in the leather displays the incomplete bating process (removal non collagen protein in the leather tanning process) that inhibit the penetration of vegetable tanning agent into the collagen fiber.

3.3. Thermogravimetric analysis

Samples were analyzed using thermogravimetric to measure the effect of percentage of different vegetable tanning agent on thermal decomposition. The thermogravimetric tests of finished calf leather were carried out in Shimadzu DTG 60 at heating rates of 10 °C min^{-1} under air atmosphere with 60 ml min^{-1} gas flow. Samples were placed in an alumina pans.

Figure 3 shows the DTG curves from each one of six samples of leathers tanned with chestnut and quebracho with different concentrations. All of the samples were degraded through three stages (Figure 3). The first stage was the initial stage between 28 and 120 °C. In this stage, for chestnut and quebracho, leather tanned with 20% of vegetable tanning agent resulted in higher residue than that of 25%. The initial stage is an endothermic phase [16] when moisture and small molecules in the leather were evaporated [15, 16].

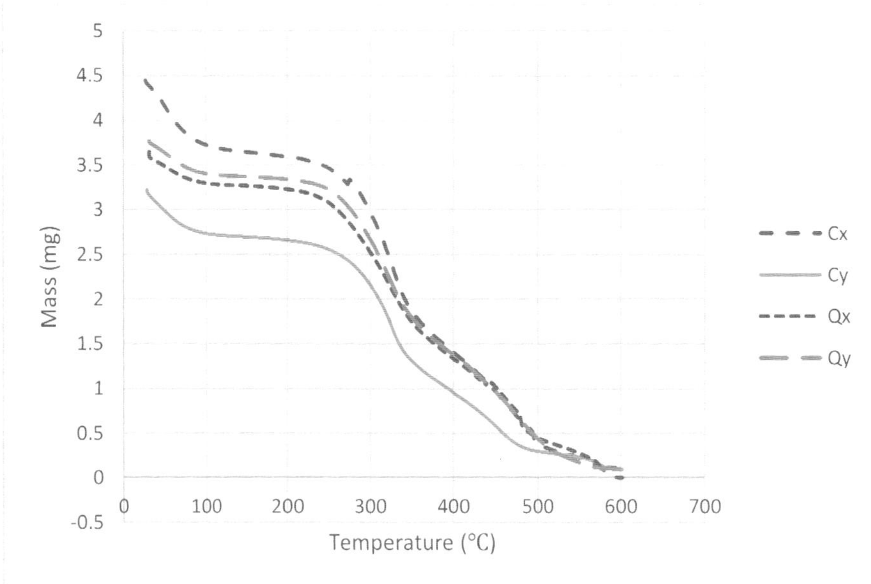

Figure 3. TG curves of leather samples.

The main phase happened in two stages: at the temperature between 205 – 509 °C and 552 – 594 °C. Those stages are more complex than the stage before. The mass loss at this stage was because of decomposition of collagen. T_{max} reflects the thermal stability of collagen of leather [7], the higher residue at the T_{max} the more stable is the collagen. From the Table 3, 20% of chestnut showed better thermal stability leather than 25% of chestnut at the initial stage. In contrast, at the first and second stages, leather with 25% of chestnut showed higher thermal stability than leather with 20% of quebracho. A previous study of Carşote, *et al.* [7] reported that the loss of water at the initial stage could affect cross-linking which increase the thermal stability because the interfibrillar water content of leather leads to thermal stability. Water molecules have been displaced by the tannin molecules when it filled the empty area of collagen at the tanning process to stabilize the collagen within leather [19].

At the second stage (Table 3), the higher content of vegetable tanning agent used the higher residue at the maximum degradation. It is clearly observed that 25% of vegetable tanning agent resulted in better leather residue than those of 20%. The second and third phase are exothermal process. Pyrolytic decomposition and thermo-oxidation were happened at these stages [20]. Due to the presence of disordered regions and the well-defined sequences of amino acid residues, leather is both crystalline and amorphous material that could show thermal properties [14]. This last phase indicates the melting crystalline area of leather or the evaporation of residual strongly bond water and continued conformational changes of super helix of collagen [16].

At 600 °C, the higher content of chestnut, the higher the residue. On the other hand, the final residue of quebracho tanned leather is almost equal (Table 4). The higher content of both chestnut and quebracho can inhibit the decomposition of the leather and therefore improve the leather thermal stability.

Table 3. Different decomposition stages of the samples.

Samples	Index	Initial stage	Stage 1	Stage 2
Cx	Temperature (°C)	28.80 – 120.71	214.336 – 376.91	552.54-593.80
	Residue at Tmax (%)	92.26	52.82	1.01
Cy	Temperature	29.37 – 108.03	205.25 – 508.63	546.61 – 593.64
	Residue at Tmax (%)	90.65	55.82	3.93
Qx	Temperature	33.90 – 96.04	215.90 – 371.26	567.25 – 578.31
	Residue at Tmax (%)	89.19	82.38	3.91
Qy	Temperature	32.57 – 107.58	218.41 – 528.61	540.32 – 583.46
	Residue at Tmax (%)	88.46	57.08	4.16

Table 4. Residue of the samples at 600 °C.

Samples	Residue at 600 °C (%)
Cx	0
Cy	3.93
Qx	2.75
Qy	2.36

3.4. Differential scanning calorimetry

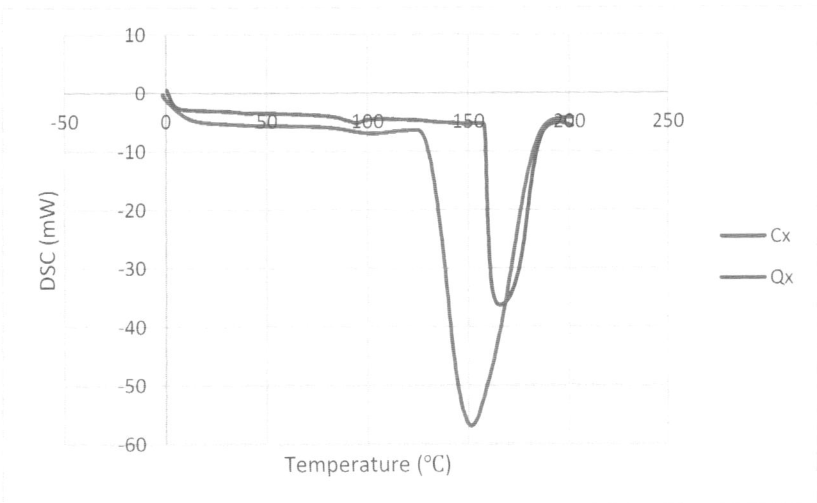

Figure 4. DSC curves of leather tanned with 20% of chestnut and quebracho, respectively.

Due to the efficiency of vegetable tanning agent, DSC was operated under the nitrogen atmosphere and 15 ml/minute flow rate, the samples were analyzed. It was conducted to determine the phase transition and denaturation temperature. DSC curves (Figure 4) showed that leather tanned with 20% of chestnut have and quebracho respectively have two endothermic processes. The maximum peak value of the endothermic phase was the denaturation temperature. From Figure 4, at the first process, chestnut tanned leather showed higher denaturation temperature (102.72 °C) than that of chestnut (94.64 °C). Meanwhile, at the primary process, the denaturation temperature of chestnut tanned leather (151.45 °C) was lower than quebracho tanned leather (165.47 °C). This results parallel with the TGA results, where the first process was the phase when the water in the leather was totally lost, and the second process was the decomposition of the materials [14, 9].

The DSC results were also in line with the conventional shrinkage temperature value, both for semi crust and finished leather. The decomposition temperature of samples from DSC method was slightly different with the shrinkage temperature (the deviation range is 6-12 °C). The higher the hydrothermal stability of leather, the higher the bond strength of the collagen [13]. From the Figure 4, all of the curves are narrow. Narrow and symmetric DSC curve indicate that the population of collagen molecules are grouped [7].

Tannin in quebracho belongs to condensed tannin, while chestnut is hydrolizable tannin. Condensed tannins consist of polymeric flavonoids, while hydrolyzable tannins are gallotannins and ellagitannins [7]. Content of gallic acid in the hydrolyzable tannin was considered as the most obviously structural difference to condensed tannins, so that it also regarded as the important factor for the formation and stabilization of tannin-metal complexes [21].

3.5. Scanning electron microscopy

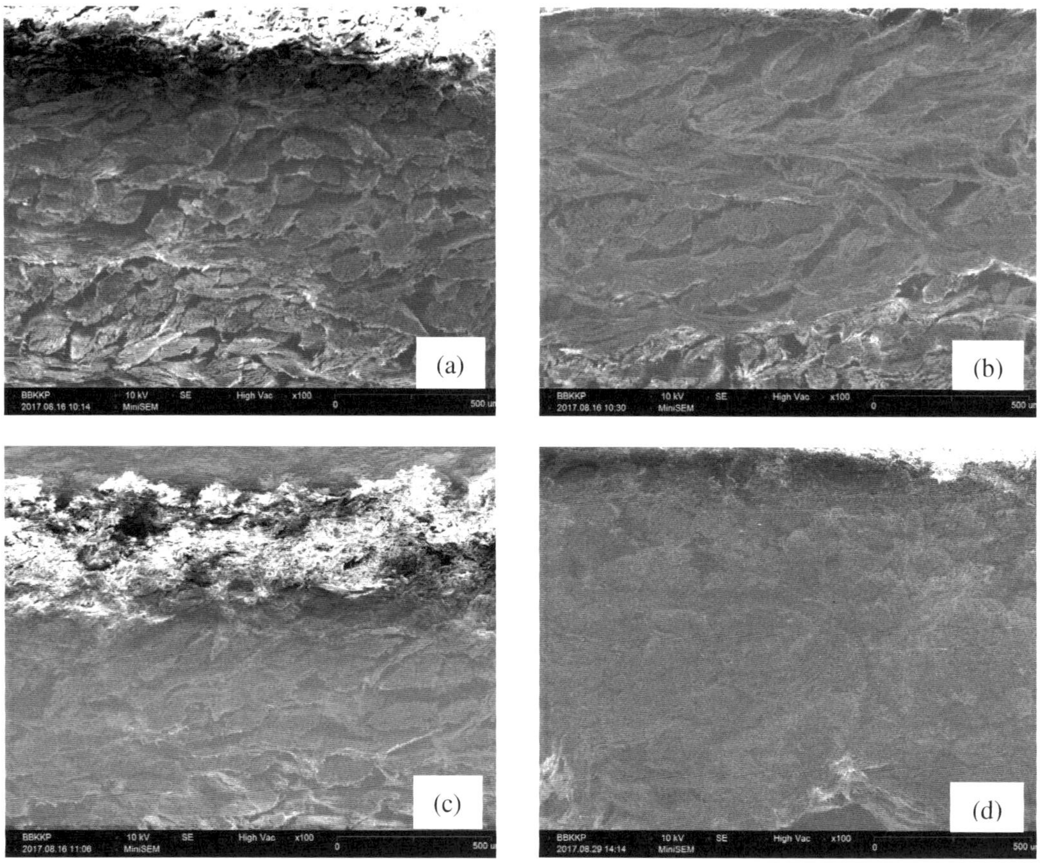

Figure 5. Leather morphology by SEM of samples:(a) leather tanned with 20% chestnut; (b) 25% of chestnut; (c) 20% of quebracho; and (d) 25% of quebracho.

Scanning Electron Microscopy SEC type SNE 3200 M was used to analyze samples at an accelerating voltage of 5 kV . The morphology of cross-section leather samples was investigated using SEM at 100X magnification are shown in Figure 5. There are no major differences in the leather morphology between 20 and 25% of chestnut, while leather tanned with 25% were tighter, bulkier, and plumper than that of 20%.

4. Conclusion

In the present work, the thermal stability of leather tanned with different tanning agent have been studied using shrinkage temperature, chemical properties, thermogravimetric analysis, DSC, and SEM. The presence of vegetable tanning agent improved the thermal stability of leather. Furthermore, it was shown that leather tanned with 20% and 25% of chestnut resulted in almost equal thermal stability. Quebracho calf leather was more resistance against heat than chestnut calf leather.

Acknowledgment

The authors would like to acknowledge the Center for Leather, Rubber, and Plastics for the support.

References

[1] Bañón E Marcilla A García A N Martínez P and León M 2016 *Waste Manag.* **48** Feb 285–99
[2] Nikonova A Andreyeva O and Maistrenko L 2016 *IOP Conf. Ser. Mater. Sci. Eng.* **111(12024)** 1-6
[3] Ali S B Haroun H E and Musa A E 2013 *J. For. Prod. Ind.* **2(5)** 25–29
[4] Khairullin A K Voznesensky E F Rakhmatullina G R Sabirov A M and Tikhonova N V 2017 *IOP Conf. Ser. J. of Phys.* **927(012025)** 1-4
[5] Mayasari H E and Yuniari A 2016 *Leather, rubb and Plast. Mag.* **32(2)** 125–34 (In Indonesia)
[6] Yang L Liu Y Wu Y Deng L Liu W Ma C and Li L 2016 *J. Therm. Anal. Calorim.* **123(1)** 413–20
[7] Carşote C Badea E Miu L and Gatta D G 2016 *J. Therm. Anal. Calorim.* **124(3)** 1255–66
[8] Mutlu M M Crudu M Maier S S Deselnicu D C Albu L Gulumser G Bitlisli B O Basaran B Tosun C C and Zengin A C A 2014 *Ekoloji* **23(91)** 83–90
[9] Carşote C Budrugeac P Miu L Yalçin F Karavana H A and Badea E 2014 *ICAMS—5th Int. Conf. on Adv. Mater. and Sys.* 505–10
[10] Sanchez-Olivares G Sanchez-Solis A Calderas F Medina-Torres L Manero O Di Blasio A and Alongi J 2014 *Appl. Clay Sci.* **102** 254–60
[11] Onem E Yorgancioglu A Karavana H A and Yilmaz O 2017 *J. Therm. Anal. Calorim.* **129(1)** 1–8
[12] Tan K W and Kassim M J 2011 *Corrosion Sci.* **53(2)** 569–74
[13] Covington A D 2009 *Tanning Chemistry: The Science of Leather* Cambridge: Royal Society of Chemistry 2009
[14] Jeyapalina S Attenburrow G E and Covington A D 2007 *J Soc Leather Technol. Chem.* **91(6)** 236–42
[15] Kasmudjiastuti E Pidhatika B Pahlawan I F and Griyanitasari G 2017 *3rdInt. Conf. on Sci. and Tech.*
[16] Budrugeac P 2015 *J. Therm. Anal. Calorim.* **120(1)** 103–12
[17] Xu W Li J Liu F Jiang Y Li Z and Li L 2017 *J. Therm. Anal. Calorim.* **128(2)** 1107–16
[18] Yang P He X Zhang W Qiao Y Wang F and Tang K 2017 *J. Therm. Anal. Calorim.* **127(3)** 2005–12
[19] Badea E Şendrea C Carşote C Adams A Blumich B and Iovu H 2016 *Microchem. J.* **129** November 158–65
[20] Budrugeac P Miu L and Souckova M 2007 *J. Therm. Anal. Calorim.* **88(3)** 693–98
[21] Teng B Jian X and Chen W 2012 *Leather Footwear J* **13(1)** 3–13

IOP Conf. Series: Materials Science and Engineering **432** (2018) 012041 doi:10.1088/1757-899X/432/1/012041

Thermal degradation and swelling behaviour of acrylonitrile butadiene styrene rubber reinforced by carbon black

H E Mayasari[1], I Setyorini[1] and A Yuniari[1]

[1] Center for Leather, Rubber and Plastics, Ministry of Industry, 9 Sokonandi Road, Yogyakarta, 55166, Indonesia

hestyeka@kemenperin.go.id

Abstract. Acrylonitrile butadiene styrene rubber (NBR) has been widely applied in the automotive industry. However, unfilled NBR has poor properties. The purpose of this publication was to study the effects of carbon black (CB) variations on the thermal degradation and swelling resistance of NBR blends. The CB used in this study were HAF, SRF, and the combination of HAF and SRF. NBR samples with different carbon black fillers were prepared by a two-roll mill. The thermal stability and swelling resistance were studied by TGA (heated in oxygen with heating rates of 10, 15, and 20 °C/min at 30-600 °C) and immersed by IRM 903 for 72 h. The degradation of NBR composites occurs in two stages. The kinetic parameters of the degradation process were studied by applying Coats and Redfern method. The filled NBR composites have better thermal and swelling resistance than unfilled NBR composite. The swelling percentage and percentage of mass decomposed at 600 °C were 6.3% and 53.34% for HAF-composite, 6.9% and 56.05% for SRF-composite, and 2.1% and 57.17% for HAF/SRF-composite. Unfilled NBR gives swelling percentage and percentage of mass decomposed of 27.1% and 78.56%. HAF is the proper filler for NBR composite to reach a good thermal and swelling resistance.

1. Introduction

Acrylonitrile butadiene rubber (NBR) is a synthetic rubber that is mostly used in automotive applications. It has a generally excellent resistance towards oil, abrasion, water, and alcohols. However, it has poor resistance to ozone because the butadiene has an unsaturated chain backbone. Unfilled NBR vulcanizate has poor mechanical properties because of the absence of strain-induced crystallization [1]. The addition of filler can improve the properties of elastomers significantly [2, 3]. One of the most commonly used filler is carbon black (CB) that plays an important role as reinforcing filler. This conductive composite material is a low cost and has high flexibility and also good mechanical properties [4-7].

As a filler, CB has various types. Based on its manufacturing process, the common CB are furnace black that has a size from 8 nanometers and thermal blacks that has a size from 200 nanometers. The common CB in the market divided into some type based on its particle size, i.e. N110 SAF (Super Abrasion Furnace), N220 ISAF (Intermediate Super Abrasion Furnace), N330 HAF (High Abrasion Furnace), N550 FEF (Fast Extrusion Furnace), N660 GPF (General Purpose Furnace), N774 SRF (Semi Reinforcing Furnace), and N990 MT (Medium Thermal) with particle sizes of 11-19, 20-25, 26-35, 40-48, 49-60, 61-100, and 200-500 nm. The different types of CB give different characteristics and CB-rubber. The important properties of CB are the structure, the particle size, and the surface activity of

CB. The lower particle size of CB, the better dispersion in the rubber molecules. The structure and surface activity of CB also related with the interaction between rubber molecules and filler [8-10].

The use of rubber products in a certain time can decrease the properties of the rubber material called degradation. The degradation of rubber material is caused by time and temperature. The rate of degradation depends on the polymer type, storage, and the application. Oil-resistant polymeric materials are also essential for the automotive application and industry. The degradation and swelling resistance analyses can give valuable information about the performance of materials and the suitable conditions of using and storage [11, 12].

Some related study about NBR composite had been done by some researcher. Khaled studied the effect of curing system on mechanic and psycho-chemical of NBR [2]. Nabil *et al.* studied the effect curing system on the thermogravimetry and swelling of NBR [13]. De Sousa *et al.* studied NBR with different clays on morphology and mechanical properties [14]. Jovanovic *et al.* found that 80 phr HAF gave the best aging resistance on NBR composite [5]. Salehi *et al.* found that HAF can improve thermal stability in fluoroelastomer composites, but not any differences with GPF and MT [9]. Manoj *et al.* found that ISAF carbon black give the better mechanical and swelling resistance than HAF and SRF on EPDM/NBR blend [15]. It is known that each type of CB has the different effect on the elastomer. However, NBR reinforcement by different types of CB has not yet been studied in detail, especially in the thermal degradation and swelling properties. In this study, the thermal decomposition and swelling of NBR blends will be studied by using thermogravimetric analysis (TGA) and by immersed the NBR composites in IRM 903 oil. The purpose of this publication was to study the effect of CB variations on thermal degradation and swelling resistance on NBR composites. So that it can optimize the use of fillers to get the required product. The fillers used were HAF (High Abrasion Furnace)/ CB N330, SRF (Semi-reinforcing Furnace)/ N774, and the combination of HAF and SRF. The degradation and swelling resistance analyses are very important to know the suitable conditions of using and storage. This is the second publication from the previous study by Yuniari *et al.* [16]. Previous publication evaluated the kinetic vulcanization and the mechanical properties of NBR composites.

2. Experimental method

2.1. Material

NBR composites were made with the formulations described in the previous publication [16] with NBR kraynac 4975 F, ML (1+4) 100 °C (75±5), ACN content of 48,5 ± 1,5%, carbon black HAF (OCI) with a particle size of 26-30 nm and SRF (OCI) with a particle size of 60-70 nm, paraffin wax (Antilux 654 A), paraffinic oil (Indrasari), zinc oxide (ZnO, Indoxide), stearic acid (Aflux 42M Rhein Chemie), 2,2,4-trimethyl-1,2-dihydroquinoline (TMQ, Kemai), n-(1,3-dimethylbutyl)-n'-phenyl-p-phenylenediamine (6PPD, Starchem), thiuram disulfide (TMTD) and 2,2-dithiobis(benzothiazole) (MBTS) (Shandong), and sulfur (Miwon).

2.2. Method

The composites were made according to the method and formulations described in the prior publication [16]. The ingredients for NBR compound as shown in table 1 were accurately weighed and mixed by a laboratory-size two-roll mill. Compounds were mixed by the sequence: NBR, ZnO and stearic acid, paraffinic oil and CB, then 6PPD, TMQ and paraffin wax, MBTS and TMTD, and the last is sulfur. After reach a homogeneous compound, approximately 25 min, the compound were stored in 23±2 °C for 24 h. Then, the curing time of the compound was tested by rheometer (Gotech 3000 A). The compounds were vulcanized in a hydraulic press (Toyoseiki A-652) at 160 °C with a pressure of 150 kg/cm^2 with time-based on the rheometer.

Table 1. NBR compound formulation.

Material	Amount, phr (per hundred rubber)			
	NBR1	NBR2	NBR3	NBR4
NBR	100	100	100	100
ZnO	5	5	5	5
Stearic acid	1	1	1	1
Paraffinic oil	10	10	10	10
HAF	0	40	0	20
SRF	0	0	40	20
6 PPD	2	2	2	2
TMQ	2	2	2	2
Paraffin wax	0.5	0.5	0.5	0.5
MBTS	1.5	1.5	1.5	1.5
TMTD	0.5	0.5	0.5	0.5
Sulfur	1.5	1.5	1.5	1.5

2.3. Thermogravimetric study

The thermogravimetric analysis was observed using Shimadzu DTG 60. The oxygen gas flowed at a rate of 30 ml/min at 30-600 °C using an alumina pan with 5-8 mg of NBR powder sample with a heating rate of 5, 10, and 15 °C/minute.

2.3.1. Kinetic parameter. This study used the Redfern Coats equation to calculate the kinetic parameters as some previous study [17,18]. TG is determined by measurement of degradation rate and fractional mass loss according to temperature change. The fraction of conversion x is defined as:

$$x = \frac{W_0 - W_t}{W_0 - W_f} \tag{1}$$

where W_t is the actual mass at any degradation time, W_0 is the initial mass, and W_f is the final mass at the end of the thermal degradation process. The rate of degradation, $\frac{dx}{dt}$, can be expressed as the product of the function of temperature and the function of conversion:

$$\frac{dx}{dt} = k(T)\, f(x) \tag{2}$$

where $k(T)$ is the rate constant and $f(x)$ is the reaction model which describes the dependence of reaction rate on the extent of reactions. If thermal degradation behavior follows the Arrhenius equation, then $k(T)$ is expressed as:

$$k = Aexp(-\frac{E}{RT}) \tag{3}$$

$$f(x) = 1 - x \tag{4}$$

For constant heating rate, $\beta = \frac{dT}{dt}$, equation (2) can be arranged:

$$\frac{dx}{dT} = \frac{A}{\beta}(1 - x)\exp(-\frac{E}{RT}) \tag{5}$$

An integration from equation (5) leads to the equation (6)

$$\ln\left[-\frac{\ln(1-x)}{T^2}\right] = \ln\left[\frac{AR}{\beta E}\left(1-\frac{2RT}{E}\right)\right] - \frac{E}{RT} \quad (6)$$

In equation (6), $\ln\left[-\frac{ln(1-x)}{T^2}\right]$ is linear with $\frac{1}{T}$ with the slope as $-\frac{E}{R}$ to calculate the activation energy. Whereas intercepts can be used to calculate the pre-exponential factor. In this study, it is assumed to take a first order reaction.

2.4. *Swelling study*

Swelling test was performed for 72 h at 23±2 °C according to ISO 1817 by immersed the NBR composites in IRM 903. The lower swelling percentage indicates the higher solvent resistance of the composite. The change in mass was calculated by equation (7).

$$Swelling\ percentage\ (\%) = \frac{(w_1 - w_0)}{w_0} x\ 100 \quad (7)$$

Where w_1 is the mass of test piece after immersion and w_0 is the mass of test piece before immersion.

3. Results and discussion

3.1. *Thermogravimetry*

Thermogravimetry analysis (TGA) is one of the common methods for thermal analysis that show a function of decreasing mass to temperature. TGA measures the rate of mass changed as a function of time or temperature. The thermogravimetric curve with different heating rate can be shown in Figure 1.

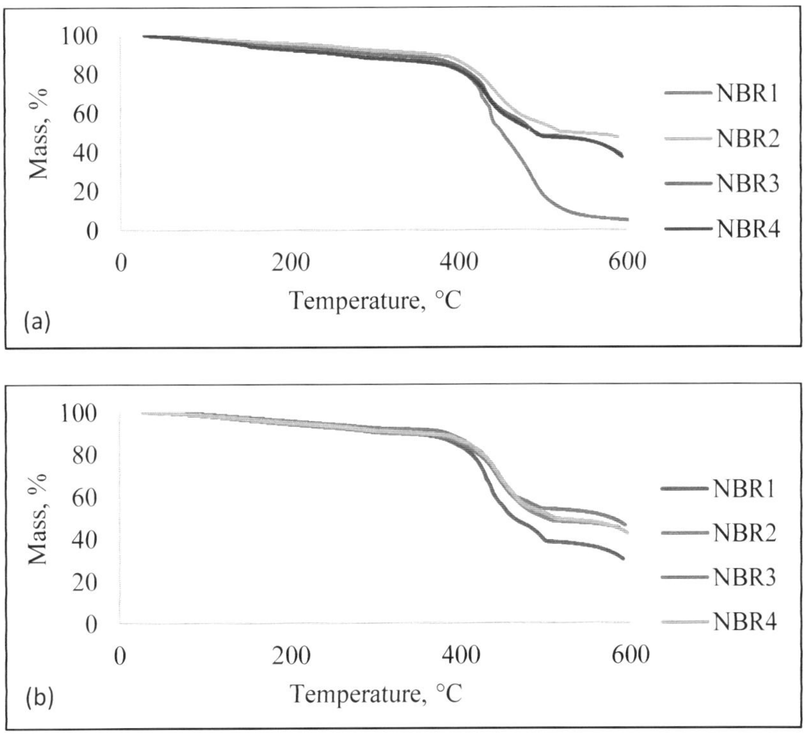

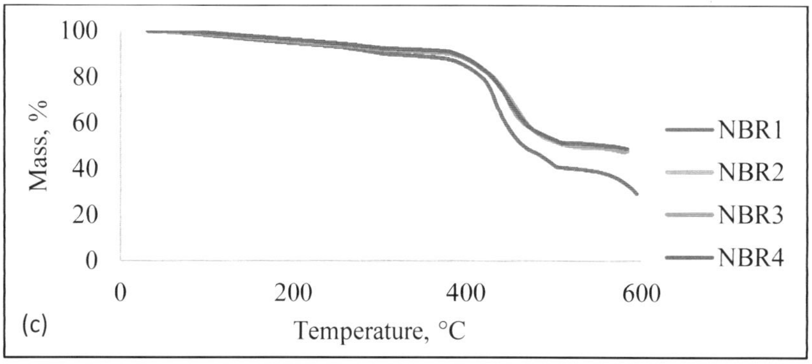

Figure 1. Thermogravimetric curves of NBR composites with the heating rate of (a) 10 °C (b) 15 °C (c) 20 °C.

Figure 1 shows that NBR1 give the lowest mass at 600 °C for all heating rate. At that temperature, the NBR composites haven't been perfectly degraded, The mass graph continues to decrease that means the burning temperature (BT) was not reached. BT is the temperature at which the mass is constant at the end of the combustion stage. At burning temperature, there has been a perfect decomposition, no more mass degradation occurs.

The degradation of NBR composites in this study occurs in two stages. The initial stage of NBR is dehydration of water content and releasing the volatile substance such as DCP, MAH (boiling point 110, 200 °C) at up to 100°C. The first stage occurs at 362 – 384 °C until 458 – 501 °C. This stage is the releasing step of non-rubber substances such as process oils, stearic acid, TMQ, TMTD, and MBTS (boiling point 302, 361, 320, 307, and 305 °C).

Table 2. Degradation of NBR composites.

Composite	Heating rate, °C/min	Stage I		Stage II	Mass decomposed at 600 °C, %
		Temp., °C	Peak temperature, °C	Peak temperature, °C	
NBR1	10	362 – 458	425	483	94.86
	15	367 – 475	432	580	69.87
	20	375 – 488	433	581	70.94
NBR2	10	376 – 489	430	563	52.75
	15	376 – 493	434	564	54.97
	20	378 – 512	456	571	52.31
NBR3	10	379 – 474	429	578	61.46
	15	380 – 486	444	579	53.85
	20	384 – 498	444	554	52.85
NBR4	10	367 – 462	428	583	62.86
	15	383 – 487	424	580	57.44
	20	383 – 501	440	569	51.19

The second stage is the process of NBR degradation. The second stage occurs at temperatures from 459 – 512 °C to more than 600 °C. At this stage, the chemical bonds of NBR start to break and the thermal resistance decrease. The degradation of NBR composites happens until more than 600 °C. At the 600 °C, the mass decomposed of filled NBR composites reach 51.19 – 63.47 %. Unfilled NBR gives higher mass decomposed as 69.87 – 94.86 %. It indicates that the filler provides good thermal resistance to NBR composites. The stages that occur in thermal degradation of NBR composites can be seen in Table 2.

Table 2 shows that for overall composites, the higher heating rate, the higher starting degradation temperature, and also the higher peak in each stage. Increasing the heating rate makes the range of temperatures increases, which may have been because of the limitations of the heat transfer. It is accordance with studied by the previous researcher [7, 8]. The increase in heating rate does not affect the degraded mass, this is in accordance with the research by Abed *et al.* [7].

The degradation temperature at first stage has no significant difference in overall composites except unfilled NBR. Unfilled NBR provides the lowest initial degradation temperature (362 °C). This is in accordance with research by Alneamah *et al.* [12] that NBR without filler is degraded at lower temperatures. The increasing of thermal resistance strongly depends on the interfacial adhesion between the filler-rubber as a result of the compatibilization [11].

Unfilled NBR provides the greatest decomposing mass (78.56%), followed by NBR4 (57.16%), NBR3 (56.05%), and NBR2 (53.34%). CB as a reinforcing filler provide high crosslink density and improve the thermal resistance. CB also provide more compatible composite than unfilled composite. This shows that CB plays an important role improving the thermal stability. HAF-composite gives the lowest mass decomposed at 600 °C. HAF has the lower particle size than SRF, this makes the HAF more easily dispersed on NBR thus providing better thermal resistance than SRF. Salehi *et al.* also get the result that CB with the lower structure has better thermal stability than CB with higher structure [3]. The higher structure shows the higher number of particle per aggregate.

CB can be a good reinforcement if the particle of CB is well dispersed in the rubber matrix. But in a certain amount of CB, it can decrease some of its mechanical properties because of the agglomeration and the poor dispersion. This agglomeration and poor dispersion may happen in HAF/SRF-composite so that the mass decomposed higher than HAF- and SRF-composite.

3.1.1. Kinetic parameter. TG data can be used to calculate the activation energy and pre-exponential factor to get the optimum design to make the best product. The kinetic parameters in stage 1 can be plotted from equation (6) and the activation energy (Ea) and pre-exponential factor can be calculated. The kinetic parameters of NBR composites in stage 1 are shown in Table 3.

Table 3. Kinetic parameters of NBR composites in stage I.

Composite	Heating rate, °C/min	Ea, kJ/mol	Average Ea, kJ/mol	A, min^{-1}	Average A, min^{-1}	R^2
NBR1	10	68.31	80.81	0.195	2.351	0.9599
	15	86.66		3.379		0.9780
	20	87.45		3.480		0.9695
NBR2	10	76.55	80.60	0.599	1.452	0.9879
	15	79.21		0.995		0.9797
	20	86.04		2.761		0.9843
NBR3	10	72.76	80.01	0.317	1.513	0.9772
	15	80.93		1.237		0.9747
	20	86.34		2.985		0.9804
NBR4	10	62.09	72.98	0.055	1.775	0.9663
	15	67.40		0.115		0.9332
	20	89.43		5.157		0.9876

From the Table 3, it is known that the correlation coefficient (R^2) at all heating rate for all composites entirely close to 1, this proves that the activation energy can be approached exactly by first-order reaction. Table 3 shows that the value of activation energy is positive. It shows that there is no transition phase at the selected temperature. Activation energy indicates the amount of energy required to start the degradation and break the chemical bond. Table 3 shows that the Ea value in stage I is NBR1> NBR2> NBR3> NBR4. In this study, the activation energy and pre-exponential factors that can be calculated just in stage I. It can't represent the function of CB as reinforcing filler in NBR, because at the first stage just step of releasing the non-rubber materials. While the NBR degradation is happening in the second stage, but in this study, the kinetic parameters of the second stage can't be calculated because the degradation is not yet done at 600 °C.

Activation energy may change depends on the composite's composition and the heating rate. Differences in the determination of the heating rate also affect the calculation of Ea. From Table 3, the higher the heating rate, the higher the Ea. This result suitable with previous research by Apay-dinvarol *et al.* [19]. Therefore, it is preferred to use low-speed heating. But, it also depends on the speed of the reaction and the type of material [13].

3.2. Swelling

Swelling occurs because there are the free volume to facilitate the mass transfer of solvent. The swelling percentage was investigated at room temperature by IRM uptake until equilibrium swelling was reached. Swelling percentage of NBR composites from equation (7) are shown in Table 4.

Table 4. Swelling percentage of NBR composites.

Composite	Swelling percentage, %
NBR 1	27.1
NBR 2	6.3
NBR 3	6.9
NBR 4	2.1

The lower swelling percentage indicated a high crosslink density [17]. Table 4 shows that unfilled NBR has the highest swelling percentage. The filled composites have higher crosslink in the composite's matrix than unfilled composite. The presence of CB proves the better interfacial interaction in NBR composites. The enhancement of interaction between NBR and filler increasing the formation of crosslink density. CB makes higher rubber network, the polymer chain is trapped or entangled in the microstructure and the porosity of CB. So, the rubber chains become highly immobilized. In addition, the high crosslink density is due to the chemical bonds from the presence of electron acceptor on the surface where macroradical rubber is formed because of the degradation during the vulcanization process [18].

HAF give better swelling resistance than SRF because HAF has a smaller particle size, that means the larger surface area so that the physical crosslink climbs up. Fillers with higher surface area provide a higher reinforcing effect on rubber [3, 6, 20]. Combination HAF/SRF give the best swelling density.

4. Conclusion

From the result, unfilled NBR gives the lowest initial degradation temperature (362 °C) and the highest decompose mass (78.56%) at 600 °C. NBR unfilled also give the highest swelling percentage (27.1%). CB makes interaction with rubber thus reinforcing the composites. HAF with lower particle size than SRF has better filler interaction with NBR. The CB filler improves the thermal stability and swelling resistance. The low-speed heating rate is proper for TG analysis. HAF give better thermal stability and swelling resistance because of the better particle interaction due to its smaller particle than SRF. HAF is proper filler for NBR to reach the best thermal stability and good swelling resistance, especially in automotive application. The TG analysis should be done at a temperature above 600 °C. Because at this temperature, the burning temperature of NBR composites hasn't been reached yet. This information is valuable to get the efficient production process and also knowing the suitable storage and application conditions.

Acknowledgment
The authors would like to acknowledge the Center of Leather, Rubber, and Plastics, Ministry of Industry, Indonesia for financial support.

References

[1] Kantala C Wimolmala E Sirisinha C and Sombatsompop N 2009 *Polym. Adv. Tech.* **20** 448–58
[2] El-Nemr K F 2011 *Mater. Des.* 2016 **32** 3361–69
[3] Ismail H Galpaya D and Ahmad Z 2009 *Polym. Plast. Tech.. Engin.* 2009 **48** 440–5.
[4] Jankovic B Marinovic-Cincovic M Jovanovic V Samarzija-Jovanovic S and Markovic G 2012 *Thermochim. Acta.* **543** 304–12
[5] Jovanović V Budinski-Simendić J Samardžija-Jovanović S Marković G and Marinović-Cincović M 2009 *Chem. Ind. Chem. Engin. Q.* **15(4)** 283–9
[6] Mostafa A Abouel-Kasem A Bayoumi M R and El-Sebaie M G 2009 *Mater. Des.* **30(5)** 1561–8
[7] Samarz S Budinski-simendic J Markovic G and Marinovic M 2013 *J. Compos.* **45** 333–40
[8] Samarzija-Jovanovic S Jovanovic V Markovic G and Marinovic-Cincovic M 2009 *J. Therm. Anal. Calorim.* **98** 275–83
[9] Salehi M M Khalkhali T Soleyman R and Barari M 2016 *Eur. J. Adv. Eng. Technol.* **3(2)** 27–30
[10] Pal K Rajasekar R Jin D Xiu Z Pal S K Das C K and Kim J K 2010 *Mater. Des.* **31(3)** 1156–64
[11] Nabil H Ismail H and Azura A R 2014 *Mater. Des.* **53** 651–61
[12] Alneamah M and Almaamori M 2015 *Int. J. Mater. Chem.* **5(1)** 1–3
[13] Jiang X Xu C Wang Y Chen Y 2015 *J. Macromol. Sci.* B **54** 58–70
[14] De-Sousa F D B Mantovani G L and Scuracchio C H 2011 *Polym. Test.* **30(8)** 819–25
[15] Manoj K C Kumari P and Unnikrishnan 2010 *J. Appl. Polym. Sci.* **120(5)** 2654–62
[16] Yuniari A Setyorini I and Mayasari H E 2016 *Leather, rubb and Plast. Mag.* 32(2) 117–24 (In Indonesia)
[17] Mayasari H E and Yuniari A 2016 *Leather, rubb and Plast. Mag.* 32(2) 125–34 (In Indonesia)
[18] Kahrizsangi R E and Abbasi M H 2008 *Trans. Nonferrous Met. Soc. China* **18(1)** 217–21
[19] Apaydin-varol E Polat S and Putun A E 2014 *Therm. Sci.* **18(3)** 833–42
[20] Li Z H Zhang J and Chen S J 2008 *Express Polym. Lett.* **2(10)** 695–704

IOP Conf. Series: Materials Science and Engineering **432** (2018) 012042 doi:10.1088/1757-899X/432/1/012042

Characterization of starch degradation during simple heating for bioethanol production from the avocado seed

C D Pratywi[1], S Marantika[1], P Dwijananti[1], Masturi[1*]

[1] Physics Department, Faculty of Mathematics and Natural Sciences, Semarang State University, Indonesia

masturi@mail.unnes.ac.id

Abstract. We have successfully identified starch degradation with the presence of glucose from the avocado seed powder by simple heating method. This work was done by varying the composition of the avocado seed powder to obtain the effect of weight fraction increase in improving the concentration of glucose. From the Luff Schoorl test method, we found that the optimal glucose concentrations reached up to 14.86% that was produced from 80 g of the avocado seed powder. Furthermore, SEM showed the crystalline structure of glucose content and swelling of granules after heat treatment. FTIR spectra also indicated the presence of glucose at 1079.6 cm^{-1} (C-O-H bending) and 1412.38 cm^{-1} (CH_2 bending modes).

1. Introduction

Bioethanol is one of the most promising alternatives to fossil fuels, which can be produced from various renewable sources rich in carbohydrates. The raw material of bioethanol makers were divided into three groups: sucrose, starchy, cellulosic fibrous (lignocelluloses) materials [1]. Meanwhile, the starch content of the avocado seed reached 85.3% can be used as raw material for bioethanol [2]. Since 2015 Indonesia has availability of the avocado seed as bioethanol source reached 76,516.40 tones with 6,208.60 tones in Central Java. In addition to being economical value, processing of the avocado seed as bioethanol will also reduce 15,303.28 tonnes of the organic waste in national region.

However, conversion of starch into glucose varies considerably depending on the process of starch degradation [3]. In this process, starch granules become hydrated, swell, and release glucose as fermentable sugar [4, 5]. In general, the degradation of complex organic compounds in starch can be done either by hydrolysis acidic or enzymatic methods [6]. During hydrolysis of starch, the amylose and amylopectin molecules undergo biochemical conversion to produce soluble sugar. Although acid hydrolysis is an effective way of generating fermentable sugars from starch, the necessity for acid recovery after hydrolysis has made it unattractive. Moreover, acid hydrolysis is often associated with the production of undesirable by-products that are toxic to microbial cells and inhibit the growth of yeast, thus affecting the overall yield of ethanol fermentation [7].

Furthermore, the pH of the solution is an important factor having significant effect on the overall process of bioethanol production [8]. Most recently, we revealed that simple heating can be used to release of glucose from starch granules without affecting pH solution [9]. Also, Wang *et al.* revealed under hydrothermal treatment, starch granules swell and lose their crystallinity and molecular organization in a process known as gelatinization [10]. Based on these backgrounds, in this work we developed and characterized starch degradation during simple heating for bioethanol production from avocado seed. We prepared the process with weight fraction variation of avocado seed powder and evaluated correlations their physicochemical and morphological properties to provide a useful information about impact of subsequent reaserch treatment parameters.

2. Experimental method

2.1. Preparation of raw material

The avocado seed waste collected from local fruits juice stalls around campus area UNNES Sekaran, Gunungpati. First, the seed was washed with plain water and dried under the sun for a week to remove excess moisture. Then, the coat of the seed was peeled to remove the outer layer. After that, the peeled seed was shredded by vegetable grater, followed by drying using microwive at 100 °C for 30 minutes and grinding until it become granulated-powdery the avocado seed.

2.2. Starch degradation by simple heating

In this work, starch degradation was done by varying weight compotition of the avocado seed powder in range 20-100 g. The powder was mixed with 200 mL of water and heated using magnetic stirer at 60 °C for 60 minutes. Then, mash of the avocad seed powder was cooled down to room temperature and filtrated to remove the slag.

2.3. Analysis and characterization of starch degradation

The filtrated solution was analyzed using Luff Schoorl method to obtain the effect of weight fraction addition in improving the concentration of glucose. In other hand, characterization of physicochemical and morphological properties before and after treatment by simple heating was analyzed using the Fourier-Transform Infrared Spectroscopy (FTIR) and the Scanning Electron Microscopy (SEM).

3. Results and discussion

SEM characterization result are shown in Figure 1 that explained the morphological effect of simple heating treatment. Figure 1 (a) shows the morphological image of the avocado seed powder before simple heating and it is obtained that the surface is composed by small granules. Otherwise, the avocado seed powder after treatment have the bigger granules (Figure 1 (b)). Both of the granules are displayed in spherical shape with smooth surface [11]. However, this change indicate that the starch granules are swell due to the water adsorption. Previous studies have shown that starch release glucose while the water adsorbed by starch [12]. It is confirmed in Figure 1 (b), where the glucose of starch perform as the reducing sugar in crystals shape [13]. An efficient starch degradation system to release the glucose also analyzed by FTIR.

Figure 1. SEM micrographs of (a) the avocado seed powder and (b) the simple heating residue.

Figure 2 shows spectral data of the avocado seed powder, residue, and filtrate after simple heating process by FTIR. Both spectral data of the avocado seed powder and treatment residue have the nearly regions in different transmittance. This is due to the content of the avocado seed powder reduced after hydrolysis but it still similar contents on the functional, chemical, and structural poperties of the starch and other. In this work, investigation of starch was performed by four main regions that can be interpretation and characterization of the key bands. This regions were as follows : below 800 cm^{-1}, 800-1500 cm^{-1} (the fingerprint region), 2800-3000 cm^{-1} (C—H stretch region), and the region between 3000-3600 cm^{-1} (O—H stretch region).

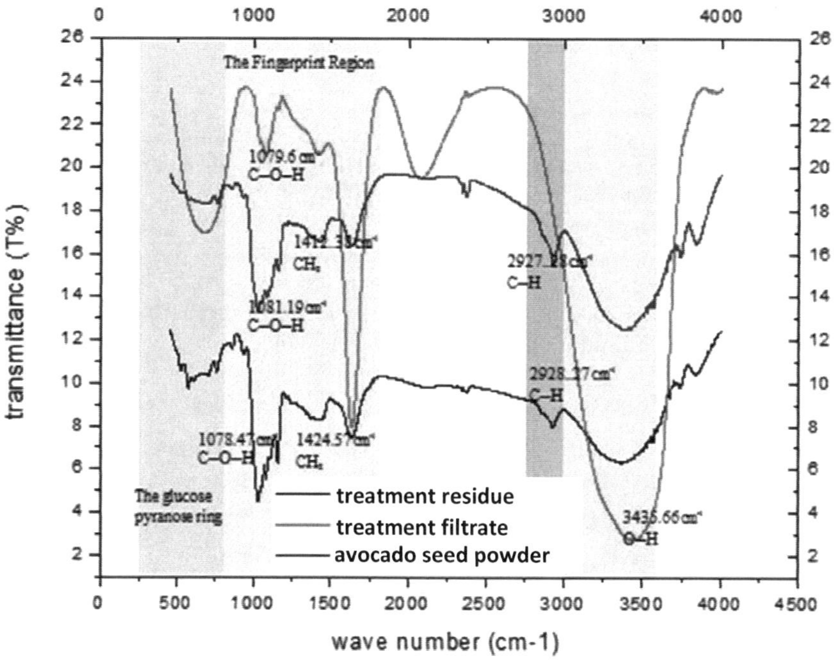

Figure 2. FTIR spectra of hydrolysis residue, hidrolysis filtrate, and avocado seed powder.

The complex vibrational modes at below 800 cm^{-1}region are the skeletal mode vibrations of the glucose pyranose ring [14]. This region of the avocado seed powder was displayed at 684.93 cm^{-1}, while the treatment residue was showed at 528.04 cm^{-1}, 576.33 cm^{-1}, and 762.97 cm^{-1}. This change shows that there is chemical modification or rupture of chemical bands in compounds such as the pyran ring or guaiacyl ring [15].

The region between 800-1500 cm^{-1} provides highly overlapping having complex spectra with difficult assignment. In this regions, the spectra of the avocado seed powder has the FTIR absorption band at 1026.71 cm^{-1} is attributed to the C(1)—H banding modes,and in other band at 1081.19 cm^{-1} is identified C—O—H bending. The emerging of 1155.94 cm^{-1} is due to coupling modes of C—O and C—C stretching, and the infrared band at 1427.83 cm^{-1}originated from CH_2 bending modes.

A band at 936.97 cm^{-1} reported by spectral of treatment residue that indicate the C—O—C vibration of the 3,6 anhydrogalactose bridge and the band also obtained to the glycosidic linkages in starches [16]. However, this band just occurs in spectra of treatment residue because there is derformation in chemical bands after simple heating [17]. In other hand, the spectra of treatment residue between 800-1500 cm^{-1} has the same of band assignment with the spectra of the avocado seed powder in different absorption band. The C(1)—H banding mode was observed at the band of 1026.97 cm^{-1} and the other band assignment could be observed at 1078.47 cm^{-1}, 1153.29 cm^{-1}, 1424,57 cm^{-1}.

The water adsorbs of the avocado seed powder and the treatment residue in the amorphous region of starch is identified as a board infrared band with peak of 1632.46 cm^{-1} and 1637.47 cm^{-1}. Both peaks indicate the water molecules vibration adsorbed in the non crystalline region of starch [14]. The region between 2800-3000 cm^{-1} in the avocado seed powder and the treatment residue was obtained at 2927.28 cm^{-1} and 2928.27 cm^{-1}. This region shows C—H streching mode observed the variations of amylose and amylopectin in the starch by the change of transmittance [18]. The O—H stretching mode of starch in the avocado seed powder and the treatment residue was obtained at 3394.41 cm^{-1}, 3567.76 cm^{-1}, 3365.38 cm^{-1}, 3525.8 cm^{-1}, and 3567.89 cm^{-1}.

An interaction between water and starch after simple heating was performed by the spectra of the treatment filtrate at the Figure 2. This is due to several peaks having the similar assignment with the spectra of the avocado seed powder and the treatment residue. This assignment shows the glucose can be bonded with molecule of water.

Furthermore, the characterization of the glucose in aqueous solution of treatment filtrate was attributed at 685.14 cm^{-1} (the glucose pyranose ring), 1079.6 cm^{-1} (C—O—H bending) and 1412.38 cm^{-1} (CH_2 bending modes). This spectra is dominated by water component, there are the O—H stretching mode and water adsorbed at 3435.66 cm^{-1} and 1634 cm^{-1}. The peak at 2078.9 cm^{-1} was obtained a combination of hindered rotation and O-H bending (water) that can explained about the vibration water molecules in aqueous solution [19].

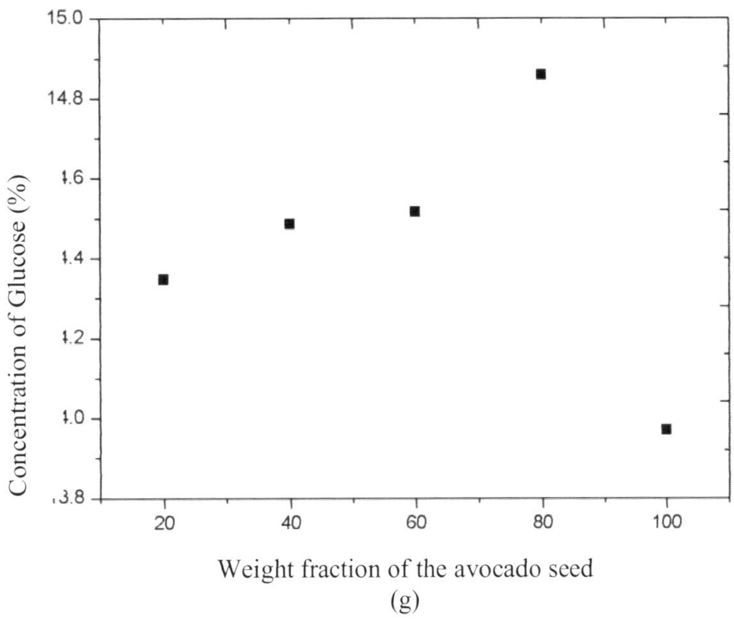

Figure 3. Concentration of glucoseas weight fraction of the avocado seed.

The starch degradation of the avocado seed powder with the water are shown in Figure 3. At 20-80% weight fraction of the avocado seed powder, it was found that the glucose concentration increased with increasing weight fraction. This is due to the increasingly of the starch that can reacted with hydrogen bond in water [20]. After 80%, the glucose concentration remarkably reduces with increasing of weight fraction, since the glucose from the granule starch unrelease into the water during simple heating [21]. The maximum amount of glucose concentration were obtained 14.86% of fermentable sugars. Furthermore, William *et al.* have studied that the starch of plant was a semicrystalline polymer and composed with 20-25% amylose [22]. However, glucose as amylose by the result of simple heating is the only one that can convert to become ethanol because it has, a mostly linear chain, typically consists of up to 3000 glucose molecules interconnected primarily by α-1,4 glycosidic linkages contain a few branched networks [14]. Additionally, the glucose yield concentration are correlated during fermentation. This due to the fact that the glucose such as nutrients of yeast to convert into ethanol [23].

4. Conclusion
Starch degradation by simple heating was a suitable process to produce the glucose for bioethanol production from the avocado seed powder. The measurement with FTIR showed that the glucose yield in aqueous solution at 685.14 cm^{-1} (the glucose pyranose ring), 1079.6 cm^{-1} (C—O—H bending) and 1412.38 cm^{-1} (CH_2 bending modes). Further, SEM images show there are changes of granules starch size due to the simple heating process. The optimal condition of the fermentable sugar was produced by 80g of weight fraction, which a concentration at 14.86% using Luff Schoorl method.

Acknowledgement
We thanks to Ministry of Research, Technology and Higher Education for supporting in finance to this work via Research Grant for fiscal year 2017-2018.

References
[1] Zabed H Sahu J N Suely A Boyce A N and Faruq G 2017 *Renew. Sustain. Energy Rev.* **71** 475–501
[2] Maryam Kasim A and Santosa 2016 *J. Food Sci. Eng.* **6** 32–7
[3] Mete A M Ülgen K Ö Kirdar B Ilsen Ö Z and Oliver S G 2002 *Enzyme Microb. Tech.* **31** 640–7
[4] Beliya E Tiwari S Jadhav S K and Tiwari K L 2013 *Energy Explor. Exploit.* **31** 771–82
[5] Wang S Li C Copeland L Niu Q and Wang S 2015 *Compr. Rev. Food Sci. Food Saf.* **14** 568–85
[6] Gundupalli M P and Bhattacharyya D 2017 *Waste and Biomass Valorization* **0** 1–8
[7] Kim K and Hamdy M K 1985 *Biotechnol. Bioeng.* **XXVII** 316–20
[8] Paulová L Patáková P and Brányik T 2013 *Engineering Aspects of Food Biotechnology* ed J Teixeira and A A Vincente (Taylor & Francis Group) 89–110
[9] Cai C Cai J Zhao L and Wei C 2014 *Food Sci. Biotechnol* **23** 15–22
[10] Wang S and Copeland L *Food Funct.* **4** 1564
[11] Yang J Xie F Wen W Chen L Shang X and Liu P 2016 *Int. J. Biol. Macromol.* **84** 268–74
[12] Ratnayake W S and Jackson D S 2008 *Adv. Food Nutr. Res.* **55** 221–68
[13] Tako M Tamaki Y Teruya T and Takeda Y 2014 *Food Nutr. Sci.* **5** 280–91
[14] Kizil R Irudayaraj J and Seetharaman K 2002 *J. Agric. Food Chem.* **50** 3912–8
[15] De Rossi E Lindino C A Cremonez P A Dos Santos K G Bariccatti R A Antonelli J and Teleken J G 2017 *Manag. Environ. Qual. An Int. J.* **28** 94–106
[16] Sekkal M Declerck C Huvenne J P Legrand P Sombret B and Verdus M C *Microchim. Acta* **112** 11–8
[17] Reyes-rivera J and Terrazas T 2017 *Xylem Methods Protoc. Methods Mol. Biol.* **1544** 46–8
[18] Movasaghi Z Rehman S and Rehman I U 2008 *Appl. Spectrosc. Rev.* **43** 134–79
[19] Sagawa N and Shikata T 2015 *J. Phys. Chem.* **119** 8087–95
[20] Mahawan M A Tenorio M F N and Gomez J A 2015 *Asia Pacific J. Multidiscip. Res.* **3** 34–40
[21] Devarapalli M and Atiyeh H K 2015 *Biofuel Res. J.* **2** 268–80
[22] Brown W and Poon T 2005 *Introduction to Organic Chemistry, 3rd Edition* (Wiley)
[23] Masturi Cristina A Istiana N and Dwijananti P 2017 *J. Renew. Natur. Mater.* **6** 56–60

The effect of $CaCO_3$ filler component on mechanical properties of polypropylene

C Budiyantoro[1], H Sosiati[1], B P Kamiel[1], M L S Fikri[1]

[1] Department of Mechanical Engineering, Universitas Muhammadiyah Yogyakarta, Lingkar Selatan St., Yogyakarta 55183, Indonesia

cahyo_budi@umy.ac.id

Abstract. The presence of filler can reduce the production cost and in certain condition it can also improve the mechanical properties. CaCO3 is one type of inorganic filler that widely used in polypropylene materials, since it can cover those two requirements above. This work aimed to study the influence of CaCO3 filler in tensile properties of polypropylene. CaCO3 was sieved through a screen to obtain uniform particles size and then filled into polypropylene with filler percentage of 5%, 15%, and 25% of weight, respectively. The mixture was shaped into dumbell specimens by using Injection Moulding Process and a universal testing machine was used for tensile testing according to ISO 527. The results showed that the mixture of 5% (w/w) CaCO3 achieved highest tensile strain and elastic modulus but had the lowest tensile strength, and that of 15% (w/w) CaCO3 attained the highest values of strength. The highest tensile properties are 24.9 MPa for tensile strength, 3.96% for tensile strain and 963 MPa for elastic modulus. Interestingly, increasing the filler content to 25% of CaCO3 caused the reduction in strength, strain and elastic modulus of the composite. Variant with 5% of filler content has the highest value of impact strength, around 168 J/cm2, whereas the lowest value of impact strength can be found in the variant with 25% of CaCO3, the hardness value decreased on the 25% CaCO3. This situation is related to the filler distribution and dispersion within the specimen due to mixing effect in injection moulding process. Scanning Electron Microscope (SEM) was used to explain this condition by observing the fractured area of specimens.

1. Introduction

Nowadays, polypropylene (PP) is well known as one of the most common polymeric materials that are used in many application sectors such as household, construction and automotive [1]. In many cases, force and load are often subjected on those products, so the products should have a good mechanical properties to carry the subjected load. To improve mechanical properties and reduce production costs, PP is often reinforced by adding fillers into the matrix. From plastic industries point of view, talc and CaCO3 are attractive filler for thermoplastic especially polypropylene (PP) because they can fulfill those two requirements, but this study was only focus on $CaCO_3$ only. $CaCO_3$ is considered as a common inorganic filler that is widely used incorporated with polypropylene materials [2]. It was observed that the tensile strengths decreased while the modulus increased with the filler content for both talc and boron nitride. The strength could decrease due to the low interfacial bond between the components [3]. This type of filler has a positive effect on the nucleation effect of PP; fine-grade filler enhances strength and stiffness [4]. Moreover, the presence of filler reduces the percentage of resin usage, so that the production cost become lower. Despite the benefit, high filler content may affect the processability, ductility, and strength of composites [5]. It necessary to know the acceptable content of filler in the polymer matrix to reach their optimum performance. When using talc as filler, 30 wt% talc-filled PP was proven to have ultimate strength in many previous studies [6].

In the polymer matrix interface, dispersion, delamination, and distribution of talc have an important effect on the properties of PP/talc composites. Talc without special treatment tends to agglomerate when it is mixed with PP, resulting in bad dispersion. In addition, low compatibility between talc surface and polypropylene matrix leads to bad interfacial adhesion and causes poor mechanical properties of the composite.

In this study, the effect of filler percentage on the tensile properties of PP were investigated. There were three variation of filler percentage added into PP resin (5%, 15% and 25%) in the form of injection-molded specimens. Then, the tensile strength, strain and elastic modulus were observed and compared.

2. Experimental method

A commercial homopolymer polypropylene PP HI10HO from Chandra Asri Petrochemical company was used as the polymer matrix. The polymer material was supplied in granulated form. Some of physical properties value of resin material is written in the Table 1; they may helpful for comparison with the properties of investigated composites.

Table 1. Properties of PP HI10HO (source: http://www.chandra-asri.com).

Properties	Typical Value	Unit
Density	0.9	g/m^3
Melt flow Index (230°C / 2.16 kg)	10	g/10 min
Tensile stress at yield (50 mm/min)	35	MPa
Tensile strain at yield (50 mm/min)	13	%
Flexural modulus (1.3 mm/min)	1500	MPa
Notched impact strength, IZOD (+23°)	30	J/m^2
Hardness Rockwell	90	R-scale
Deflection Temperature (@ 0.455 MPa)	104	°C
Vicat Softening Temperature	152	°C

Considering material availability, a local grinded $CaCO_3$ was used as the filler material. It is normally a hygroscopic material and need to be dried at 30 °C – 55 °C in order to minimize the moisture content. After drying, the calcium carbonate was sieved in a such of manner to split the agglomeration and to get smaller particle size. The filler were mixed together with PP using three variation:

- 1st variation, 5% filler content: 50 g of $CaCO_3$ added into 950 g of PP
- 2nd variation, 15% filler content: 150 g of $CaCO_3$ added into 850 g of PP
- 3rd variation, 25% filler content: 250 g of $CaCO_3$ added into 750 g of PP.

Table 2. Optimum setting paramater of injection molding process.

Parameter	Typical Value	Unit
Injection Pressure	105	Bar
Injection Time	6.5	Second
Holding pressure	90	Bar
Holding time	4	Second
Melting temperature	230	°C
Cooling temperature	35	°C

The dumbell specimens were fabricated by the 70-ton MEIKI Injection Molding machine. The standard of processing procedures were based on ISO 294, whereas the shape of the dumbell specimen followed ISO 527 type 2. The high quality of injection molded specimens are indicated by several points: ideal and uniformly shaped as per drawing, minimum sinkmark and warpage, flatness etc. Those points can be achieved by optimum parameters that obtained from trials. Table 2 shows the optimum setting parameter of injection molding process.

To fulfill the testing standard, at minimum 5 specimens were prepared for each filler variations as can be seen on Figure 1.

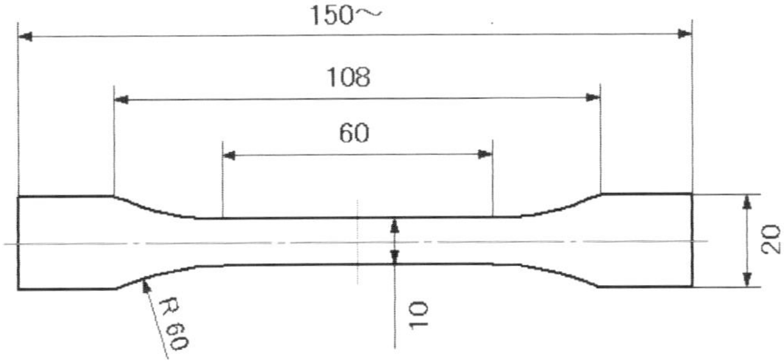

Figure 1. Dumbell specimen according to ISO 527-2 IB as result of injection molding process.

The specimens were tested by using a Zwick Roell Z020 Universal Testing Machine according to ISO 527 to get tensile properties data. Tensile testing was made at speed of 50 mm/min. Some other specimen were cut into standard dimension of test specimen (80 x 10 x 4 mm), as showed on Figure 2, and tested by Charpy Impact testing according to ISO 179 to get impact properties data.

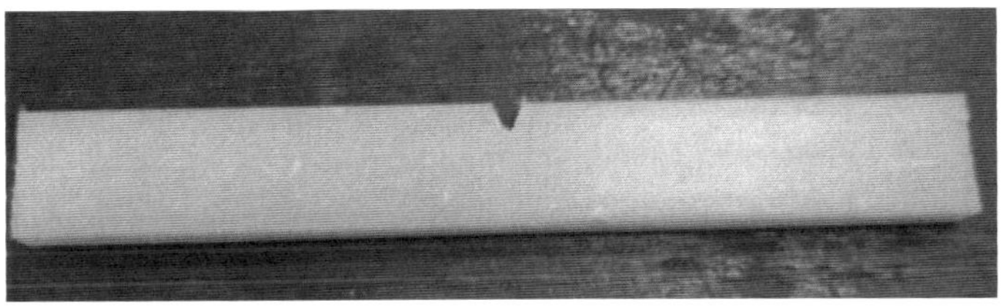

Figure 2. Impact test specimen.

The Charpy impact test with three-points support were performed on notched specimens to evaluate the toughness behaviour of specimen. According to the standard, the support span is 62 mm. The pendulum hammer has nominal impact energy of 4 J and an impact velocity of 2.9 m/s. The sample were notched by angular single lip milling cutter, with the angle of 60° and base radius of 0.25 mm. The Shore D hardness testing according to ASTM D2240 - 15 was also perfomed on the specimen. There were three samples for each variations and for each of them, the indentation was applied. The hardness value of materials was determined by the change of depth of conical indentor when penetrating the surface of the samples. The results were also compared with the hardness of unfilled PP.

Material characterization was visually done by using Scanning Electrone Microscope (SEM); it was used to provide information in microstructures related to mechanical properties. The type of SEM is Hitachi TM3030 plus. Additionally, Energy Dispersive X-Ray Spectroscopy (EDS) was used to get semi-quantitative elemental results on the fractured area. Those two devices were provided by PT. Fajar Mas Murni.

3. Results and discussion

3.1. Tensile test

The tensile test result of PP/$CaCO_3$ with filler content variation of 5%, 15% and 25% gives three mechanical properties such as tensile strength, tensile strain and elastic modulus.

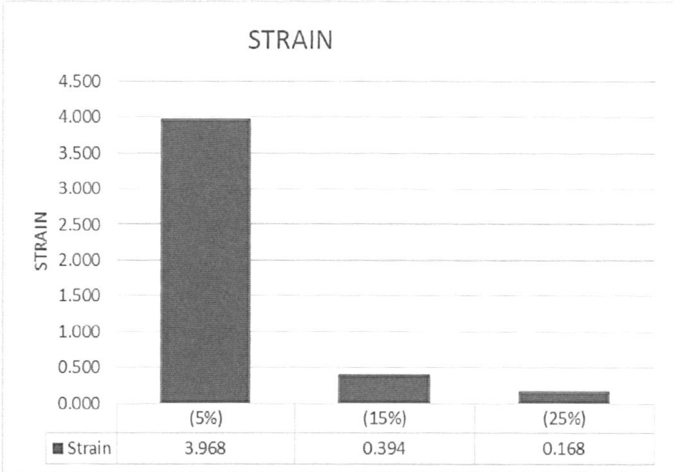

Figure 3. Strain result of three filler content variation.

According to material data, the strain of unfilled PP is 13%, Figure 3 shows that addition of filler reduces the strain of material significantly. Filler content contributes the brittleness of PP/CaCO3 composite material.

Figure 4 shows that the average value of elastic modulus of polypropylene material with filler content of $CaCO_3$ 5% is 963,2 MPa, then with 15% CaCO3 is 388,9 MPa and the lowest elastic modulus is given by 25% $CaCO_3$ which has the value at 333 MPa. It can be said that the addition of $CaCO_3$ to the polypropylene matrix reduce the elastic modulus. The first variant with 5% filler has the largest surface contact area which makes its elastic modulus higher, as described in some previous research that elastic modulus of polymer depending on the filler surface contact area [7].

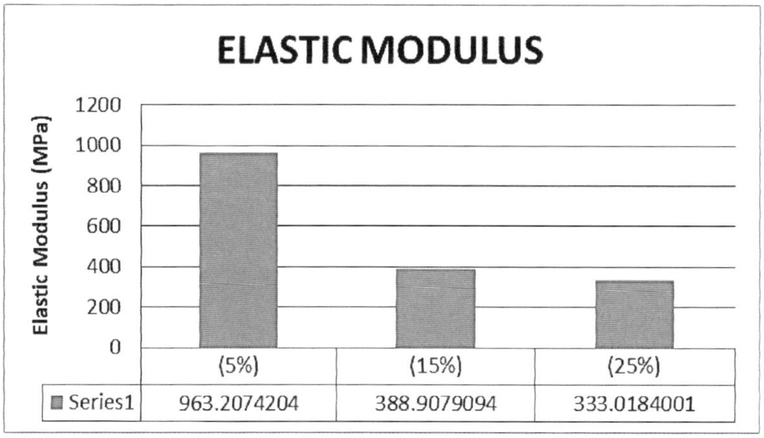

Figure 4. Elastic Modulus of three filler content variation.

Figure 5 explains that the polypropylene material with filler content of $CaCO_3$ 5%, 15% and 25% attained the tensile strength of 21.19 MPa, 24,9 MPa, and 22,13 MPa, respectively. The highest value of yield tensile strength is in $CaCO_3$ 15% filler. Interestingly, the larger contents of $CaCO_3$ (i.e. 25%)

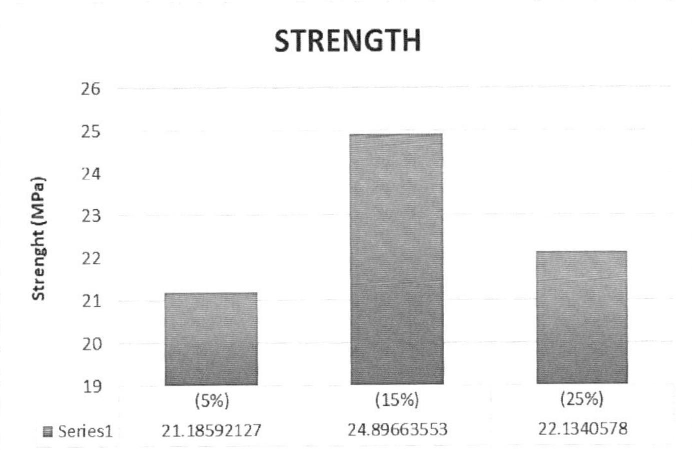

Figure 5. Yield tensile strength results.

did not bring to successive increase in the tensile strength. In contrast to the elastic modulus, yield strength depends on both surface contact area and interfacial strength [8]. Those factors were strongly related to the distribution and dispersion of filler particles contained inside matrix. Distribution is the way the particles fill the matrix, whereas dispersion is the way these particles are agglomerated or not. In order to get clear explanation of this issue, structure investigation was done on the cross section area of fractured specimen for both second and third variant by using Scanning Electron Microscope (SEM). Figure 6 shows that the filler in the second and the third variant are not fully distributed. Moreover, the dispersion on the 2nd variant is better than the 3rd variant. In general, the size of agglomerated filler (white spots) of third variant are bigger than the 2nd variant, so that the the ability to withstand pulling force is reduced, then the yield strength become lower. By increasing the magnification into 100x, as seen in Figure 7, it is clearly seen that the filler powder itself cannot blend properly. Inhomogeneous filler dispersion can lead to a considerable reduction in the mechanical strength of the composites.

Even at a smaller particle size the agglomeration was still occurred (Figure 8), indicating that the sieving process of filler was not optimal and the mixing effect in the barrel as well.

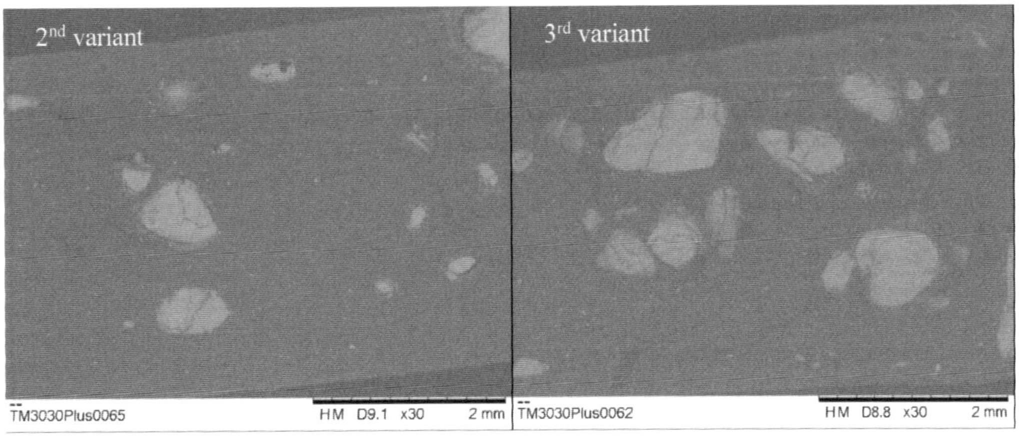

Figure 6. Filler distribution on fractured surface of 2nd variant (15% filler) and 3rd variant (25% filler).

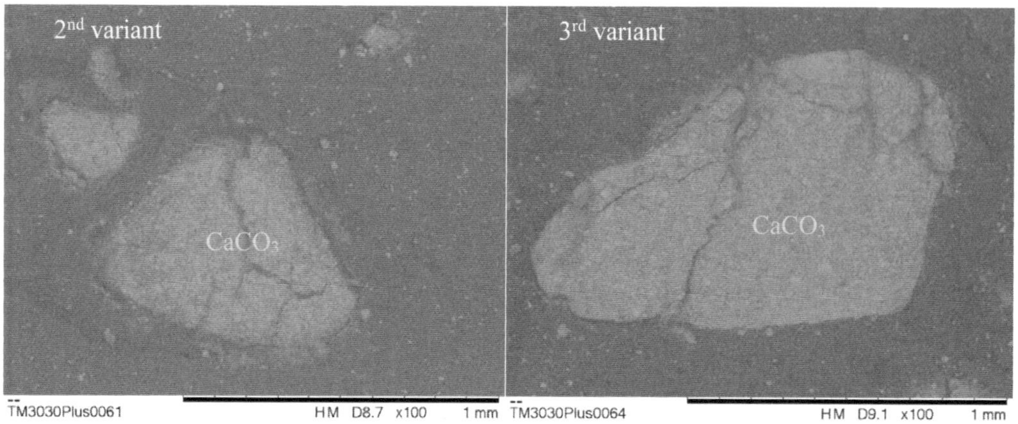

Figure 7. Filler agglomeration on fractured surface of 2nd variant (with 15% filler) and 3rd variant (with 25% filler).

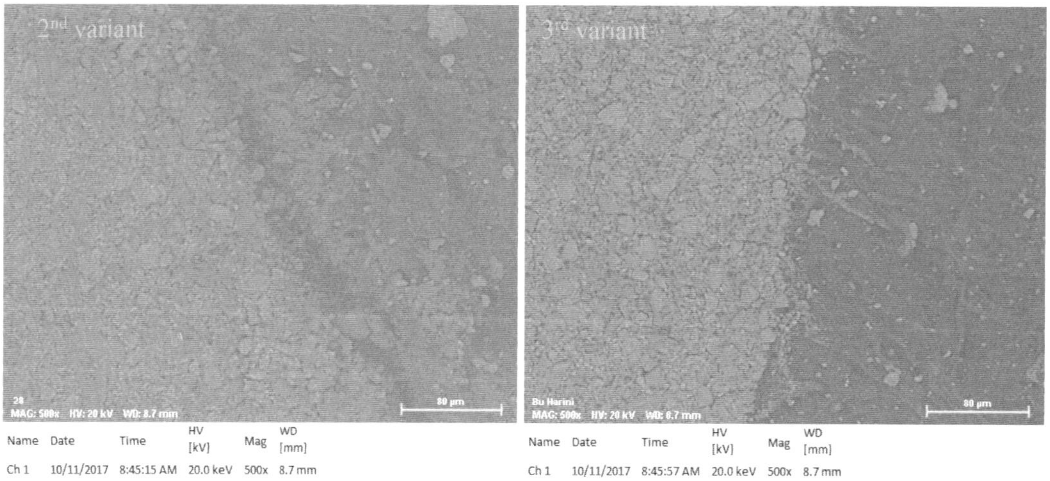

Figure 8. SEM images using 500x magnification.

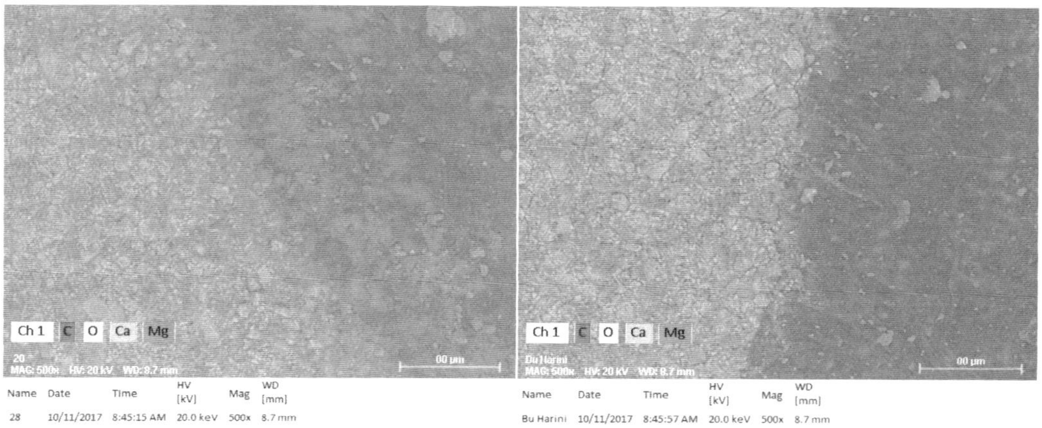

Figure 9. EDS images confirming the elemen of $CaCO_3$.

IOP Conf. Series: Materials Science and Engineering **432** (2018) 012043 doi:10.1088/1757-899X/432/1/012043

As the SEM show compositional contrast that results from the atomic distribution and different atomic number of elements, Energy Dispersive Spectroscopy (EDS) was used to identify the certain elements in the fractured area and their relative proportions. As shown in Figure 9, it is certain that the elements present on the surface of the fractured specimen consist of: C, Ca and O, which are the constituents of $CaCO_3$.

It is clear the dispersion and distribution of filler is strongly affected by two factors: first, powder preparation (sieving and drying) and second mixing effect on the injection molding process. To achieve better filler dispersion and distribution, it is necessary to have higher back pressure and faster screw rotation to get better mixing effect, unfortunately there is no screw rotation adjustment on the MEIKI 70-ton Injection Molding Machine.

3.2. Impact test

Figure 11 shows the relationship between filler content and the value of impact strength. The variant with 5% of filler content has the highest value of impact strength, around 168 J/cm2, whereas the lowest value of impact strength can be found in the variant with 25% of $CaCO_3$. It is understandable that the addition of $CaCO_3$ reduces the impact strength of PP, it become more brittle as shown in Figure 10. This confirmed the results of research conducted by J. Schoene *et. al.* [9].

Figure 10. Brittle Fracture of PP/$CaCO_3$.

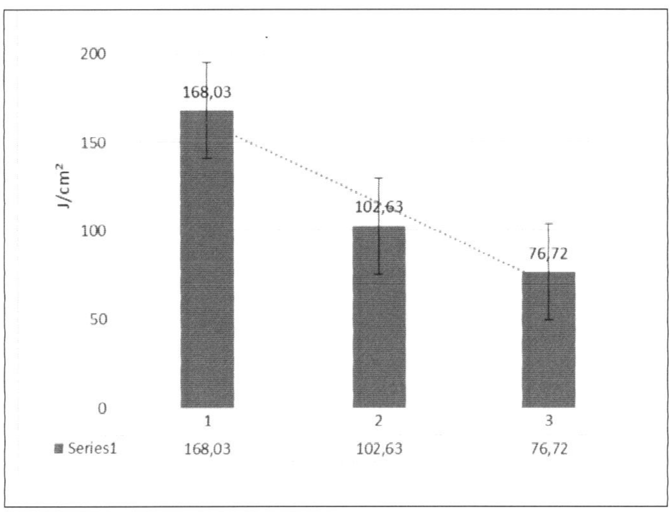

Figure 11. Impact test results.

3.3. Hardness

The hardness values (Shore D) of the filled PP composites with different percentage of $CaCO_3$ are shown in Figure 12. Normally, CaCO3 will reduce the molecular chain activity of the PP matrix, resulting in increasing its stiffness and and hardness correspondingly [10]. Here the situation is similar to the result of yield strength where the hardness value decreases for the 25% $CaCO_3$ content.

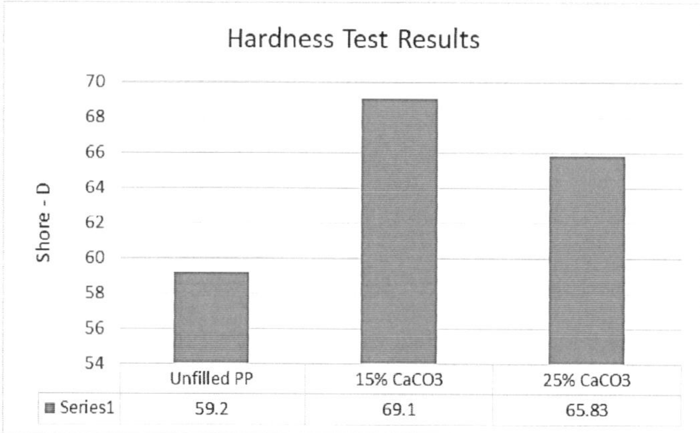

Figure 12. Shore D hardness comparison.

4. Conclusion

An investigation was carried out to study the effect of filler content on the elastic modulus, tensile strength, strain, hardness and impact strength of $CaCO_3$ filled PP. The results can be concluded as follows;

1) Elastic modulus and yield strain were decreased with the increasing of filler content
2) Tensile yield strength seems to be affected by the dispersion of the filler. This situation is related to the mixing effect in injection molding process.
3) Addition of $CaCO_3$ reduces the impact strength of PP.
4) The value of hardness increase with addition of filler but uneven distribution and dispersion impaired this properties.

Acknowledgement

The authors would like to thank Universitas Muhammadiyah Yogyakarta for the financial and facilities support and Mr. Galuh Setyanto Pradana from PT. Fajar Mas Murni for providing SEM and EDS data.

References

[1] Hargitai H Rácz I 2012 *Polymer Composites* 747-90
[2] Dey K Sharmin N Khan R Nahar S Parsons A Rudd C 2011 *J. of Thermoplast. Composite Mater.* **24** 695-711
[3] Hargitai H Török D 2014 *Acta Technica Jaurinensis* **7(4)** 345-55
[4] Castillo L Barbosa S Capiati N 2012 *Society of Plastics Engineers (SPE)*
[5] Rai U Singh R 2004 *Materials Letters* **58(1-2)** 235-40
[6] Leong Y Abu B M Ishak Z Ariffin A Pukanszky B 2004 *J. of App. Polymer Sci.* **91(5)** 3315-26
[7] Laura D Keskkula H Barlow J Paul D 2002 *Polymer* **43(17)** 4673-87
[8] Turcsányi B Pukánszky B Tüdõs F 1988 *J. of Mater. Sci. Lett.* **7(2)** 160-2
[9] Schoene J Kotter I 2012 *J. of Plastic Tech.* **8**
[10] Liang J Tang C Li R Wong T 1988 *Metals and Materials* **4(4)** 616-9

Characteristics of hydroxyapatite from buffalo bone waste synthesized by precipitation method

I Pawarangan[1] and Y Yusuf[1]*

[1] Physics Department, Universitas Gadjah Mada, North Sekip BLS 21 Yogyakarta 55281, Indonesia

*yusril@ugm.ac.id

Abstract. Hydroxyapatite (HA), $Ca_{10}(PO_4)_6(OH)_2$, is the most widely used ceramic biomaterials. It is a mineral found in natural teeth, bone and hard tissue. In this paper, we present a synthesis and characterization for obtaining HA by wet precipitation. This study mainly focuses on the use of bio-waste materials, *i.e.*, buffalo bones to obtain HA. In order to synthesized HA from buffalo bones, we use $Ca(OH)_2$, H_20 and H_3PO_4 as the starting precursors. The material was sintered at temperatures of 650 °C, 850 °C, and 1050 °C for 7 hours to improve its particle size distribution. Identification of functional group using FTIR shows absoption namely at wavenumber values of 3699.20, 3787.92 and 3706.91 cm^{-1} which indicate hydroxyl group (OH). Phosphate band were identified by peak 1022.20 and 1118.63 cm^{-1} and 1215.06 to 1459.31^{-1}cm which indicates carbonate group. Ratio of Ca/P for synthesized HA is about 1.67. XRD pattern results indicated that the synthesized buffalo bone HA closely correspond to commercial HA phase.

1. Introduction

Bones naturally consist of inorganic (70 %), organic compounds (20 %) and water (10 %). HA with the chemical formula $Ca_{10}(PO_4)_6(OH)_2$ is a main organic phase of human hard tissues with structure similar to natural teeth and bone [1]. Additionally, HA is an osteoconductive, non-toxic [2], non-inflammatory and non-immunogenic agent, has bioactive properties and due to these properties, HA becomes a promising candidate material for human hard tissues substitute, and therefore it is extensively used in biomedical applications [3, 4].

The preparation of bio-waste materials is an interesting procedure to produce HA, the advantage is not only because it uses bio-waste, but both enviromental and economic benefits are obtained through waste recovery. In addition, HA generated from bio-waste materials, such as buffalo bones, may be less likely to be rejected by living organs due to its physicochemical similarity to natural human bone [5]. Comparative studies involving the synthetic and the natural bone of HA have been performed [6]. The result indicated that their chemical compositions were similar. The use of bio-waste materials to produce HA powders using various synthesis method has been widely studied using such materials as animal bone, seashells, corals and shells. Buffalo bones is one of the animal bone bio-waste in Indonesia. Studies concerning buffalo bones as a potential HA source material are scarce.

Different methods have been explored for converting materials into HA powder such as solid state reaction, sol-gel synthesis, microwave processing, hydrotermal methods and wet synthesis. The different methods are used to synthesis HA, the different morphologies, grain sizes, stoichiometries and molar Ca/P ratio can be obtained. The wet synthesis is recoqnized to be the most simple, cheap and suitable

IOP Conf. Series: Materials Science and Engineering **432** (2018) 012044 doi:10.1088/1757-899X/432/1/012044

method available [7]. This methods would lead to obtain the type of HA which is characterized by a high specific surface area, small size distribution including small particle agglomeration, high purity and crystallinity. The HA powders are obtained from the reaction of inorganic oxide chemical solutions.

The fabrication of bone into ceramic must be done through a sintering process, where the powder is compacted and sintered. The biological properties can be controlled by sintering at high temperatures and also the potential applications can be restricted [8]. Phases of cattle bone and bone waste sintered at different temperatures have been reported. Distinct structural changes were detected on sintering 600 °C and the formation of HA was identified after sintering as low as 600 °C [9].

The objective of this study is to obtain and to analyze the HA powders by using wet precipitation method and the phase of buffalo bones at elevated sintering temperatures of 650 °C, 850 °C, and 1050 °C which are the approximate sintering temperature ranges of synthetic HA attain.

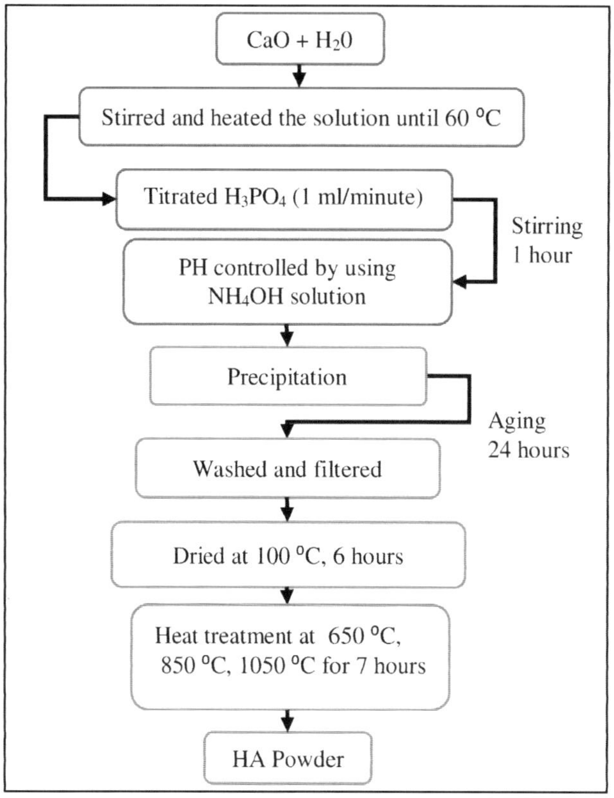

Figure 1. Wet precipitation procedure for buffalo bone HA synthesis.

2. Experimental method

2.1. *Preparation material*

Femur from buffalo bones were first cut into pieces, then the bones were cleaned, boiled for an hour, rinsed in water and destilled water, soaked in NaOH for 24 hours, and further soaked in C_3H_6O for 24 hours and then dried. The bones were subsequently sintered at 1000 °C for 4 hours.

2.2. *HA shynthesis*

HA was synthesized using the wet pecipitation method [1]. As the starting precursors, 120 ml $Ca(OH)_2$, 167.73 ml H_2O (distilled water), 12.27 ml H_3PO_4 were used. The chemical reaction is :

$$10Ca(OH)_2 + 6H_3PO_4 \rightarrow Ca_{10}(PO_4)_6(OH)_2 + 18H_2O$$

Ammonium hydroxide, NH_4OH, were used to adjust the solution to raise the pH greater than 10 in order to obtain a stoichiometric HA. The whole procedure is shown in the flow diagram in Figure 1.

2.3. Characterization of HA

Microstructure and crystal size analysis of HA were performed by using Scanning Electron Microscopy (SEM). The powder samples were coated using a carbon evaporation coating unit. To identify the phases present before and after the heat treatment and to determine the particle size distribution, X-ray diffractometer was used with CuKα radiation. FTIR spectroscopy was used to analysis the characteristic phase of the presence of phospate groups and surface hydroxyl groups.

3. Results and discussion

3.1. Morphological analysis

Figure 2 shows the morphology of the HA powders sintered at 650 °C, 850 °C and 1050 °C for 7 hours. It could be seen from Figure 2(a), that at 650 °C the HA particles are partially agglomerates. This is due to HA particle surface area being quite larger. Increasing the sintering temperatures to 850 °C and 1050 °C has made the surface area of HA particle to become smaller, then the agglomeration area decreases. The uniform grain size with a narrow size distribution of HA particles could also be seen. This correspond to the crystallinity improvement of the HA, and this narrow size is clearly visible after sintering at 1050 °C for 7 hours as shown in Figure 2(a). The interconnection of HA particles has also improved by increasing the sintering temperature as shown in Figure 2(b) and 2(c).

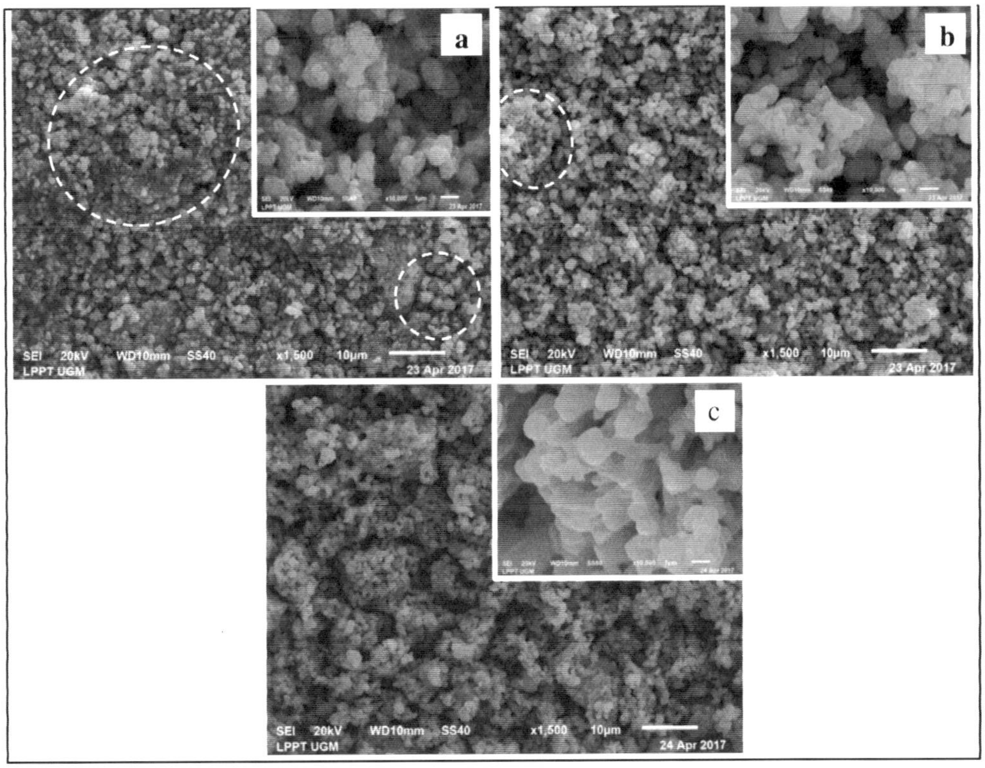

Figure 2. SEM images of HA sintered at (a) 650 °C, (b) 850 °C and (c) 1050 °C for 7 hours. Insets: magnification 10.000x. Dashed line is agglomeration area.

The main chemical components of the synthesized natural buffalo bone could be expressed in term of percent mass as CaO (54.03 %) and P_2O_5 (45.97 %). Other components of HA for different sintering temperatures are shown in Table 1.

Table 1. Chemical components of HA.

Elements (% mass)	Temperature (°C)		
	650	850	1050
Calcium	39.11	34.69	38.44
Posphate	18.67	16.13	19.33
Magnesium	0.63	0.62	0.64
Natrium	0.58	0.28	0.64

Chemical component of HA is dominated by calcium and phosphorus as a main component of natural bone. The existences of magnesium and natrium [10, 11] in small amounts are corresponding to the inorganic compounds in natural bone [12].

Commercial HA has a chemical formula $Ca_{10}(PO_4)_6(OH)_2$ with a calcium and phosphorus molar ratio of 1.67. Table 2 shows the molar ratio calcium/phosphorus (Ca/P) of HA based on buffalo bone.

Table 2. Molar ratio calcium and phosphorus at different sintering temperature of buffalo bone HA.

Temperature (°C)	Ca/P molar ratio
650	1.61
850	1.67
1050	1.52

As shown in Table 2 for sintering temperature at 650 °C, 850 °C and 1050 °C, Ca/P molar is ~1.67. At 850 °C, the Ca/P molar ratio is 1.67. Theoretically, the stoichiometric HA molar ratio is 1.667 for pure HA [12, 13]

Figure 3 shows the chemical elements of buffalo bone HA with molar ratio 1.67 sintered at 850 °C. The composition consist of 34.69 % mass calcium and 16.13 % mass phosphorus.

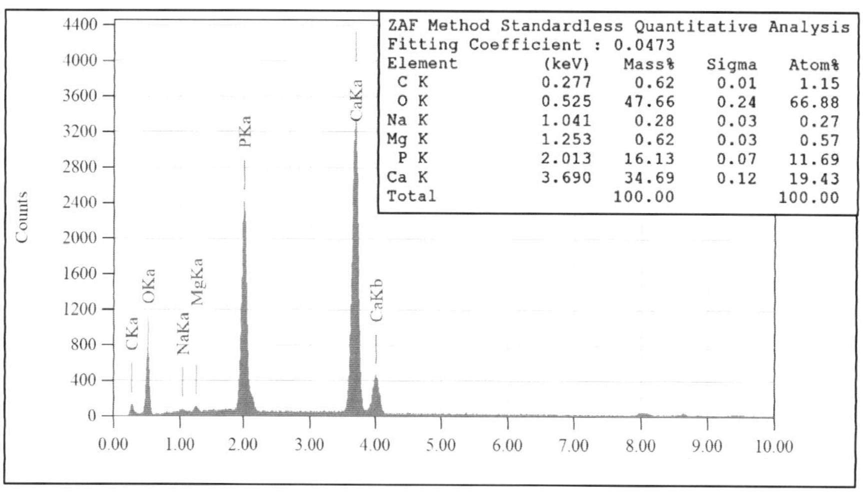

Figure 3. Chemical elements of buffalo bone HA with molar ratio 1.67.

3.2. FTIR analysis

FTIR spectral analysis of HA sintered at 650 °C, 850 °C and 1050 °C for 7 hours is shown in Figure 4.

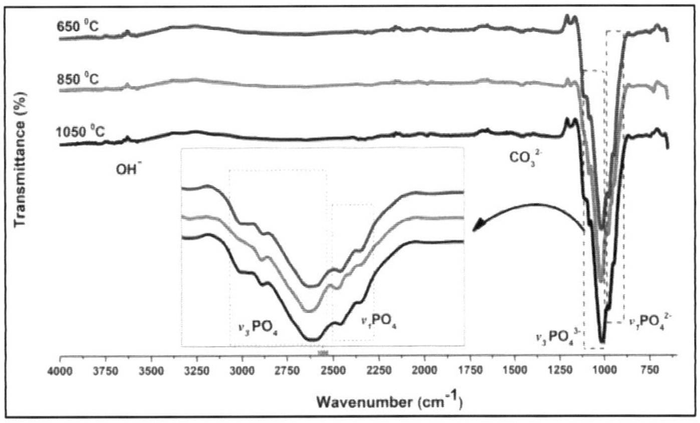

Figure 4. FTIR spectrum of the sintered HA at sintering temperature of 650 °C, 850 °C and 1050 °C for 7 hours.

The characteristic structural OH peaks appeared at 3572 and 633 cm^{-1} [14, 15] for the sample sintered 1050°C indicates the formation of HA. Phosphate band were identified by a peak at 1022.20 cm^{-1} [15]. This band is due to the sintering process, in which increasing the temperature would also increase the crystallinity of the sample which then also increases the purity of HA, as analyzed in XRD result (Figure 5). The existence of the C-O at 1442.65 cm^{-1} [16] is not surprising because human natural bones have this compound.

3.3. Structural analysis

Figure 5 presents an X-ray diffraction pattern of sintered HA at 650 °C, 850 °C and 1050 °C for 7 hours.

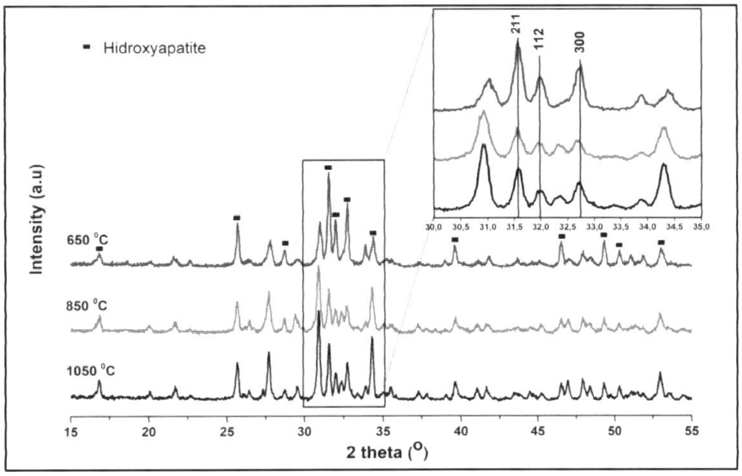

Figure 5. XRD Pattern of sintered at 650 °C, 850 °C and 1050 °C for 7 hours.

Characteristic of the HA powder is clearly identified by comparing the experimental XRD patterns to the Joint Committee on Powder Difraction Standards (JCPDS) by no. #090432. There are three main peaks of HA, they are 2θ =31.85°; 32.31°; 32.95° respectively. The XRD pattern of the synthesized HA at 650 °C, 850 °C and 1050 °C for 7 hours indicates the presence of the HA phase which are evidenced by the diffraction lines located at 2θ=31.77°; 32.23°; 32.94°. This indicates the reflection corresponding to the (211), (112), and (300) *hkl* indexes. The additional peaks between 112 and 300 indicate the impurity of samples. Sintering process using conventional furnace and then Ca ions in sample is in contact with carbon dioxide to form calcium carbonate. The existence carbonate phase in the sample has been confirmed by FTIR result.

A narrow XRD peak observed at high sintering temperature is an indication of the high crystallinity of the sample [13, 18]. The structural analysis using XRD pattern showed that the HA was crystalline in all samples.

4. Conclusions

Hydroxyapatite is successfully synthesized from buffalo bones by wet precipitation method and elevated sintering temperatures of 650 °C, 850 °C, and 1050 °C for 7 hours respectively. The calcium-phosporus ratio of 1.67 is obtained at 850 °C. Uniform grain size with a narrow size distribution corresponding to the cystalinity improvement of the HA obtained after the sintered at 1050 °C for 7 hours. The presence of HA phase is evidenced by the diffraction lines located at 2θ=31.77°; 32.23°; 32.94° indicates the reflection from 211, 112, and 300 crystal planes. All XRD peaks patterns of synthesized HA are correspond solely to commercial HA phase.

Acknowledgments

I. Pawarangan gratefully acknowledge Indonesia Endowment Fund for Eduation (LPDP) Scholarships (No. :PRJ-57/LPDP/2016) for supporting this study and PUPT Grant for supporting this conference. Y. Yusuf would like to thank PUPT Grant (No. 2456/UN1.P.III/DIT-LIT/LT/2017), Indonesia. The author also gratefully acknowledge LPPT UGM, Indonesia for many wise discussions and assistance.

References

[1] Alfar A Ghorbani M Ehsani N Saeri M R Sorrell C C 2003 *Materials and Design* **24** 197-202
[2] Zhao Q Shen Y Ji M Zhang L Jiang T Li C 2014 *J. Ind. Eng. Chem.* **20** 544–8
[3] Barakat N A M Khil M S Omran A M Sheikh F A Kim H Y 2009 *J. of Mater. Proces. Tech.* **209(7)** 3408–15
[4] Larry L Hench 199 *Biomaterials* 8(19) 1419-23
[5] Laonapakul T 2015 *Synthesis of hydroxyapatite from biogenic wastes* Khon Kaen University, Thailand
[6] Galutvina M M Semochkina L S Glagoley A N Osanov D P 1973 *Biol. Nauki* **16**
[7] Narasaraju T S B Phebe D E 1996 *J. Mater. Sci.* **31** 1–21
[8] Jarcho M 1963 *Clinical Orthopaedics* 157 259-78
[9] Fattah A Salim Ghazi 1982 *Sprechsaal* 115
[10] Kamalanathan P Ramesh S Chandran H Bang L T Niakan A Tan C Y Purbolakson J Teng W D 2014 *Ceramics Int.* **40** 16349-59
[11] Ramirez-Gutierrez C F Londoño-Restrepo S M del Real A Mondragón M A Rodriguez-García M E 2017 *Ceramics Int.* **43** 7552-9
[12] Murugan R Ramakrishna S 2006 *Acta Biomaterialia* **2** 201-6
[13] Niakan A Ramesh S Ganesan P Tan C Y Teng W D 2015 *Ceramics Int.***41** 3024-9
[14] Haberko K Bucko M M Brzezinska-Miecznik J Haberko M Mozgawa W Panz T Pyda A Zarebski J 2006 *J. of the Uropian Ceramic Soc.* **26** 537-42
[15] Jahandideh R Behnamghader A Rangie M Youzbashi A Joughehdoust S Tolouei R 2009 *Key Engin. Materi.* **396–398** 607–10
[16] Pal A Paul S Choudhury A R Balla V K Das M Sinha A 2017*Materials Letters* **203** 89-92
[17] Ramesh S Chandran H Natasha A N Tan C Y Bang L T Ramesh S Ching C Y 2016 *Cer. Int.* **42** 7824-9
[18] Paul S Pal A Choudhury A R Bodhak S Balla V K Sinha A Das M 2017 *Cer. Int.* **43** 15678-84

IOP Conf. Series: Materials Science and Engineering **432** (2018) 012045 doi:10.1088/1757-899X/432/1/012045

Porous-structure engineering of hydroxyapatite-based scaffold synthesized from *Pomacea canaliculata* shell by using polyethylene oxide as polymeric porogen

V J Mawuntu[1] and Y Yusuf[1]

[1] Department of Physics, Universitas Gadjah Mada, North Sekip BLS 21, Yogyakarta 55281, Indonesia

yusril@ugm.ac.id

Abstract. Polymeric porogen leaching method was used to fabricate hydroxyapatite (HAp)-based scaffolds. The starting material was prepared from golden apple snail (*Pomacea canaliculata*) shells. The synthesized material exhibit nanostructured HAp/β-TCP composite. Pore engineering is carried out by removing the polyethylene-oxide (PEO) as porogen from the mixture in sintering process. By varying the concentration of weight of PEO in the mixture, different porosity of scaffolds were obtained for each treatment. The weight concentration of PEO plays an important influence on the morphological porous structure of the HAp-based scaffolds. The FTIR spectra of HAp-based scaffolds did not exhibit PEO functional groups which means that the porogen was completely leached out from scaffold. The thermal analysis of porogen during heating confirms that PEO has leached in sintering process. The mechanical strength of the scaffold decreased as the pore size increased.

1. Introduction

Bioceramic materials such as calcium phosphate become a new alternative material in orthopedic applications, due to their advantages over commonly used metal materials. Calcium phosphate has excellent material properties in terms of biocompatibility as well as quality of integration with bone [1]. Bone tissue engineering has been widely developed to overcome the problem of bone defects by restoring and maintaining the natural function of bone tissues. One type of artificial bone tissue engineering developed today is the fabrication of bone scaffold both polymer-based and ceramic-based scaffold.

Ceramics hydroxyapatite (HAp) is a part of the apatite mineral compound having a $Ca_{10}(PO_4)_6(OH)_2$ chemical formula with a calcium-phosphate (Ca/P) molar ratio of 1.67. Hydroxyapatite itself has biocompatibility properties similar to human bone tissue that compatible with bone and can improve bone regeneration process. Natural materials with high amounts of calcium carbonate are highly potential to be utilized as calcium source of synthetic HAp. The golden apple snail (*Pomacea canaliculata*) shell has an aragonite phase composition, which is the phase generally present in calcium carbonate ($CaCO_3$) [2].

Porosity is an important property that needs to be taken into account in the fabrication of bone scaffolds to allow growth and vascularization of bone or in other words enhance the properties of the bioactivity of the scaffold. By conditioning the composition and pore structure, we can make it possible to obtain a scaffold that acts as an osteoconductor in order to increase the scaffold's ability to support bone regeneration [3]. Porosity will enhance the osteoconductivity of biomaterials, the properties that allow attachment, migration, proliferation, and phenotypic expression of bone cells that lead to new bone formation of biomaterials [4].

Bone biocompatibility is related to the material's ability to support bone growth by paying attention to the body's acceptance response to implant materials and toxic properties of implantable materials in the body. To be used as a site for the formation of new bone, the implant material must function as a media of stem cells and osteoblasts to adhere to and develop well [5]. The bone implant material should have good osteointegration with bone tissue. The biocompatibility of a material can be evaluated on the physical function of the material, including the biological response that may occur.

The porogen leaching method is a common method used to fabricate scaffolds with controlled porosity. The main concept of this method is to disperse porogens such as salts and waxes in both solvent and powder form through the process of porogen evaporation of the scaffold material [6]. Through this method, pores and interconnectivity of pores can be produced in the scaffold. The structure of pore geometry depends on the type of porogenic agent used [7]. This study were used polymeric porogen (*i.e.,* PEO) to fabricate the porous HAp scaffolds. PEO polymers has been reported to be utilized as biomedical materials due to its low toxicity and biocompatibility, and approved by the FDA for several medical application [8].

2. Experimental method

2.1. Synthesis of HAp

Pomacea canaliculata shells were used to prepare the starting material of porous HAp-based scaffold. Shells that have been identified were then cleaned from the macro impurities. Shell samples that have been cleaned were dried in air with temperature 100 °C. The dried shell was then milled using a ball mill in order to obtain a homogeneous shell powder. Shell powder was then tested for its elemental composition by EDS, and its functional group by FTIR. These test were performed to ensure that the shell is calcium carbonate then can be used as a source of calcium in the synthesis process. Shell powder was then calcined at high temperature to decompose the calcium carbonate into calcium oxide (CaO). The sample was then hydrated by reacting CaO with H_2O to obtain calcium hydroxide ($Ca(OH)_2$).

$Ca(OH)_2$ solution was reacted with H_3PO_4 by wet precipitation method, with 1 mL/min flow rate of H_3PO_4. The mixture was stirred in a temperature of 60 °C. The reaction was controlled at pH 10 by using ammonium solution (NH_4OH). The synthesis result was then characterized by using EDS to ensure that the molar ratio of Ca/P met the ideal molar ratio of HAp. In addition, the crystallinity of the synthesized HAp material was also characterized using XRD to see if the crystallinity of the material was in accordance with the standard HAp crystallinity. The functional groups of HAp were also characterized by using FTIR analysis. The synthesized material that meets the properties of HAp then used as the starting material in porous-structure HAp-based scaffold engineering.

2.2. Porous HAp-based scaffold engineering

The fabrication of HAp scaffold was carried out by using additional porogen with the intention of modifying the synthesis results to produce a controlled porous material [9,10]. To analyze the influence of the concentration of polymeric porogen to the porous structure of HAp scaffold, then the concentration of the porogen was varied. Polyethylene oxide (PEO 400000, Aldrich) was used as porogen by varying the concentration (i.e. 0, 5, 10, and 15% wt.). The preferred porogen does not have negative reaction with the body in case of interaction between body tissues and the materials [8].

The variation of porogen concentration was intended to observe the effect of porogen concentration treatment on the scaffold porosity. Porous-structure scaffold is expected to increase the osteoconductivity of bone tissues and support the growth of new bone tissues in the scaffold. Nonporogen HAp scaffold was prepared to compare with the porous HAp scaffold. The mixing results were heated at high temperatures to produce crystalline HAp. The scaffold was sintered at 1000 °C.

High temperatures were used to obtain the crystallinity of the materials. The morphology, crystallinity, and functional groups of HAp scaffolds were characterized by using Scanning Electron Microscope (SEM-EDS), X-ray Diffractometer (XRD), and Fourier Transform Infrared Spectrophotometer (FTIR), respectively. The thermal behavior of the porogen was characterized by using Differential Thermal Analyzer/Thermogravimetric Analyzer (DTA/TGA), and the mechanical properties by using Knoop Microhardness Tester.

3. Results and discussions

3.1. Synthesis of HAp

The SEM image shows the interconnected granular shape of HAp. Figure 1(a) shows the microstructure morphology of sintered HAp. The starting HAp powder consists of agglomerates formation. The molar ratio of synthesized HAp confirmed is 1.67 and it correspond to the molar ratio classification of HAp. The elemental composition of HAp is shown in Figure 1(b).

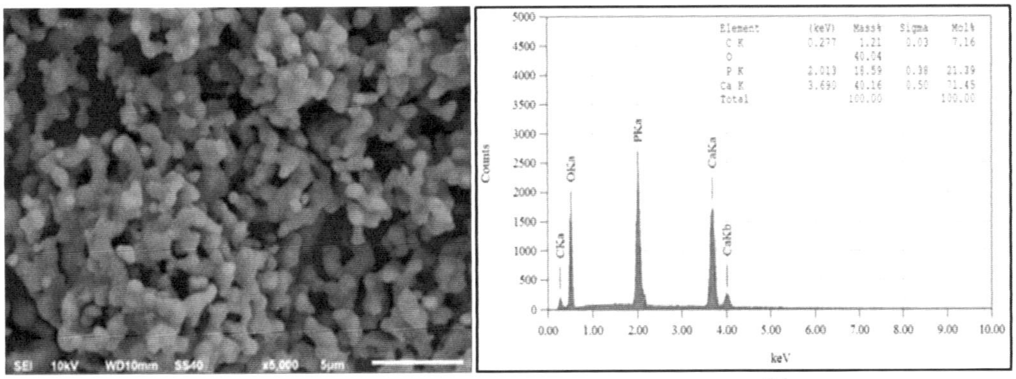

(a) (b)

Figure 1. SEM-EDS result: (a) image of synthesized HAp (powder); (b) elemental composition of synthesized HAp (powder).

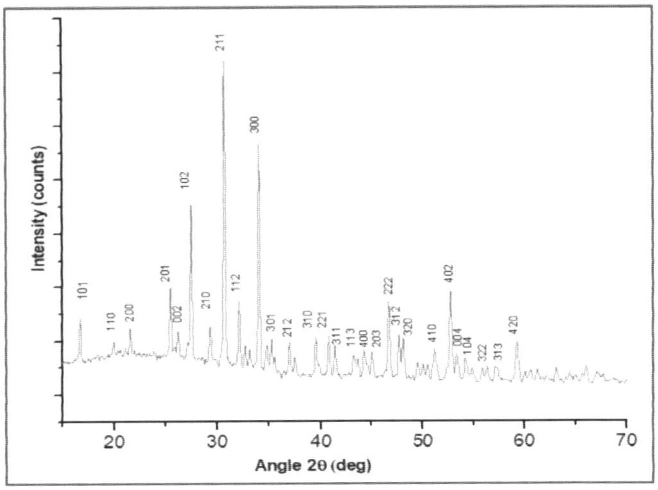

Figure 2. XRD pattern of synthesized HAp.

The X-ray diffraction peaks of the synthesized HAp is shown in Figure 2. The sharp and distinct peaks of the HAp indicate that the HAp was highly crystalline. The peaks were identified by referring

IOP Conf. Series: Materials Science and Engineering **432** (2018) 012045 doi:10.1088/1757-899X/432/1/012045

JCPDS (No. 09-0432) for HAp. The diffraction patterns of XRD correspond solely to HAp phase. The crystallite size of HAp is 49 ± 2 nm (nano-grained HAp), with lattice parameter a is equal to 0.954 nm, and lattice parameter c is equal to 0.698 nm. The XRD pattern of HAp exhibits characteristic of the highest peak at 2θ that is closed to lattice plane 211. The other peaks around 2θ is represent the HAp typical structure.

3.2. Porous HAp-based scaffold engineering

The evaporate particles of PEO from the HAp+PEO mixtures at sintering process left porous-structured morphology to HAp scaffold. From the SEM results, it is seen the dense sintered HAp has pore size distribution below 50 μm (Figure 3a). The HAp scaffold with 5% porogen concentration (wt.) was not sufficient to make interconnect the pores. The average pore size ranged below 100 μm. The morphology of the scaffold surface is mostly crater-shaped with a non-uniform pore distribution (Figure 3b). The addition of 10% (wt.) of PEO porogen in mixture was able to create porous structure on the scaffold with macro-pore sizes ranging from about 100 μm to 450 μm, but there appeared to be some craters that had not yet turned to pores (Figure 3c). The microporous structure of HAp scaffold shows the pore size ranges from ~1 to ~3 μm and it did not exhibit significant differences of pore size for each treatment.

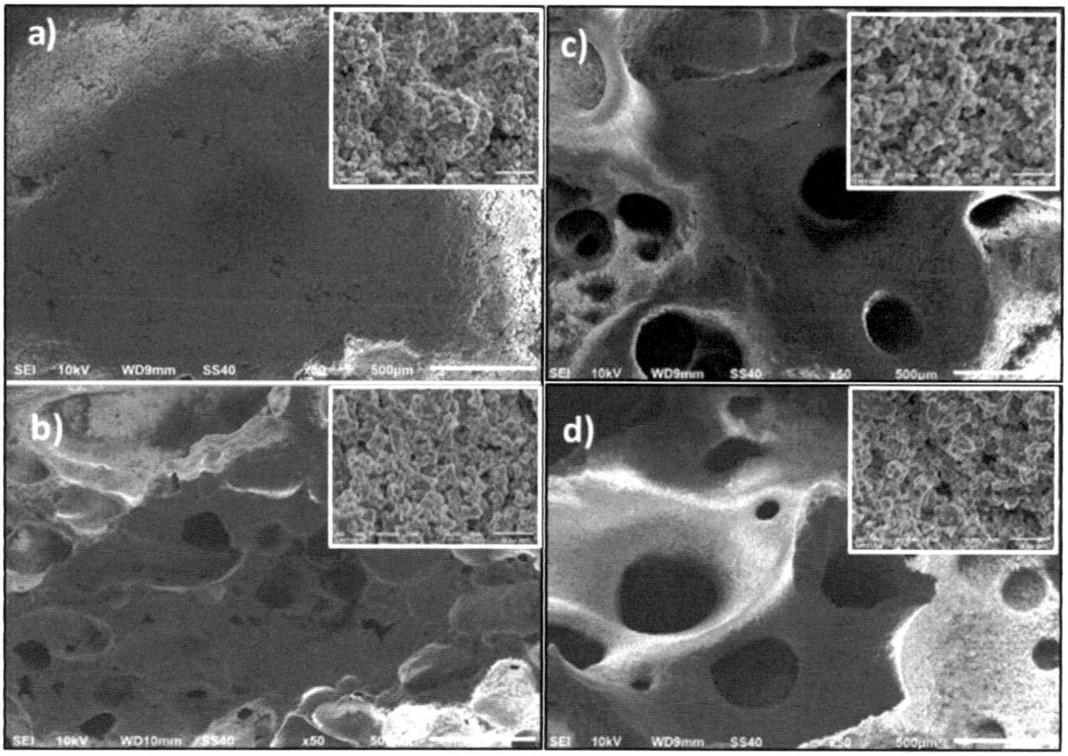

Figure 3. SEM image of HAp-based scaffold: (a) non-porogen (sintered); (b) HAp + 5 % wt. PEO (sintered); (c) HAp + 10 % wt. PEO (sintered); (d) HAp + 15 % wt. PEO (sintered). Inset: microstructure pores of HAp (magnification 5000 times).

The porogen in the suspension slowed down the densification process of the HAp particles, causing the particles to be less interconnected by the higher porogen concentrations in the suspension. Pore interconnectivity has been seen to result from using of porogen by 15% (wt.) in mixture (Figure 3(d)). The surface craters were much reduced and the pore distribution of range size more than 400 μm was varied more than 5% areas of the scaffold's surface. The pore size distribution of the scaffold fulfilled the requirements of bone tissue engineering, where the 100-1000 μm pore size of the scaffold has the biological function for cell penetration, cellular growth, and bone ingrowth [3].

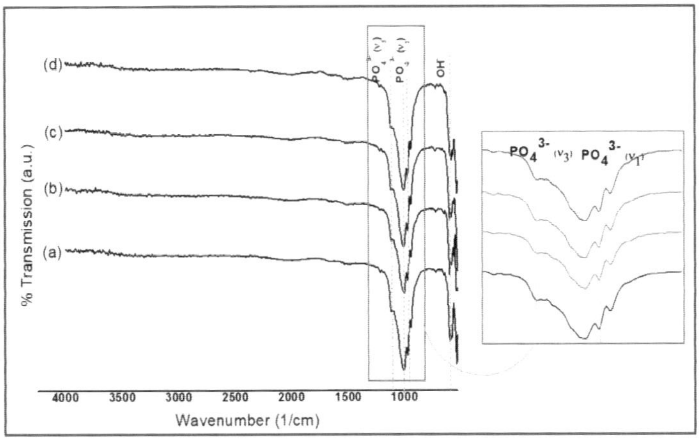

Figure 4. FTIR spectra of synthesized HAp (sintered); (a) non-porogen HAp; (b) HAp + 5 % wt. PEO; (c) HAp + 10 % wt. PEO; (d) HAp + 15 % wt. PEO.

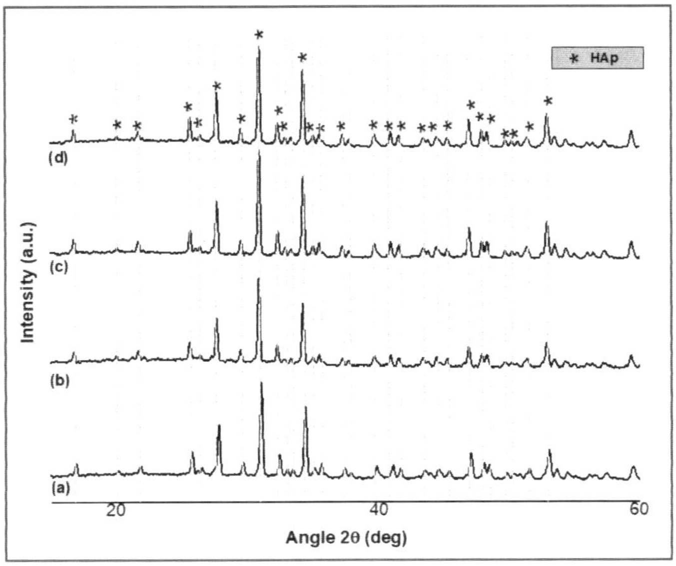

Figure 5. XRD pattern of synthesized HAp: (a) non-porogen HAp; (b) HAp + 5 % wt. PEO; (c) HAp + 10 % wt. PEO; (d) HAp +15 % wt. PEO.

The IR spectra (Figure 4) shows the characteristic spectra of HAp. The FTIR band at wavenumber 1014.48 cm^{-1} corresponds to the presence of a PO_4^{3-} bending vibration. The band at 972.05 is attributed to phosphate stretching vibration. The band at 551.60 cm^{-1} corresponds to vibration of hidrogel ions (OH^-) [11, 12]. The carbonate ion (CO_3^{2-}) does not appear at the spectra.

The FTIR spectroscopy also was used to study the chemical changes during the porous structure engineering process. The IR spectras of HAp scaffolds compared to the synthesized HAp confirm that no chemical decomposition occured in the porous structure engineering process. There was no significant difference peak between the synthesized HAp and the porous-structure HAp scaffold.

The XRD patterns of HAp scaffolds treated with the addition of porogen shows that there are no significant shifting and changing of crystalline peaks (Figure 5). The peaks were also identified by referring to JCPDS data. The diffraction patterns of the scaffolds correspond solely to HAp phase with relatively small intensity of β-TCP. The crystallite size of the HAp-based scaffold tended to increase as porogen was added to the mixture (Figure 6). The addition of porogen tended to increase the agglomeration of crystallite size of the grains.

The synthesized HAp crystal occurs uniformly, the unit cell dimension change isotropically, which cause changes in lattice parameters and peak shifts. There was no peak widening indicates peak shift (211) was caused by uniform strains and changes in *d*-spacing values [13].

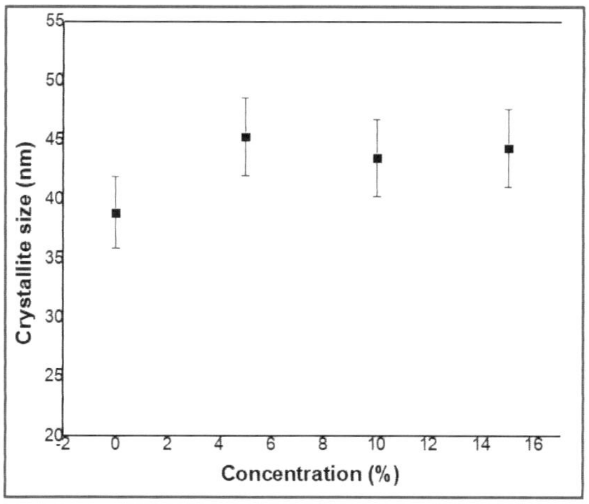

Figure 6. Crystallite size of HAp-based scaffold fabricated from various concentration of porogen PEO.

The lattice parameters *a* and *c* of the scaffold expanded compared to the lattice parameters of the HAp powder. The change of lattice parameters is likely due to the sintering process performed during the HAp densification process. This also indicates a lattice contraction of HAp. The difference in density of the HAp scaffold from each variation of porogen concentration appears to have no significant difference. However, compared to the density of the synthesized HAp powder, the density of the scaffold seemed significantly reduced.

The DTA/TGA graph shows the degradation of PEO mass in sintering process by 32 °C/min heating rate, and it corresponds with the sintering rate of the scaffold. There are some weight loss regions, corresponding to the moisture evaporation, and mass degradation of PEO (Figure 7). The broader temperature range indicates the lower thermal stability of the porogen. The weight of the PEO was drastically reduced at 412 °C by 99.910 % weight loss. The melting point of PEO is 68.77 °C. Mass degradation is associated with an exothermic event on the DTA chart. This process is followed by an endothermic event indicating that the sample in this process had decreased in enthalpy to a more stable structure. It was completely leached-out from the scaffold about 412 °C for about 10 minutes after the sintering begin from the ambient temperature. The remaining temperature and time of sintering was proposed to produce crystalline material of the scaffold.

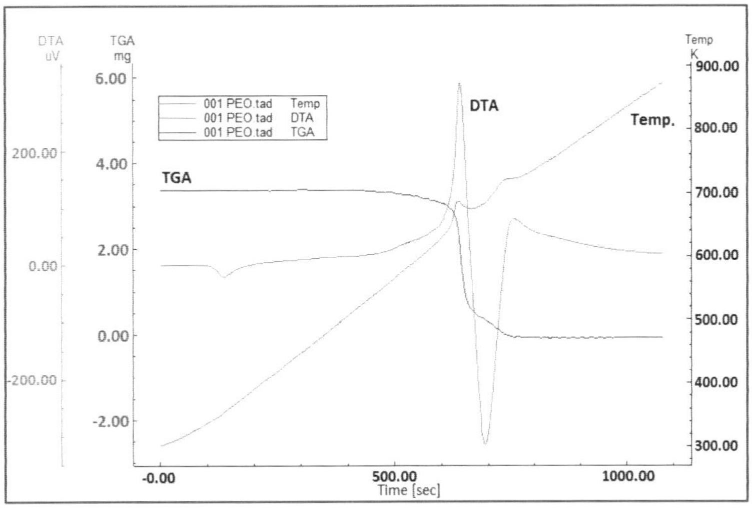

Figure 7. Thermal analysis result of PEO.

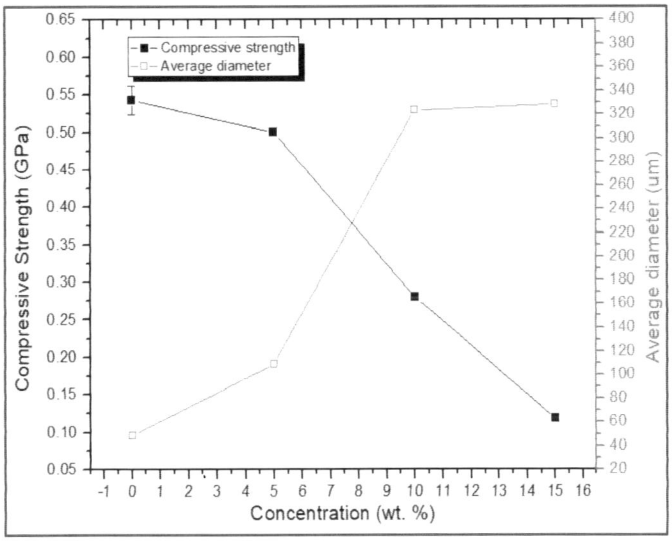

Figure 8. Effect of PEO weight concentration to the mechanical strength and pore size of HAp-based scaffold. Mechanical strength is inversely proportional to the pore size of scaffold.

The mechanical strength of the scaffold decreased with the addition of porogen (Figure 8). It was inversely proportional to the increase of pore size as shown in Figure 3. The microhardness strength of the scaffold is 0.54 GPa, 0.50 GPa, 0.28 GPa, and 0.12 GPa for concentration 0, 5, 10, and 15 % (wt.) respectively. Increase of pore distribution of scaffold caused the mechanical ability of the material to decrease. Nano-bioceramic HAp that was successfully synthesized in this study is an effective option for improving the mechanical properties of materials, where nano-sized ceramics shows better mechanical properties than micro-size bioceramics [14].Ceramic materials with large porosity have large exposed surface areas as well, so the risk of decreasing the mechanical strength of materials by the environment will increase.

4. Conclusion

The results showed that the golden apple snail (*Pomacea canaliculata*) shells can be utilized as a source of calcium for HAp synthesis. Hydroxyapatite with Ca/P molar ratio 1.67 was successfully fabricated. The crystallite size of synthesized HAp is nano-structure HAp. The polymeric porogen leaching method by using PEO as the porogen agent is succeeded to fabricate scaffolds with controlled porosity. Pores appear to be homogeneous at 10% (wt.). Pore interconnectivity has occurred with the addition of 15% (wt.) porogen. The microhardness of the scaffold is getting decrease while the pore size increase. The porogen was completely leached-out from the scaffold in the sintering process. The synthesized HAp that was treated using this method succeeded in fabricating porous HAp-based scaffold.

Acknowledgment

Vicky J. Mawuntu would like to express profound gratitude to Indonesia Endowment Fund for Education (LPDP). Yusril Yusuf is immensely grateful to Indonesian Ministry of Research, Technology, and Higher Education through PUPT Grant (2456/UN1.P.III/DIT-LIT/LT/2017) for the financial support in this research. They would also like to thank LPPT UGM for the technical assistance.

References

[1] Maria V and Gonzalez-Calbet J M 2004 *Progress in Solid State Chem.* **32** 1-31
[2] Leelatawonchai P and Laonapakul T 2014 *Adv. Mater. Research* **931-32** 370-4
[3] Sanchez S S and Arcos D 2008 *Key Engineering Mater.* **377** 19-42
[4] LeGeros R Z and Ben-Nissan B 2014 *Adv. in Calcium Phosphate Biomaterials* ed Ben-Nissan B (Heidelberg: Springer) 1-17
[5] Mankin H J Hornicek F J Tomford W W and Gebdhardt M C 2005 *Bone Regeneration and Repair: Biology and Clinical Applications* ed Lieberman J R and Friedlaender G E 241-61
[6] Subia B Kundu J and Kundu S C 2010 *Tissue Engineering* ed Eberli D 141-57
[7] Mano J F Silva G A Azevedo H S Malafaya P B *et al* 2007 *J. R. Soc Interface* **4** 999-1030
[8] Srichana T and Domb A J 1972 *Biomedical Materials* 83-122
[9] Hou Q P Grijpma D W and Feijen J 2003 *Biomaterials* **24** 1937-47
[10] Lee S H Kim B S Kim S H and Kim Y H 2004 *Macromol. Biosci.* **4** 802-10
[11] Tripathy G and Bikramjit B 2012 *Ceramics International* **38** 341-349
[12] Murphy W L Kohn D H Mooney D J 2000 *J. Biomed Mater Res.* **50** 50–8
[13] Weidenthaler C 2010 *Nanoscale* **3** 792-810
[14] Wang J and Shaw L L 2009 *Biomaterials* **30** 6565-72

Synthesis and characterization of hydroxyapatite based on green mussel shells (*perna viridis*) with the variation of stirring time using the precipitation method

M Sari[1] and Y Yusuf[1]

[1] Department of Physics, Universitas Gadjah Mada, North Sekip BLS 21, Yogyakarta 55281, Indonesia

yusril@ugm.ac.id

Abstract. Hydroxyapatite (HA) from green mussel shells (*Perna viridis*) has been successfully synthesized with a variation of stirring time using the precipitation method. The green mussel shells were calcined in furnace at temperature of 950°C for 2 h. AAS result shows that the level of Ca was 49.5757%. X-Ray diffractometer result shows that the crystallization of CaO was high because it has big crystallite size and small microstrain. The analysis of XRD shows that for HA with the stirring time of 30 min and 45 min have the same crystallization with the crystallite size of (74.91 ± 4.9) nm and (74.91 ± 4.7) nm. From DTA/TGA analysis, HA samples of stirring time 45 min and 15 min are more stable than HA with the stirring time of 30 min. The FTIR spectra show that HA with the stirring time of 30 min and 45 min could have lower transmittance value. From SEM result, HA with the stirring time of 15 min has small agglomerate shape and thick structure of particles. Therefore, HA with the stirring time of 15 min becomes the best synthesized from this research which it can be used as a coating material in implanted applications.

1. Introduction

So far, the treatment to overcome bone damage cases has used the heavy metals to replace the damaged bones in the body. However, this treatment still inflicted a problem, namely the metal stored in the body had a low biocompatibility level that caused sick and bruise in the flesh around the metal. Therefore, it taken a material that it can be used to coat the metals to be implanted into the body, so the metal attain a high biocompatibility with the body. The approach is taken using the materials that are already present in the body, constituents of bone tissues such as apatite compounds. Presently, there is an artificial bone similar to the original bone constituent, namely hydroxyapatite [1].

Hydroxyapatite ($Ca_{10}(PO_4)_6(OH)_2$, HA) is a major component of human bones and teeth [2]. It is commonly used in orthopedic, dental and maxillofacial applications [3-5]. HA has most stable crystal phase of potassium phosphate with the hexagonal structure that has the lattice parameters: a = 9.433Å c = 6.875Å and a variable Ca/P mol ratio is 1.67 [6-7]. The advantages of hydroxyapatite are porous, bioactive, non-corrosive and wear-resistant. Hydroxyapatite has a weight of 69% of the weight of pure bone and it is the most stable compound in body fluids and dry for up to the temperature of 1200 °C [8].

A variety of synthesis techniques of hydroxyapatite have been developed such as sol-gel procedure [9], precipitation from aqueous solution [10], hydrothermal [11-13] and solid-state reactions [14]. In this study, we use the precipitation method to synthesis HA. The selection of precipitation methods is done on several considerations such as many hydroxyapatite relatively synthesized without used the organic solvents (relatively small cost); it is simple process with large product (87%), so it is suitable for large-scale (industrial) production, it needs inexpensive reagents and Ca/P products with the phase compositions can be obtained. Although this process depends on variables such as pH, aging, and temperature, it is effective and inexpensive compared to sol-gel [15].

Hydroxyapatite made by chemical synthesis is called synthetic hydroxyapatite. Synthetic hydroxyapatite can be obtained not only through the reaction of synthetic compounds, but also it can be reacted the synthetic compound with a natural compound. HA can be synthesized from the materials that contain high calcium such as cow bones, fish bones, cuttlebones, eggshells and mussel shells [16].

Here, we use waste mussel shell in Indonesia that have not been processed in Indonesia. The production of mussel shell in Indonesia from 2002 has continued to increase [17]. The waste mussel shells have a high calcium carbonate content of 95-99% [18] so it can be used as a source of calcium in HA synthesis. The analyses of Energy Dispersive X-Ray Fluorescence (EDXRF) show that the composition of minor minerals in the green mussel shells are Ca 99.5%, Sc 0.24% and Sr 0.47% [19]. In this study, the waste green mussel shells (*Perna viridis*) were taken from the Depok beach, Yogyakarta, Indonesia.

2. Experimental method

2.1. Preparation of calcium oxide

The waste green mussel shells (*Perna viridis*) were cleaned in boiling water for 30 min and washed using distilled water to remove attached materials such as shell meat and algae. Then, they were dried using the furnace at the temperature of 100°C. It was refined using ball mill to obtaining the smaller particle size. The powder of green mussel shells was heated at the temperature of 950 °C for 2 h.

2.2. Synthesis of hydroxyapatite

An amount of 2 gram of calcium oxide was mixed with 50 ml distilled water. Then, $(NH_4)_2$ HPO_4 solution (2.8285 gram in 50 ml distilled water) was slowly added dropwise at rate 1 ml/min to the CaO powder, the liquid mixture was stirred with the velocity at 300 rpm and done the variation of stirring time of 15 min, 30 min, and 45 min while heated at temperature of 70°C. The pH of the mixture was controlled above 9 by adding ammonium hydroxide (NH_4OH (25 %) 3M. The mixture was stirred by magnetic stirrer for 50 min and heated at temperature of 70 °C. The solution was subjected to aging treatment for 24 h and washed using distilled water. The solution was filtered to obtain the precipitate of HA. The precipitate of HA was dried at temperature of 100 °C for 2 h. Finally, HA was calcined at temperature of 950 °C for 3 h using the furnace to obtain the pure hydroxyapatite.

2.3. Characterization

In the preparation step, the green mussel shells sample was characterized using the atomic absorption spectrophotometer (AAS) to determine the level of Ca in the green mussel shells. The calcium oxide was characterized using X-Ray Diffractometer (XRD) to determine crystallite size and microstrain. We use Fourier Transform Infrared Spectroscopy (FTIR) to determine the functional groups of calcium oxide. We use Scanning Electron Microscopy (SEM) to know the morphology of calcium oxide [20].

In the synthesis step, the HA samples was characterized using X-Ray Diffractometer (XRD) to determine the crystal structure of HA, Differential Thermal Analysis/Thermogravimetric Analysis (DTA/TGA) to analyse the thermal and stability properties. We use Fourier Transform Infrared Spectroscopy (FTIR) to determine the functional groups of OH^-, PO_4^{3-}, and CO_3^{2-} in the HA samples. We use Scanning Electron Microscopy (SEM) to know the morphology of HA samples [20].

3. Result and discussion

3.1. Calcination

The green mussel shells need to be calcined before it used as calcium precursors (Ca). The calcination was done in order to decomposition reaction of $CaCO_3$ to CaO occurred. In this condition, all organic components of green mussel shells will burn exhausted to CaO and H_2O. As the reaction occurring during the calcination process is as follows:

$$CaCO_{3(s)} \rightarrow CaO_{(s)} + CO_{2(g)} \quad (1)$$

The enthalpy change for the decomposition reaction of $CaCO_3$ was 177.7 kJ / mol [21]. Therefore, the enthalpy change for the decomposition reaction of $CaCO_3$ was positive, so the reaction was occurred on an endothermic [21]. In this study, the calcination of the samples was done at 950 °C because the decomposition reaction of $CaCO_3$ to CaO can be occurred at the minimum of 750 °C [22]. According to calculations using the enthalpy concept, parameter of increasing temperature was 1426 °C so that the variations of the calcination temperature can ideally be carried out at ~ 1400 °C. Giving the heat greatly helped the optimization of the decomposition reaction. When the $CaCO_3$ compounds received heat, the moving atom would be faster and this movement broke the chemical bonds of $CaCO_3$ into CaO and CO_2. Therefore, increasing the calcination temperature tended to make the breaking of chemical bonds $CaCO_3$ into CaO faster. Besides that, the analysis of AAS shows that the level of Ca for the calcination temperature of 950 °C is 49.5757%.

Table 1. Comparison the level of Ca from green mussel shells.

Researchers	Year	Level of Ca (%)
Yulianti	2009	28.97% and 39.55%
Hidayat	2013	45.62%
Siswanto	2013	44.53 %
Mona Sari	2017	49.575 %

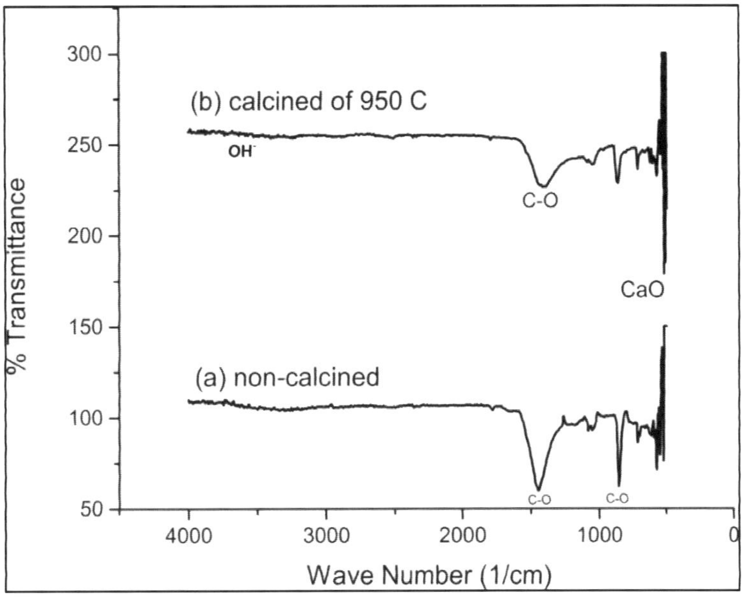

Figure 1. FTIR spectra of CaO samples (a) non-calcination, (b) calcination of 950°C.

According to Table 1, the level of Ca from researcher is greater than the others. Thus, the calcined green mussel shells at 950°C have become potential to precursors in production of hydroxyapatite.

The calculation of crystallite size is 184.8 ± 7.9 nm and the small value of microstrain is 0.00302. This indicates that calcium oxide has a small crystal defect. Microstrain can become one of the forms of crystal imperfections due to distortion or dislocation. The small value of microstrain indicates the small defect of crystal.

FTIR spectra analysis in CaO is performed to know the functional groups of C-O, OH^-, and CaO formed. According to Figure 1, the non-calcined of green mussel shells do not show the stretching mode of OH^- and the functional group of CaO. The calcined of green mussel shell at temperature of 950 °C indicates the stretching mode of OH^- at 3641.34 cm^{-1}. The samples of CaO have shown the C-O functional groups of calcium carbonate contained in the green mussel shells. The functional bands of calcium carbonate indicate close to the commercial calcium carbonate functional group at 1400 cm^{-1}, 877 cm^{-1} and 700 cm^{-1} [23]. The functional group of CaO for the calcination temperature of 950 °C is 871.76 cm^{-1} even for the calcined CaO of 950 °C also shows CaO bond formed at 667.32 cm^{-1}.

SEM characterization was done to determine the morphology of the green mussel shells and calcium oxide from the green mussel shells. In this study, the images were taken with 5000x magnification. According to Figure 2(a), the green mussel shells have a large particle shape with more heterogeneous particle distribution. The morphology of green mussel shells also shows like producing cavities. In Figures 2(b), it can be seen that there are some relief lines on the surface of the particles resulting from calcination at 950 °C.

Figure 2. Results of SEM Characterization for samples (a) green mussel shells, (b) calcination temperature of 950 °C.

3.2. XRD analysis of the HA samples

The XRD pattern of the HA from calcined green mussel shells are shown in Figure 3. From Figure 3, HA with stirring time of 15 min, 30 min and 45 min were obtained the peak of 31.54°, 31.63° and 31.57°, respectively with *hkl* index of close to 211. These agreed with the data Joint Crystal Powder Diffraction Standard (JCPDS) No.09-0432. According to the calculation using the Bragg equation, we obtained the distance between the crystal planes of HA for stirring times 15 min, 30 min and 45 min are 2.83 Å, 2.82 Å and 2.83 Å, respectively. This result is close to the crystal plane of the HA at 2.88 Å appropriate with international standards (ISO 13779-3, ISO 13175-3) for hydroxyapatite implants [24]. Therefore, the results of XRD analysis in this study are not expected to produce any other crystalline phase other than the HA crystalline phase.

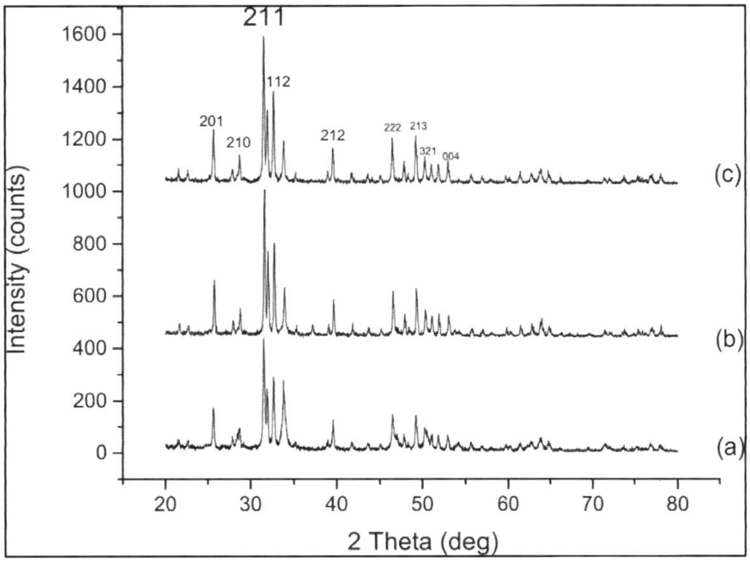

Figure 3. XRD pattern of HA samples with the stirring time (a) 15 min, (b) 30 min, (c) 45 min.

As shown in Table 2, there is an inverse relation between the crystallite size and micro strain. Microstrain is a form of crystal imperfection that appears as a distortion or dislocation. A small microstrain value indicates a small amount of defect in the crystal [25]. The crystallite size and microstrain of HA for stirring times 30 min [(74.91 ± 4.9) nm] and 45 min [(74.91 ± 4.7)nm] are almost same. This means, they have same crystal defect.

Table 2. XRD results for hydroxyapatite.

Name of Samples	Crystallite Size (nm)	Lattice Parameter (Å)	X-Ray Density (g/cm^3)	Microstrain
HA with the stirring time of 15 min	66.31 ± 4.5	8.65	10.31	0.0077
HA with the stirring time of 30 min	74.91 ± 4.9	8.63	10.38	0.0067
HA with the stirring time of 45 min	74.91 ± 4.7	8.65	10.31	0.0068

According to Table 2, the X-ray density value corresponds to the parameter lattice of samples. From samples of HA, HA with the stirring time of 30 min has the greatest X-ray density value from the others. This result indicates that the sample has larger atomic density than the others. Thus, the crystallization of HA depends on the stirring time. If the stirring time was longer than 15 min, no HA were produced [26]. This result can be seen in crystallite size, small microstrain value at HA with the stirring time of 30 min and 45 min.

3.3. *FTIR spectra of the HA samples*

According to FTIR spectra (Figure 4), all samples have HA functional group. HA with the stirring times 15 min, 30 min and 45 min show stretching mode of OH^- at 3641.34 cm^{-1}, 3645.19 cm^{-1}, 3645.19 cm^{-1} and 3571.90 cm^{-1}, respectively. All samples show the bending modes of OH^- at 632.61 cm^{-1} and stretching $_v$(P-O) modes of PO_4^{3-} at 960.48, 1026.05 and 1087.77 cm^{-1}. HA with the stirring time 15 min shows the bending (P-O) PO_4^{3-} at 567.89 and 597.89 cm^{-1}, however HA with the stirring times 30 min and 45 min show the bending (P-O) PO_4^{3-} at 594.03 and 563.17 cm^{-1}, respectively. According to these data, HA with the stirring times 30 min and 45 min could have lower transmittance value. The spectra bands at 1419.50 and 875.62 cm^{-1} represent CO_3^{2-} groups. The existence of CO_3^{2-} groups in FTIR spectra can be derived from the reaction of calcium oxide with carbon dioxide in free air during the synthesis. In addition, CO_3^{2-} groups were possibly existed in raw green mussel shells before the synthesis process. Carbon dioxide will come into contact with the distilled water that are solvent in this reaction and produce CO_3^{2-} and into the crystal lattice of HA.

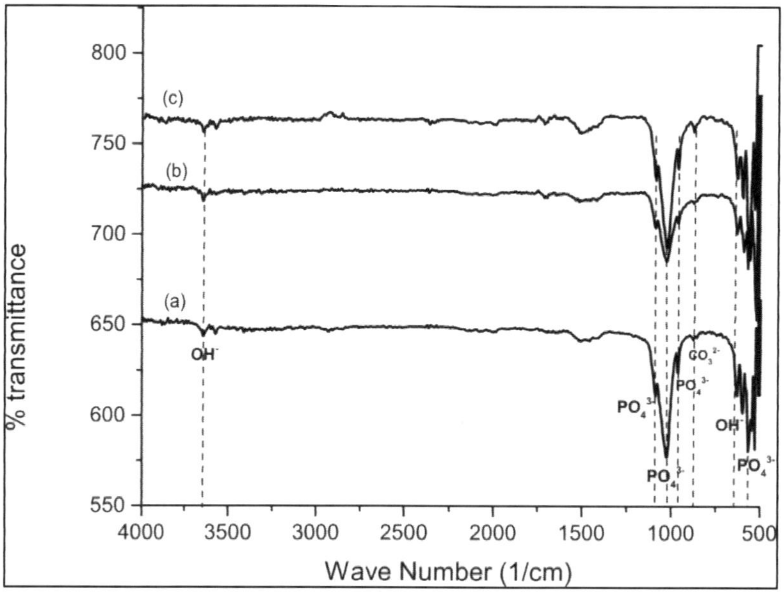

Figure 4. FTIR spectra of HA samples with the stirring time (a) 15 min, (b) 30 min, (c) 45 min.

The presence of carbonate in the HA structure is assessed to reduce the thermal stability to be confirmed from the DTA/TGA results. If the calcium carbonate content is high, there will be a significant decrease in the sample. The presence of CO_3^{2-} groups is common because of human bones have also them. This is the natural substitution for PO_4^{3-} according to formula of molecules Ca_{10} $(CO_3)x(PO_4)_6$ $(2/3)_x$ (OH) or called carbonated HA. The existence of carbonate is unavoidable if HA synthesis is conducted in the open air. There needs to be environmental innertization (reactor) by passing inert gas such as nitrogen (N_2) so that the process of mixing precursor is free from outside air contaminant

3.4. DTA/TGA analyses of the HA samples

According to DTA result (Figure 5), the enthalpy with the exothermic process for the HA with the stirring times of 15 min, 30 min and 45 min are 186.41 J/g, 150.66 J/g, and 72.49 J/g, respectively. HA with the stirring time of 45 min removed the least energy than HA for stirring times of 15 min and 30 min. The result of TGA, HA with the stirring time of 45 min shows a faster weight loss at 427°C with a weight loss of 0.897%. For HA with the stirring time of 15 min and 30 min, the weight loss are 2.171 % and 1.745%, respectively. The weight loss was suggested due to the evaporation of moisture water as the result of CaO and $(NH_4)_2$ HPO_4 reactions. Therefore, HA with the stirring time of 15 min contains much carbonate. HA with the stirring time of 45 min shows the strong elimination of water because it has the fastest weight loss at 427°C. HA samples of stirring time 45 min and 15 min are more stable than HA with the stirring time of 30 min.

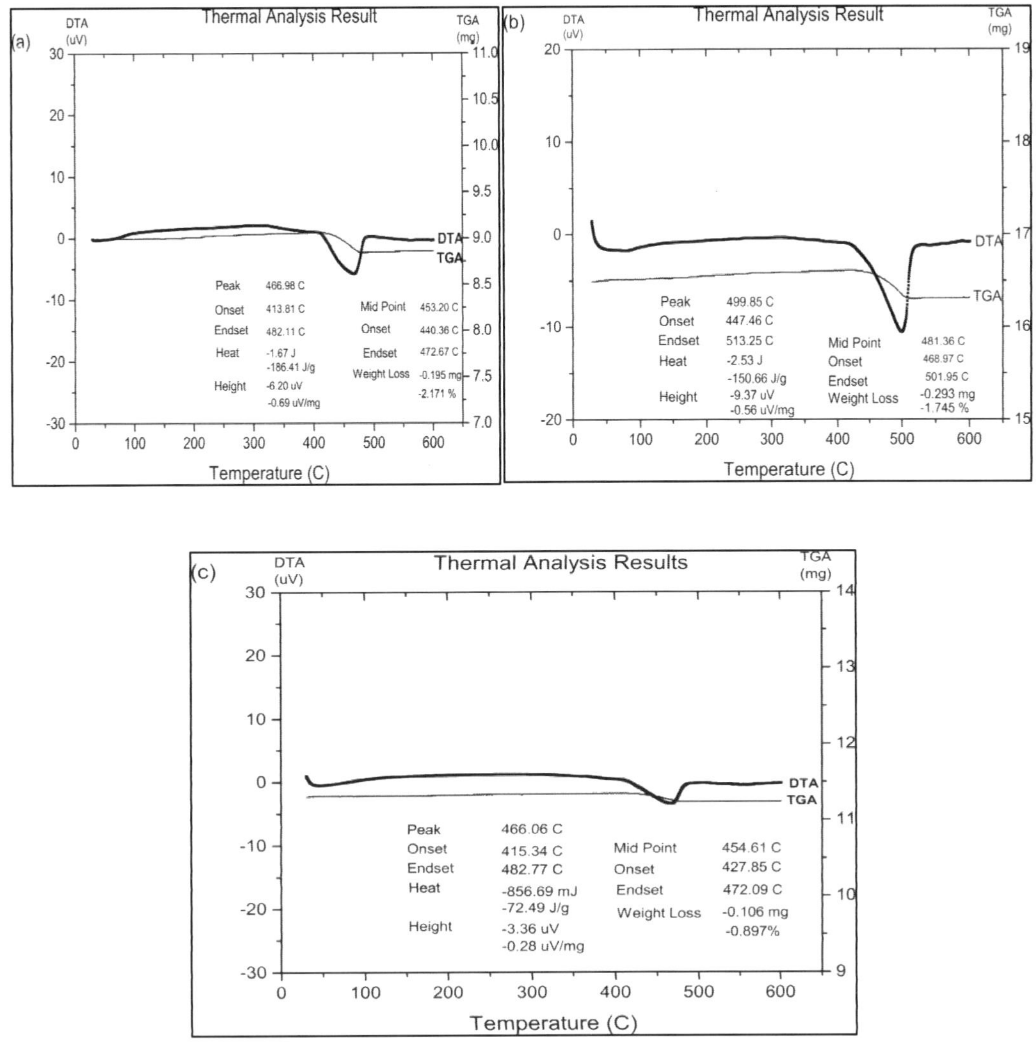

Figure 5. Analysis of DTA/TGA for HA samples with the stirring time (a) 15 min, (b) 30 min, (c) 45 min.

3.5. Morphology of HA

Figure 6 shows the morphology of HA with the stirring time of 15 min, 30 min, and 45 min. HA with the stirring time of 15 min has small agglomerate shape and solid structure. The SEM results show that HA synthesized by using precipitation method is fine grain of uniform size. HA with the stirring time of 30 min and 45 min attain a form with large and uneven clumps. Therefore, by increasing the stirring times, the agglomeration will increase.

IOP Conf. Series: Materials Science and Engineering **432** (2018) 012046 doi:10.1088/1757-899X/432/1/012046

Figure 6. SEM images of HA sample with the stirring time of (a) 15 min, (b) 30 min, (c) 45 min.

4. Conclusion

In this experiment, HA was synthesized successfully from green mussel shells through use of the precipitation method with variations of stirring time. The results of the AAS analysis showed that the Ca level at a calcination temperature of 950 °C was 49.5757%. Per the XRD results, the calcium oxide calcined at 950 °C exhibited the largest crystallite size, meaning it had high crystallinity and a shortened amorphous phase. According to the results of production of hydroxyapatite, HA with the stirring time of 30 min and 45 min have the same crystal defect because the crystallite size and microstrain of HA for stirring times 30 min are almost same. They could have lower transmittance value. HA samples of stirring time 45 min and 15 min are more stable than HA with the stirring time of 30 min. From SEM result, HA with the stirring time of 15 min has small agglomerate shape and thick structure of particles.

Acknowledgement

Mona Sari acknowledges the LPDP Scholarship, Indonesia (No. PRJ-62/LPDP/2016) for financial support in this research and she acknowledges the Ministry of Research, Technology and Higher Educations, Indonesia through PUPT Grant (2456/UN1.P.III/DIT-LIT/LT/2017) for supporting this conference. Yusril Yusuf acknowledges the Ministry of Research, Technology and Higher Educations, Indonesia through PUPT Grant (2456/UN1.P.III/DIT-LIT/LT/2017) for financial support in this research. The authors acknowledge the facilities as well as technical assistance from staff at LPPT UGM, Indonesia.

References

[1] Mulya N Fadli A and Amri A 2016 *Jom FTEKNIK* **3** 1-7
[2] Rujitanapanich S Kumpapan P and Wanjanoi P 2014 *Proc. 11th Eco-Energy and Mater. Sci. and Engin.* 112-7
[3] Akram M Ahmed R Shakir I Ibrahim W and Hussain R 2014 *J. Mater. Sci.* **49** 1461-75
[4] Gao Y Cao W-L Wang X-Y Gong Y-D Tian J-M Zhao N-M and Zhang X-F 2006 *J. Mater. Sci. Mater. Med.* **17** 815-23
[5] Huber F-X Berger I McArthur N Huber C Kock H-P Hillmeier J and Meeder P 2008 *J. Mater. Sci. Mater. Med.* **19** 33-8
[6] Kalita S J and Verma S 2010 *Mater. Sci. Eng. C* **30** 295-303
[7] Mohandes F Niasari M S Fathi M H and Fereshteh Z 2014 *Mater. Sci. Eng. C* **45** 29-36
[8] Saleha Halik M Annisa N Sudirman and Subaer 2015 *Proc. Scientific Meeting XXIX HFI Jateng & DIY* Yogyakarta 124-7
[9] Peng H Wang J Lv S Wen J and Chen J F 2001 *Ceram. Int.* **41** 14340-9
[10] Cox S C Jamshidi P Grover L M and Mallick K K 2014 *J. Mater. Sci. Mater. Med.* **25** 37-46
[11] Kumar R Prakash K H Yennie K Cheang P and Khor K A 2005 *Key. Eng. Mater.* **284** 59-62
[12] Liu F Wang F Shimizu T Igarashi K and Zhao L 2006 *Ceram. Int.* **32** 527-31
[13] Wu S-C Tsou H K Shu H C. Hsu S–K Liou S P and Ho W-F 2013 *Ceram. Int.* **39** 8183-8
[14] Wu S-C Hsu H-C Shu H C Hsu S–K Chang Y-C and Ho W-F 2016 *J. Asian Ceram. Soc.* **4** 85-90
[15] Suryadi 2011 *Synthesis and Characterization of Hydroxyapatite Biomaterials by Wet Chemical Precipitation Process* Thesis University of Indonesia
[16] Sari M and Yusuf Y 2017 *Proc. The Biomaterials Int.* Fukuoka Japan
[17] Muntamah 2011 *Synthesis and Characterization of Hydroxyapatite from the Blood Mussel Shells (Anadara granosa, sp)* Thesis Bogor Agricultural University Indonesia
[18] Shavandi A Bekhit A E-D Ali A Sun Z and Ratnayake J T 2015 *Powder. Tech.* **273** 33-9
[19] Siriprom W Chumnanvej N Choeysuppaket A and Limsuwan P 2012 *Proc. Eng.* **32** 1123-6
[20] Mc Mahon G 2007 *Analytical Instrumentation A Guide to Laboratory Portable and Miniaturized Instruments* edited by J. Willey and Sons Ltd England 130-70
[21] Zhang D Lu S Gong L L Cao C–Y and Zhang H-P 2016 *Energy. Conv. Management.* **109** 94-102
[22] Shavandi A Bekhit A E-D Ali A and Sun Z 2015 *Mater. Chem. Phys.* **149-150** 607-16
[23] Shariffuddin J H Jones M I and Patterson D A 2013 *Chemic. Eng. Research. Design* **91** 1693-704
[24] Gross K A Berndt C C Stephens P and Dinnebier R 1998 *J. Mater. Sci.*, **33** 3985-91
[25] Sari M and Yusuf Y 2018 *Int. J. of Nanoelectronics and Mater.* **11**
[26] Kalita S J and Verma S 2010 *Mater. Sci. Eng. C* **30** 295-303

The effect of clay and mixing conditions to fire resistant and mechanical property of unsaturated polyester – glass fiber composites by a vacuum bagging method

D A Priyani[1], S Prabowo [2], A L Juwono[1] and O Ujianto[2]

[1] Department of Physics, Faculty of Mathematics and Natural Sciences, University of Indonesia, Depok, 16424, Indonesia
[2] Center of Polymer Technology. BPPT Puspitek Area, Setu, South Tangerang, 15314, Indonesia

deanandepriyani@gmail.com

Abstract. Glass fiber reinforced polymer (GFRP) composites have a weakness in fire resistant property. The addition of clay can improve the fire resistant property of GFRP. However, a good mixing is needed to disperse the clay. This research aims to observe the effect of clay and mixing conditions on fire resistance and impact strength of GFRP/clay composites. Unsaturated polyester (UP), glass fiber and 3-Aminopropyltriethoxysilanewere used as a matrix, a reinforcement and a compatibilizer, respectively to build the composites. Unsaturated polyester-clay-silane were mixed using a magnetic stirrer with a variation of rotation speed and mixing duration. The composites were fabricated using a vacuum bagging method. The composition of clay was varied from 1, 2, and 3 wt%. The rotation speed and mixing time were varied at 100, 150, 200 rpm and 60, 90, and 120 minutes, respectively. UP-glass fiber was used as a comparison material. The results showed that composites with 1 wt% clay content was not burned with the burning rate value of 0 mm/min. There were no significant effects of variables on impact strength in between the boundaries that attributed to different mechanisms such as shear and diffusion that worked together to disperse the clay. The shear action should work at high speed rotation and short mixing time, while diffusion at low speed rotation and long mixing time.

1. Introduction

Glass Fiber Reinforced Polymer (GFRP) composites are widely used in many fields, such as aircraft, automotive and construction materials [1-3]. It was evidence that unsaturated polyester (UP) has been used as a matrix for GFRP composites, because the composites had good mechanical properties and UP had lower price compared to others thermoset polymers [4]. However, GFRP is not resistant to fire. From previous researches it turned out that glass fibers reinforced UP had poor fuel resistance properties [5]. Therefore, the addition of a filler as a flame retardant was necessary.

One filler that had been proved could improve GFR-UP fire resistant is clay [5]. Although, the hydrophilic property of the surface of clay caused poor interaction or interface bond with the matrix. Therefore, a compabilizer, such as silane, was needed to facilitate clay dispersion in polymer [6]. Moreover, the addition of silane could improve the mechanical properties of GFR-UP [7].

The clay dispersion in polymer also depends on processing method [8]. Mixing process was one of commonly used method to disperse clay in polymer. Rotation speed and mixing time are process parameter in mixing process which could affect nanocomposite performance [9]. Magnetic stirrer was one of tools which have rotation speed and mixing time parameter. The rotation speed produces shear tension. The higher speed rotation could produce a higher shear tension which could facilitate clay dispersion. However, a high shear tension could make the composites degradation [10, 11]. Therefore, the varieties of process parameters were required to find the effect of mixing process on GFR-UP fire resistant and mechanical properties.

The specimen preparation can also affect GFR-UP property according to fiber and nanocomposite laminating. Vacuum bagging is one of laminating process which has the following advantages; has comprehensive pressure distribution, the amount of resin content that can be controlled, having customized mold, having low amount of void, and being not dependent on operator [12].

The objective of this research was study the effect of clay addition and mixing conditions on fire resistance and impact strength of GFR-UP/clay nanocomposites. The processing variables were clay content, rotation speed and mixing time.

2. Experimental method

2.1. Materials

Unsaturated polyester (UP) Yukalac 123 B - EX (IV) and E-glass fiber with chopped strand mat type (CSM) as first filler were supplied by PT Justus Kimia Raya. Montmorillonite (Cloisite 20A) or MMT C20A as second filler was provided by Southern Clay and silane (3-Aminopropyltriethoxysilane) as compatibilizer was produced by Sigma Aldrich.

2.2. Preparation of nanocomposites

Firstly, MMT C20A was sieved and dried at 80 °C for 24 hours. UP, clay, and silane were stirred manually for 4 minutes. The clay composition was varied as low, medium, and high (1, 2, and 3 wt%). The composition of silane was 0.5 wt%. Then, the solution was mixed by magnetic stirrer method. Methyl Ethyl Ketone Peroxide (MEKP) was also added, and mixed manually for ± 40 secs. The materials were processed at eleven conditions that are shown in Table 1.

Table 1. Settings conditions applied on sample preparations.

Sampel	Clay (wt%)	Rotation speed (RPM)	Time (minute)
Control	0	0	0
1	1	100	60
2		100	120
3		200	60
4		200	120
5	3	100	60
6		100	120
7		200	60
8		200	120

After mixing, specimens were prepared by vacuum bagging. At first, hand lay-up technique was applied, with nine layers CSM, then the composite was covered by peel ply, perforated film, breather and vacuum bag respectively. Vacuum method was applied with -1 bar pressure at room temperature for 3 hours. Then, the samples were cut according to test standard (UL 94H and Impact Charpy ISO 179).

3. Characterization

Flammability and impact strength test were conducted to analyze fire resistance and mechanical properties with ASTM D635 (UL 94H) and Impact Charpy ISO 179 standard. The flammability test was carried out 5 specimens using methane gas Impact strength test was carried out to 5 specimens using the Ceast Resil Impactor Junior with speed of 3.7 m / s, pendulum energy 15 Joule, and correction energy 0.05 Joule. In this research, the composites of contained 1 wt% clay and 3 wt% clay were identified as "low" "high" compositions respectively.

4. Results and discussion

4.1. Effect of clay, mixing rotation speed, and time on fire resistant

In this research, it was found that UP with 1 and 3 wt% organoclay can improve the fire resistant up to 100%. Table 2 shows that the low clay composition samples at low speed and high speed, generate the similar burning rate which is 0 in result. In compared to previous study, polyamide 6 with 2 wt% organoclay could improve 2.4% in fire resistant property [13].

Table 2. Burning rate and impact strength GFRP / clay composites.

Sample	Clay (wt%)	Rotation speed (RPM)	Time (minute)	Burning rate (mm/min)	Impact Strength (kJ/m^2)
Control	0	0	0	8.5	129 ± 12
1	1	100	60	0	121 ± 12
2		100	120	0	128 ± 12
3		200	60	0	153 ± 17
4		200	120	0	133 ± 10
5	3	100	60	8	143 ± 21
6		100	120	0	137 ± 13
7		200	60	0	123 ± 15
8		200	120	6.7	13711

4.2. Effect of clay, mixing rotation speed, and time on fire resistant

Figures 1(a) and (b) show that the result of impact test value selection. Samples with "low" and "high" clay composition were selected to study the comparison. In low composition clay samples, the diffusion and shear mechanism worked well. Meanwhile, in high composition clay samples, the diffusion and shear mechanism did not work well because probably there was an agglomeration by visual looking and it decreased the impact value. In the previous research diffusion and shear worked well together [11]. The difference value between in low and high composition clay samples was not substantial. It was proven that diffusion and shear mechanism had no significant effect in impact value.

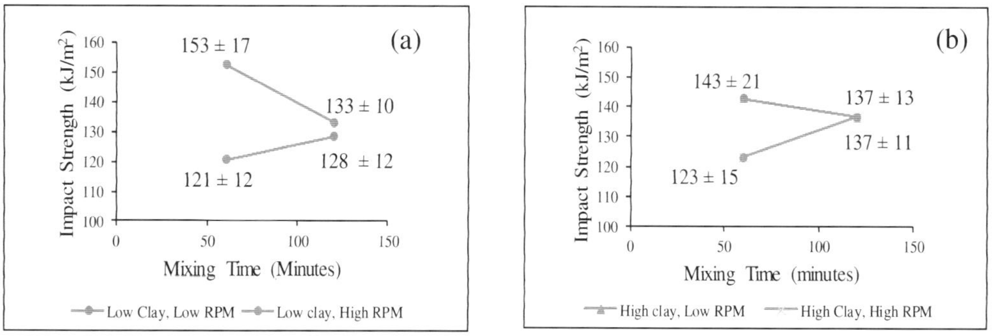

Figure 1. Impact strength GFRP/clay composites (a) low level of clay, (b) high level of clay.

There are two variables which affect the process of polymer mixing, namely diffusion and shear. Shear mechanism would work at the high-speed rotation and short time. In the other hand, diffusion mechanism would work at the low speed rotation and longer time [10, 11]. In this research shear and diffusion did not show positive result.

5. Conclusion

It could be concluded that additional clay composition with percentage of 1wt% improved the fire resistant property by producing unburned GFR-UP composites. With the burning rate value of 0 mm/min. In the other hand, the shear and diffusion mechanism did not affect significantly the impact strength values for both composites 1 wt% and 3 wt% clay compositions.

Acknowledgment

This work was supported by University of Indonesia for providing finance from HIBAH PITTA and Center of Polymer Technology for providing the facilities during the research.

References

[1] Monti M Tsampas S A Fernberg S P Blomqvist P Cuttica F Fina A Camino G 2015 *Polym. Degradat. and Stability* **121** 1-10
[2] Baljinder K Kandola Bhatti W Kandare E 2012 *Polym. Degradat. and Stability* **97** 2418-27
[3] Michaela R Petersen Chen A Roll Mark S J Jung Yossef M 2015 *Composites B* **78** 109-21
[4] Varga Cs Miskolczi N Bartha L Lipoczi G 2010 *Mater. and Design* **31** 185–93
[5] Nazare S Kandola B K Horrocks A R 2016 *Polym. Adv. Technol* **17** 294–303
[6] Lee G-W Lee N-J Kang J Lee K-J Nam J-D 2002 *Composites Sci. and Tech.* **62** 9-16
[7] Romanzini D Frache A Zattera A J Amico S C 2015 *J. of Phys. and Chem. of Solids* **87** 9-15
[8] Ray S S Okamoto M 2003 *Prog. Polym. Sci.* **28** 1539–641
[9] Bashir M A 2008 Effect of nanoclay dispersion on the processing of polyester nanocomposite (McGill Uiiversity, Montreal, QC, Canada, H3A 2K6)
[10] Moghri M Shahabadi S I S Garmabi H 2013 *J. of Reinforced Plast. & Compos.* **32(24)** 1956-65
[11] Ujianto O Jollands M Kao N 2015 *Adv. Mater. Research* **1105** 46-50
[12] Boey F Y C Lye S W 1992 Void reduction in autoclave processing of thermoset composites *Composite* **23** 261-65
[13] Song L Hu Y Lin Z Xuan S Wang S Chen Z Fan W 2004 *Polym. Degradat. & Stab.* **86** 535-40

IOP Conf. Series: Materials Science and Engineering **432** (2018) 012048 doi:10.1088/1757-899X/432/1/012048

Effect of multi walled carbon nanotube (MWCNT) and mixing conditions on unsaturated polyester (UP) physical properties

V Prawitasari[1], O Ujianto[2] and A L Juwono[1]

[1] Department of Physics, Faculty of Mathematics and Natural Sciences, University of Indonesia, Depok, 16424, Indonesia
[2] Center of Polymer Technology. BPPT Puspitek Area, Setu, South Tangerang, 15314, Indonesia

venty.prw19@gmail.com

Abstract. Unsaturated polyester resins (UP) are widely used for thermoset composite application, but to increase its mechanical and physical properties, fillers are required. In this study, UP was mixed with multi-walled carbon nanotube (MWCNT) using mechanical stirrer to improve the mass change after water absorption (WA) and the value of coefficient of friction (COF). The MWCNT was varied at 0.3 and 0.6 wt%, the stirring parameters were also varied with the mixing time at 30 amd 90 minutes, with mixing speed at 500 and 1500 rpm, respectively. The results showed that the mass change after WA test giving 41% increase and COF improved 100%. The improvement on WA and COF might be influenced by high shear energy density, produced by high mixing speed, which might facilitate complete separation of MWCNT.

1. Introduction

UPs are extremely versatile in properties, applications and have been a popular thermoset used as polymer matrix in composites [1]. However, thermoset can not cure before heated. Therefore, an accelerator or promoter is needed to accelerate curing process [2]. UP has lower water absorption (WA) value than epoxy [3], but if it mixed with wrong filler, the WA property can be decreased [4]. Therefore the right filler must be used to get the best WA property.

One filler that can improve the WA in UP is multi-walled carbon nanotube (MWCNTs) better than glass fiber, plant fiber and fur [4-7]. Moreover, previous studies reported that MWCNT could improve UP's coefficient of friction (COF) values [5]. The improvement COF caused by its sliding interface mechanism [8] and the improvement WA caused by its barrier performance (hydrophilic behavior) [9]. The improvement of these properties also depends on the preparation parameters [10].

There are various methods to prepare MWCNT composites, such as solution mixing, bulk mixing, melt mixing and in situ polymerization [10]. The solution mixing method is more efficient than others because multi-walled carbon nanotubes dispersing in composite can be done directly. Moreover, the solution processing method is the best method when the product is in the form of paste on solution. There are two ways that are considered in solution mixing, namely mechanical stirrer and sonication

[10]. In mechanical stirrer, the parameter, time and speed, could be set, while in sonication, only the time could be set [11]. The mixing process is important to discuss, because the dispersion of MWCNT can influence the composite properties.

Due to this reasons, the research of UP/MWCNT composite is necessary to be done. In this research the effect of composition percentage of MWCNT and mixing condition will be investigated by using mechanical stirrer to produce the UP 150 HRN-EX/MWCNT composites with the best properties.

2. Experimental method

2.1. Materials

UP 150 HRN-EX, *Methyl Ethyl Ketone Peroxide* (MEKP) and *Cobalt Naphtenatel* (Co-NAP) were used as matrix, catalyst and accelerator/promoter respectively and purchased from PT. Justus Sakti Raya, Indonesia. MWCNT with outer diameter 30 - 50 nm and length 10 - 20 μm, purity >95wt% and ash <1.5wt% was supplied by Cheaptube.

2.2. Preparation of nanocomposites

The variation of MWCNT that used was 0.3 wt% and 0.6 wt%; and they were identified as low and high composites respectively. Initially, UP and MWCNT were mixed manually for 1 minute, then continued by solution mixing in mechanical stirrer. Mechanical stirrer was set at 500 rpm as the lowest speed and 1500 rpm as the highest speed and mixing time for 30 minutes as the low duration and 90 minutes as the high duration with various compositions (Table 1) [12, 13]. Co-NAP was then added and mixed manually for 10 seconds. MEKP was also added after that, and mixed manually for 10 seconds. The solution was poured into mold with a dimension 30 x 30 x 0.3 cm.

2.3. Testings

WA test was carried out in Water Bath Gesellschaft fur Labortechnik mbH D-30938 Burgwedel typ 1004 no 10966099 I made in Germany based on ISO 62 standard. The temperature and soaking duration were set at 23°C and 24, 48, 96, 192 hours. COF test was conducted in an OTS FOOR societa per azioni vignate–mi–italy s/n: 20006498 according at ASTM D 1894 standard.

3. Results and discussion

Table 1 shows the WA and COF test results related to MWCNT compositions and stirring conditions (duration and stirring speed). The lower value of the increased weight represents the better property of WA and the higher value of COF exhibits the better anti-slip property that could be found in the nanocomposites [3-14]. The lowest value of weight change is (0.48 ± 0.08) %, or 41% reduction on WA compared to the control value. Moreover, in COF testing, the highest value COF is 0.29 ± 0.010 or 100% increasing compared to the control value.

The addition of increasing CNT composition is related to the higher filler in the materials, as a result, the dispersion might not spread evenly. That phenomenon did not cause any effect on increasing property of WA and COF, which could be observed in WA and COF test results that the increasing CNT composition which is caused by the immersion does not indicate the significant decreasing value. Therefore, the CNT composition is not a significant factor to increase the water resistant and anti-slip property. The results of this study agreed well with previous studies [15, 16]. However, there are another factor that can affect WA and COF property, such as mixing speed and duration [17].

The higher the speed and the higher the mixing time, the lower the WA and COF value. The value of stirring condition given depends on the value of CNT compositions that were applied. The higher amount of CNT composition which was used exhibited the higher amount of stirring condition given [18].

Figure 1 shows MWCNT composition decreases corresponding to the increased stirring duration, the increased weight which shows the reduction indicates the cause of shear stress given that could disintegrate the bonds of CNT agglomeration and disperse the CNT by visual looking. Otherwise, in a high condition, shear stress cannot disintegrate the bonds of CNT agglomeration, however the higher amount of CNT composition causes the agglomeration that resists the water to be absorbed. That condition also occurred in a high stirring speed in both CNT composition, which could be observed from the test result that was not overly distinct. The increased water absorption is caused by the reduction of CNT length or can be identified as the damage of the CNT bonds [19].

Table 1. Water absorption and COF test results.

MWCNT Composition [wt%]	Speed [rpm]	Time [min]	Sample (MWCNT_Speed_Time)	Weight Change (%)	COF
Control			UP	0.81	0.007
0.3	500	30	Low_Low_Low	0.71 ± 0.10	0.029 ± 0.010
		90	Low_Low_High	0.48 ± 0.08	0.005 ± 0.003
	1500	30	Low_High_Low	0.52 ± 0.07	0.014 ± 0.005
		90	Low_High_High	0.57 ± 0.01	0.031 ± 0.001
0.6	500	30	High_Low_Low	0.70 ± 0.10	0.004 ± 0.005
		90	High_Low_High	0.64 ± 0.07	0.012 ± 0.011
	1500	30	High_High_Low	0.63 ± 0.10	0.002 ± 0.004
		90	High_High_High	0.68 ± 0.21	0.012 ± 0.005

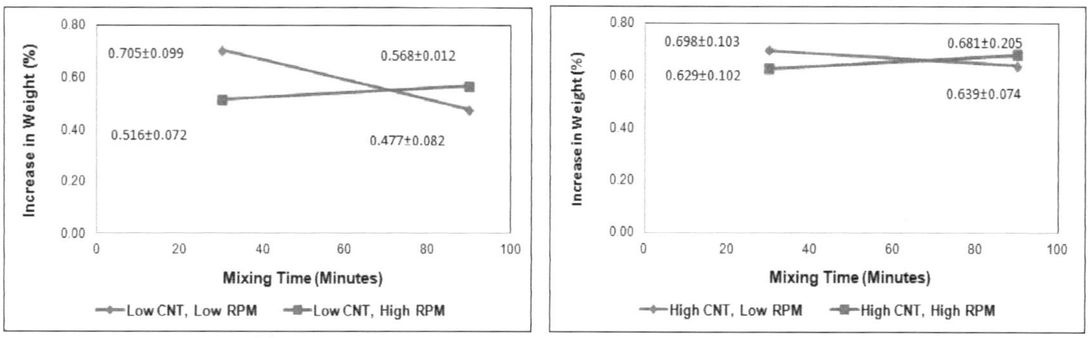

(a) (b)

Figure 1. Weight change vs mixing time during water absorption test, (a) low CNT composition, (b) high CNT composition.

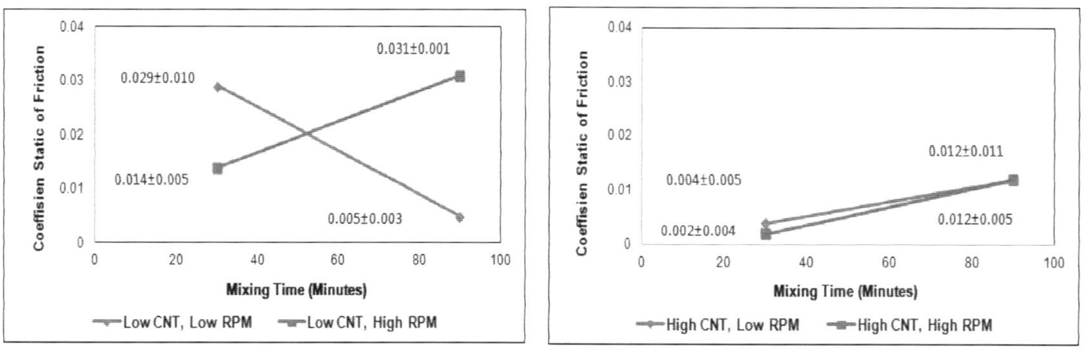

(a) (b)

Figure 2. COF value vs mixing time, (a) low CNT composition, (b) high CNT composition.

Figure 2(a) represents the similar condition with the Figure 1(a), which shows the CNT has been evenly dispersed that can increasing WA but decreasing COF. The other factor which can affect COF value is the rubbed surface, the more slippery the surface that rubs on the sample causes the lower COF value. The second factor which affects COF value is CNT direction towards the direction of friction, if the CNT direction is perpendicular to the direction of friction, then the COF value will increase [20].

Figure 2(b) exhibits the increasing result and does not show any significant difference. That phenomenon was indicated by the higher number of added filler that causes the molecular weight increased by the agglomeration. The external power of blade rotation and stirring duration does not generate shear stresses which can disintegrate filler aggregates, so that the dispersion does not evenly spread and may cause the homogeneous size [19]. When the CNT composition and stirring duration are similar, but the stirring speed is dissimilar, that causes the decline at the high stirring speed. That phenomenon is caused by the agglomerated CNT which binds the bigger energy than the shear energy density which generates activation energy barrier so that the exfoliation occurs temporary and decreasing that value.

4. Conclusion

The fabrication of UP/MWCNT composite could be mixed together properly using a mechanical stirrer. To obtain a good property, it may need high supply energy density, low binding energy aggregates, and high fracture resistance individual of MWCNT. By increasing amount of MWCNT the stirring condition needs to be higher, so that MWCNT can be well dispersed. It decreased the water absorption, then increased COF value. It is concluded that UP/MWCNT nanocomposites have complex properties which depend on stirring conditions.

Acknowledgment

Authors would like to thank you to University of Indonesia for financial support from HIBAH PITTA 2017 and Center of Polymer Technology for facility supports.

References

[1] Leo L 2016 *Scientific Journal of Engineering Students* **1(1)** 7–12 (in Indonesia)
[2] Hernandar W 2004 *Effect of Fiber Volume Faction on Mechanical Properties of Unsaturated Polyester Composite* (in Indonesia)
[3] Sugiman Putra I Gozali M and Setyawan P 2015 *Paper* 1–7
[4] Farias M d Farina M Pezzin A and Silva D 2009 *Mater. Sci. and Engin. C* **29** 510–3
[5] Murarescu M Dima D Andrei G and Ciupagea L 2012 *Assessment of Carbon Nanotubes Functionalization Through Bending and Wear Tests on Polyester Nanocomposites* 1–5
[6] Agrawal S Singh K K and Sarkar P K 2016 *Tribology International*
[7] M S K Raghul R Balaguru I and Ganesh K M M 2015 *Development and Testing of Coir Fiber Reinforced Sandwich Panel* 209–15
[8] Song H-J Zhang Z-Z and Men X-H 2007 *European Polym. J.* **43** 4092–102
[9] Ruban Y J V Mon S G and Roy D V 2012 *Plastics Engin. & Tech.* **15(2)** 133–49
[10] Spitalsky Z Tasis D Papagelis K and Galiotis C 2010 *Progress in Polym. Sci.* **35(3)** 357–401
[11] Nguyen Q T Ngo T D Tran P Mendis P and Bhattacharyya D 2015 *Composites Part A: Applied Sci. and Manufact.* **74** 26–37
[12] Makki M S I Abdelaal M Y Bellucci S and Salam M A 2014 *Fullerenes, Nanotubes and Carbon Nanostructures* **22(9)** 820–33
[13] Shokrieh M M Saeedi A and Chitsazzadeh M 2013 *Composite Structures* **452-453(24)** 3593–9
[14] Golchin A Wikner A and Emami N 2016 *Tribology International* **95** 156–61
[15] Singh T Patnaik A and Satapathy B K 2013 *NANO: Brief Reports and Reviews* **8(3)** 1350025
[16] Liu C Lei W Cai Z Chen J Hu L Dai Y and Zhou Y 2013 *Industr. Crops and Prod.* **49** 412–8
[17] Afrinaldi B 2016 Thesis: University of Indonesia (in Indonesia)

[18] Liu C-X and Choi J-W 2012 *Nanomaterials* **2(4)** 329–47
[19] Huang Y Y and Terentjev E M 2012 *Polymer* **4** 275-95
[20] Dickrell P *et al.* 2005 *Tribology Letters* **18(1)** 59-62

Haematological and inflammatory responses of intramedullary tibial iron based implant in sheep

S F Siallagan[1], R E Nurjannah[1], H Raudlowi[1], H Maheshwari[1], M Silalahi[2], A Boediono[1], S Estuningsih[1] and D Noviana[1,*]

[1] Faculty of Veterinary Medicine, Bogor Agricultural University, Bogor, Indonesia
[2] Center for Science and Technology of Advanced Materials BATAN Kawasan Puspiptek, Serpong, Tangerang, Indonesia

deni@ipb.ac.id

Abstract. Iron-based implants have been widely researched as biodegradable orthopaedic implants although non-degradable metallic materials are still the golden standard as orthopaedic implant to avoid removing implant surgery when bone healing achieved. The aim of this study was to observe haematological and inflammatory responses of iron-based intramedulary biodegradable implants. Eight sheep were divided into two groups according to the material implanted, which were the control group using stainless steel Steinman pin and treatment group using degradable iron-based pin. The pin was inserted into a transverse fracture of sheep tibial bone medullary cavity. Blood sampling was performed on jugular vein on day 0, 1, 3, 7, and 14 after implantation. The examination includes the number of erythrocytes, haematocrite values, hemoglobin levels, platelet values, the number of leukocytes, and leukocyte differential cell. Iron-based bone pin implantation did not cause significant changes on erythrocyte profile, leukocyte count and leukocyte differential cell when compared to stainless steel bone pin. The interaction between iron-based implant and bone tissue, acute and chronic inflammatory responses shows a dynamic variation. It can be concluded the iron-based implant is still tolerated sub-chronically in sheep.

1. Introduction

Fracture cases are quite common in Indonesia. According to the regional health research (RISKESDAS) of Indonesia in 2013, the incidence of fracture was ranked 4th with a rate of 5.8% due to injury [1]. Fracture bone could be immobilized using biomaterial implant. Inert biomaterial such as titanium-based is still the golden standard as orthopaedic implant [2]. To avoid implant surgery, iron-based as a biodegradable metallic material is widely developed as orthopaedic bone implants [3, 4]. Iron used as a bone implant is still debated because of its toxicity [5, 6], but iron also had proven has minimal response in mice [7].

Iron is an essential mineral for heme biosynthesis in developing erythroblasts [8] and also known as an element required for cognitive development in human [9]. Implantation procedures in fracture cases also could affect the number of erythrocyte. Because iron is poorly acquired through diet [10], then in most cases iron supply could be obtained from erythrocyte disposal and iron recycling require macrophage in the liver as primary organ supporting [11]. The use of iron-based biomaterial as orthopaedic biodegradable materials will provide the source of iron ion due to it will gradually be degrade by the time [12]. Several studies were conducted to explore the potential of iron as a bone implant, but just a few of them discussed the response of iron as a bone implant to the systemic response of the body. Therefore, the aim of this study was to observe systemic effects of iron-based biodegradable implant using blood examination in sheep.

2. Experimental methods

2.1. *Implantation preparation procedures*

All of the experimental methods were already approved by animal care and use committee of Bogor Agricultural University number 12-2015 RSHP FKH-IPB. Eight male sheep with 21±4 kg range weight were divided into iron-based and control groups. Control group was implanted with 5 mm of diameter stainless steel based intramedullary implant (Steinmann Pin, sklar®, USA) while an iron-based group was implanted with 5 mm of diameter of iron-based implant with (90.02% of iron, BjPT 6, Tunggal Jaya Steel®, Indonesia). Before implantation, all animal had 2 weeks acclimation and had blood examination at the end of acclimatization period to determine the health condition.

One hour before implantation procedures started, all of the animal was injected with anti-tetanus serum (anti-tetanus serum 1.5®, 1 mL/sheep, intramuscularly) and general antibiotic combination of gentamycin (Bio-genta®, Biopharchemie, Vietnam) and amoxicilin (intramox-150 LA®, Interchemie, Holland). Implantation procedures was performed under general anaesthesia influenced. General anaesthesia we used was xylazin 2% (Xyla®, Interchemie, Holland) and ketamine 10% (Ketamin®, Kepro BV, Netherland) accompanied with Atropin sulfate (Atropine®, PT Indofarma, Indonesia) as premedication.

2.1.1. *Implantation and post implantation procedures.*

The implantation procedure was performed in right lateral recumbency. The skin incision was applied to the lateral of tibia-fibula. The pin was inserted into a transverse fracture of left-tibial sheep bone intramedullary. The retracted muscle was sutured using polyglactin 910 3/0 (Vicryl®, Ethicon, USA) and wounded skin was sutured using silk 3/0 (PT Inti sumber Hasil Sempurna, Indonesia). The wound was bandaged using antibiotic gauze pad (Sofra-tulle®, Sanafi-aventis, Thailand) and hypafix adhesive (Hypafix®, PT. BSN Medical, Sweeden).

Post-implantation treatment was using the combination of gentamycin and amoxicilin general and tramadol. Combination of gentamycin and amoxicilin general antibiotic was administered intramuscularly to prevent secondary infection. Tramadol (Tramadol®, PT Indofarma, Indonesia) also administered intravenously as an analgesic. The bandage was replaced every two days until day 14 using antibiotic gauze pad.

2.1.2. *Blood examination.*

Blood sampling was performed on jugularis vein at day 0, 1, 3, 7, and 14 after implantation. The whole blood sample was examined to have the number of erythrocytes, hematocrite values, hemoglobin levels, platelet values, the number of leukocytes and leukocyte differential cell.

2.1.3. *Data analytics.*

Data were conducted using SPSS for Windows with a P value of $P \leq 0.05$ as statistically significant and expressed as mean ± standard deviation. Data were analyzed using analysis of variance (ANOVA) and then post-hoc DUNCAN.

3. Result and discussion

3.1. *Erythrocyte profile*

Erythrocytes, also known as red blood cells, are the most common type of blood cell that plays the main role to deliver oxygen to the soft tissue. The importance of the role of erythrocyte then the observation of erythrocyte profile is the major concern to know the health condition. The blood examinations of male sheep include erythrocyte counts, hematocrit counts, hemoglobin level, and thrombocyte counts (Table 1, Figure 1).

Table 1. Erythrocyte profile of control and iron-based implant in tibial bone of male sheep on different observation day.

Group	Day 0	Day 1	Day 3	Day 7	Day 14
Erythrocyte count ($10^6/\mu L$)					
Control	3.43± 0.10^{x,a}	3.33±0.10^{x,a}	3.18±0.26^{x,a}	3.10±0.20^{x,a}	3.18±0.30^{x,a}
Iron-based	3.48±0.21^{x,a}	3.53±0.29^{x,a}	3.10±0.22^{x,a}	3.20±0.36^{x,a}	3.10±0.42^{x,a}
Haematocrit (%)					
Control	30.50±1.73^{x,a}	31.00±2.45^{x,a}	28.25±2.75^{x,a}	28.00±2.94^{x,a}	29.75±1.50^{x,a}
Iron-based	30.75±2.63^{x,a}	31.25±2.87^{x,a}	28.50±2.08^{x,a}	28.50±3.42^{x,a}	28.25±4.03^{x,a}
Haemoglobin (g/dL)					
Control	10.10±0.52^{x,b}	10.20±0.83^{x,b}	9.40±0.89^{x,ab}	8.83±0.36^{x,a}	9.73±0.67^{x,ab}
Iron-based	10.43±1.03^{x,a}	10.38±0.95^{x,a}	9.38±0.67^{x,a}	9.28±1.05^{x,a}	9.30±1.32^{x,a}
Thrombocyte ($10^6/\mu L$)					
Control	0,27±0,11^{x,a}	0,22±0,14^{x,a}	0,23±0,13^{x,a}	0,312±0,14^{x,a}	0,37±0,14^{x,a}
Iron-based	0,12±0,03^{x,a}	0,19±0,12^{x,a}	0,20±0,10^{x,a}	0,15±0,04^{x,a}	0,15±0,09^{x,a}

Desription: data presented in the average ± standard deviation. The same superscript letters (x,y) in different columns showed no significant differences (P>0.05). The same superscript letters (a,b,c) in different row showed no significant differences (P>0.05).

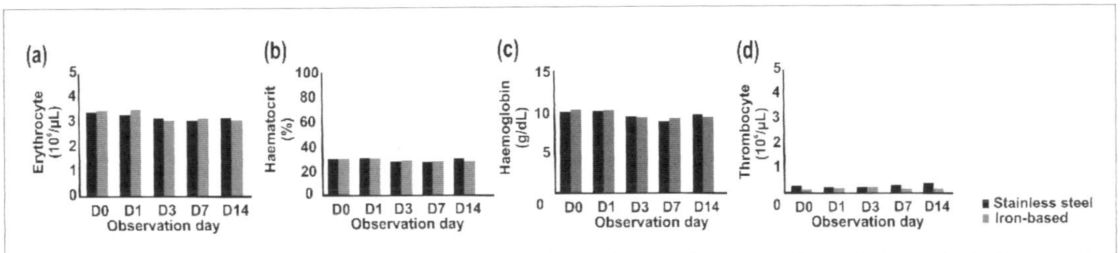

Figure 1. Erythrocyte profile of stainless steel and iron-based implan in tibial bone of male sheep on different observation day. (a) Erythrocyte ($10^6/\mu L$), (b) Haematocrit (%), (c) Haemoglobin (g/dL), (d) Thrombocyte ($10^6/\mu L$).

The number of erythrocytes fluctuates insignificantly different at each observation time (Table 1, Figure 1a). In the first two weeks after bone pin implantation, the erythropoiesis response has not been stable. The decrease in erythrocyte number can be due to the cytotoxicity of the material. The erythrocyte membrane ruptured and the hemoglobin released into the blood plasma that induced by the implant design and the mechanical properties of the implanted surface [7].

Haematocrit is the percentage of erythrocyte in the blood volume so that the capacity of oxygen that is carried in the blood can be calculated [13]. Based on observations of acute hematocrit values (Table 1, Figure 1b), there was an increase in hematocrit value on day 1 and decrease on day 3, 7, and 14 when compared with day 0 before implantation. Increased hematocrit values on the first day after implantation correspond to an increased erythrocyte number. Low hematocrit values can be caused by anemia or blood loss, hemorrhage, increased erythrolysis, anemia, myeloma, lymphoproliferative hepatocellular carcinoma, and excessive hydration. High hematocrit values can be caused by spleen contractions, hypoproteinemia, erythrocytosis, and dehydration [14].

Hemoglobin is a protein formed from the symmetrical pair of polypeptide chains to carry oxygen from the lungs which play an important role in the biological cycle [15]. Observation of acute hemoglobin levels after implantation of the iron-based implant are listed in Table 1. Based on observation, hemoglobin values insignificantly decreased on day 1, 3, and 7 in the iron-based implant group. The hemoglobin increment may be due to blood loss during implantation and the new hemoglobin has not formed. Hemoglobin accumulates in 3-4 days along with the formation of reticulocytes. Otherwise, on day 14 the value of hemoglobin has insignificantly increased. Iron is also known as a major component of hemoglobin [14], therefore iron-based implantation could accelerate the formation of hemoglobin.

Thrombocyte or platelet value is insignificantly increasing in the first two weeks due to the process of wound healing after implantation (Table 1, Figure 1d). The main role of platelets is to repair the vascular damage and proven hemorrhage by forming a hemostatic plug [13]. The correlations between iron and platelet count were negligible [16].

3.2. Leucocyte profile

Leucocyte and its differentials have the primary role to defend the body through foreign body phagocytic process and antibodies proceed, transportation, and distribution in immune response [17]. The leucocyte count and differentiation are presented in Table 2 and Figure 2.

Table 2. Leucocyte profile of control and iron-based implant in tibial bone of male sheep on different observation.

Groups	Day 0	Day 1	Day 3	Day 7	Day 14
Leucocyte Count (10^3/μL)					
Control	$16.67\pm3.39^{x,a}$	$18.37\pm2.08^{x,a}$	$19.32\pm5.45^{x,a}$	$16.47\pm2.54^{x,a}$	$15.67\pm1.08^{x,a}$
Iron-based	$20.4\pm2.30^{x,y,a}$	$27.72\pm8.82^{y,a}$	$18.90\pm7.16^{x,y,a}$	$15.52\pm5.92^{x,a}$	$17.32\pm4.29^{x,a}$
Eosinophil (%)					
Control	$0.25\pm0.5^{x,a}$	$0.25\pm0.5^{x,a}$	$1.25\pm0.95^{x,a}$	$0.5\pm0.57^{x,a}$	$0.25\pm0.5^{x,a}$
Iron-based	$0.75\pm0.95^{x,a}$	$1.25\pm0.95^{x,a}$	$0.25\pm0.5^{x,a}$	$1.5\pm0.57^{x,a}$	$1\pm0.81^{x,a}$
Band Netrophil (%)					
Control	$0.25\pm0.5^{x,y,a}$	$0.25\pm0.5^{y,a}$	$1.25\pm0.95^{x,y,a}$	$0.5\pm0.57^{x,ya}$	$0.25\pm0.5^{x,a}$
Iron-based	$0.75\pm0.95^{x,a}$	$1.25\pm0.95^{x,a}$	$0.25\pm0.5^{x,a}$	$1.5\pm0.57^{x,a}$	$1\pm0.81^{x,a}$
Segmented Netrophil (%)					
Control	$22.5\pm6.75^{x,a}$	$27.75\pm14.17^{x,a}$	$30\pm7.11^{x,a}$	$27.5\pm8.34^{x,a}$	$21.75\pm1.26^{x,a}$
Iron-based	$22.25\pm10.04^{x,a}$	$36\pm29.4^{x,a}$	$27.5\pm6.45^{x,a}$	$25.25\pm4.92^{x,a}$	$28.75\pm10.11^{x,a}$
Lymphocyte (%)					
Control	$70.25\pm8.18^{x,a}$	$67.75\pm12.92^{x,a}$	$63.75\pm9.7^{x,a}$	$69.5\pm7.59^{x,a}$	$72.75\pm2.21^{x,a}$
Iron-based	$73\pm10.13^{x,a}$	$59.25\pm33.5^{x,a}$	$67.5\pm7.93^{x,a}$	$68.5\pm4.65^{x,a}$	$64.75\pm9.56^{x,a}$
Monocyte (%)					
Control	$4\pm1.82^{x,a}$	$1.5\pm1.73^{x,a}$	$4\pm1.41^{x,a}$	$3.25\pm1.25^{x,a}$	$3.75\pm1.5^{x,a}$
Iron-based	$2.75\pm0.95^{x,a}$	$3.25\pm1.25^{x,a}$	$3.25\pm1.89^{x,a}$	$4\pm1.41^{x,a}$	$3.75\pm1.7^{x,a}$
Basophil (%)					
Control	0 ± 0	0 ± 0	0 ± 0	0 ± 0	0 ± 0
Iron-based	0 ± 0	0 ± 0	0 ± 0	0 ± 0	0 ± 0

Desription: data presented in the average ± standard deviation. The same superscript letters (x,y) in different columns showed no significant differences (P>0.05). The same superscript letters (a,b,c) in different row showed no significant differences (P>0.05).

Leucocyte count increment in both groups (Table 2, Figure 2a) indicates the presence of a post-implantation response caused by the procedure or the implanted implant. Leukocytosis occurs as a normal protective response to the physiologic stressor [18]. The highest total number of leucocyte in control and the iron-based group were on day 3 and day 1 post-implantation, respectively. After the bone injury, leucocyte rapidly infiltrates the fracture site, which marks the inflammatory phase of the fracture healing. Insignificantly changes in this study indicated the inflammatory response was still tolerable to the sheep.

Eosinophils count data showed insignificantly fluctuated results and still in normal range (Table 2, Figure 2b). Eosinophils are a variety of white blood cells and one of the immune systems responsible for combating multicellular parasites and associated with the allergic disorder [18]. The iron administration is known have no exacerbate response in allergic experimental [19]. It indicated both implants did not trigger an allergic reaction.

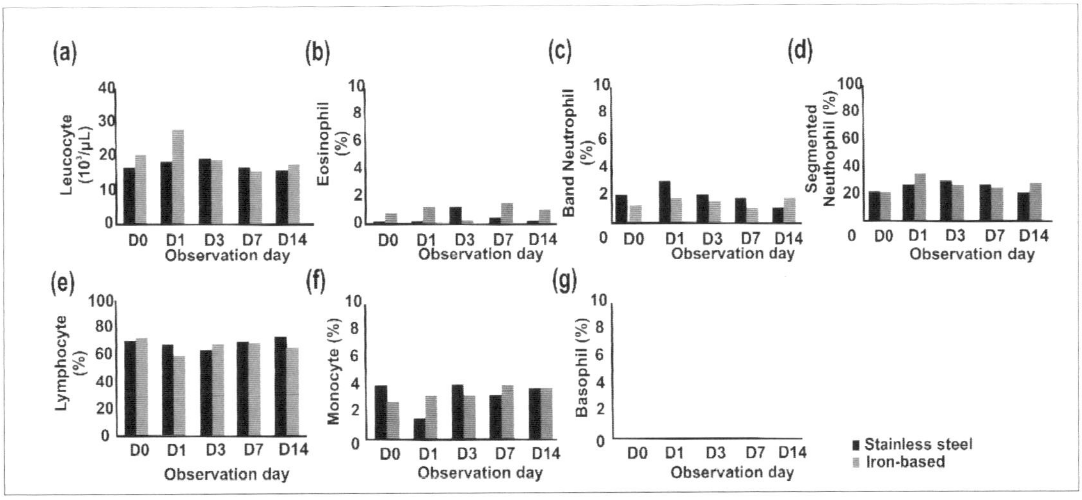

Figure 2. Leucocyte profile of stainless steel and iron-based implan in tibial bone of male sheep on different observation day. (a) Leucocyte (10^3/μL), (b) Eosinophil (%), (c) Band neutrophil (%), (d) Segmented neutrophil (%), (e) Lymphocyte (%), (f) Monocyte (%), (g) Basophil (%).

The inflammation process could activate neutrophils and macrophages and lead to phagocytosis of foreign materials, which may occur in the early phase of bone healing. Table 2 shows the highest neutrophil number in control and iron-based implant are day 3 and 1, respectively. The highest monocyte number in control and iron-based implant are occurred on day 3 and 7, respectively. The results of this study in accordance with previous research that neutrophils and then followed by macrophages appeared in the early stages of inflammation [20]. These insignificant neutrophils and monocytes changes between both groups indicated there was no excessive inflammatory response at various observation times.

Basophils counts were zero in both groups (Table 2, Figure 2g). An increase in the circulating number of Basophils is a rare and generally, is a response to inflammation and immediate hypersensitivity reactions [15]. Basophils play a role in inflammatory reactions during an immune response, as well as in the formation of acute and chronic allergic disease. Based on the basophils count, both implanted implant groups have no hypersensitive reaction.

4. Conclusion

Generally, iron-based implant did not cause significant changes in systemic effects of an iron-based biodegradable implant through erythrocyte and leucocyte profile observation. It can be concluded the iron-based implant is still tolerated sub-chronically in sheep.

Acknowledgment
The authors acknowledge the support of Indonesia Ministry of Research, Technology and Higher Education in international research collaboration and scientific publication scheme.

References

[1] Agency for Health Research and Development, Basic Health Research 2013 Jakarta (ID): Agency for Health Research and Development Ministry of Health RI 104
[2] Kopova I Strasky J Harcuba P Landa M Janecek M Bacakova L 2016 *Mater Sci Eng C.* **60** 230-8
[3] Ulum M F Arafat A Noviana D Yusop A H Nasution A H Kadir A Hermawan H 2014 *J. Mater. Sci. Eng.* **36** 336-44
[4] Orinakova ROrinak A Buckova L M Giretova M Medvecky L Labbanczova E Kupkova M Hrobovcakova M Koval K 2013 *Int. J. Electrochem. Sci.* **8** 12451-65
[5] Rajabian F Mohri M Heidarpour M 2017 *Vet. Rec.* **181** 265
[6] Zhang D Wong C S Wen C Li Y 2017 *J. Biomed. Mater. Res.* **105** 148-58
[7] Paramitha D Noviana D Estuningsih S Ulum M F Nasution A K Hendrawan H 2015 *AIP Conf Proc.* **1677** 070022
[8] Kafina M D Paw B H 2017 *Metallomics* **9** 1193-203
[9] More S Shivkumar V B Gangane N Shende S 2013 *Anemia* **2013** 819136
[10] Hentze M W Muckenthaler M U Galy B and Camaschella C 2010 *Cell.* **142** 24
[11] Theurl I Nairz M Tymoszuk P *et al.* 2016 *Nat Med.* **22** 945-51
[12] Yusop A H Daud N M Nur H Kadir M R A Hermawan H 2015 *Scientific Reports* **5** 11194
[13] Stockham S L Scoot M A 2008 Fundamentals of Veterinary Clinical Pathology (Missouri: Blackwell) 54-102
[14] Rote N S McCance K L 2014 Structure, and Function of the Hematologic System (Missouri: Elsevier) Chapter 27 945-80
[15] Schechter A N 2008 *Blood* **112** 3927-38
[16] Holbro A Volken T Buser A Single J P Halter J P Passweg J R Tichelli A Infanti L *Vos Sanguinis* **112** 87-92
[17] Nse Abasi N Etim Mary E Williams Akpabio U Edem EA Offiong E 2014 *Agricul. Sci.* **2** 37-47
[18] Schwartz A Rote N S 2014 Alteration of leucocyte, lymphoid, and hemostatic function (Missouri: Elsevier) 1008-1051
[19] Maazi H Shirinbak S Bloksma N Nawijn M C van Oosterhout AJ 2016 *Clin Exp Immunol.* **166** 80-6
[20] Bastian O W Koenerman L Alblas J Leenen L P Blokhuis T J 2016 *Clin Immunol.* **164** 78-84

IOP Conf. Series: Materials Science and Engineering **432** (2018) 012050 doi:10.1088/1757-899X/432/1/012050

Effect of fibre sequence on the mechanical properties of woven fan palm/glass fabric reinforced polymer hybrid composites

H Hestiawan[1,2], Jamasri[2], Kusmono[2]

[1] Department of Mechanical Engineering, Faculty of Engineering, Universitas Bengkulu, Jln. WR. Supratman, Kandang Limun, Bengkulu 38122, Indonesia

[2] Department of Mechanical and Industrial Engineering, Faculty of Engineering, Universitas Gadjah Mada, Jln. Grafika No. 2, Yogyakarta 55281, Indonesia

hestiawan1@yahoo.com

Abstract. The aim of this research is to investigate the influence of fibre sequence on the mechanical properties of woven fan palm/glass fabric reinforced polymer hybrid composites. Research specimens were varied in fibre sequence between woven fan palm and glass fabric. Vacuum bagging method was used to manufacture the specimens (at pressure of -70 cmHg). The results of tensile, bending, and impact tests show that the addition of glass fabric can increase the mechanical properties of the hybrid composite. Based on the fibre sequence, the glass fabrics on surface of the hybrid composite have the best mechanical properties.

1. Introduction

In the last decade, the development trend of composite materials has changed in the use of natural fibres as a substitute for synthetic fibres. These are supported by several advantages possessed by natural fibres, such as light weight, high flexibility, and reduced engine usage during processing, recyclable and biodegradable. They are also renewable and have relatively high strength and stiffness and cause no skin irritations. On the other hand, natural fibres also have limitations, such as moisture uptake, quality variations as well as poor moisture resistance when compared to synthetic fibres [1-4].

One way to improve the mechanical properties of composites can be done by reinforcing using two or more variations of fibres in the same matrix material so as to take the best advantage of the properties of both fibres. This results in a hybrid composite and having increased mechanical properties than natural fibre composite. Tensile, bending and impact strength increases with increasing amounts of fibres up to 30%. There are a significant decrease in mechanical strength if the number of fibres is added [5]. Synthetic fibres can be used as a reinforcement together with natural fibre. Synthetic fibre commonly used in industry is glass fibre because it is light and strong. Although the strength of glass fibre is lower and more brittle than carbon fibre, but glass fibre is relatively cheaper [6].

The mechanical properties of the woven hybrid composite are better than the woven composite due to the different structures of the cross section. The hybrid composite shows a significant increase in tensile strength over the pure woven banana and knead fibre reinforced polyester composites [7]. The hybrid composites with glass fabric have better strength, toughness, and out-of-plane stiffness, as well as easier handling in production than unidirectional composites [8]. Glass fabrics offer superiority in terms of good dimensional stability and high packaging density. Mechanical properties of woven coir-glass fabrics/polyester composites [9] and woven jute-glass fabric/polyester composites [10] were evaluated with the effect of woven fabric reinforcement, which was observed to improve mechanical properties of hybrid composites. By varying the stacking sequences of reinforcing layers in the composites, higher mechanical properties of hybrid composites can be achieved [11].

However, there are few studies have been done on the woven natural fibres. John and Naidu [12] reported that the addition of 25, 50 and 75% glass fibres on sisal fibre reinforced composites increased their tensile strength and strain respectively by 28% and 26%. In the present study, woven fan palm and glass fabric composites were chosen due to its ability to be produced in a continuous form, hence easier in producing a woven mat fabric. The fibre sequences are varied from 1 to 4 glass fibres and their effects on tensile, flexural, and impact properties are examined.

2. Experimental method

2.1. Materials

The following materials are used for hybrid composite including woven fan palm (WFP), E-glass fabric 200 g/m^2, unsaturated polyester BQTN 157, and methyl ethyl ketone peroxide (MEKP) as catalyst. Hestiawan *et al.* [13] describe the extraction of fan palm fiber from the young leaves of gebang plant (Figure 1). Prior to form FPF, the fibers were soaked in fresh water for 24 hours. Then they were rinsed and followed by drying at room temperature for 2 days. The Fan palm fibers were made by traditional weaving machine and formed WFP as shown in Figure 2. The weight of the woven FPF was approximately 140 g/m^2. WFP was supplied by Jogjavanesia Handicraft, Indonesia. Unsaturated polyester and MEKP catalyst were provided from by Justus Kimiaraya, Indonesia.

Figure 1. Gebang plant.

2.2. Fabrication of composites

Prior to fabricating the composite laminates, the WFP was placed in an oven at 105 °C for 1 hour in order to remove the moisture contents of fibres. Nine different kinds of laminates were prepared 22 mm x 22 mm size with total seven plies maintains. The sample designation were summarized in Table 1. Five specimens were fabricated for each group used the vacuum bagging technique as shown in Figure 3. The vacuum bagging technique provides a number of improvements over traditional methods such as better fibre-to-resin ratio, very consistent resin usage, and unlimited set up time, and it is cleaner. The process of making specimens is carried out at a pressure of -70 cmHg for 2 hours. Comparison of

IOP Conf. Series: Materials Science and Engineering **432** (2018) 012050 doi:10.1088/1757-899X/432/1/012050

hardener and polyester resins was 1: 100 (v/v). Demoulding the specimens was done after curing at room temperature for 2 hours, then the specimens were stored on a flat surface under room conditions. The next step the specimen is cut and formed into a tensile, flexural, and impact test specimen according to ASTM standard.

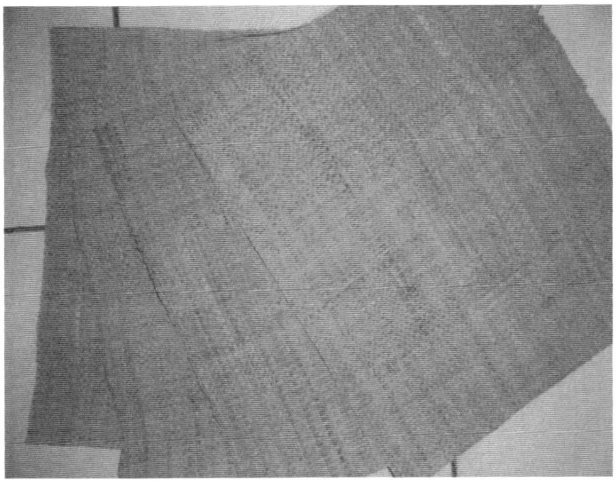

Figure 2. Woven fan palm (WFP).

Table 1. Identification of sample used for each composite.

Composite sample code	Number of WFP	Number of glass fabric	Fiber sequence
F	7	0	$[F_7]$
G1	6	1	$[F_3/\bar{G}]_S$
G2A	5	2	$[G/F_2/\bar{F}]_S$
G2B	5	2	$[F/G/F/\bar{F}]_S$
G3A	4	3	$[G/F_2/\bar{G}]_S$
G3B	4	3	$[F/G/F/\bar{G}]_S$
G4A	3	4	$[G/F/G/\bar{F}]_S$
G4B	3	4	$[G_2/F/\bar{F}]_S$
G	0	7	$[G_7]$

G : Glass fabric F : WFP

2.3. Mechanical properties of composites

The effect of fibre sequence on the mechanical properties of woven FPF/UPE composites was investigated in the materials laboratory of the Mechanical and Industry Engineering Department, Universitas Gadjah Mada, Indonesia. The tensile tests of hybrid composites were formed according to ASTM D 638 standard. Tests were conducted using Shimadzu testing machine at a cross-head speed of 5mm/min as per ASTM D 638 standard. All specimens were cut to the 165 mm x 20 mm x actual thickness and were finished by grinding machine (Figure 4). Five specimens with identical dimensions for each composite material were tested and average result was determined. The flexural tests were investigated using the three-point bending fixture according to the ASTM D790 standard, with a crosshead speed of 5 mm/min. The test was conducted using the Torsee's Universal. The rectangular shape specimens were prepared with dimensions of 80 mm x 10 mm x actual thickness. For each case, five specimens were tested and the average values reported. The impact strength of the samples was measured using Charpy impact test machine, according to the ASTM 5942 standard. All impacted specimens were un-notched. The standard specimen size was 75 mm x 10 mm x actual thickness. The average values of five specimens for each sample have been reported.

Figure 3. Vacuum bagging technique.

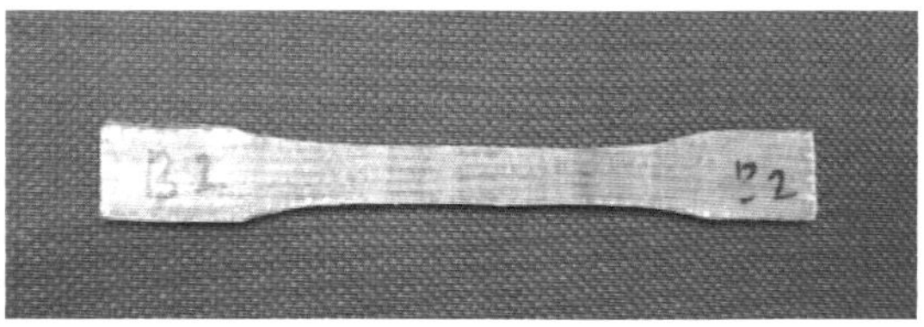

Figure 4. Tensile specimen of hybrid composites.

3. Result and discussion

3.1. Tensile properties

The effect of fibre sequence on the tensile properties of hybrid composites using are studied and shown in Figure 5. From Figure 5, it can be seen that the adding glass fabrics improve the tensile strength of hybrid composite. The highest tensile strength was obtained in the addition of four glass fabrics with sequential fibre arrangement (G4A) of 130 MPa, but still lower than pure glass fabric reinforced composites. It is attributed to the reason that glass fabrics are stronger than WFP.

In addition, the tensile strength of specimens increases due to fibre sequence in the hybrid composite. The G2A vs G2B, G3A vs G3B, and G4A vs G4B sample have different tensile strength approximately 9%, 14%, and 17% respectively. The glass fabric contained in the hybrid composite surface has a higher tensile strength than the glass fabric contained inside the hybrid composite. It is observed, in this study, that the effect of tensile strength not only the number of glass fabric but also fibre sequence in the hybrid composites. Tensile properties are higher in the composites that keep the glass fabrics as outer layers.

The similar behaviour was observed in previous work based on fibre sequence in hybrid composites [14, 15]. Fractures photos of the tensile test specimens are presented in Figure 6. The type of fibre fracture, both WFP and glass fabric composites is brittle fracture. The cross section of glass fabric looks sharper than WFP. This indicates that the glass fabric has a higher fibre tensile strength than WFP so that the strength of the hybrid composite is highly dependent on the presence of glass fabric.

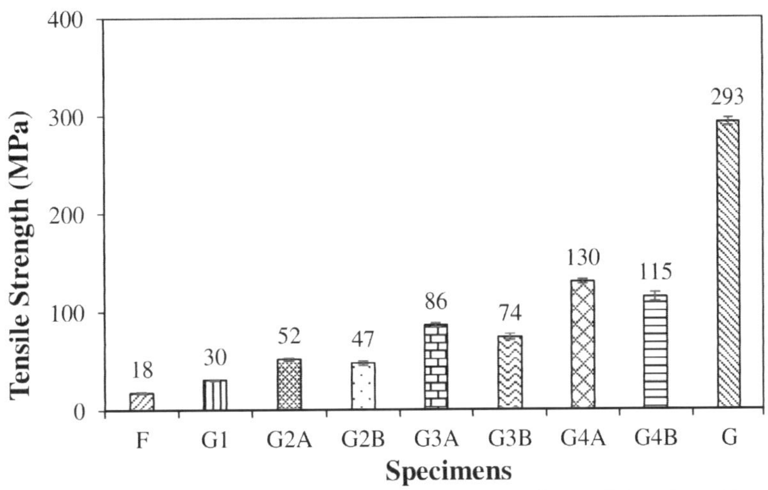

Figure 5. Tensile strength of hybrid composites with different fibre sequences.

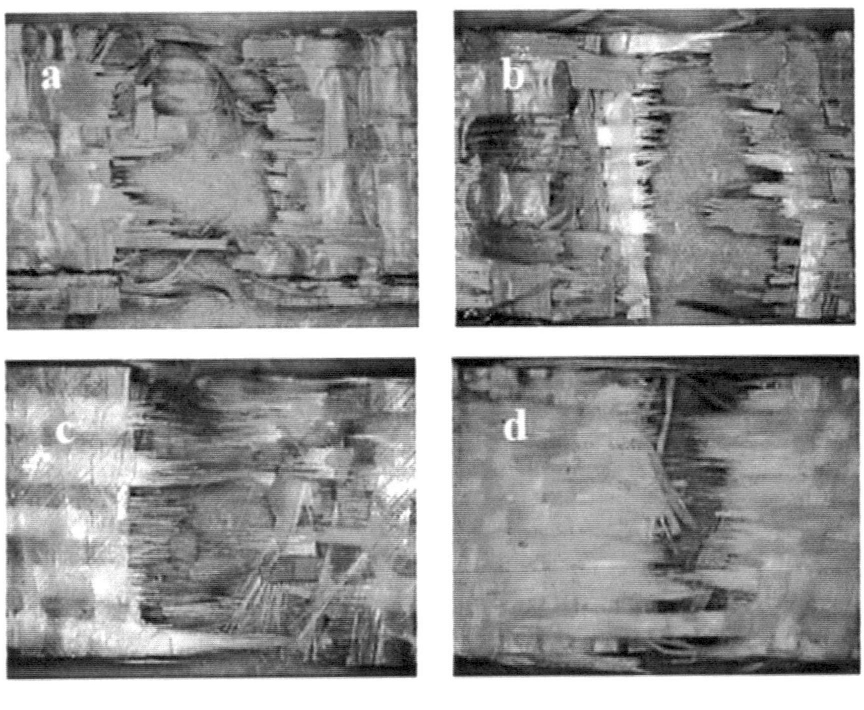

Figure 6. Fracture photograph of tensile test of (a) WFP composite (F); (b) Hybrid composites (G2B); (c) Hybrid composite (G4A); (d) Glass fabric composite (G).

3.2. Flexural strength

Figure 7 shows the flexural strength of hybrid composites with respect to fibre sequence. The flexural strength of the WFP reinforced composite is found to be 52 MPa. The addition of glass fabric enhances flexural strength. The flexural strengths of hybrid composites were obtained to be greater for fibre sequence with four glass fabric (G4A and G4B) having 168 MPa and 154 MPa, respectively. Although the flexural strength of hybrid composites still lower than that of pure glass fabric reinforced composites being 284 MPa. Previous reports also showed improvement of the flexural properties of composites due to incorporation of glass fabric [4, 16]. The same trend of data as like as tensile testing was observed for the case of flexural properties. The glass fabrics on the outside of the bamboo fibre on the hybrid composite have better flexural strength and modulus than the other fibre sequences [17].

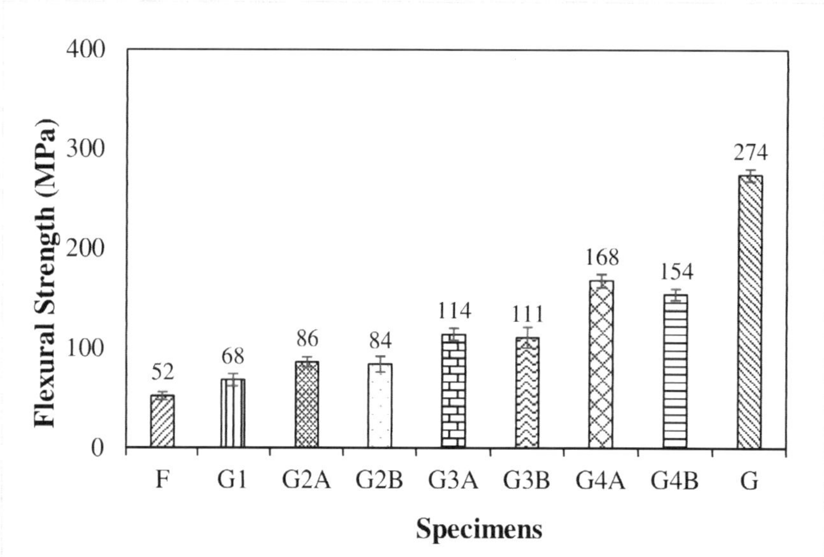

Figure 7. Flexural strength of woven hybrid composites with different fibre sequences.

3.3. Impact properties

Unnotched specimens were not broken during Charpy tests. Figure 8 shows impact strength of hybrid composites with different fiber sequences. The increase in the impact strength by increasing the number of glass fabrics can be attributed to the higher energy dissipation at the fiber/matrix interfaces to release fibers from the matrix. The impact strengths of hybrid composites were found to be greater for fiber sequence with four glass fabric (G4A and G4B) having 49 kJ.m^{-2} and 46 kJ.m^{-2}, accordingly. In this study it was found that, the impact strength of hybrid composites is in similar trend as the tensile and flexural properties. Similar results were obtained by Yahaya *et al.* [18] where the impact strength of hybrid composite was higher than that of woven kenaf composite. The specimen photos after impact test are shown in Figure 9. Delamination is not only occurred on WFP reinforced composites but also composite hybrids (9(a-c)). While the glass fabric reinforced composite, delamination was not observed so it has better impact strength (9(d)). It meant that there was a good interfacial adhesion between the unsaturated polyester matrix and glass fabric. On the other hand, poor interaction between the woven fan palm and the polyester resin can be observed in Figure 9(a-c) leading to delamination.

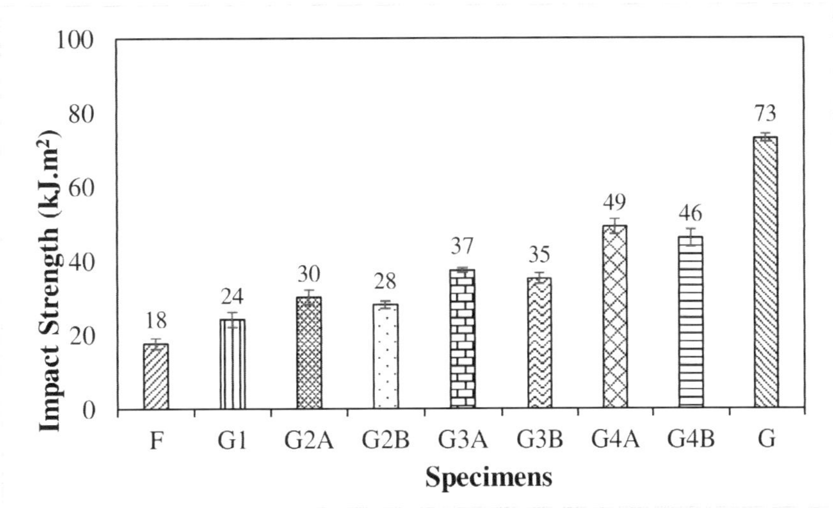

Figure 8. Flexural strength of woven hybrid composites with different fiber sequences.

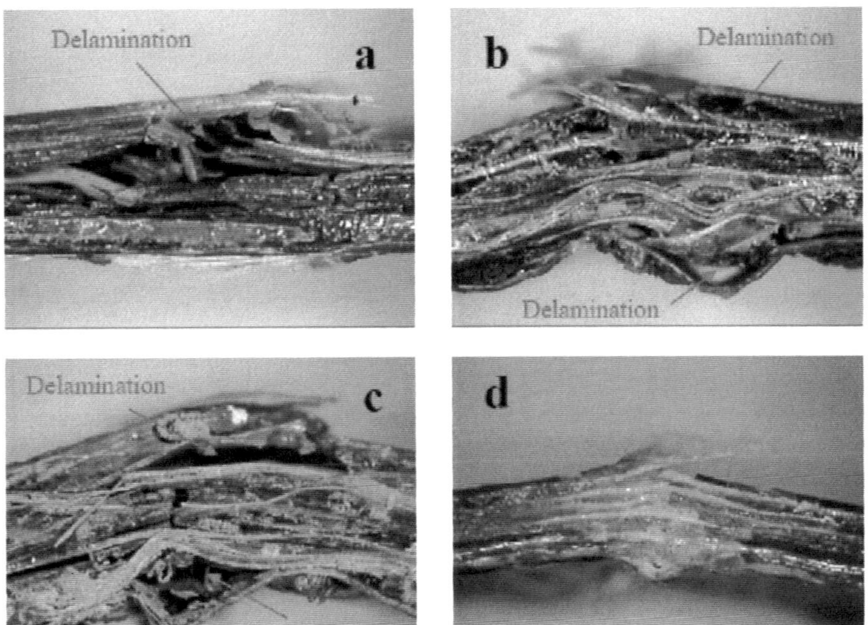

Figure 9. Fracture photograph of impact test of (a) WFP composite (F); (b) Hybrid composites (G3B); (c) Hybrid composite (G4A); (d) Glass fabric composite (G).

4. Conclusion

Woven fan palm/glass fabric reinforced polymer hybrid composites were successfully prepared. It was found that that fibre sequence strongly influenced the mechanical properties of the hybrid composites. The mechanical properties increased with increasing glass fabric content in hybrid composites with the highest tensile, flexural, impact strength being 130 MPa, 168 MPa, and 49 kJ.m^{-2}, respectively for hybrid composites reinforced with 4 glass fabrics. Based on the fibre sequence, the glass fabrics on surface of the hybrid composite have the best mechanical properties.

Acknowledgement
This work was carried out under research grant Research Grants Doctoral Dissertation (Contract no. 901/UN30.15/LT/2017), DRM Kemenristek DIKTI. The authors acknowledge the support of Kemenristek DIKTI, Universitas Bengkulu, and Universitas Gadjah Mada, Indonesia.

References

[1] Bledzki A K *et al* 1996 *J. Appl. Pol. Sci.* **5** 1329
[2] Oksman K M *et al* 2003 *Compos. Sci. Technol.* **63** 1317
[3] Ahmad F *et al* 2015 *Macromol. Mater. Eng.* **300** 10
[4] Jarukumjorn K and Suppakar N 2009 *Compos. B* **40** 623
[5] Mohanty S *et al* 2006 *Compos. Sci. Technol.* **66** 538
[6] Dwivedi U K and Navin C 2009 *J. Mater. Proc. Technol.* **209** 5371
[7] Alavudeen A *et al* 2015 *Mater. Des.* **66** 246
[8] Mariatti M *et al* 2008 *J. Compos. Mater.* **42** 931
[9] Jayabal S *et al* 2011 *J. Compos. Mater.* **45** 2279
[10] Ahmed K S and Vijayarangan S 2008 *J. Mater. Proc. Technol.* **207** 330
[11] Cheong S K and Lee S H 1997 *J. Kor. Soc. Mach. Tool. Eng.* **6** 96
[12] John K and Naidu S V 2004 *J. Reinf. Plast. Comp.* **23** 1815
[13] Hestiawan H *et al* 2016 *App. Mech. Mater.* **842** 61
[14] Salman S D *et al* 2016 *Malay. J. Civ. Eng.* **28** 10
[15] Khalil H P S A *et al* 2009 *J. Reinf. Plast. Compos.* **28** 1123
[16] Amico S C *et al* 2010 *J. Reinf. Plast. Compos.* **29** 179
[17] Latha P S *et al* 2015 *J. Ind. Text.* **46** 1
[18] Yahaya R *et al* 2016 *Def. Technol.* **12** 52

Energy adsorption of carbon-based material doping boron with Ti and Ni metal ion decoration for hydrogen storage application using density functional theory

V V R Repi [1,2,3], S R Wicaksono[1,2], F Hidayanti[2]

[1] Computational Material Laboratory, Engineering Physics Department, Universitas Nasional, Jakarta, Indonesia 12520

[2] Engineering Physics Department, Universitas Nasional, Jakarta, Indonesia 12520

[3] Centre for Sustainable and Resources Management (CSERM), Universitas Nasional, Jakarta, Indonesia 12520

vekky_repi@civitas.unas.ac.id

Abstract. The adsorption of hydrogen and the gravimetric capacity of boron-doped graphene decorated with transition metals, titanium and nickel as candidates for hydrogen storage material have been investigated earlier by using Density Functional Theory. Our calculations based on Perdew-Burke-Ernzerhof function of the Generalized Gradient Approximation method and the van der Waals interaction calculations correction using Grimme DFT + D2 method were implemented in Quantum ESPRESSO. Our result showed that from 6 of the 27 proposed material designs were tested and had adsorption energy in accordance with the criteria range of hydrogen storage material to obtain storage that can react at room temperature and achieve hydrogen gravimetric capacity of 8.927% to 9.568% which was corresponding to the target of the US Department of Energy for Light Fuel Cell Vehicles by 2020 which is only 4.5%. The six materials had average hydrogen adsorption energy of -0.397 eV to -0.223 eV per hydrogen molecule.

1. Introduction

There are at least three main reasons why hydrogen has the potential to be the ideal energy carrier to replace fossil fuels [1, 2]. First, the availability of hydrogen in nature is abundant because hydrogen is a forming element of water molecule. The next reason is that hydrogen has high efficiency for energy conversion with its energy density of 142 MJ / kg which is four times as big as its energy density contained in liquid hydrocarbon products. The last reason is that hydrogen fuel is one of the environmentally friendly energy because the emission of hydrogen energy is water vapor.

In hydrogen economy, there are two crucial aspects of its implementation namely the activity to produce hydrogen and the hydrogen storage activity. However, in this study, we focused solely on the problem of hydrogen storage to propose the design of hydrogen storage material that meets the requirement which is 4.5% of hydrogen weight for storage purposes of Light-Duty Vehicle which becomes the target of the US Department of Energy by 2020 [3]. From many methods in hydrogen storage, we have chosen hydrogen storing method with solid-state material which is safer and potentially containing more hydrogen mass with less storage space than other storage methods [4].

IOP Conf. Series: Materials Science and Engineering **432** (2018) 012051 doi:10.1088/1757-899X/432/1/012051

There are many materials that have been studied as hydrogen storage material candidates. One of the many materials that interest us is carbon-based hydrogen storage material. This is because the availability of carbon in nature is as much as the availability of hydrogen. The reason why we conducted a preliminary study of hydrogen storage using carbon-based materials is that carbon-based nanomaterials have been extensively studied for its lightweight characteristics and high surface-volume ratio [5, 6]. Unfortunately, pure carbon-based hydrogen storage material, in its use according to the adsorption, has a very small adsorption energy and cannot achieve a corresponding adsorption energy level of about -0.7 to -0.2 eV to produce an alternating process in hydrogen storage [7, 8].

In this study, we conducted a preliminary study to observe the adsorption of hydrogen molecules on carbon-based materials by comparing pure carbon materials in graphene form, boron-substituted graphene carbon material, and boron-substituted graphene carbon material with additional decoration of titanium and nickel transition-metal materials. According to the substitution of carbon atoms with boron atoms, it was found that boron opened the band gap on graphene carbon material that made the material suitable to perform atomic switches and device formation which made the material better in its use [9]. In those substitution, there was also chemical optimization of electron transmutation that obtained extra electrons from metal atoms, and boron had a property similar to carbon [10-12]. Furthermore, decorating using transition metals had been successful on improving and meeting the criteria of the gravimetric efficiency of hydrogen storage material in computation study [13]. The purposes of this research is to conduct a study on the effects of transition-metal atoms on both substituted and unsubstituted graphene carbon materials and to design the material meets the criteria that have beenrequired.

2. Experimental method

The Density Functional Theory calculations which were implemented in the Quantum ESPRESSO software package [14] and used XCrysDen [15] were conducted to visualize the material designs. In this calculations, the Ultrasoft potential with Generalized Gradient Approximation (GGA) in Perdew-Burke-Ernzerhoft (PBE) [16] format for the potential of correlations was used to ensure the accuracy of calculation for electron exchange and Grimme DFT + D2 empirical correction scheme to calculate van der Walls interaction effects [17-19]. The Plane-wave cutoff energy has been set to 25 Ry and the convergence of the force of each atom was less than 0.05 eV / Å.

The modeling of the carbon material as the material base of the surface taken from the graphene form by taking eight atoms which were expected to represent the whole graphene (bulk graphene) is shown in Figure 1. The modeling of the surface material with the parameter of a and c lattice length of 4.926 Å, based on the results of the experiment which was conducted [20, 21], used a vacuum space of 20 Å to avoid interaction. The Monkhorst-Pack matrix [22] with k-mesh was used to perform Brillouin zone sampling.

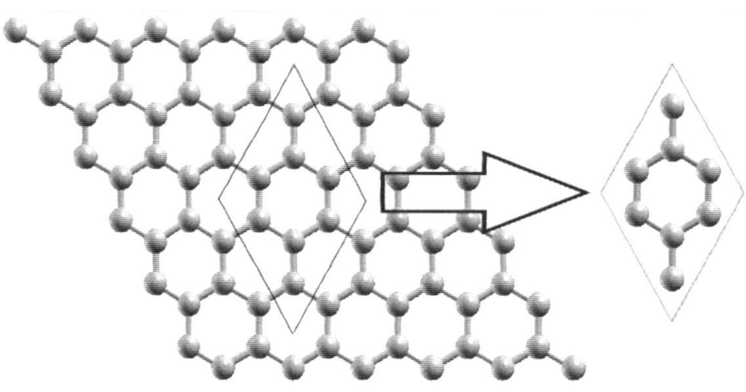

Figure 1. Simplifikasi permodelan material karbon grafena.

The relaxation simulation which was performed on all models was conducted by calculating the application of the self-consistent field to find the most stable condition of the system geometrically, and in order that energy obtained from the system represents the amount of system energy in a stable state. The energy was obtained from the relaxation calculations of each model to be used to calculate the energy of the transition-metal bond and the adsorption energy of the hydrogen molecule to obtain hydrogen storage material candidate.

In our study, there were three equations involved. Firstly, it was to calculate the transition-metal bonding energy at the surface. Secondly, it was to calculate the average adsorption energy of hydrogen. And the last was the equation to calculate the hydrogen gravimetric of the overall material designs.

The first equation is E_b, Energy bond of the transition metal on the surface material was calculated by using the equation:

$$E_b = E_d - (E_s + E_{TM}) \tag{1}$$

where E_d is the energy when the surface has transition-metal decoration or the total energy of the material surface, E_s is the energy of the surface or the energy of the material surface, and E_{TM} is the total energy of the free transition-metal atoms. Furthermore, our study focused on hydrogen storage capacity. There were two natural calculations in this section, the calculation of hydrogen adsorption energy and gravimetric or the weight of hydrogen contained in the cell unit.

For the calculation of hydrogen adsorption energy, it can be calculated by using equation (2).

$$E_a = (E_T - (E_d + nE_{H_2}))/n \tag{2}$$

where E_T is the total energy of hydrogen adsorption in the surface material with or without transition metals, Ed has the same terminology with equation (1). E_{H_2} is the total energy of free hydrogen molecules, and n represents the sum of the adsorbed hydrogen molecules.

Next, we used equation (3) to calculate the weight percentage or gravimetric of the hydrogen molecule.

$$\mathrm{H}_2(wt\%) = \left[\frac{m_{H_2}}{m_{H_2} + m_{host}}\right] \times 100 \tag{3}$$

where m_{H_2} is the mass of a total number of the adsorbed H_2 molecule and m_{host} is the mass of the surface.

3. Result and discussion

This study focused on the calculations which involve those three equations above divided into three parts of the research. The first calculation was to calculate the bond energy of nickel and titanium transition metals of the surface material either substituted by boron atoms or not. The next calculation was the calculation of the adsorption energy of hydrogen molecules on the overall surface designs, either with or without transition-metal decoration. The last calculation applied if in the previous stage, the surface design obtained the adsorption energy of the corresponding hydrogen molecule or about -0.7 eV, then the surface will be recalculated by the amount of hydrogen that meets the needs of the United States Department of Energy.

3.1 Bond energy of transition-metal decoration

The bond energy of transition-metal decoration can be obtained by using equation (1). It was the

IOP Conf. Series: Materials Science and Engineering **432** (2018) 012051 doi:10.1088/1757-899X/432/1/012051

calculation of the bond energy of surface material which was substituted by boron or not towards nickel and titanium transition metals. There were three surfaces to be tested, namely the surface of eight carbon atoms (C8), one boron-substituted atom on carbon (BC7), and two boron-substituted atoms on carbon (B2C6), as shown in Figure 2. Then there were three possible locations for placing the transition metals such as the bridge site that was on the carbon-carbon or boron-carbon bond center, the hollow site which was located on the hollow or the hexagon ring of the surface, and the top site where the position of decorating atoms located directly above carbon or boron atoms.

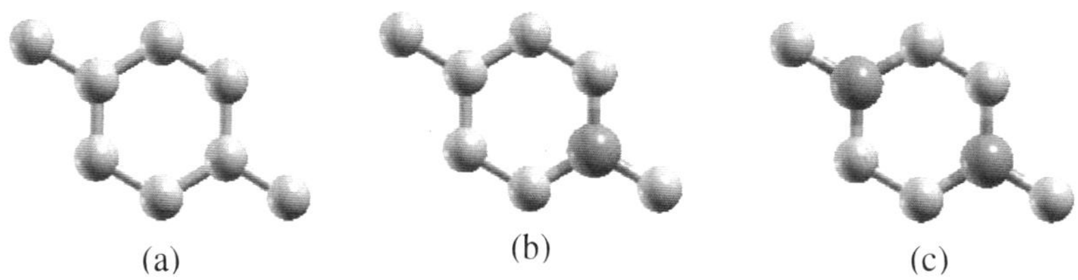

Figure 2. The model surface of, (a) C8, (b) BC7, and (c) B2C6, where the carbon show as the yellow atom and the boron show as the grey atom.

Table 1 is the summary of the bond energy calculations of nickel transition-metal decoration of the surface material. Based on the results of the calculations, it can be seen that the increase in the number of boron atoms in the substitution of carbon results in the increase of bond energy (more negative).

Table 1. Summary calculation energy binding (E_b) of nickel atom on the surface.

Location	E_b on C8 (eV)	E_b on BC7 (eV)	E_b on B2C6 (eV)
Bridge	-1.736	-2.593	-2.863
Hollow	-1.877	-2.921	-3.260
Top	-1.685	-2.594	-2.554

Like nickel decoration, titanium transition-metal decoration of the surface also applied the same thing, the increase of bond energy occurred was in direct proportion to the increase of the amount of boron atoms substitution of the surface material at each bonding site. In the case of the best sites of each material type was not like nickel decoration, on C8 and BC7 surface materials, they had the best site at hollow sites and for B2C6 surface material, titanium decoration had a preference at the bridge site. A summary of the calculation of the bond energy of the titanium transition-metal decoration can be seen in Table 2.

Table 2. Summary calculation energy binding (E_b) of titanium atom on the surface.

Location	E_b on C8 (eV)	E_b on BC7 (eV)	E_b on B2C6 (eV)
Bridge	-1.761	-3.431	-4.053
Hollow	-2.489	-3.775	-4.047
Top	-1.525	-3.141	-3.813

3.2 Adsorption energy of the hydrogen molecule

The purpose of this preliminary study is to observe the adsorption of hydrogen on carbon-based material either with boron atoms substitution or not and the presence of nickel or titanium decoration atoms at three points of symmetry on the surface (bridge, hollow, and top site). We tested all three types of material surfaces at all points of symmetry. Firstly, the test was performed on the surface materials

IOP Conf. Series: Materials Science and Engineering **432** (2018) 012051 doi:10.1088/1757-899X/432/1/012051

without transition-metal decoration. The second test was performed on surface materials with nickel decoration. And the last was on the surface materials with titanium decoration. The test involved equation (2) to obtain the energy adsorption of the hydrogen molecule.

Based on the calculation of the adsorption energy of hydrogen molecules on surfaces which were not decorated with transition metals, we found that nothing can reach the range of adsorption energy in accordance with the literature [6, 7]. On the surface of C8, it could achieve adsorption energy of -0.066 eV at the hollow site, -0.062 eV at hollow sites of the surface of BC7, and -0.066 eV at the top site of the B2C6 surface. The summary of the calculations is shown in Figure 3.

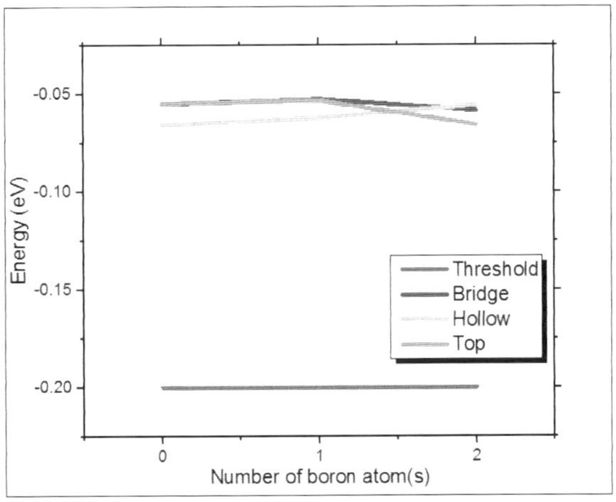

Figure 3. Summary adsorption energy calculation on surface without decoration.

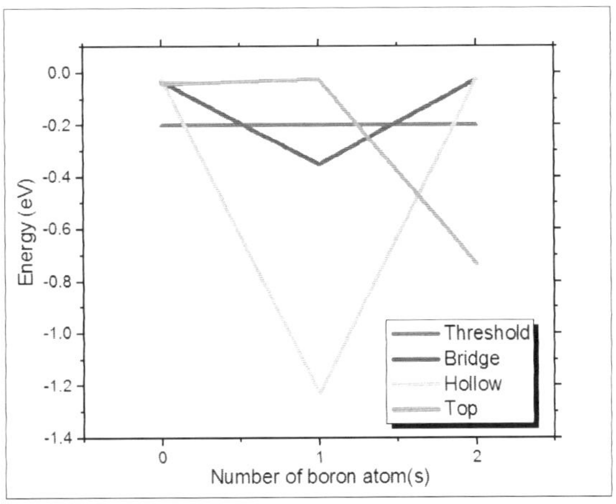

Figure 4. Summary adsorption energy calculation on surface with Nickel decoration.

The presence of nickel transition-metal material decoration improved the material's ability to adsorb hydrogen molecules. In our calculations, a material design, nickel which was decorated at the bridge site on the surface of the BC7 material, reached the recommendation range for adsorption energy of hydrogen storage with an adsorption energy of -0.352 eV. The other two designs gained the adsorption energy value which was under the recommendation range or below -0.7 eV namely nickel-decorated

BC7 material at the hollow site and the nickel transition-metal decorated B2C6 surface material at the top site which had bond energy with one hydrogen molecule of -0.732 eV. Both designs were indeed more negative than the recommended value as hydrogen storage material because the hydrogen molecules will be bonded more strongly with greater energy that greater energy to break the bond was also needed. However, we believed in adding the number of hydrogen molecules to our material designs. A summary of these overall calculations was depicted in the gradient in Figure 4, where the calculations below the recommendation line will be further calculated by adding the number of hydrogen molecules.

The titanium decoration at the top sites was successful in improving the ability of surface materials to achieve the range of adsorption energy of hydrogen molecule for the recommended hydrogen storage application. We obtained three design materials that meet the needs. Firstly, C8 material surface which had an adsorption energy of -0.362 eV. Secondly, BC7 material surface with an energy of -.0337 eV. The last was B2C6 material surface which had adsorption energy of -0.308 eV. The summary of the calculations we performed is presented in the graph shown in Figure 5.

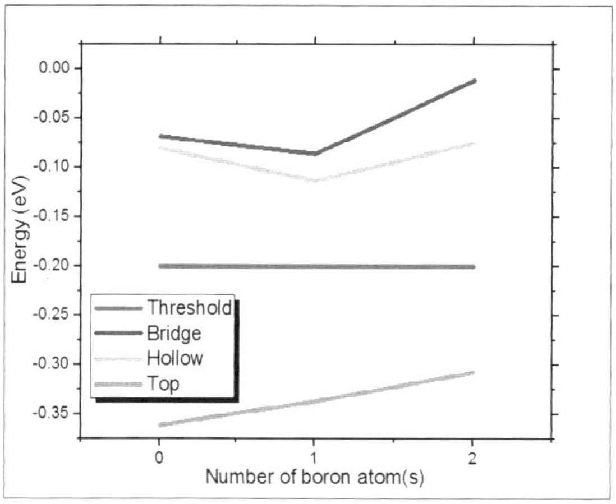

Figure 5. Summary adsorption energy calculation on surface with Titanium decoration.

3.3 Further calculation with more hydrogen molecule

Those six materials that had met the requirements for further calculations will be recalculated by adding the number of hydrogen molecules. For surfaces with nickel transition-metal atomic decoration, we added seven more hydrogen molecules to all candidate material designs, and added six hydrogen molecules to the surface with titanium decoration.

Table 3. Hydrogen adsorption energy calculation result with gravimetric above US Department of Energy target.

Material Combination	nH_2	Total adsorption energy (eV)	Average adsorption energy (eV)	Gravimetric (%)
bridge-Ni@BC7	8	-1.938	-0.242	9.502
hollow-Ni@BC7	8	-1.783	-0.223	9.502
top-Ni@B2C6	8	-2.737	-0.342	9.568
top-Ti@C8	7	-2.779	-0.397	8.927
top-Ti@BC7	7	-2.926	-0.366	8.995
top-Ti@B2C6	7	-2.429	-0.347	9.064

The summary of the results can be seen in Table 3, all material designs have reached the recommendation range of hydrogen adsorption energy for hydrogen storage to obtain back and forth storage in the temperature room and to obtain gravimetric value over the target of US Department of Energy for light-duty fuel cell vehicle for 2020. From the results of this adsorption energy, the material designs we proposed could be good candidates for the hydrogen storage.

4. Conclusion

In this research, the Density Functional Theory calculations for our pilot study has been performed concerning on hydrogen adsorption of our material designs combined with transition-metal decoration on carbon-based material both with and without boron substitution. It is found that the addition of transition metals can increase the adsorption energy of carbon-based materials so that the addition of hydrogen molecules can produce the average adsorption energy that reaches the range of adsorption energy which is suitable for hydrogen storage. Our results showed that the adsorption energy has fulfilled the needs of the United States Department of Energy on carbon-based material with boron doping and transition-metal decoration. The average hydrogen adsorption energy of -0.397 eV to -0.223 eV with gravimetric of 8.927% to 9.568% is close to the energy required to obtain back and forth storage at room temperature. Therefore it would be a good candidate for hydrogen storage.

Acknowledgment

Thanks to the GRID@LIPI team, the Indonesian Institute of Sciences who has granted permission to use LIPI's High Performance Computing facility to help researchers complete this study.

References

[1] Schlapbach L and Zuttel A 2001 *Nature* **414** 353–8
[2] Tozzini V Pellegrini V 2013 *Phys. Chem. Chem. Phys.* **15** 80–9
[3] Departement of Energy 2017 *Target Explanation Document: Onboard Hydrogen Storage for Light-Duty Fuel Cell Vehicles* United States of America
[4] Dixit M Maark T A Pal S 2011 *Int. J. Hydrogen Energy* **36** 10816–27
[5] Liu C Fan Y Y Liu M Cong H T Cheng H M and M S 1999 *Science* **286** 1127-9
[6] Dillon A C Jones K M Bekkedahl T A Kiang C H Bethune D S and Heben M J 1997 *Nature* **386** 377–9
[7] Beheshti E Nojeh A Servati P 2011 *Carbon N. Y.* **49** 1561–7
[8] Ao Z Dou S Xu Z Jiang Q Wang G 2014 *Int. J. Hydrogen Energy* **39** 16244–51
[9] Sharma R Khan S Goyal V Sharma V and Sharma K S 2017 *FlatChem* **1** 20–33
[10] Andrey T A A V Henrietta L and B D G 2015 *Int. J. Energy Res.* **39** 524–8
[11] Alexandrova A N Birch K A and Boldyrev A I 2003 *J. Am. Chem. Soc.* **125** 10786–7
[12] Olson J K and Boldyrev A I 2012 *Chem. Phys. Lett.* **523** 83–6
[13] Bhattacharya A Bhattacharya S Majumder C and Das G P 2010 *J. Phys. Chem. C* **114** 10297–301
[14] Giannozzi P Baroni S Bonini N Calandra M *et al.* 2009 *J. of Phys.:Conden. Matt.* **21** 395502-20
[15] Kokalj A 2003 *Comput. Mater. Sci.* **28** 155–68
[16] Perdew J P Burke K Ernzerhof M 1996 *Phys. Rev. Lett.* **77** 3865–8
[17] Grimme S 2006 *J. Comput. Chem.* **27** 1787–99
[18] Bučko T Hafner J Lebègue S Ángyán J G 2010 *J. Phys. Chem. A.* **114** 11814–24
[19] Luo Z Fan X Pan R An Y 2017 *Int. J. Hydrogen Energy.* **42** 3106–13
[20] Novoselov K S Geim A K Morozov S V Jiang D 2004 *Science* **306** 666–9
[21] Cervenka J Flipse C F J 2007 *J. Phys. Conf. Ser.* **61** 190
[22] Monkhorst H J Pack J D 1976 *Phys. Rev. B.* **13** 5188–92

Position accuracy evaluation of multileaf collimator (MLC) static in linear accelerator

Y Rahmat[1], S A Pawiro[1], Y A Baihaqi[1], M Fadli[1,2] and A Nainggolan[2]

[1] Department of Physics, Faculty of Mathematics and Natural Sciences, University of Indonesia, Depok, Indonesia
[2] Departement of Radiotherapy, MRCCC Siloam Hospital, Jakarta, Indonesia

supriyanto.p@sci.ui.ac.id

Abstract. This study was aimed to evaluate the accuracy of static MLC positions using gafchromic BT3 film. EBT3 gafchromic film was scanned using Vidar dosimetry Pro Advantage and was analyzed using Matlab algorithm. Matlab algorithm which determined leaf edge of MLC was using Edge Detection Function with Laplacian of Gaussian Method. The precision of MLC leafs on Linac 1 and Linac 2 was measured using the Matlab application code and being compared to early data in TPS setting. The measurement showed that average value of the best precision was obtained at line 6 and the lowest at line 2 both in Linac 1 and 2. MLC leaf measurement error using Matlab was still within tolerance limit AAPM TG 142 which is ± 1 mm. An Error value of the MLC was increased if the leaf movement over the normal position. This effect was caused by an influence of vertical line scattering (VL) and inaccuracy positioning of gafchromic film EBT3 when scanned. The result of this study showed that MLC position accuracy in LINAC 1 and 2 was still within tolerance limit based on AAPM PG 142. It could be an alternative measurement of MLC position accuracy more easily.

1. Introduction

Radiotherapy is one of the main choices for people in the world and in Indonesia for the treatment of cancer. Approximately,50 percent of all cancer patients around the world use radiotherapy as the one of the method for their cancer treatment [1]. Radiotherapy techniques use ionizing radiation that causes damage to cancer cells and healthy tissue.

Therefore, there are several types of radiation techniques used to achieve the optimum dose for cancer cells and a minimum dose for the surrounding target or the normal tissue. Radiotherapy techniques have developed starting from 2D (conventional) techniques to 3-dimensional techniques such as Three Dimensional Conformal Radiotherapy Technique (3D-CRT), Intensity Modulated Radiotherapy (IMRT), and Volumetric Arc Radiotherapy (VMAT). Furthermore, IMRT techniques are able to achieve the conformal doses better than the 3D-CRT technique, especially for irregular shapes target volumes which employed the multileaf collimator. The variation in radiation intensity or fluence is shaped by the shift of MLC leaf which attenuated and collimated the radiation beam according to the required dose distribution.

In 3D-CRT and IMRT, MLC of linear accelerator was used to modulate the fluence in the same field to increase the conformity distribution of doses in the area tumor [2]. There are two kinds of MLC, they are static MLC (SMLC) and dynamic MLC (DMLC) which have been implemented in a clinical setting. Technically, the static MLC is not moving during beam on (discrete radiation) whereas the dynamic MLC allow the MLC moving continuously during radiation, taking place in each direction on the gantry angle [3]. From several segments, the radiation beam was formed into a fluence map consisting of a set of segments from all radiation fields of different intensities so that the dose distributions given to the target will be conformal.

The accuracy of the MLC position is one of the most important factors determining a patient's dose [4]. It was proposed by Mu *et al.*, that a small systematic error in the inaccuracy of the MLC position could have a major effect on the doses which are given to the target as well as on the normal tissue [5]. The American Association of Physicist in Medicine (AAPM) Task Group (TG) number 142 recommended positioning accuracy for the position of MLC with tolerance ± 1 mm [6]. A standard measurement is carried out by using the LINAC apparatus with Electronic Portal Imaging Devices (EPID). In the contrary, there are many LINAC apparatus which have no EPID yet. This study was designed in order to obtain another way of measuring MLC position precision accurately without using EPID.

Similar studies had ever been held before. El-Maraghy *et al.* (2014) used 1 Linac Varian with MLC 80 leaves. Each leaf had a thickness of 1 cm and the study field size was 1 cm. Measurement the file used film Kodak EDR2 and being analyzed by using Matlab. Average of the result was 9.95 ± 0.09. Julia Gotstedt (2015) used 1 Linac with MLC 120 leaves and the thickness was 0.5 cm. Field shape was irregular and accuracy was measured by using EPID [8]. Matlab was also used to analyze the accuracy as the first study did.

2. Materials and methods

2.1 Scanning slab water fantom

This study employed the solid water phantom with density 1.19 gr/cc. Slab water phantom was scanned by using CT simulator Philip Brilliance 16 slices with 120 kV energy, 50 mAs and a slice thickness of 2 mm. Scanning of slab water fantom by using CT simulator was to obtain the image of the slab water phantom which was sent to TPS for MLC patterns simulation. The gafchromic EBT3 film was positioned at a depth of 1.5 cm in the build-up area and the base thickness 10 cm to avoid backscattering. At the time of scanning was done by a marker on slab water phantom to determine the reference point. After acquiring the image of slab water phantom then the data could be sent to TPS for planning.

2.2 Planning MLC pattern

This study was performed by using two Linacs namely Varian Clinac iX at Siloam MRCCC Hospital Semanggi with 120 MLCs millennium consisting of 80 leafs with a thickness of 0.5 cm and 40 leaf MLC with a 1 cm thickness. This study only used 80 leaf MLC with a thickness of 0.5 cm contained in Bank A and Bank B. Besides MLC, Linac has equipped jaws with length 36.7 cm and height 7.8 cm. MLC pattern was created in Treatment Planning System (TPS) Eclipse version 11 with Anisotropic Analytical Algorithm (AAA) algorithm. The inverse planning method was used to determine the radiation parameters for initial treatment such as collimator and gantry angle, MU value and number of the field. This MLC pattern was created based on the previous study which provided 10 MLC patterns [7]. This study was performed by using 3D-CRT with static MLC. An overlapping gap was made between the first and second pattern with 1 cm distances and 0° collimator angle. The gap was aimed to get a flat profile in order to see the pattern accuracy. The third to sixth as well as first and second pattern but being different in MLC position depending to film size. The seventh and eighth patterns were also set to be overlapping and collimator angle was 0°. However, the MLC opening was 10 cm wide both in Bank A and Bank B. They formed a horizontal pattern called horizontal line (HL). The ninth and tenth patterns were made similar to the seventh and eighth one, but with the collimator angle was 90°. These last patterns were vertical called vertical line (VL). Horizontal line and vertical line were perpendicular to each other, the intersection of these two lines was used as a reference point to measure the position accuracy of each MLC leaf on every line [7]. The MU value assigned to the first to the sixth pattern were 200 MU while seventh to the tenth pattern were 250 MU. The next step of planning is the simulation of MLC patterns using TPS and sent to Linac machine for irradiation of the gafchromic EBT3 film. Figure 1 showed 10 MLC patterns created in TPS.

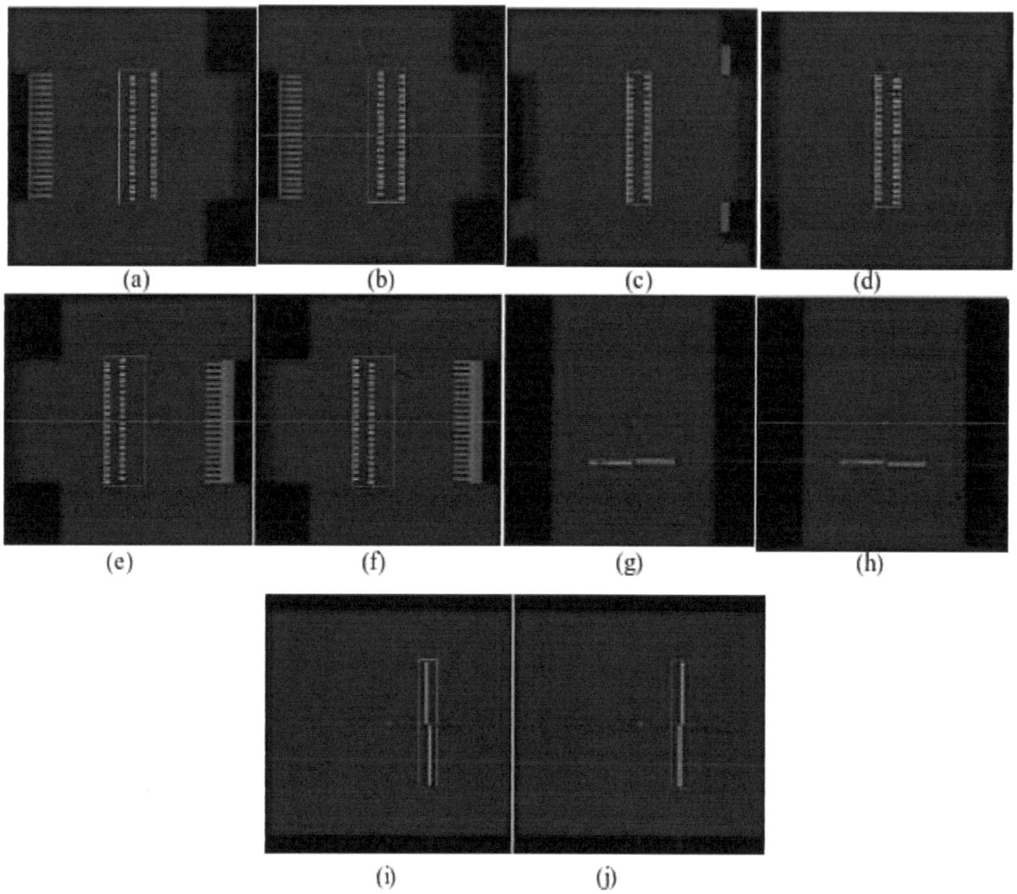

Figure 1. The MLC pattern design created on TPS Exclipse (a) the first pattern, (b) the second pattern, (c) the third pattern, (d) the fourth pattern, (e) the fifth pattern, (f) the sixth pattern, (g) seventh pattern, (h) eighth pattern, (i) the ninth pattern, and (j) the tenth pattern.

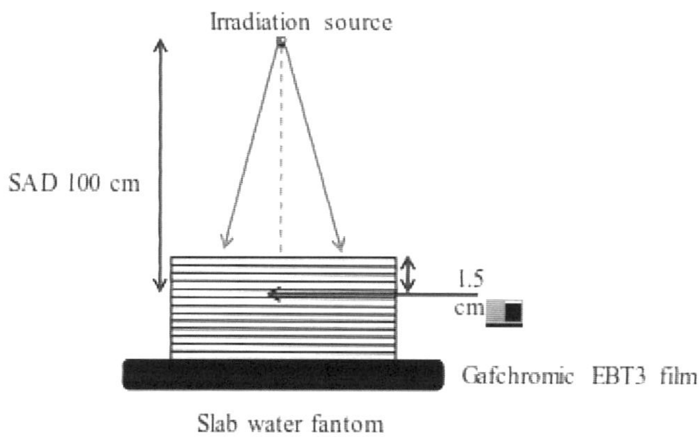

Figure 2. Setting of gafchromic EBT3 film irradation position on slab water phantom.

2.3 Irradiating of gafchromic EBT3 film

The MLC pattern created at the TPS was sent to Linac for irradiation with 6 MV photon energy. Gafchromic EBT3 film was inserted in the slab of water phantom Adapter Plate RW3-FC65 with size 30cm x 30cm size at maximum depth of 1.5 cm, source to axis distance (SAD) of 100 cm and backscatter materials with 10 cm thickness. The size of EBT3 gafchromic film used in the experiment was 25.4cm x 20.32cm. The position of gafchromic EBT3 film when irradiating 10 MLC patterns was not changed, but the MLC pattern did change. Furthermore, each irradiation on the EBT3 gafchromic film was repeated three times to evaluate precision and accuracy of the MLC leaf position. After irradiation on gafchromic EBT3 film with as long as 2 x 24 hours, then it was scanned using Vidar dosimetry Pro scanner. The scanned image of gafchromic EBT-3 film was converted into Tagged Image File Format (TIFF) and being analyzed by using Matlab. Figure 2 shows the setting of irradiation of gafchromic EBT3 film on slab water phantom.

2.4 Matlab code

Matlab algorithm was employed to evaluate the images. The evaluation of the images was performed using the Matlab function Edge Detection with Laplacian of Gaussian Method. This method was found to be the most suitable one for this experiment with a zero threshes value and a standard deviation of sigma 6 After the edge of leaf position was detected, the mark "+" was created. The MLC aperture was determined with a vertical line as the reference. From the reference position, we can automatically calculate how far the distance is between Bank A and Bank B. Matlab reading results are named from right to left with line 1 to line 6 and giving leaf numbers from bottom to top starting from leaf 11 to leaf 50.

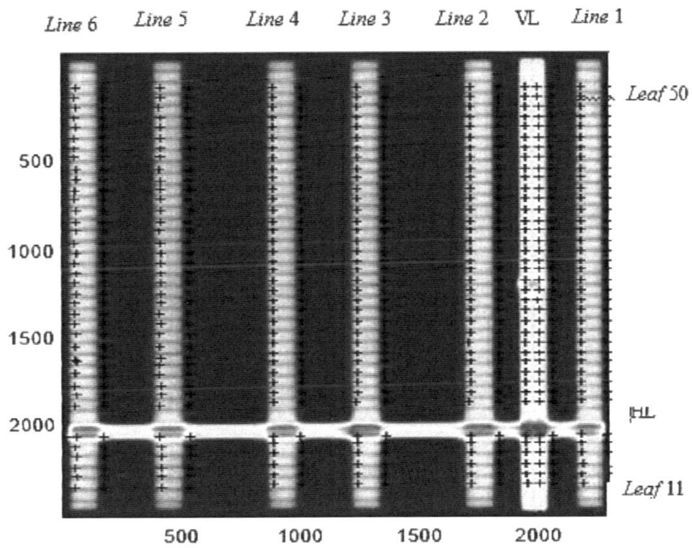

Figure 3. The reading results of MLC pattern in gafchromic EBT 3 film used Matlab algorithm.

3. Results and discussion

3.1 Accuracy test of MLC aperture

3.1.1 Linac 1. MLC aperture accuracy test was aimed to determine the error value of each MLC leaf. It used gafchromic EBT3 film three times. After the test, the film was analyzed using Matlab algorithm. The result showed average error value of MLC aperture on line 1 to line 6 respectively were (0.34 ±

0.12) mm, (0.29 ± 0.17) mm, (0.13 ± 0.07) mm, (0.14 ± 0.07) mm, (0.18 ± 0.11) mm, and (0.12 ± 0.07) mm. These results showed that line 1 had the highest MLC aperture value and line 6 had the lowest. The high error value was resulted by scattering effect of the vertical and horizontal line that affected the test result. However, aperture error values for each MLC leaf were still within the tolerance limits recommended by AAPM TG 42 namely ± 1 mm. They showed that the performance of MLC Linac 1 was still in good condition but a periodic check is necessary to ensure that the MLC works optimally. Figure 4 shows the average error value of TPS data MLC aperture on Linac 1 using Matlab algorithm.

3.1.2 Linac 2. The average error value of the MLC aperture using Matlab on Linac 2 was still below the tolerance limit value. They were respectively (0.14 ± 0.08) mm, (0.53 ± 0.13) mm, (0.11 ± 0.07) mm, (0.10 ± 0.07) mm, (0.47 ± 0.16) mm, and (0.21 ± 0.09) mm. It could be seen that line 2 had the highest aperture error value and the lowest was in line 4. These result described linac 2 had good MLC aperture accuracy. Figure 5 shows the TPS data MLC aperture average error value using Matlab algorithm on Linac 2.

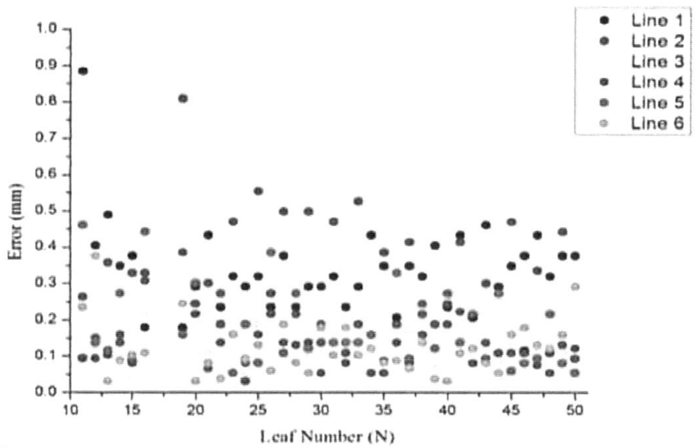

Figure 4. The average graph of the MLC aperture error of the six lines between the Matlab measurement results and the TPS data on Linac 1.

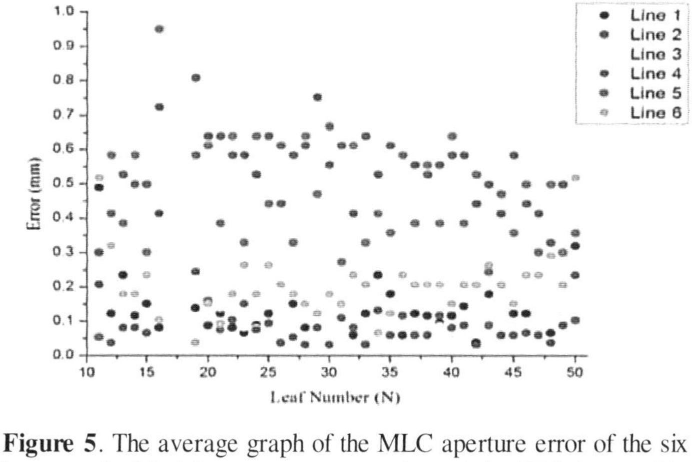

Figure 5. The average graph of the MLC aperture error of the six lines between the Matlab measurement results and the TPS data on Linac 2.

3.2 Leaf position accuracy evaluation

3.2.1 Linac 1 at bank A. MLC position accuracy test was aimed to ensure the accuracy of each MLC leaf forming irradiation field according to TPS data. The test reference point was at the vertical line (VL) and the horizontal line (HL) intersection. When the reference point was obtained, test position measurements could be done to end of each leaf at Bank A and Bank B horizontally. The values on line 1, line to line 6 respectively were (0.32 ± 0.14) mm, (0.59 ± 0.15) mm, (0.31 ± 0.14) mm, (0.61 ± 0.14) mm, (0.42 ± 0.17) mm, and (0.50 ± 0.15) mm. The results showed that line 4 has the highest error value, this was due to leaf movement effect from Bank A passing the normal position. These values showed that Linac 2 was still within the tolerable range recommended by AAPM TG 42 namely ± 1 mm. Although it is not significant, the average value of MLC position accuracy shows that the further movement of leaf MLC passing the normal position the higher the error value becomes. It could be seen by the condition being closer to line 6, the higher error value. It was due to the position of the leaf at Bank A was moving away from the normal position.

3.2.2 Linac 1 at Bank B. In Bank B, the MLC error position average value of each line was relatively smaller than Bank A. The values on line 1 to line 6 respectively were (0.15 ± 0.07) mm, (0.34 ± 0.19) mm, (0.26 ± 0.14) mm, (0.53 ± 0.10) mm, (0.32 ± 0.14) mm and (0.50 ± 0.16) mm. They show that the highest error value was at line 4 and the lowest was at line 1. The cause was the movement of the leaf at Bank B not far away passing the normal position. Overall, error value indicates the error position occurring in bank B was still within the tolerance limits recommended by AAPM TG 142 namely ± 1 mm.

3.2.3 Linac 2 at Bank A. MLC leaf error position evaluation average value on line 1 to line 6 respectively were (0.15 ± 0.08) mm, (0.88 ± 0.18) mm, (0.30 ± 0.14) mm, (0.96 ± 0.17) mm, (0.77 ± 0.19) mm, and (0.82 ± 0.16) mm. The values showed that line 4 has the highest error average value and line 1 was the lowest. As many as 14.91% of the total leaf had error value exceed the tolerable limit of AAPM TG 142 namely ± 1 mm overlimit values were at line 2, line 4, line 5 and line 6. In general, the values showed position error in Bank B still being tolerable according to AAPM TG 142 namely ±1 mm average error position of six line in Bank B Linac 1.

3.2.4 Linac 2 at Bank B. The average value of MLC error position on line 1 to line 6 were (0.21 ± 0.11) mm, (0.33 ± 0.15) mm,(0.31 ± 0.12) mm, (0.94 ± 0.13) mm, (0.31 ± 0.12) mm, and (1.02 ± 0.13) mm. From the average error value of MLC leaf position showed line 6 has the highest average value. The percentage of the total leaf was 14.91%. The cause was due to scattering from the horizontal line and vertical line.

4. Conclusion

Overall the average error value of MLC aperture measurements using Matlab on Linac 1 and Linac 2 was still within the AAPM tolerance limit of TG 142. The average value of leaf position accuracy on Linac 1 was also within tolerable limits.

Acknowledgment

The authors would like to express special thanks to the medical physicist of Department of Radiotherapy MRCCC Siloam Hospital Semanggi (M. Fadli and A. Nainggolan) who helped with the measurements and other preparations along this research; also to the Laboratory of Medical Physics, University Indonesia which provided equipment needed in this research. This study was supported by Hibah Pascasarjana University of Indonesia 2017 with contract number 688/ UN2.R3.1/ HKP.05.00/ 2017.

References

[1] International Atomic Energy Agency 2007 Vienna: International Atomic Energy Agency 430

IOP Conf. Series: Materials Science and Engineering **432** (2018) 012052 doi:10.1088/1757-899X/432/1/012052

[2] Schlegel W 2006 *New Tech. in Radiation Oncology* chapter 11 New York: Stranger 1-11
[3] American Association of Physicist in Medicine 2001 American Association Physical in Medical
[4] Yan G Liu C Simon T A Fox C and Li J G 2009 *J. Appl. Med. Phys.* **10(1)** 120-8
[5] Mu G Ludlum E and Xia P 2007 *Phys. Med. Biol.* **53(1)** 77-88
[6] American Association of Physicist in Medicine 2009 American Association Physicist in Medicine's Sciences 36 4197-212
[7] El-maraghy K A Metwaly M El-sayed E S M and Mohamed A 2014 *J. of Radiation Research and Appl. Sci.* **7(2)** 230-40
[8] Gotstedt J Hauer A K and Back A 2015 *Med. Phys.* **42(7)** 3911-21

Modification of ZIF-8 using direct mixing method at a room temperature

A Mustaqim[1], B Yuliarto[1,2*], Nugraha[1,2]

[1] Advanced Functional Materials Laboratory, Engineering Physics Department, Bandung Institute of Technology, 10 Ganesha St., Bandung 40132, Indonesia
[2] Research Center for Nanoscience and Nanotechnology (RCNN), Bandung Institute of Technology, 10 Ganesha St., Bandung 40132, Indonesia

brian@tf.itb.ac.id

Abstract. ZIF-8 modified with CNT doping have been successfully synthesized using direct mixing methods at room temperature. The characteristics of ZIF-8 and ZIF-8 with CNT were studied using XRD and SEM. The structural integrity and phase purity of both materials were obtained through X-ray diffraction patterns and was found comparable to that of sodalite-type ZIF-8. The peaks were found at 2θ = 7.4°, 16.5°, 22.1°, 25.7°, 26.7°, and 32.53°. When ZIF-8 was doped with CNT, the peaks of ZIF-8 decreased. Scanning electron microscopy of pure ZIF-8 showed hexagonal nanocrystals, and the particle sizes were distributed at approximately 50-140 nm. CNT-doped ZIF-8 showed that ZIF-8 attached to CNT was uniformly with diameters of approximately 80-120 nm. It was concluded that ZIF-8 modified using CNT had been succesfully synthesized.

1. Introduction

Zeolitic imidazolate frameworks (ZIFs) are a sub-family of porous metal-organic frameworks (MOFs) built of tetrahedrally coordinated metal ions bridged by imidazolates [1]. The crystal structure of zeolitic imidazolate frameworks (ZIFs) has the same topologies as aluminosilicate zeolites. In zeolites, the framework is built by tetrahedral silicon or alumunium bridged by oxygen atoms. Whereas in ZIF-8, the tetrahedral frameworks are built by transition metal (Zn) as nodes and imidazolate linkers as bridges [2, 3]. Similiar to zeolites, the frameworks of ZIF-8 has pores and channels that allow access of guest molecules and can differentiate species on the molecular level. ZIF-8 have more flexibility in surface and structure modification, rigidity, and better thermal, hydrothermal, and chemical stabilities. In 2017, Zhang has succesfully modified ZIF-8 structure with rod-like structures [4]. That is advantages ZIFs then zeolites.

ZIF-8 can be synthesized using several methods such as hydrothermal, solvothermal, microwave assisted, and direct mixing. Among those methods, the easiest way to get ZIF-8 is using direct mixing methods. Because it can be prepared in room temperature. Surface area ZIF using this method is 1073 m^2g-1 [3].

ZIFs have many advantages on the morphology and properties. It has a uniform porosity and an exceptional chemical and thermal stability. Its made ZIFs have a wide range of promising applications such as supercapacitor using ZIF-67 [5], gas storage [6], drug delivery [7], and chemical sensors [8].

To enhance the ZIFs capabilities several treatment have been done by several treatments such as modified structure, solvent, and modification by another materials. This work reports the use of CNT on ZIF-8 and the influence CNT as modification material on ZIF-8 also investigated on this study.

2. Experimental method

First, 132 mg of Zinc nitrate hexahydrate was dissolved in 12 ml methanol. Second, H-MIM (imadizolate) 1320 mg was dissolved in 60 ml methanol. Then, the two precursor were mixed together drop by drop and stirred for about 30 minutes. The mixtures were washed three times using DI water and methanol. Then, the precipitate was dried at 80 ^{o}C for two hours to obtain ZIF-8 powders.

The CNT dopping was performed by mixing modified CNT on Zinc nitrate hexahydrate using methanol as solvent. A total of 30 mg CNT and 120 mg PVP were dissolved in 60 ml methanol and stirred 30 minutes and sonicated 1 hour, then centrifuged. The precipitate of CNT was redispersed in 15 ml of methanol. 132 mg of zinc nitrate hexahydrate dissolved on 12 ml methanol. Then redispersed CNT and Zinc nitrate hexahydrate mixed drop by drop and then mixed with H-Mim 1320 mg that disolved in 60 ml methanol drop by drop under stirred condition afterwards stirred about 30 minutes then centrifuges and washed 3 times using DI Water. Then dried 80°C to get ZIF-8 doped CNT powders. The crystallinity and morphology of ZIF-8 and ZIF-8 doped CNT were characterized by X-Ray Diffraction (XRD) and Scanning Electron Microscope (SEM), respectively.

3. Results and discussion

The diffraction patterns of ZIF-8 and ZIF-8 doped CNT are shown in Figure 1. The XRD pattern of ZIF-8 (in Figure 1) is identical to the presented patterns [1, 9, 10] confirming the succesful synthesized. The peaks was found at 2θ = 7.4, 16.5, 22.1, 25.7, 26.7, 32.53. We found the peak of ZIF-8 doped CNT still formed. That indicate the dopping of CNT did not disorganize or destroy the assembly of metal (Zn) as nodes and ligands (H-methylimidazole) as frameworks to form ZIF-8 and we also found decreasing the peak of ZIF-8 when dopping with CNT (b in Figure 1) and lower than pure ZIF-8. It because when ZIF-8 doped with CNT decreasing concentration of MOF (ZIF-8). As can be seen, The peak of CNT were not appeared. It can be attributed to the high dispersion of CNT in ZIF-8.

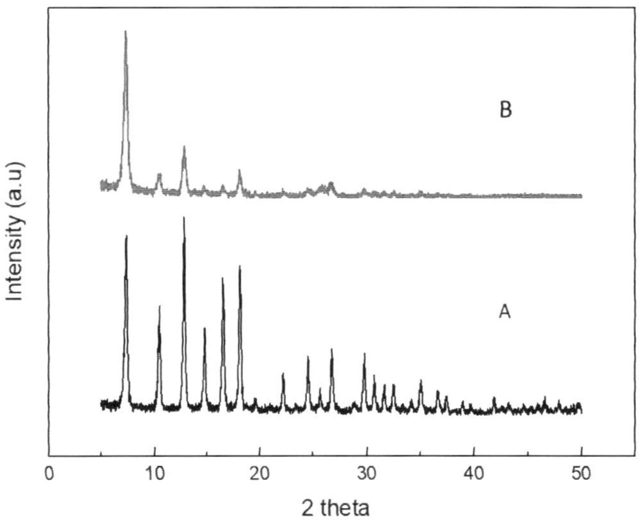

Figure 1. XRD pattern of a) ZIF-8, b) ZIF-8 doped CNT.

Figure 2 (a and b) shows the morphology and size using scanning electron microscopy (SEM) of ZIF-8 and ZIF-8 doped CNT. From Figure 2, it can be seen that the morphology of ZIF-8 has diameter of around 50-140 nm while for ZIF-8 doped CNT it is about 80-120 nm. The presence of CNT made

ZIF-8 attached on it. It can be seen on Figure 2 (b) and also made diameter size of ZIF-8 more uniformly. It confirmed the XRD pattern of ZIF-8 doped CNT lower than pure ZIF-8 because the CNT has high dispersed with ZIF-8.

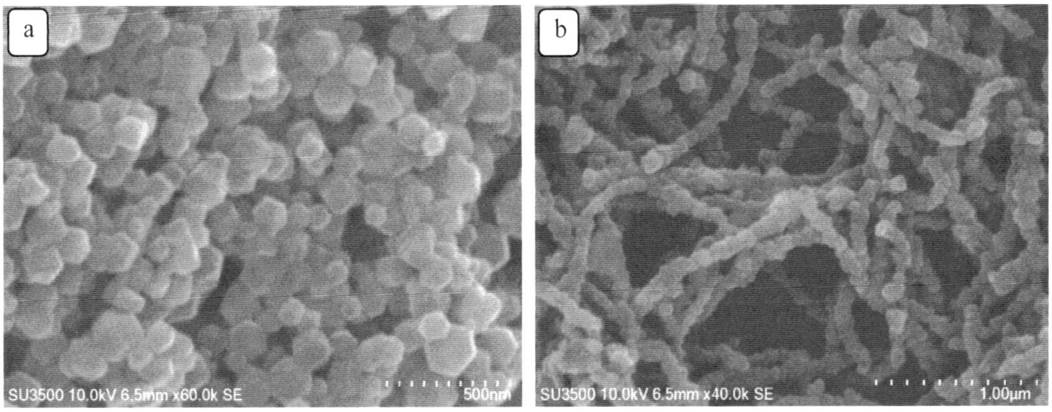

Figure 2. Morphology images of a) ZIF-8, b) ZIF-8 modified CNT.

4. Conclusion

The modification of ZIF-8 using CNT has been succesfully synthesized. The hexagonal morphology of ZIF-8 attached on CNT and made diameter size of ZIF-8 more uniform. The high dispersion of CNT decreases the peaks of XRD patterns of pure ZIF-8.

References

[1] Jafar A Manouchehr V Niyaz M M and Iran A 2017 *Chem. Eng. J.* **326** 1145-58
[2] Guan Y Shi J Xia M Zhang J Pang Z Marchetti A Wang X Cai J and Kong X 2017 *Appl. Surf. Sci.* **423** 349–53
[3] Pan Y Liu Y Zeng G Zhao L and Lai Z 2011 *The Royal Soc. of Chemis.* 2071–3
[4] Zhang Y Zhao Y Ke X and Zhang L 2017 *Mater. Letters* **206** 233–6
[5] Wang L Yang H Pan G Miao L Chen S and Song Y 2017 *Electrochimica Acta* 16–23
[6] Jiang M He H Yi W J Huang W Pan X Wang M Y and Chao Z S 2017 *Electrochem. Comm.* 5-9
[7] Kaur H Mohanta G C Gupta V Kukkar D and Tyagi S 2017 *J. of Drug Deliv.Sci. and Tech.*106-12
[8] Lu G and Hupp J T 2010 *America Chem. Soc.* 1406-10
[9] Cravillon J Schroder C A Bux H Rothkirch A Caro J and Wiebcke M 2012 *Cryst. Eng. Comm.* **14** 492–8
[10] Gross A F Sherman E and Vajo J J 2012 *Dalton Trans* **41** 5458–60

Computer aided diagnosis (CAD) for thorax radiography children patient with segmentation deformable models method

R Seftina[1], Prawito[1], D S Soejoko[1]
[1] Department of Physics, Faculty of Mathematics and Natural sciences, University of Indonesia, Depok, West Java 16332, Indonesia

rahmiseftina@gmail.com

Abstract. This study developed a correlation test Computer Aided Diagnosis (CAD) radiographic of children pulmonary using segmentation Deformable Models method to detect abnormalities. Deformable Models method searched abnormalities by value of the image pixel. Deformable models method used two variations, namely deformable models with median filter and deformable models with wiener filter. Abnormal result lung pixel values of segmentation wiener deformable models with wiener filter is 186-255 and deformable models with median filter is 190-255. Deformable Models used wiener filter have ROC result relatively higher than Deformable models used median filter with value of accuracy 78.5%, sensitivity 74.5%, specificity 80.0%, precision 90.0% and overall error of 21.0%. ROC result of segmentation using Deformable models method evince that Deformable models method can separated normal and abnormal tissue from radiographic of children pulmonary. However, Deformable models method can not definitively determine lung infection from radiographic of children pulmonary but this method can help radiologist to separated normal and abnormal tissue automatically with computer.

1. Introduction

Computed radiography (CR) is an imaging modality with digital system that widely used for the examination of the thorax and abdomen. Contrast and brightness from image of CR can be modified, so the image of CR becomes more clear and sharp. In addition, dose effect of CR is lower than CT so as not to endanger patients with unfavorable conditions and also patients who are still too young or too old. Sometime the abnormalities on thorax examination using CR, covered by other anatomic structures or the brightness of nodules is similar with background (normal tissue) and that's make difficult to see abnormalities direct from the image [1]. Especially in lung examinations of children who have dimensions of small lung size and fewer number of alveoli. Because of that, in this study Computer Aided Diagnosis (CAD) used to detection of abnormalities by separated normal and abnormal tissue on image from CR. The success of such a separation will be easier to read, so it can be an additional input for radiologists in the detection automatically with computer [2].

Medical image segmentation with CAD is one of the most challenging research topics in image processing and computer vision. A variety of different methods have been proposed [5]. Segmentation method widely used in CAD is Markov Random Field (MRF) models, thresholding approaches, region growing approaches, artificial neural network, atlas-guide approaches, and Deformable Models [6]. In this study, CAD for radiography of children pulmonary with segmentation deformable models method is used to detect child's lung abnormalities with infection trend. This is because Deformable Models have brought tremendous impact to classical problems in image processing such as providing ways to devise efficient computational algorithms for automatic segmentation. Other than that, the segmentation result provides by Deformable Models are usually dependent on the contour initialization and can provide partial objects boundaries [7].

2. Experimental method

This study was conducted by CR image of pediatric lung, 1-15 years old, from Rumah Sakit Anak dan Bunda Harapan Kita. The image was taken from November 2014 until January 2016 with inspection standard PA, AP, and supine. There was 136 image was equipped with a physician evaluation result used in this study. From 136 images, 66 abnormal images used as the model data and 70 images with 20 normal image and 50 abnormal image used as test data.

CAD program developed used MatLab 2016 with some commands like the active contour, filter, Fuzzy Clustering Means (FCM) segmentation and deformable models segmentation. CAD development of children CR lung is divided into preparation, image segmentation, and evaluation. The preparation is done by change format of sample images from DICOM to TIFF, grouping sample images be normal and abnormal image, proceed image with normalization, and image cropping. Image segmentation is done by Fuzzy Clustering Means (FCM) segmentation and deformable models segmentation with two methods of deformable models, namely deformable models with median filter and deformable models with wiener filter. Evaluation is performed by calculating the success rate is expressed by the ROC (Receiver Operating Characteristics).

Sample images are used in this study have different kV and mAs. Variation of kV and mAs produces sample images have different contrast and brightness levels for each image. Other than that, sample images also have sections that can interfere with segmentation process. Therefore, on the preparation performed normalization and image cropping. Normalization aims to equalize contrast and brightness levels of sample image with range pixel value 0-255. Image cropping aims to remove unwanted part of image, so the time required to run segmentation become shorter.

On the image segmentation, each image was processed by FCM segmentation and deformable models segmentation. FCM segmentation was done with the minimum point search method histogram automatically. The minimum point of the histogram is used to find the number of regions/clusters contained in the image and determine the threshold value [3]. In this study deformable models was done with active contour model driven by Globally Signed Region Pressure Force (GSRPF). GSRPF is done in this study because GSRPF is robust to contour initialization and has the ability to stop the curve evolution close to evenill-defined (weak) edges [4]. Parameter deformable models used in this study are iterations and sigma to control the smoothness of the level set function in each iterations.

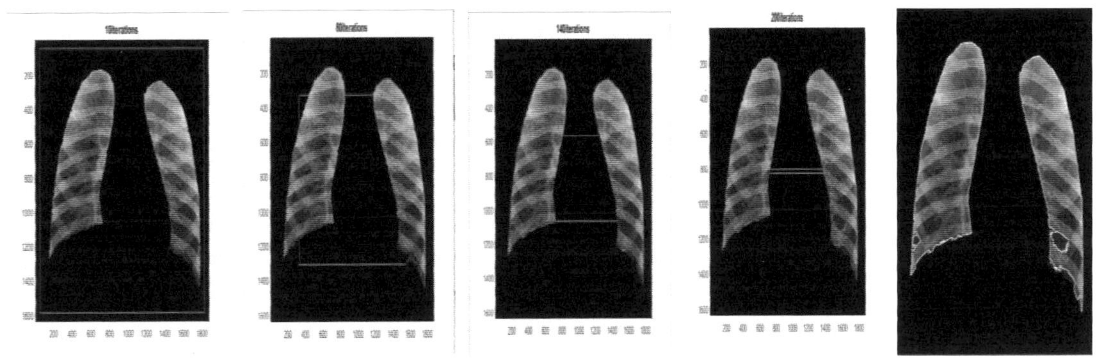

(a) (b) (c) (d) (e)

Figure 1. The segmentation result of deformable models (a) iterations 10, (b) iterations 80, (c) iteration 140, (d) iteration 200, and (e) final result of deformable models.

Evaluation was done by calculating the ROC (Receiver Operating Characteristics) image segmentation methods against a reference image from physician evaluation results. ROC counts is the accuracy, sensitivity, specificity, precision, and overall error. ROC uncertainty calculation is repeated five times by crossing data/cross validation. Cross validation is useful to determine the stability of the method.

3. Results and discussion

Medical image segmentation is the process of labeling each voxel in a medical image data set to indicate its tissue type and provide information about the anatomical structure. The various confrontations in medical image segmentation are poorly defined boundaries, blur or weak edges, intensity inhomogeneity, inconsistency in image quality while doing a scan and variable object shapes in medical images [10, 12]. Because of that, using segmentation method can be useful to detection of abnormalities by separated normal and abnormal tissue.

In Other research, MRF segmentation method used to detection abnormalities from radiographic of children pulmonary. MRF is a probabilistic which captures such contextual constraints. Like as in Deformable models method, MRF method search abnormalities by value of the image pixel. In n that research, MRF method used four variations like MRF without a filter, median filter+MRF, wiener filter+MRF, and adaphisteq filter+MRF. ROC segmentation result wiener filter relative higher than others. ROC wiener filter results show the value of accuracy 81.4%, sensitivity 82.0%, specificity 80.0%, precision 91.1% and overall error of 18.6% [3].

Like as in MRF method research, in this study Deformable Models method use to detection abnormalities from radiographic of children pulmonary. Deformable Models are curves or surfaces that deform two-dimensional (2D) or three-dimensional (3D) digital images under the influence of both internal and external forces and user defined constrains [13]. In this study, Deformable Models method used to detection abnormalities from radiographic of children pulmonary by separate normal and abnormal tissue.

Lung image has a lot of noise. Noise the most disturbing is the ribs. To reduce the influence of the ribs to the image of the lung treated with FCM segmentations and various filter. FCM segmentation used to reduce ribs from the lung image and to clarify the cluster to distinguish between bone and abnormalities. The difference result of deformable models segmentation with FCM and deformable models segmentation without FCM can be seen in Figure 2.

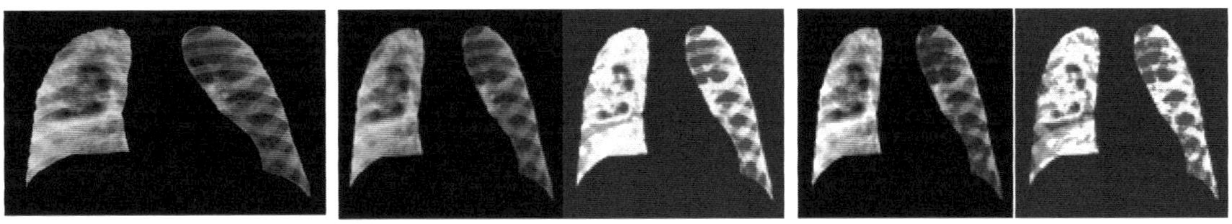

(a) (b) (c)

Figure 2. FCM segmentation image result (a) Original image, (b) Deformable models without FCM, (c) Deformable models with FCM

To reduce influence of the ribs and another noise in lung image was done with various filters including a gaussian filter, median filter, a standard deviation filter, wiener filter, histeq filter, histeq + wiener filter, adaphisteq filter and adaphisteq + wiener filter. Criteria of filter used in this study is a filter that can reduce noise without creating a blurry image and can keep the details of the sharpness of the corners of the object. Figure 3 show result of deformable models segmentation with various filter. From Figure 3 median filter and wiener filter can meet criteria of filter that used in this study.

Abnormal cluster search done by looking at cluster pixel values from segmentation result of 66 abnormal images that used as the model data. Through the observation of 66 abnormal image, the range of values obtained abnormal image pixel in the image that is subsequently used as the pixel value of data models. Pixel value of model data will be compared to pixel value of the test data obtained to determine the normal and abnormal image based on the pixel value. Table 1 is a cluster of range pixel values obtained by averaging the pixel values of the image.

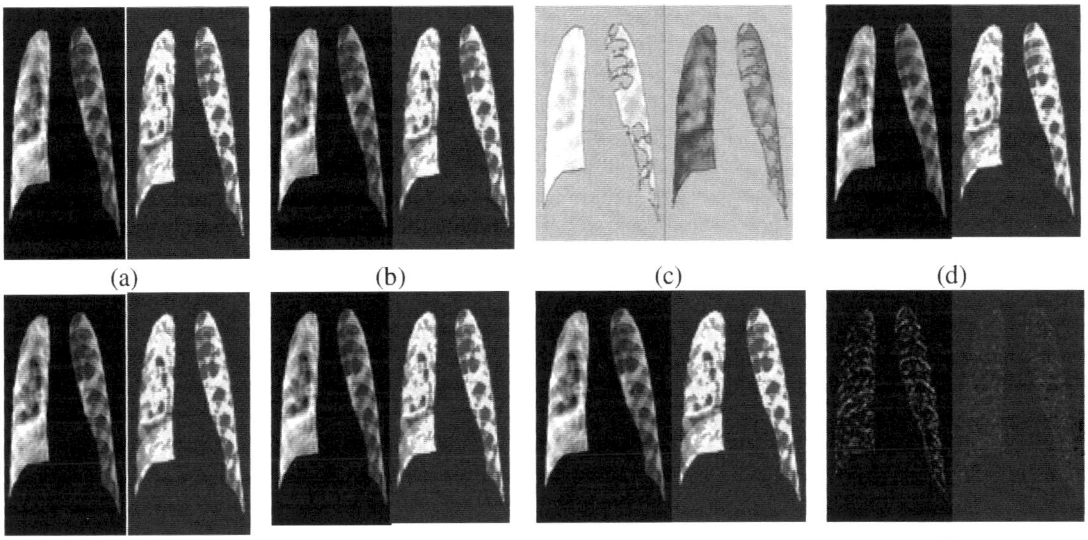

(a) (b) (c) (d)

(e) (f) (g) (h)

Figure 3. X-ray image with several treatment filter (a) median filter, (b) wiener filter, (c) histeq filter, (d) adaphisteq filter, (e) adaphisteq+wiener filter, (f) histeq+wiener filter, (g) Gaussian filter, and (h) standar deviation filter.

Table 1. Range pixel value of the model data used as a reference for the test data.

Method	*Range of* pixel value
Median filter + *Deformable models*	190-255
Wiener filter + *Deformable models*	185-255

Normal and abnormal images based on pixel values will be compared with doctors' evaluation results to determine whether the image is segmented in accordance with the results of the physician's evaluation. Based on the results of the evaluation that has been done, then calculated ROC value shown in Table 2.

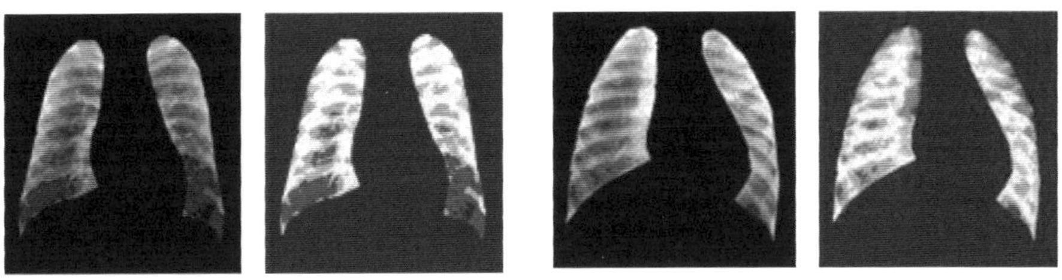

(a) (b)

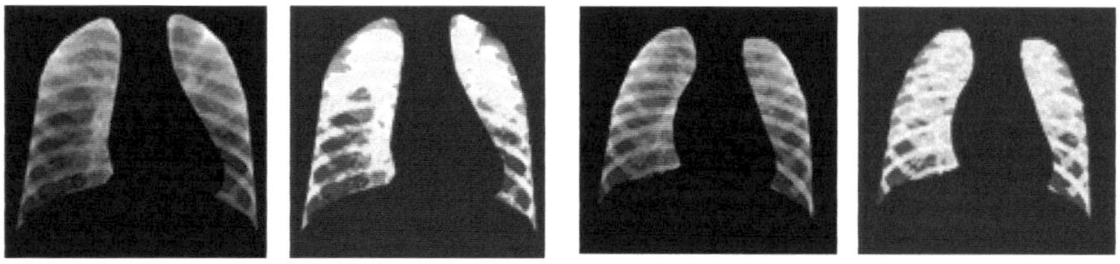

(c) (d)

Figure 4. ROC result of abnormal image used deformable models segmentations with median (a) normal image segmented as normal image (TN), (b) abnormal image segmented as abnormal image (TP), (c) abnormal image segmented as normal image (FN), and (d) normal image segmented as abnormal image (FP).

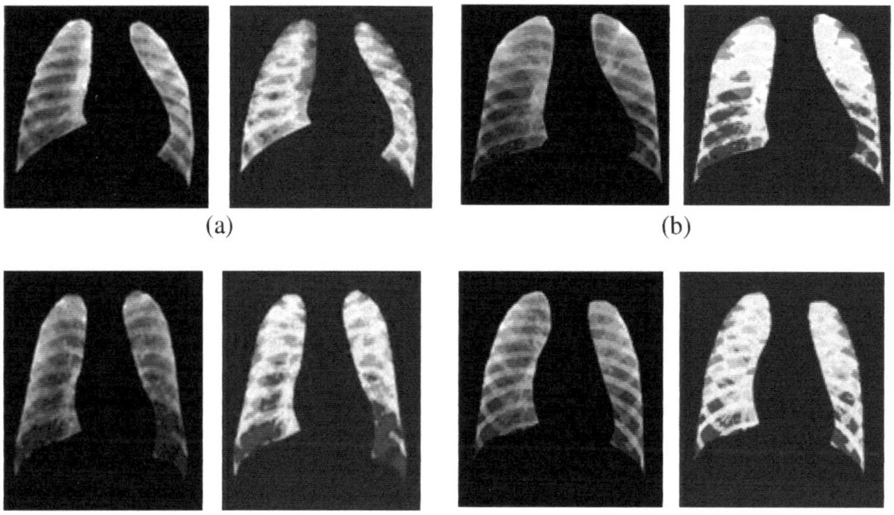

(a) (b)

(c) (d)

Figure 5. ROC result of abnormal image used deformable models segmentations with wiener filter (a) abnormal image segmented as abnormal image (TP), (b) abnormal image segmented as normal image (FN), (c) normal image segmented as normal image (TN), and (d) normal image segmented as abnormal image (FP).

Table 2. ROC value of Deformable models method with median filter and Wiener filter.

	Deformable models	
ROC	Median filter	Wiener filter
Accuracy	70.0% ± 2.8%	78.5% ± 3.1%
Sensitivity	66.0% ± 4.2%	74.5% ± 4.1%
Specificity	80.0% ± 0.0%	80.0% ± 0.0%
Precision	89.2% ± 2.1%	90.0% ± 2.8%
Overall Error	30.0% ± 2.9%	21.0% ± 3.0%

ROC segmentation results wiener filter is relatively higher than median filter. ROC wiener filter results show the value of accuracy 78.5%, sensitivity 74.5%, specificity 80.0%, precision 90.0% and overall error of 21.0%. However, differences of ROC for any kind of success rate is not more than 5% that means Deformable Models can be implemented to detect abnormalities.

4. Conclusion

Deformable models method searched abnormalities by value of the image pixel. Deformable models method used two variations, namely median filter + Deformable Models and Wiener filter+Deformable Models. Abnormal result lung pixel values with segmentation Wiener filter+Deformable models was 186-255 and median filter + Deformable Models was 190-255. Wiener filter+Deformable models method showed ROC result relatively higher than median filter+Deformable models with value of accuracy 78.5%, sensitivity 74.5%, specificity 80.0%, precision 90.0% and overall error of 21.0%. Differences of ROC value using Deformable Models method was not more than 5% that means Deformable models could be implemented to detect abnormalities. However, Deformable models method cannot definitively determine lung infection from radiographic of children pulmonary.

Acknowledgment

At the outset we would to thank Rumah Sakit Anak dan Bunda Harapan Kita for allowed to use the data of CR image of children patient from Rumah Sakit Anak dan Bunda Harapan Kita as the research sample in this study. The acknowledgment would be incomplete without thanking Septia Ardiani who get data from Rumah Sakit Anak dan Bunda Harapan Kita.

References

[1] Shiraishi J Li Q Suzuki K Engelmann R and Doi K 2006 *Med Phys.* **33** 2642-53
[2] van Ginneken B 2012 *5th Int. Workshop on Pulmonary Image Analysis* 9-21
[3] Adiani S Prawito Soeharso D S 2017 *Proc. of The 7th Annual Basic Sci. Int. Conf.* 132-5
[4] Abdelsamea MM and Tsaftaris SA 2013 18th Int. Conf. on Dig. Sign. Proces. (DSP)
[5] Zhang S Wang L Jiang L and Liu S 2010 3rd Int. Cong. on Image and Signal Proces. (CISP2010)
[6] Dzung Xu L C & Jerry L 2000 *Current Methods in Medical Image Segmentation* Baltimore; Departement of Electrical and Computer Engineering Johns Hopkins University
[7] Yang B 2005 *Geometric Deformable Models Using the Level Set Method* Canada
[8] Demirkaya O Asyali M S and Sahoo P K 2009 *Image Proc. with MATLAB, App. in Medicine and Biology* London: CRC Press Taylor & Francis Group
[9] Dougherty G 2011 *Medical Image Proces, Tech. and App.* NewYork: Springer
[10] Jayadevappa D Kumar S S and Murty D S 2009 *Int. J. of Comp. Sci.* **36(3)**
[11] Liu Brian and Ashish R 2013 *Hindawi Pub. Corporat. Comput. and Math. Methods in Medicine*
[12] McInerry T Terzopoulos D 1996 *Medical Image Analysis* **1(2)** 91 – 108
[13] Kass M Witkin A and Terzopoulus D 1988 *Int. J. Comp. Vision* **1** 312-33

IOP Conf. Series: Materials Science and Engineering **432** (2018) 012055 doi:10.1088/1757-899X/432/1/012055

Preparation of $LiFePO_4$ added PPy as cathode materials for li-ion battery

I Gunawan [1], J Ginting [1], P Yustinus [1], Sudaryanto [1]

[1] Centre of Advanced Materials Science and Technology, National Nuclear Energy Agency, South Tangerang 15310, Indonesia

irgun.IG@gmail.com

Abstract. $LiFePO_4$ is an alternative cathode material for lithium-based battery cells. $LiFePO_4$ demonstrates many advantages such as non-toxicity, low cost, and stable structure. One of the main problems of $LiFePO_4$ is its low electronic conductivity, which results in its poor rate capability due to the poor kinetics of lithium intercalation/deintercalation process. The objective of this study is to increase the electrical conductivity of $LiFePO_4$. An in-situ carbon source such as polypyrrole PPy was applied to prepare $LiFePO_4$/C composites by heat treatment. The nature of polypyrole in the system can be considered not only as carbon source, but also as a structure director. The $LiFePO_4$/C composites were prepared with the addition of PPy which were varied between 2.5, 5, 7.5, 10 and 15 % weight, and graphite also added in the amount of four times the PPy content. The crystal structure, surface morphology, elemental analysis, and conductivity of $LiFePO_4$/C composites were characterized by X-ray diffraction and scanning electron microscopy combined with energy dispersive spectrometry and use of LCR meter, respectively. The XRD patterns of $LiFePO_4$/C composites showed that the $LiFePO_4$/ C particles belong to a pure orthorhombic system of olivine-type structure (Pnma, JCPDS: 40-1499). Carbon phase was not found apparently from the XRD pattern, which indicated that it may be present in amorphous form. The particles morphology were composed of primary particles approximately 400 nm in diameter, with homogeneous pores and were agglomerated, while carbon deposition was observed on the surface. The addition of PPy increased the conductivity of materials. The highest conductivity of $LiFePO_4$/C composite was 9.74×10^{-2} S cm^{-1} reached by the addition of 15% PPy.

1. Introduction

This research is part of Advanced Materials Science and Technology Centre, National Nuclear Energy Agency's program in new and renewable energy research for sustainable energy sources. $LiCoO_2$ and $LiMnO_2$ cathode materials preparation has been done [1, 2]. In this research, lithium ferophospate ($LiFePO_4$) will be developed as an alternative to cathode materials for lithium-based battery cells. The lithium battery is expected to be a solution to energy issues related to environmental issues. The use of lithium battery cathodes materials in phosphate form will be readily paired with solid lithium electrolyte, glass, or polymer. Since the pioneering work of Padhi [3], orthophosphate $LiFePO_4$ has been intensively studied as cathode material for Li-ion batteries. As one of the most promising cathode materials, $LiFePO_4$ has been intensively investigated for large lithium ion batteries in electrical vehicles (EVs and HEVs). $LiFePO_4$ demonstrates many advantages such as non-toxicity, low cost, and stable structure that makes it is safe to use as a large lithium battery cathode material in electric and hybrid vehicles with a high theoretical capacity (170 mAh/g). However, $LiFePO_4$ has an inherently low electrical conductivity, which results in its poor rate capability due to the poor kinetics of lithium intercalation/deintercalation process [4]. Also reported the $LiFePO_4$ phase did not have a high capacity was due to some lithium or iron disorder with around 7 % iron on the lithium sites [5-7]. The electronic conductivity of $LiFePO_4$ at room temperature is about 10^{-9} S / cm [8], and electrochemical capacity of materials is 85 mAh/g [4]. Such low conductivity makes it difficult to attain the theoretical capacity of $LiFePO_4$ as the Li-ion battery cathode. Nowadays many approaches have been successfully adopted to improve its electrical conductivity, such as the control of particle size, and metal cation doping such as carbonaceous materials and some metals enhance the charge/discharge performance [9-12]. These factors (particle size and conductive coating) are largely affected by the method of preparation. Therefore, to produce commercially available $LiFePO_4$ a proper preparative method is an urgent research target [13]. The cathode conductivity is primarily determined by iron content. The choice of phosphate salts is due to the molecular size that allows the lithium ion to move from and to the electrolyte in the process of charged/discharged. The lithium content in the cathode can be expected as an ion source of a lithium-based battery component. Lithium iron phosphate has a strong covalent bond between oxygen and phosphate to form a strong polyanion unit that has a better stability than its oxide. The formation of the polyanion also enlarges the free space available for lithium. The P-O-metal bond helps to stabilize the redox energy of the metal cations and allows for faster ion transfer. In addition, the bonds between oxygen atoms in phosphate are stronger than cobalt, so some $LiFePO_4$ cells have heat resistance properties in their use [4, 14].

$LiFePO_4$ has a discharge voltage of about 3.4 V versus lithium. It shows no obvious fading even after several hundred cycles, and shows power capabilities comparable to that of the mixed layer oxides [15]. Typical compositions of cathode films are metal substrates $LiFePO_4$ 72 wt. %, acetylene black 8 wt. %, and 20 wt. % polyethylene oxides as binder. Acetylene black ensures electrical conductivity throughout the film [16], and the binder makes the film intact during processing. These materials are dissolved on a metal substrate using acetonitrile.

In the previous works, we have studied the addition of polyacryllic acid, (PAA) [17] and polylactic acid (PLA) to $LiFePO_4$ [18], and the addition of organic matters as reducing agents as well as a carbon sources [19, 20]. In this paper we will report the preparation of $LiFePO_4$/C composites by addition of PPy as carbon source to to increase the electrical conductivity of $LiFePO_4$. The characterization focused on exploring the relationship of structure and properties of materials.

2. Experimental

2.1 Materials
The experiment was carried out by using materials of lithium ferrophosphate, $LiFePO_4$ (MTI China, purity = 99%), polypyrrole, PPy (Aldrich, purity = 99 %), and graphite, C (Aldrich, purity = 99%).

2.2 Methodology
The starting materials were $LiFePO_4$, PPy, and C. They were mixed in the mortar. The $LiFePO_4$ was purchased from MTI China as obtained without further treatment. The weigth ratio of the PPy in the mixture to $LiFePO_4$ were varied as 0, 2.5, 5, 7.5, 10 and 15 %, with samples code in a row is K0, K1, K2, K3, K4 and K5 respectively. The graphite added were 4 times of PPy content. The mixture then heated in furnace at 300 °C for 1 h. The products then ready for characterizations.

2.3 Characterizations
To determine the crystal phase composition of the $LiFePO_4$/C composites, XRD analysis was carried out at room temperature using a Pananalytical-Empyrean diffractometer (Japan) with Cu Kα radiation (λ = 0.15406 nm), over the 2θ collection range of 5–80°. The accelerating voltage of 40 kV, emission current of 300 mA and the scanning speed of 4°/min were used. The particles size and elemental analysis

were measured using SEM combined with EDS (JEOL JSM 6510 LA, Japan) working on a 20 kV voltage. Electrical conductivity measurements were performed using LCR meter HIOKI 3532-50, China.

3. Results

The morphologies of the $LiFePO_4$/C composites were investigated using SEM. The surface of $LiFePO_4$ coated with a carbon layer can inhibit the particle growth and help obtain a smaller particle [21]. While surface coatings and nanosized syntheses have proved to improve the power density, they are usually associated with lower tap density for the cathode-anode material as the particle size decreases [22].

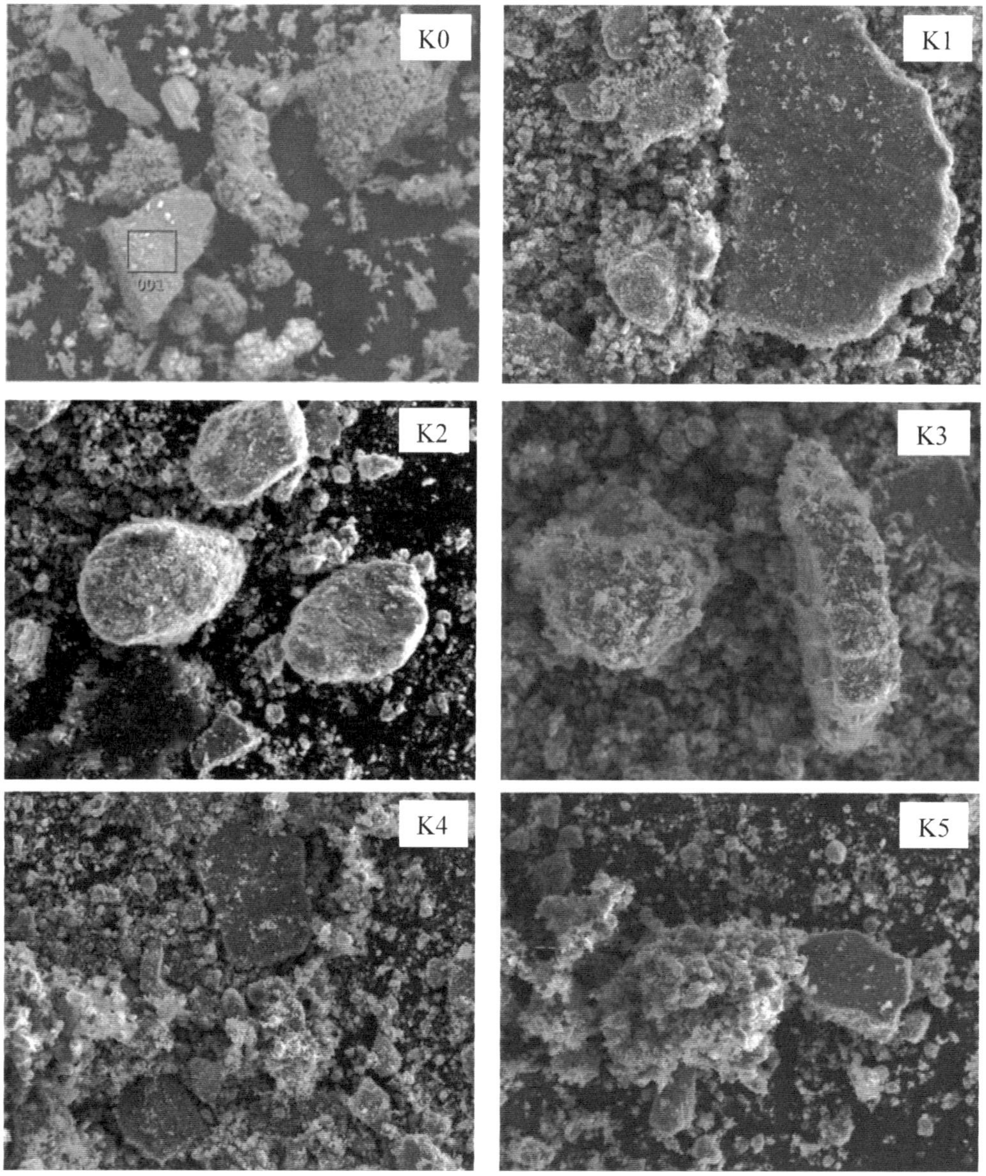

Figure 1. The morphologies of the $LiFePO_4$/C composites for LFP/PPy 0, 2.5, 5, 7.5, 10 and 15 % as sample code of K0,K1, K2, K3, K4 and K5 respectively. Investigated using SEM with a magnification of 5000X.

Figure 1 (K0) shows the arrangement of $LiFePO_4$ particles without the addition of PPy. The arrangement of these particles is agglomerated which its distance between the particles is a wide vacancy. Figure 1 shows the panoramic morphologies for LFP/PPy 2.5 % (K1); this image informs $LiFePO_4/C$ particles with a size distribution (1-5 μm). Detailed investigations of the particles showed that they were composed of primary particles approximately 400 nm in diameter, and homogeneous pores with carbon deposition were observed on the surface. Figure 1 (K2, K3, K4 and 5) shows the spherical morphology for the LFP/PPy 5, 7.5, 10 and 15 % respectively. Similar to LFP/PPy 2.5 %, all the $LiFePO_4/C$ particles comprised 200 nm-diameters, and homogenous pores were formed in these samples. The particles are less agglomerated and tend decreased in size with increasing PPy content compared those in K1 sample.

Table 1. Elemental analysis of the $LiFePO_4/C$ composites samples.

Name of sample/PPy content (wt. %)	Elemental analysis (mass %)			
	C	O	P	Fe
K1/2.5	24.20	31.57	15.96	28.27
K2/5	24.43	40.48	13.71	21.38
K3/7.5	26.60	32.66	13.91	26.83
K4/10	30.11	33.50	13.86	22.54
K5/15	36.24	34.75	11.55	17.46

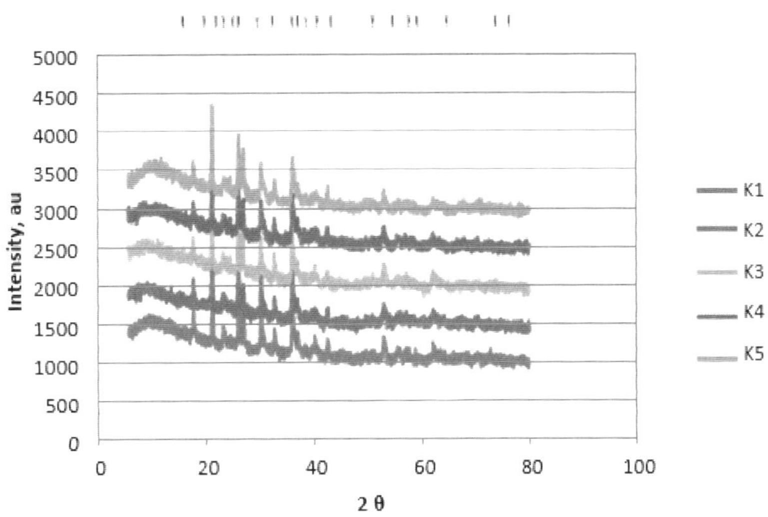

Figure 2. The X-ray Diffraction (XRD) patterns obtained for $LiFePO_4/C$ composite.

The XRD patterns of samples for K1, K2, K3, K4 and K5 can be seen in Figure 2. There are insignificant difference patterns for these samples. The majority of diffraction lines can be indexed as a single phase belongs to an ordered olivine structure with the orthorhombic Pnmb space group. The main diffraction peaks of $LiFePO_4$ occurs at angles 2θ = 17.04°, 20.7°, 24°, 29.6°, 30.98°, 35.5°, and 42.2° correspond to the crystal planes [020], [011], [101], [200], [210], [201], and [112] mark as bar in Figure 2. The phases of $LiFePO_4$ are in the sharp diffraction peaks illustrate the highly crystalline. The nature of polypyrole in the system can be considered not only as structure director, but also as a carbon source.

Carbon phase was apparently not found from the XRD pattern, which indicates it may be present in amorphous form. These results correspond to the literature [23].

Analysis of the complex permittivity is measured and usually done with the technique of curve-fitting based on normally, one of the following models: Debye model, Cole-Cole, Cole-Davidson, a model of Havriliak-Negami and KWW. An analysis of curve-fitting is usually used to find the relaxation time. In addition, Cole-Cole plot of dielectric constants have also been used to analyse parameters of the dielectric [24]. As shown in the Figure 3, Cole-Cole plot for composite material $LiFePO_4/C$ straight line at low frequency range in the initial state and a half circle in the middle and high frequency range. It was found that a linear line appears at low frequencies, reveals that the passivation film stable.

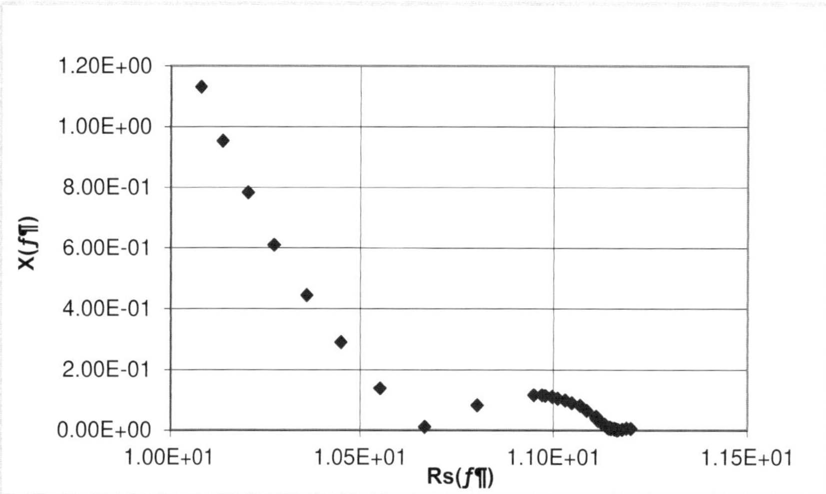

Figure 3. Cole-Cole plot from the results of measurements of electrical conductivity of $LiFePO_4/C$ for a sample of K5.

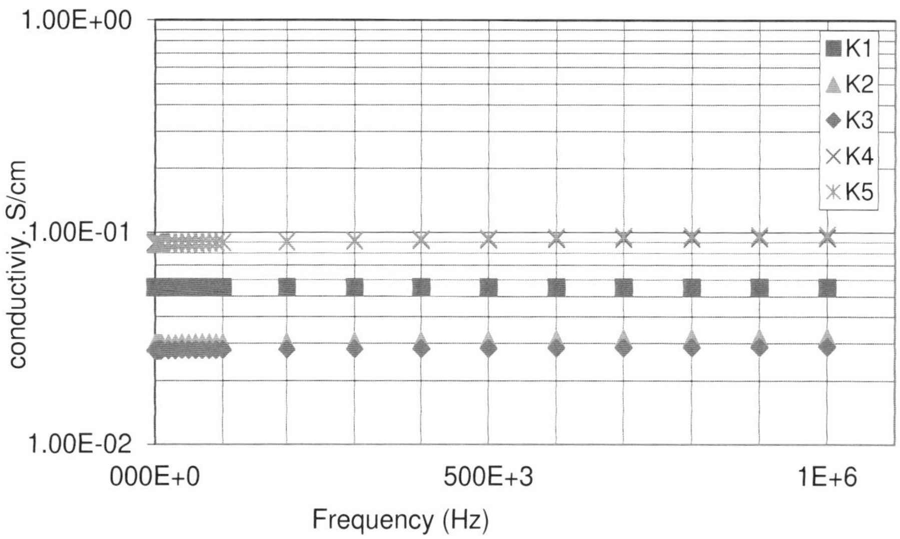

Figure 4. The electrical conductivity graph of $LiFePO_4/C$ composites for sample of K1, K2, K3, K4 and K5 as a function of frequency.

The radius of a half circle appears in the medium to high frequency range with the voltage increasing, revealed that happens a decrease in resistance transfer. Conductivity measurement is carried out using High Precision LCR-meter which is connected to a computer for data acquisition. The addition of PPy is increasing the conductivity of materials. In addition of PPy 5% and 7.5% i.e. a sample of K2 and K3, the conductivity tends to decrease compared to samples of K1. This may be caused by the limited mobility of the charge carriers and the establishment of a neutral ion pairs and ion aggregates. On the larger content of PPy i.e. 10 and 15% the particle distribution may be spread more evenly, forming a network of $LiFePO_4/C$ in an amount more so the conductivity then increases. The highest conductivity of $LiFePO_4/C$ composite was 9.74×10^{-2} S cm^{-1} reached by addition of 15% PPy. The $LiFePO_4$ conductivity before added of PPy is 4×10^{-4} S cm^{-1} [25] increase to 9.74×10^{-2} S cm^{-1} by addition of 15% PPy, or an increase in 243.5 time.

4. Conclusion

There is insignificant difference for diffraction pattern of samples K1, K2, K3, K4 and K5 using XRD. Diffraction patterns showed that $LiFePO_4$ dominate the main peaks are marked by 3 strongest diffraction peaks at angles $2\theta = 20.73°$, 24°, 33°, and 33.7°. The nature of polypyrrole in the system can be considered not only as a director of the structure, but also as a source of carbon. But the peak corresponding to the carbon is not visible, due to the amorphous structure.

Morphology of $LiFePO_4/C$ composite surface was observed using SEM showed particles of $LiFePO_4/C$ with a varied size distribution (1-5 μm). A detailed investigation of the particle consists of primary particles with a diameter less than 5 μm, the homogeneous pore and carbon deposition observed on the surface.

The addition of PPy is not immediately increase the conductivity of materials acquired. In addition of PPy 5% and 7.5% i.e. a sample of K2 and K3, the conductivity tends to decrease compared to samples of K1. This may be caused by the limited mobility of the charge carriers and the establishment of a neutral ion pairs and ion aggregates. On the much more content of PPy i.e. 10 and 15%, the particle distribution may be spread more evenly, forming a network of $LiFePO_4/C$ in an amount more so the conductivity then increases.

Acknowledgements

We would like to thanks to Dr. Abu Khalid Rivai, as the head of BSBM Division BATAN, for supporting the research and writing of this paper.

References

[1] Honggowiranto W Jahja A K and Kartini E 2014 *Proc.of The Int.Conf. on Materials Sci.& Tech.* ICMST 497–504
[2] Kartini E Honggowiranto W Jodi H and Jahya A K 2014 *14th Asian Conf. Solid State Ionics* 2 ACSSI 978–81
[3] Nanjundaswamya K *et al.* 1996 *Solid State Ionics* **92** 1–10
[4] Sun X Sun K Chen C Sun H and Cui B 2013 *Int. J. Mater. Chem.* **2(5)** 218–24
[5] Pereirinha P G and Santiago A 2012 Coimbra Portuguese
[6] Luo J Y Cui W J He P and Xia Y Y 2010 *Nat. Chem.* **2(9)** 760–5
[7] Hua N Wang C Y Kang X Y Tu E D and Han Y 2010 *J. Inorg. Mater.* **25(8)** 887–92
[8] Yin Y Li X Mao X Ding X and Yang S 2013 *J. Mater. Sci. Technol.* **29(10)** 937–42
[9] Chen R Wu Y and Kong X Y 2014 *J. Power Sources* **258** 246–52
[10] Meng L Y Wang B Ma M G and Lin K L 2016 *Mater. Today Chem.***1–2** 63–83
[11] Wang G Ma Z Shao G Kong L and Gao W 2015 *J. Power Sources* **291** 209–14
[12] Hu J *et al.* 2009 *J. Mater. Sci. Tech.* **25(3)** 405
[13] Palomares V *et al.* 2007 *J. Power Sources* **171(2)** 879–85
[14] Zhang Y Wang C Y and Tang X 2011 *J. Power Sources* **196(3)** 1513–20
[15] Lassi U Hu T Pohjalainen E Kallio T Kordas K and Jantunen H 2014 *Int. J. Mater. Sci.* **4(1)** 1

[16] Gabrisch H Wilcox J and Doeff M M 2008 *Electrochem. Solid-State Lett.* **11(3)** A25
[17] Gunawan I and Sudaryanto 2011 *Proc. Sem. Nat. Mat. & Met.* 385–92
[18] Gunawan I Sugiantoro S and Deswita 2011 *Proc. of The Int. Conf. on Innovation in Polymer Sci. & Tech.* IPST 2011 53–7
[19] Honggowiranto W and Gunawan I 2015 *Chem. Industry and Envy.* 57–64 (In Indonesia)
[20] Gunawan I and Honggowiranto W 2015 *Macromol. Symp.* **353** 225–30
[21] Jugović D and Uskoković D 2009 *J. Power Sources* **190(2)** 538–544
[22] Örnek A and Efe O 2015 *Electrochimica Acta* **166** 338–49
[23] Miao C Bai P Jiang Q Sun S and Wang X 2014 *J. Power Sources* **246** 232–8
[24] Shukla N Thakur A K Shukla A and Marx D T 2014 *Int. J. Electrochem. Sci.* **9(12)** 7644–59
[25] Gunawan I 2012 P*roc. Scientific Meet. Mat. Sci. and Tech.* 119–23 (In Indonesia)

IOP Conf. Series: Materials Science and Engineering **432** (2018) 012056 doi:10.1088/1757-899X/432/1/012056

Effect of samarium in Al-Zn-Cu as a micro-alloying element for low voltage sacrificial anode

D Ferdian[1, a], Y Pratesa[1, b], F Y Ramadhan[1c]

[1] Department of Metallurgy and Materials Engineering, University of Indonesia, 16424 Depok, Indonesia

[a)] deni@metal.ui.ac.id
[b)] yudha@metal.ui.ac.id
[c)] fajar.yusya@gmail.com

Abstract. Development of a low-voltage sacrificial anode for offshore structure has attracted increasing attention in recent years. The low-voltage operation can prevent the structure from overprotection and the resulting stress corrosion cracking. One of the efforts made to create this low-voltage sacrificial anode is by replacing indium as a conventional alloy with other elements such as copper, silicon, or gallium. Previous research showed that copper has the potential for low voltage properties, but the presence of interdendritic corrosion reduces the efficiency of the anode. The addition of samarium is aimed at a grain refinement that is expected to reduce the interdendritic phases' size and disperses evenly. In this research, Al-Zn-Cu alloy with a variation of samarium composition by 0.1% Sm, 0.3% Sm, and 0.5% Sm was produced and it was followed by characterization which include metallographic test, differential scanning calorimetry, scanning electron microscopy, and energy dispersive spectroscopy. The results showed the dendrite size was reduced, in proportion with the addition of samarium, from 383 μm to 261 μm. Intermetallic Al-Cu-Sm phases were found in many interdendritic areas. Samarium also produced undercooling phenomena during the solidification process. Samarium contents below 0.3% acted as a grain refiner. Above that level, it will transform into a new phase. These findings indicate the Al-Zn-Cu-Sm alloy is a potential alloy to be developed into a low-voltage sacrificial anode.

1. Introduction

Presently, Indonesia is no longer a major oil exporting country. Decline in oil production from the proven reservoir is the main reason for this condition. It is more difficult to find the possibility of new resources, as the possible resources are located in deep sea regions. Most of the exploration resulted in the coastal and offshore area. Nevertheless, the marine environment is more corrosive than the onshore for the metallic structures of an oil platform. Therefore, it requires better protection to support the offshore infrastructure, one of which is a sacrificial anode.

From several types of available protection methods, the sacrificial anode is one of the corrosion protection mechanisms that can be used in a seawater environment. The sacrificial anode offers advantages because it is easier for installation and has a lower maintenance costs. Currently, the most widely used sacrificial anode is an aluminum-based anode of aluminum-zinc-indium alloy (Al-Zn-In). Al-Zn-In anodes produce excellent efficiency in its applications. Zinc addition is intended to make the beta phase of aluminium (β aluminum) which makes the passive layer of alumina oxide easy to penetrate by the environment [1]. Meanwhile, the addition of indium will prevent the alumina's passive coating to readily re-form.

However, in some cases of applications of Al-Zn-In, failures as a result of stress corrosion cracking were found. This damage was due to a miscalculation of the sacrificial anode design that causes the steel structure, which should be protected, attacked by stress corrosion cracking. The addition of indium makes the potential of protected metals very low and reaches the evolution of hydrogen gas area in Pourbaix diagram. Several studies have examined if the replacement of indium with other elements could prevent the stress corrosion cracking. Other researchers has shown that gallium, silicon, and copper are the elements that made low-voltage aluminum sacrificial anode [2]. Aluminium-gallium (Al-Ga) operated within or near the desired potential range of -750 to -850 mV with anode efficiencies greater than 95%. Ga was found at elevated levels below the ground surface of uncorroded samples, with varying levels of Ga through the depth of corroded specimens [3].

A prior study by Ferdian *et al.* [4, 5] shows that copper is more efficient than silicon as an alloy for the low voltage sacrificial anode. The corrosion rate of this alloy increased with the addition of copper as its alloy. Furthermore, there is an inductance curve in the EIS test results indicating a decrease in oxide resistance of the aluminum. Copper makes the alloy potential more positive and makes the width of the loop even larger which indicates it takes longer time for re-passivation. Al-Zn-Cu alloy is chosen to be developed in this research due to the economical price and ease of obtaining it in the market. Based on literature review, several elements were already tried as an alloying element for the sacrificial anode. Gallium, silicon, and copper in aluminum anode showed increasing of steel potential protection and qualify the criteria to be a low voltage sacrificial anode as illustrated in Figure 1.

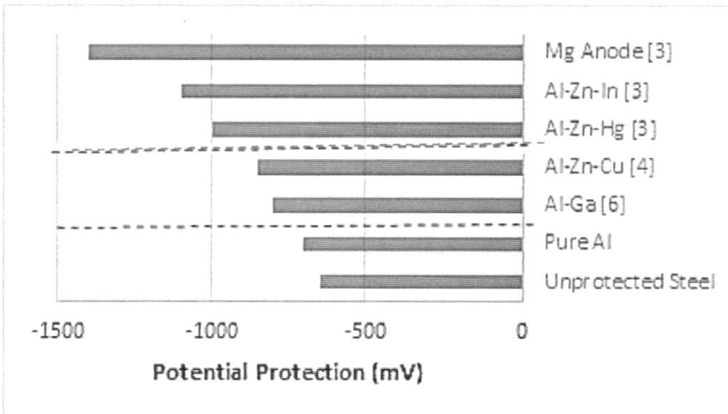

Figure 1. Comparison of potential protection from various aluminum sacrificial anodes [3-5].

Samarium is classified as a rare earth element and is known to have the ability as a grain refiner in aluminum alloys. According to Hu *et al.* [7], samarium can work as grain refinement by altering the morphology of the α-Al dendrite into a smaller grain. The changes also modified the morphology of eutectic silicon and iron-rich phase. Based on that finding, this research is expected to multiply the interdendritic area formed to make the intermetallic phase more evenly distributed in smaller sizes than without the addition of samarium. Another work by Hongxu *et al.* [8], shows that the addition of Sm elements could affect the morphology of eutectic silicon. The addition of 0.3% samarium in the alloy made the eutectic smoother than without samarium and made the phase more rounded. The effect of samarium in silicon eutectic modification is optimum until the addition of 0.6% Sm. A higher percentage of samarium addition can make eutectic silicon the original shape.

2. Methodology

The sample for sacrificial anode was made by a casting process. A resistant furnace was used with graphite crucible which heated at a temperature of 750 °C for 90 minutes. The stirring process of the aluminum was carried out using graphite bar, and the remnant of slag and impurities on the surface of molten metal was removed before pouring process. In order to achieve the target alloying composition, binary Al-10Sm was made as a master alloy and melted in a graphite crucible and then mixed with Al-5Zn-0.5%Cu. The alloying element is weighed to achieve the target for a quarternary alloy of Al-5Zn-0.5Cu-xSm. The casting product was characterized by several tests such as metallographic, Energy Dispersive Spectroscopy, and Differential Scanning Calorimetry.

3. Result

3.1 Differential scanning calorimetry

DSC experiments were carried out under a low nitrogen flow and cooling rate of 10 °C min^{-1}. They consisted of heating the sample up to 700 °C, holding it at that temperature for 10 min for homogenization, and finally cooling it down to room temperature. Figure 2 shows the DTA thermogram upon cooling for different samarium addition. Two solidification reactions occur successively at decreasing temperature. The nucleation of (a-Al) dendrites (refer to the first peak in Figure 2) and followed by the formation of rich samarium intermetallic phase as the second peak. The peak becomes distinctively clearer as the samarium level increases, as seen in thermogram for Al-5Zn-0.5Cu-0.5Sm. The addition of samarium also slightly shifted the precipitation temperature of α-Al solidification into the lower temperature.

Based on thermogram, two-phase transformation occur at a temperature of approximately 620 °C for α-Al solidification and 580 °C for samarium rich intermetallic phase. Analysis of the ternary phase diagram of Al-Zn-Sm showed no indication of high-temperature intermetallic phase precipitation, while analysis with ternary Al-Cu-Sm, the second peak could be associated with $SmCu_4Al_{11}$ phase. This temperature is near with the isothermal temperature of Al-Cu-Sm as shown in Figure 3 and conforms with $SmCu_4Al_{11}$ formation.

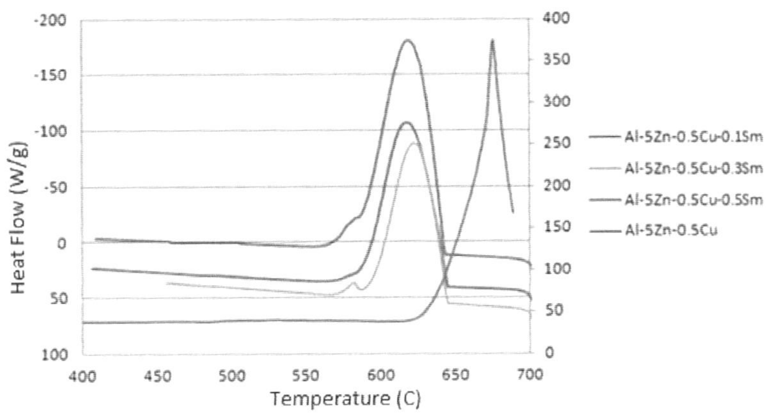

Figure 2. DSC result from Al-Zn-Cu and Al-Zn-Cu-Sm.

3.2 Microstructure

3.2.1 Metallography. The microstructural observation was conducted using a metallographic method using optical emission spectroscopy and scanning electron microscope (SEM). Microstructural observations were performed using optical microscopy (OM) for as cast and controlled cooling rate (DSC). All of the metallographic examinations is taken by the non-etchant process. It can be seen clearly, grain boundaries revealed with brighter colors. The intermetallics appear on grain boundaries signed by white color. The intermetallic is formed at the grain boundary because the grain boundary is the last area that solidified during the dendrite formation as seen in Figure 4.

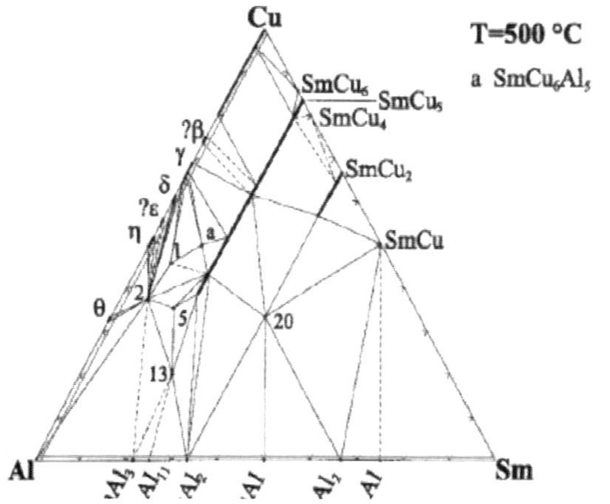

Figure 3. Isothermal section at 500 °C of ternary Al-Cu-Sm phase diagram [10].

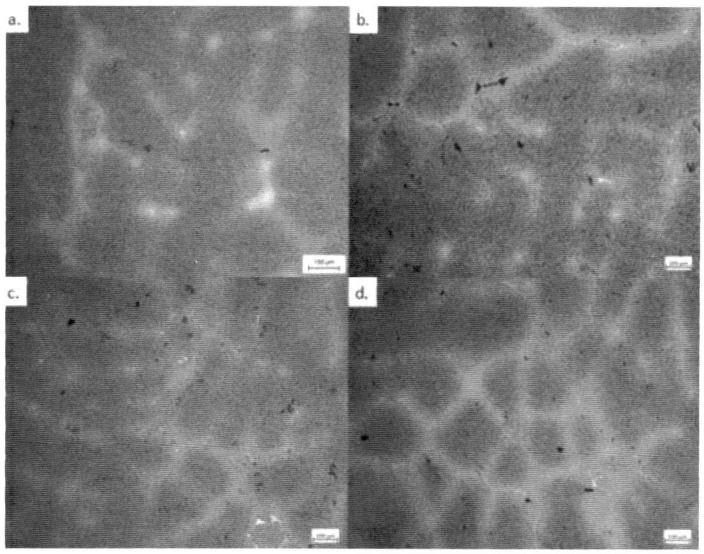

Figure 4. Effect of samarium in microstructure of Al-Zn-Cu with addition (a) 0% Sm, (b) 0.1% Sm, (c) 0.3% Sm and (d) 0.5% Sm.

The alloy that contain 0.5 wt% samarium has many intermetallics in the grain boundary compared with 0.1 wt% and 0.3 wt% Al-Zn-Cu-Sm alloy. This indicates more samarium addition made a new intermetallic phase in the grain boundary with the other element. Zhi Hu *et al.* also found the similar result after 1% addition of Sm in AlSiCu alloy. They found two type of precipitate at 1% Sm addtion. It indicated most of the samarium will bound with the other element rather than dissolved in the matrix [7].

Metallographic observations also show a decrease in the length of secondary dendrite arm spacing (SDAS). The average length of grain size that was formed before the addition of samarium was 383 µm while the observation after the addition of 0.5% wt samarium became 245 µm. Based on Figure 5, the most optimal addition is 0.3% samarium to produce finer SDAS. Above these concentration, DSC and metallography analysis showed that Sm started to form a new intermetallic phsse. The same phenomenon was also found in Al-Si-Mg alloy which showed a silicon phase modification at 0.6% Sm [11]. This result indicated the potential for better intermetallic dispersion from the as-cast product.

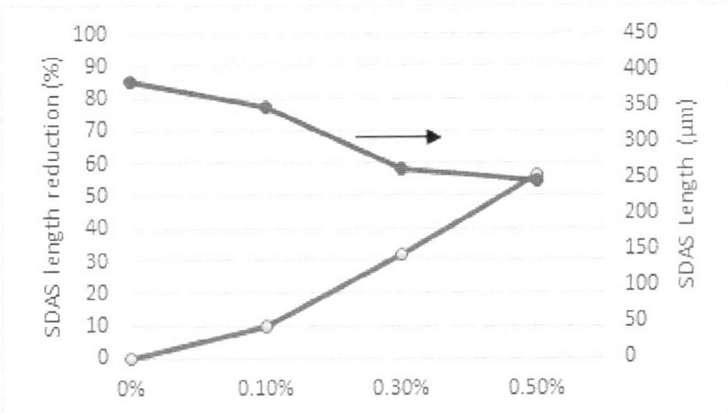

Figure 5. Effect of samarium addition in Al-Zn-Cu-xSm to SDAS.

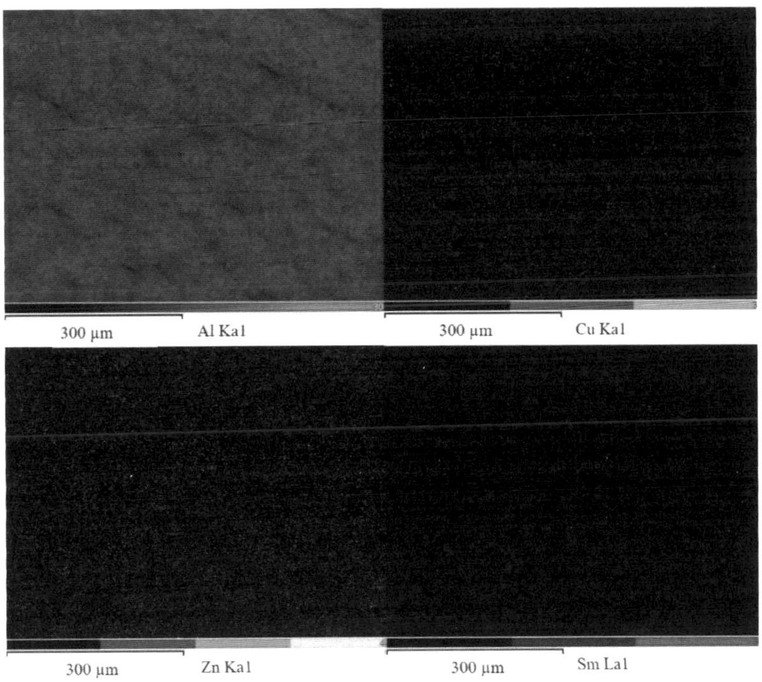

Figure 6. Elemental mapping of Al (red), Zinc (blue), Yellow (copper) and Samarium (purple).

3.2.2 Energy dispersive spectroscopy: element mapping. Energy dispersive spectroscopy test performed to the sample with the highest samarium content (0.5% Sm). All items of the alloy are spread evenly on the microstructure as seen in the elemental mapping results in Figure 6. Several areas in Al element mapping result showed less aluminum content in the interdendritic region. On the contrary, these area has a more abundant Sm and Cu than aluminum. Zinc is spread evenly in an aluminum matrix. This condition is caused by the high solubility level of zinc in aluminum. Copper still gathered around, and bound with, the samarium in the interdendritic area form an intermetallic. It proved the possibility of $SmCu_4Al_{11}$ in this alloy as predicted by DSC result.

4. Conclusion

Samarium acted as grain refiner in Al-5Zn-0.5Cu-xSm alloy. Differential scanning calorimetry showed an intermetallic $SmCu_4Al_{11}$ formed at temperature 580 °C and the α-Al solidification formation shifted to the lower temperature after an addition of 0.1% Sm in the alloy. The most efficient of samarium addition for grain refinement is 0.3% Sm. It could reduce the secondary arm spacing size by approximately 25%. Secondary arm spacing reduced from 383 μm to 261 μm after addition of 0.3% Sm. A higher samarium content increases the intermetallic formation in the grain boundary of Al-Zn-Cu-Sm alloy.

References

[1] Muazu A and Yaro S 2011 *J. Miner. Mater. Charact. Eng.* **10(2)** 185–98
[2] Monzel M William J Druschitz Alan P 2014 Maxfield *Corrosion* 4284
[3] Baker D S 2015 Virginia Polytechnic Institute and State University
[4] Ferdian D Pratesa Y Togina I and Adelia I 2017 *Proc. Eng.* **184** 418–22
[5] Ferdian D Pratesa Y Togina I 2017 *IOP Conf. Ser. Mater. Sci. Eng.* 12026
[6] Ma-Li H B Weili L I Hongjie Z Yonggui Y 2010 *J. Chinese Soc. Corros. Prot.* **4** 2–5
[7] Hu Z Yan H and Rao Y S 2013 *Trans. Nonferrous Met. Soc. China Eng.***23(11)** 3228–34
[8] Qiu H Yan H and Hu Z 2013 *J. Alloys Compd.* **567** 77–81
[9] Liang S M and Schmid-Fetzer R 2016 *Calphad Comput. Coupling Phase Diagrams Thermochem.* **52** 21–37
[10] Riani P Arrighi L Marazza R Mazzone D Zanicchi G and Ferro R 2004 *J. Phase Equilibria Diffus.* **25(1)** 22–52
[11] Qiu H Yan H Hu Z 2013 *J. of Alloys and Compd.* **567** 77-81

Development of battery performance data acquisition system for monitoring battery performance inside solar cell system

A E Mulyono[1], T Mustika[1], H P Sulaikan[1], E Kartini[2]

Center for Materials Technology, Agency for Assessment and Application Technology, Puspiptek Serpong, South Tangerang 15314, Indonesia.

Center for Science and Technology for Advanced Materials National Nuclear Energy Agency, Puspiptek Serpong, South Tangerang 15314, Indonesia.

aditya.eka@bppt.go.id

Abstract. Indonesian battery consortium for solar cell battery has made several battery prototypes and ready to be tested in a real system. Monitoring performance of battery prototype, while it is running in real system is necessary in order to study its behavior. The results are important to be used for improvement of the battery prototype itself. Accurate monitoring of battery performance can be obtained by fitting the instrumentation with the application and condition which are monitored. Generally the system for monitoring battery performance only measures electrical current and voltage, meanwhile for further development, it is also important to measure another variables like internal and ambient temperature of the battery pack. Therefore development of battery performance data acquisition system for monitoring battery performance inside solar cell system is acquired. Battery voltage and current will be measured using a DC voltage sensor based on voltage divider, and using non invasive current sensor WCS1800, respectively. Temperature inside battery pack and ambient temperature are measured using thermocouple type K, and BME280, respectively. All measured variables are pooled to the microcontroller and then they are transmitted to the display screen. This system is expected to be used for monitoring battery performance accurately.

1. Introduction

In battery research and development, monitoring performance of battery prototype while it is running in real system is necessary. This because its behavior in real system is important to be known for further improvement of the battery prototype itself. Accurate monitoring of battery performance can be obtained by fitting the instrumentation with the application and condition which are monitored. Nowadays several companies have already provided battery monitoring product or data loggers. But their product cannot fulfill the requirement from the consortium (Indonesian battery consortium for solar cell battery), for example temperature monitoring is not included in their system. To fulfill the team requirement, the system should be customized.

Several researcher has also made their own customized battery or solar monitoring system [1–16]. However all their system were made to communicate with PC (Personal Computer with windows as Operating system). This paper will describe a new approach for data acquisition system that replacing the role of PC in another researcher's system by using android-based device as display screen and data logger. Android device based application can be easily made by using app Inventor while making PC-based application cannot using app inventor. Moreover Android-based device offering more flexibility in data flow management than PC-based device.

2. Experimental method

Generally the system consists of two main sub-system, namely, a system for measuring the variable which is called "Measurement System" and a system for displaying the data on display screen and for logging the data into internal memory of Android-based device which is called "Display and Logging System". Block diagram for this system can be seen in Figure 2. Measurement System is placed inside the panel of Solar Cell system, while Display and Logging System can be placed remotely from Measurement System, for example, can be placed inside the building. Physical illustration of the system can be seen in Figure 1.

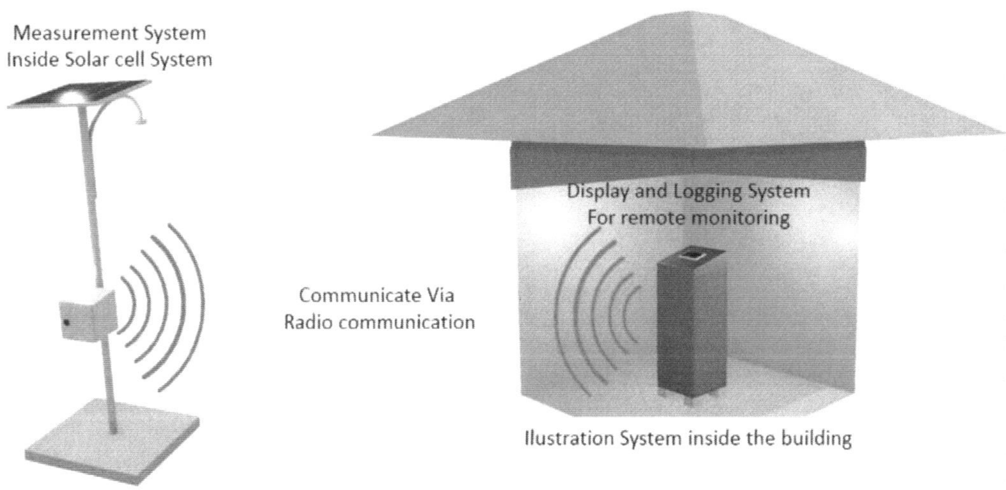

Figure 1. Illustration of the integrated system for solar street lamp.

Variables that measured in this system are voltage, current and temperature. Voltage will be measured using simple voltage divider with the ratio 1:5. Current will be measured using WCS 1800 which output is already in the voltage. Temperature inside the battery is measured using thermocouple type K that will be interfaced using module MAX6675. Ambient temperature is measured using BME280.

Figure 2 shows that voltage and current is interfaced to microcontroller (arduino Nano) *via* module ADC (Analog to Digital converter) ADS1256 which has 24 bit full resolution (23 bit if used single ended mode) and this module using SPI (Serial Peripheral Interface) protocol to communicate with Arduino Nano. Module MAX6675 for thermocouple and BME280 are also communicate through SPI protocol with arduino Nano. All the data that has been collected by arduino Nano then will be transmitted to the other side of the system through module HC12 (radio communication module), Nano communicate with HC12 via UART protocol.

On the other side of system, another HC12 receive the information and then forward the information to microcontroller (arduino Mega), based on the datasheet, the communication range between a pair of HC12 is maximum 1.8 km Line of Sight. From this microcontroller information were forwarded to display screen via bluetooth communication. Finally the information would be displayed on display screen, 10 inch Android-based device, Chuwi Hi10 Pro. The Graphical User Interface (GUI) was made using App Inventor, beside displayed, information are also logged by this device.

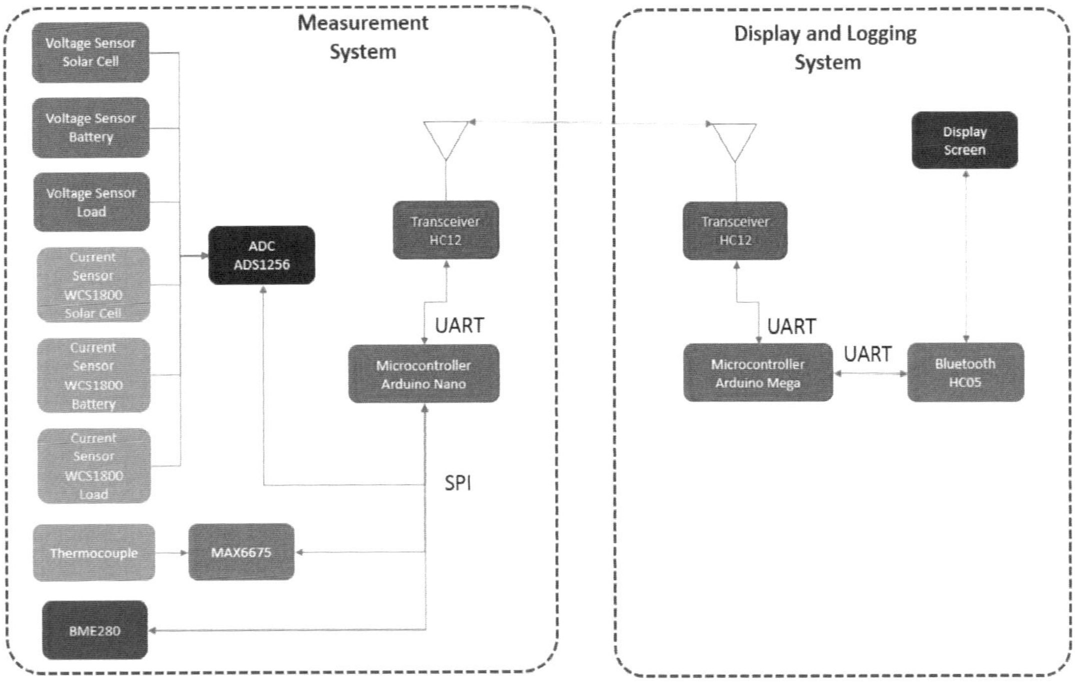

Figure 2. Block diagram for battery monitoring system for solar cell system.

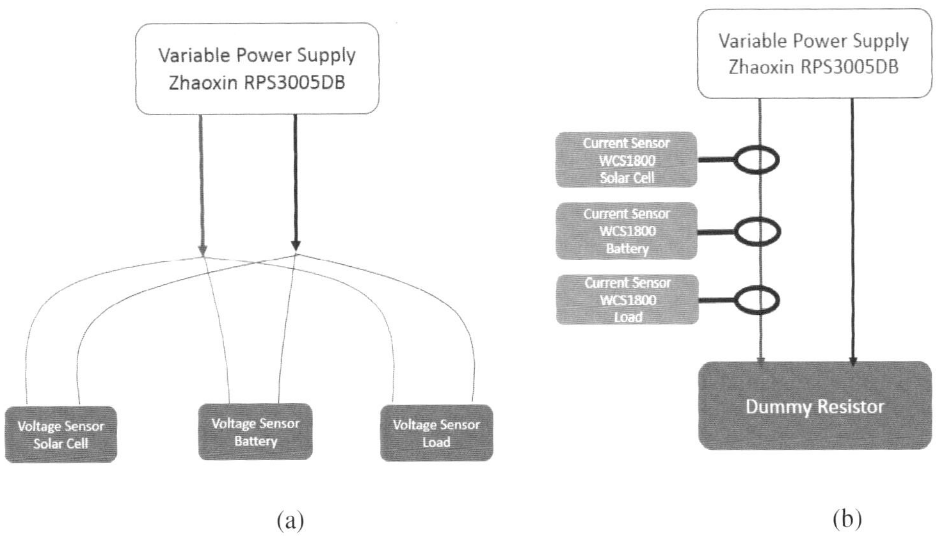

Figure 3. (a) Voltage sensor verification experiment, (b) current sensor verification experiment.

(a) (b)

Figure 4. (a) Photo of voltage sensor verification experiment, (b) Photo of current sensor verification experiment.

In this paper, the system was tried to be applied in lab scaled, not in real solar cell system yet because it was still under development. Voltage and current measurement verification was done using Variable Power Supply Zhaoxin RPS3005DB. Voltage and Current verification experiment was done separately to make sure the value that measured by the sensor did not affect each other. Figure 3 below illustrate how the experiments was done, Figure 4 is the documentation of experiment. Temperature Measurement data validity was referred from its datasheet.

After passed the verification step, the system was tested on the simulation of solar cell system which is replaced solar cell module with variable power supply in solar cell system. The other part of solar cell system is still same which is consist of solar charger controller, battery and load. The illustration of this experiment can be seen in Figure 5.

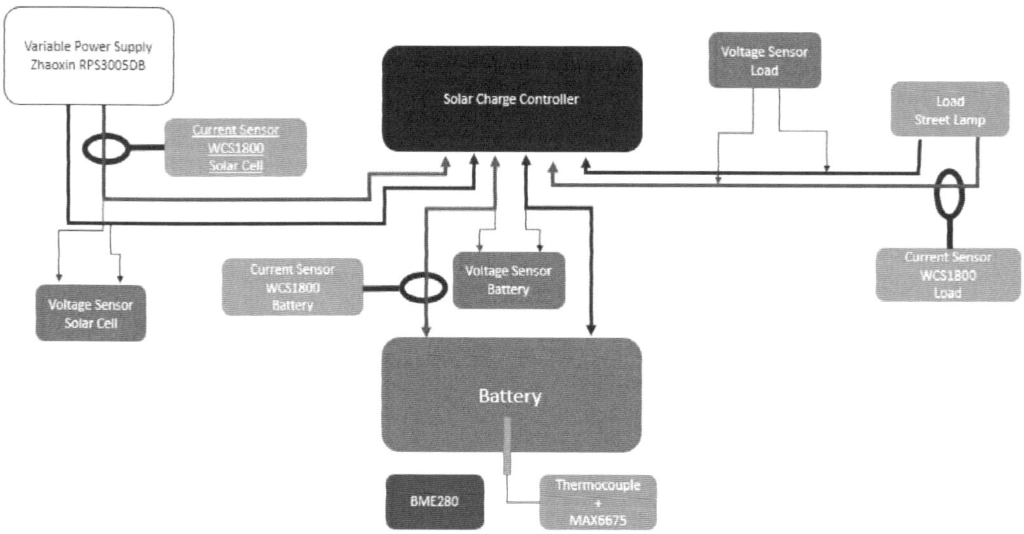

Figure 5. Simulation of solar cell system experiment.

3. Result and discussion

Results from voltage and current sensor verification step can be seen in Figures 6(a) and 6(b). Figure 6 describes the relationship between voltage and current that was measured by the prototype *vs.* value that was givenby verification device (Zhaoxin RPS3005DB). It showed that relationship between input value from verification device and the value that read by prototype was tend to be linear. Specifically, mathematical relationship that obtained from this experiment can be written as in equation [1–5].

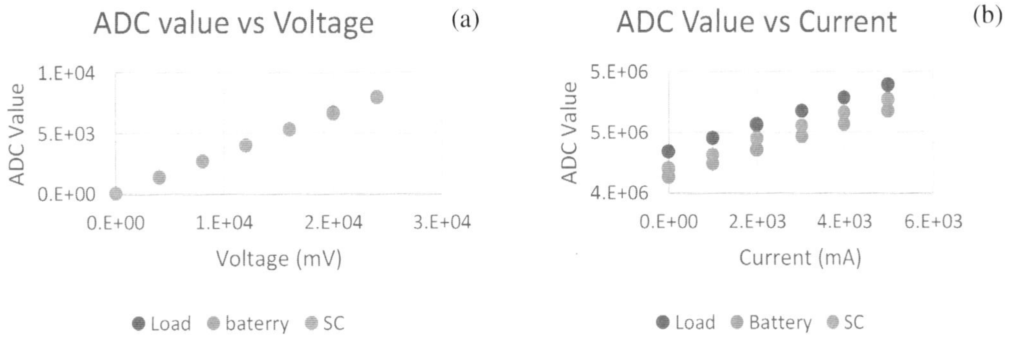

Figure 6. (a) Data plot for voltage sensor verification experiment (b) Data plot for current sensor verification experiment.

$$\boldsymbol{P} = \boldsymbol{a}.\boldsymbol{X} + \boldsymbol{b} \tag{1}$$

$$\boldsymbol{P} = \begin{bmatrix} Load\ Voltage \\ Battery\ Voltage \\ Solar\ Cell\ Voltage \\ Load\ Current \\ Battery\ Current \\ Solar\ Cell\ current \end{bmatrix} \tag{2}$$

$$\boldsymbol{a} = \begin{bmatrix} 0{,}003 & 0 & 0 & 0 & 0 & 0 \\ 0 & 0{,}003 & 0 & 0 & 0 & 0 \\ 0 & 0 & 0{,}003 & 0 & 0 & 0 \\ 0 & 0 & 0 & 0{,}009 & 0 & 0 \\ 0 & 0 & 0 & 0 & 0{,}0091 & 0 \\ 0 & 0 & 0 & 0 & 0 & 0{,}0087 \end{bmatrix} \tag{3}$$

$$\boldsymbol{X} = \begin{bmatrix} ADC\ value\ for\ Load\ Voltage \\ ADC\ value\ for\ Battery\ Voltage \\ ADC\ value\ for\ Solar\ Cell\ Voltage \\ ADC\ value\ for\ Load\ Current \\ ADC\ value\ for\ Battery\ Current \\ ADC\ value\ for\ Solar\ Cell\ current \end{bmatrix} \tag{4}$$

$$\boldsymbol{b} = \begin{bmatrix} -96{,}865 \\ -121{,}8 \\ -128{,}65 \\ -38.980 \\ -37.703 \\ -36.434 \end{bmatrix} \tag{5}$$

IOP Conf. Series: Materials Science and Engineering **432** (2018) 012057 doi:10.1088/1757-899X/432/1/012057

When ***P*** are physical value that want to be read, ***X*** are ADC value that readed by the prototype, ***a*** and **b** are constant that obtained from experiment. This equation was used for translating ADC value to information that shown by display system.

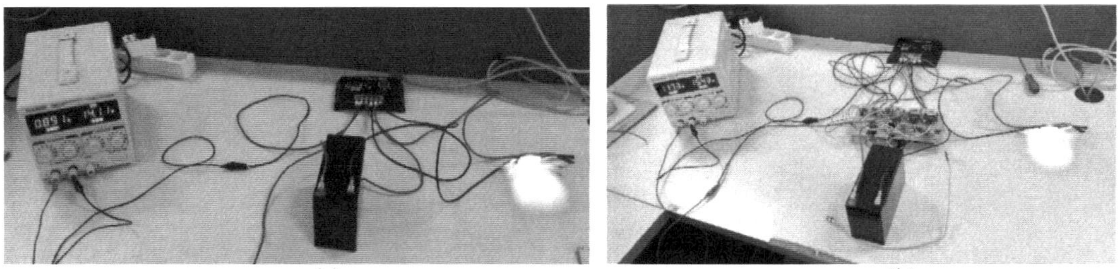

(a) (b)

Figure 7. Documentation from simulation of solar cell system experiment (a) without data acquisition system (b) with data acquisition system.

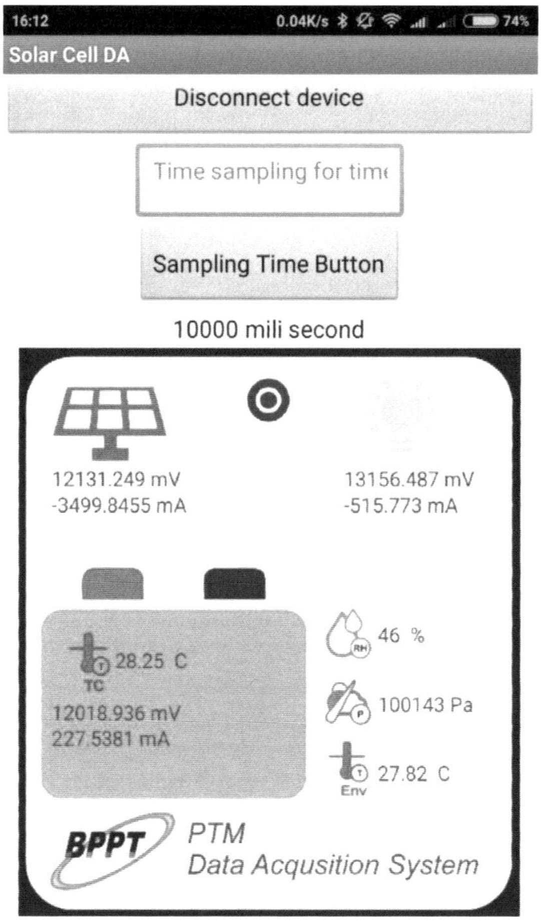

Figure 8.Screenshot of display screen.

	A	B	C	D	E	F	G	H	I	J	K	L	M	N	O	P	Q	R	S	T	U	V	W	X
1061	4-October-2017	16.17.00	1937241	1937255	4386165	4272790	4167591	3804321	4025476	4063323	27580	28250	45000	100158000	13061.63	524.89	2.220.781	-33.364.073	11.954.628	12.061.319	27.58.00	28.25.00	45	100158
1062	4-October-2017	16.17.10	1937233	1937179	4385100	4272112	4166431	3802334	4024837	4062646	27560	28000	46000	100159000	13.058.435	530.992	2.115.221	-33.536.942	11.952.711	12.059.288	27.56.00	28	46	100159
1063	4-October-2017	16.17.20	1937262	1937207	4384565	4271732	4167121	3804050	4024183	4062087	27590	28250	45000	100161000	13056.83	534.412	2.178.011	3.338.765	11.950.749	12.057.611	27.59.00	28.25.00	45	100161
1064	4-October-2017	16.17.31	1937254	1937214	4383475	4273457	4166835	3804329	4023575	4061498	27600	28250	45000	100160000	13053.56	-518.887	2.151.985	-33.363.377	11.948.925	12.055.844	27.06.00	28.25.00	45	100160
1065	4-October-2017	16.17.41	1937290	1937328	4382687	4272637	4165250	3803165	4023007	4060871	27630	28500	45000	100162000	13.051.196	-526.267	200.775	-33.464.645	11.947.221	12.053.963	27.63	28.05.00	45	100162
1066	4-October-2017	16.17.51	1937306	1937174	4382153	4272341	4167091	3803786	4022418	4060310	27630	27750	45000	100161000	13.049.594	-528.931	2.175.281	-33.410.618	11.945.454	12052.28	27.63	27.75	45	100161
1067	4-October-2017	16.18.01	4534280	7868160	1937252	4273672	4167043	3803432	4021685	4059650	27630	28250	45000	100157000	5.714.891	-516.962	2.170.913	-33.441.416	11.943.655	12050.3	27.63	28.25.00	45	100157
1068	4-October-2017	16.18.11	1937235	1937224	4379913	4272936	4168244	3802658	4021227	4059029	27620	28000	45000	100161000	13.042.874	-523.576	2.280.204	-33.508.754	11.941.881	12.048.437	27.62	28	45	100161
1069	4-October-2017	16.18.21	1937222	1937200	4379334	4273352	4167059	3803148	4020702	4058323	27630	28250	45000	100156000	13.041.137	519.832	2.172.369	33.466.124	11.940.306	12.046.319	27.63	28.25.00	45	100156
1070	4-October-2017	16.18.31	1937230	1937231	4378710	4273086	4167288	3803766	4020241	4057750	27610	28250	45000	100161000	13.039.265	522.226	2.193.208	33.412.358	11.938.923	12044.6	27.61	28.25.00	45	100161
1071	4-October-2017	16.18.41	4534280	7868160	1937263	4272866	4166444	3803061	4019315	4057225	27630	28250	45000	100161000	5.714.924	-524.206	2.116.404	-33.473.693	11.936.145	12.043.025	27.63	28.25.00	45	100161
1072	4-October-2017	16.18.51	1937239	1937163	4376704	4273269	4167571	3804187	4018934	4056698	27590	28000	45000	100160000	13.033.247	-520.579	2.218.961	-33.375.731	11.935.002	12.041.444	27.59.00	28	45	100160
1073	4-October-2017	16.19.02	4534280	7868160	1937245	4274261	4167816	3805040	4018403	4055955	27580	28250	45000	100157000	5714.87	-511.651	2.241.256	-3.330.152	11.933.409	12.039.215	27.58.00	28.25.00	45	100157
1074	4-October-2017	16.19.12	1937212	1937133	4375378	4273394	4167137	3804646	4017775	4055537	27570	28250	45000	100161000	13.029.269	-519.454	2.179.467	-33.335.798	11.931.525	12.037.961	27.57.00	28.25.00	45	100161
1075	4-October-2017	16.19.22	1937221	1937141	4374461	4273609	4167717	3804836	4017227	4054876	27560	27500	46000	100164000	13.026.518	-517.519	2.232.247	-33.319.268	11.929.881	12.035.978	27.56.00	27.05.00	46	100164
1076	4-October-2017	16.19.32	1937242	1937221	4373598	4273140	4167902	3804842	4016468	4054320	27570	28500	46000	100158000	13.023.929	321.74	2.249.082	33.318.746	11.927.604	12034.31	27.57.00	28.05.00	46	100158
1077	4-October-2017	16.19.42	1937177	1937161	4373578	4274188	4166638	3804416	4016240	4053674	27560	27750	45000	100163000	13.023.869	512.308	2.134.058	33.355.808	11926.92	12.032.372	27.56.00	27.75	45	100163
1078	4-October-2017	16.19.53	1937174	1937128	4372479	4273840	4167462	3804351	4015595	4053154	27580	28000	46000	100165000	13.020.572	515.44	2.209.042	33.361.463	11.924.985	12.030.812	27.58.00	28	46	100165
1079	4-October-2017	16.20.03	1937150	1937105	4371949	4273132	4167772	3804650	4015183	4052563	27590	28500	45000	100169000	13.018.982	-521.812	2.237.252	-3.333.545	11.923.749	12.029.039	27.59.00	28.05.00	45	100169
1080	4-October-2017	16.20.13	1937130	1937143	4371113	4272958	4167566	3804268	4014625	4052173	27590	28250	45000	100167000	13.016.474	-523.378	2.218.506	-33.368.684	11.922.075	12.027.869	27.59.00	28.25.00	45	100167
1081	4-October-2017	16.20.23	1937225	1937183	4370218	4272763	4167349	3803550	4013988	4051564	27580	28250	45000	100164000	13.013.789	-525.133	2.198.759	-3.343.115	11.920.164	12.026.042	27.58.00	28.25.00	45	100164
1082	4-October-2017	16.20.33	1937162	1937213	4375457	4273603	4166417	3803945	4041663	4035193	27590	28250	45000	100168000	13.029.506	-517.573	2.113.947	-33.396.785	12.003.189	11.976.929	27.59.00	28.25.00	45	100168
1083	4-October-2017	16.20.43	1937191	1937203	4374327	4272898	4166170	3804305	4041047	4034728	27600	28250	45000	100166000	13.026.116	-523.918	209.147	-33.365.465	12.001.341	11.975.534	27.06.00	28.25.00	45	100166
1084	4-October-2017	16.20.53	1937130	1937157	4373920	4272493	4167257	3805115	4040634	4034129	27600	27750	45000	100164000	13.024.895	527.563	2.190.387	33.294.995	12.000.102	11.973.737	27.06.00	27.75	45	100164
1085	4-October-2017	16.21.03	4534280	7868160	1937182	4273518	4166315	3804444	4040068	4033539	27580	28250	45000	100165000	5.714.681	518.338	2.104.665	33.353.377	11.998.404	11.971.967	27.58.00	28.25.00	45	100165
1086	4-October-2017	16.21.13	1937143	1937041	4372690	4272687	4167375	3806036	4039487	4033095	27560	28500	45000	100165000	13.021.205	525.817	2.201.125	-33.214.868	11.996.661	11.970.635	27.56.00	28.05.00	45	100165
1087	4-October-2017	16.21.24	1937054	1937044	4371723	4273642	4166123	3804968	4039097	4032505	27560	28500	45000	100165000	13.018.304	-517.222	2.087.193	-33.307.784	11.995.491	11.968.865	27.56.00	28.05.00	45	100165
1088	4-October-2017	16.21.34	1937121	1937074	4371402	4272565	4166223	3805146	4038462	4032102	27550	28000	45000	100172000	13.017.341	-526.915	2.096.293	-33.292.298	11.993.586	11.967.656	27.55.00	28	45	100172
1089	4-October-2017	16.21.44	1937079	1937002	4370668	4272094	4166758	3804454	4037966	4031560	27530	28250	45000	100167000	13.015.139	-531.154	2.144.978	-33.352.502	11.992.098	11966.03	27.53.00	28.25.00	45	100167
1090	4-October-2017	16.21.54	1937027	1937057	4369931	4272962	4165876	3804125	4037559	4030994	27490	27750	45000	100165000	13.012.928	-523.342	2.064.716	-33.381.125	11.990.877	11.964.332	27.49.00	27.75	45	100165
1091	4-October-2017	16.22.04	1937131	1937075	4369159	4273615	4166416	3804221	4036882	4030664	27500	28250	45000	100171000	13.010.612	-517.465	2.113.856	-33.372.773	11.988.846	11.963.342	27.05.00	28.25.00	45	100171
1092	4-October-2017	16.22.14	1937090	1937066	4368949	4272923	4167126	3805826	4036445	4030145	27480	28250	45000	100171000	13.009.982	523.693	2.178.466	33.233.138	11.987.535	11.961.785	27.48.00	28.25.00	45	100171
1093	4-October-2017	16.22.24	1937051	1937050	4368125	4272715	4166871	3805506	4035912	4029581	27500	28000	45000	100173000	13007.51	525.565	2.155.261	33.260.978	11.985.936	11.960.093	27.05.00	28	45	100173
1094	4-October-2017	16.22.34	4534280	7868160	1937103	4272532	4166631	3804830	4035531	4029092	27490	28500	45000	100166000	5.714.444	-527.212	2.133.421	-3.331.979	11.984.793	11.958.626	27.49.00	28.05.00	45	100166
1095	4-October-2017	16.22.44	1937039	1937073	4367260	4273181	4167067	3805203	4035033	4028579	27470	28000	45000	100171000	13.004.915	-521.371	2.173.097	-33.287.339	11.983.299	11.957.087	27.47.00	28	45	100171
1096	4-October-2017	16.22.55	1937024	1937008	4366312	4273424	4166478	3805359	4034431	4028123	27430	28250	45000	100172000	13.002.071	-519.184	2.119.498	-33.273.767	11.981.493	11.955.719	27.43.00	28.25.00	45	100172
1097	4-October-2017	16.23.05	4534280	7868160	1937072	4272878	4165929	3806539	4034008	4027623	27390	28250	46000	100173000	5.714.351	-524.098	2.069.539	-33.171.107	11.980.224	11.954.219	27.39.00	28.25.00	46	100173
1098	4-October-2017	16.23.33	1937030	1937012	4365188	4272775	4166493	3804828	4033539	4027040	27410	28500	46000	100176000	12.996.699	-525.025	2.120.863	-33.319.964	11.978.817	11952.47	27.41.00	28.05.00	46	100176

SCAD2-4-October-2017

Figure 9.Screenshot of data logging results.

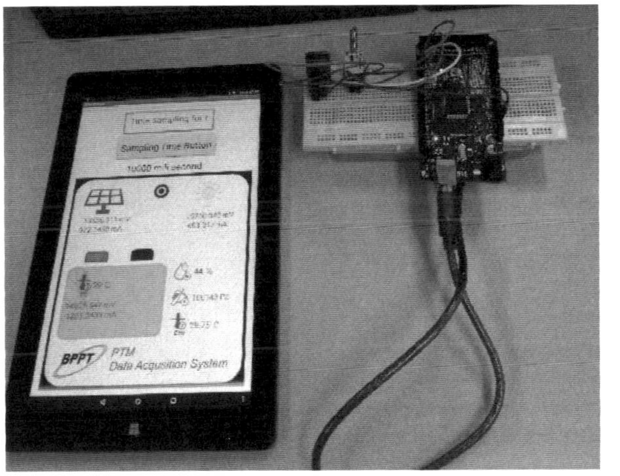

Figure 10. Documentation of Display and Logging system while running.

Results from simulation of solar cell system experiment can be seen on Figure 7. Screenshot for display and data that had been logged by the logging system during experiment can be seen on Figure 8 and 9. Documentation of Display and Logging system while running can be seen in Figure 10. It can be seen from Figure 7 until 10 that the system worked well. The data that were measured from Measurement System could be successfully displayed and logged by Display and Logging System.

Voltage measurement range from this device is 0-25 Volt, current measurement range is-35 until 35 Ampere. Thermocouple in this prototype has range from 0-1024 degree Celsius, and BME280 for environment temperature can measure from -40 until 85 degree Celsius. The advantage of this prototype compared to other reference is in ADC unit, and its software is in Display and Logging System. The ADC that used in this device is 24 bit which is give more resolution than other reference that used

ADC less than 24 bits [5, 7, 14, 16–18]. Software that was used to build an application in Display and Logging System is MIT App Inventor, which is free and easy to use, meanwhile software that was used by reference are not free [2, 4, 6, 8–10, 13, 14, 19]. This experiment had successfully proved that data acquisition using Android-based device can be implemented for monitoring battery performance inside solar cell system.

4. Conclusion

The system for monitoring battery performance which used an Android-based device has been developed. Voltage, current, and temperature were successfully measured by the prototype. The data from measurement system to display system can also successfully be transmitted. The Graphical User Interface for displaying information also works well. The logging system is also works fine. Overall the system works fine, but it is still have several deficiency such as, its prototype casing, its capacity algorithm verification, the determination of real maximum distance, etc.

Acknowledgment

We would like to express my gratitude to PTM-BPPT and Indonesian Consortium for Solar Cell Battery for their sponsorship and loaning of the equipment. This work is partially supported from the grant for the New and Renewable Energy Research Consortium for Lithium Battery Development as Energy Storage of Solar Electricity Power funded by the Indonesian Ministry Research and Technology.

References

[1] Nagalakshmi R Kishore B Babu D P 2014 Design and Development of a Remote Monitoring and Maintenance of Solar Plant Supervisory System **3**
[2] Kumar C M A and Manivanan G N *IOSR J. Comput. Eng.* 33–6
[3] Harid N Bogias A C and Haddad A 2012 *Recent Res. Commun. Electron. Signal Process. Autom. Control* 181–6
[4] Murugan H B Nagaraju S V S Kumar K S Ramakrishnan K C and Sowjanya V E 2007 *Int. J. Adv. Res. Electr. Electron. Instrum. Eng. An ISO Certif. Organ.* **3297**
[5] Mukaro R and Carelse X F 1999 *IEEE Trans. Instrum. Meas.* **48** 1232–8
[6] Dogaru V Cepisca C Andrei H and Ivanovici T 2011 Data Acquisition in Photovoltaic Systems *Renew. Energy - Trends and Applic.* (InTech)
[7] Krishna A M Rao K P and Prakash M B 2012 *Int. J. Eng. Res. Tech.*. **1**
[8] Zaw M and Min W 2014 Design and Implementation of a data-acquisition system to monitor and control a photovoltaic power generation system
[9] Sousa S Onofre M Antunes T Branco C Maia J Rocha J-I and Pires V F 2013 *Int. Conf. on Renewable Energy Research and Applic. (ICRERA)* (IEEE) 1084–8
[10] Joshi M Mishra A Rakhasiya R and Balpande S 2017 *Int. Res. J. Eng. Techl.* 2440–2
[11] Sharma M Bansal K and Buddhi D 2015 *Procedia Comput. Sci.* **48** 582–8
[12] Vyas M Chudasama K Bhatt M and Gohil B 2016 *Int. J. Curr. Eng. Int. J. Curr. Eng. Techl.* **6**
[13] Sharma M Bansal K and Buddhi D 2015 *Procedia Comput. Sci.* **48** 582–8
[14] Koutroulis E and Kalaitzakis K 2003 *Renew. Energy* **28** 139–52
[15] Faizal A Abidin Z Jusoh M H James E Mutalib S A Junid A Ihsan A and Yassin M *IOP Conf. Series: Mater. Sci. and Engin.* **99** 70
[16] Jucá S C S Carvalho P C M and Brito F T 2011 *Sensors (Basel)* **11** 743–56
[17] Jamaluddin A Sihombing L Supriyanto A Purwanto A and Nizam M 2013 *Proc. of the 2013 Joint Int. Conf. on Rural Inform. and Commun. Techn. and Electric-Vehicle Tech. rICT and ICEV-T 2013* (IEEE) 1–4
[18] Jamaluddin A Nur'aini A Jumari A and Purwanto A 2017 *Int. J. Power Electron. Drive Syst.* **8** 926–34
[19] Sureshgoud S and Rao N 2014 *Int. J. Mag. Eng. Technol. Manag. Res.* **1** 630

IOP Conf. Series: Materials Science and Engineering **432** (2018) 012058 doi:10.1088/1757-899X/432/1/012058

Synthesis of $Na_2Ti_3O_7$–based anode for sodium-ion battery using solid state reaction method

Slamet P [1]*, W B Widayatno [1], A Subhan[1], B Prihandoko[1]

[1] Research Center for Physics – Indonesian Institute of Sciences, PUSPIPTEK Area Build. 442, Serpong, South Tangerang 15314, Banten, Indonesia Telp. 021-7560570, Fax. 021- 7560554

slam013@lipi.go.id

Abstract. $Na_2Ti_3O_7$-based anode for sodium-ion battery was synthesized through solid-state reaction of high purity Na_2CO_3 and TiO_2. The dry and wet mixing by using ethanol assistance were used to compare the effects on their properties. Simultaneous Thermal Analysis (STA) was carried out to determine the temperature of sintering process. Precusors, both from dry and wet mixing, were sintered at 900 °C for 4 h in air. The structure and phase of the sintered samples were characterized by X-ray diffraction (XRD), while the particle size was determined using a particle size analyzer (PSA). The samples were then assembled into coin cells for performance test purposes. The electrochemical processes of all of samples were examined using an automatic battery cycler system. It is shown that ethanol assisted-wet process did not decrease the sintering temperature but it decreased the particle size of samples. Phase analysis showed that the phases for both treatments are $Na_2Ti_3O_7$ and $Na_2Ti_6O_{13}$, and they had a high crystalinity. Cyclic voltammetry shows that all samples have an unbalanced pair of redox peaks and show formation SEI layers.

1. Introduction

Energy is the most important thing in our live because everybody in entire world needs energy to do one's activity. Almost all energy which is consumed every day is from the dirty fossil fuel energy. The fossil fuel energy has been used in some economic sectors like industry, transportation, and home. Burning fossil fuel energy every time will increase the amount of carbon dioxide in the atmosphere and will lead to global warming and climate change [1]. The way to combat these impacts to our live is by decreasing the fossil fuel energy use and replacing it with renewable energy from wind, solar, tidal, and hydro energy. To apply these renewable energy sources to the grid, electrochemical energy storage device is required. Electrochemical energy storages like lithium ion batteries are seen to be the most favorable candidate because they have high energy densities, ability to demonstrate flexible response times, portability, pollution-free operation, long cycles, and high efficiencies [1, 2].

Batteries are a collection of electrochemical cells which can convert chemical energy into electrical energy and its reverse. The three main constituents of a lithium ion cell are: an anode (which is oxidized), a cathode (which is reduced), and the electrolyte. These electrochemical devices convert electrical energy into chemical energy when the devices are charged and convert chemical energy into electrical energy when the devices are used [3].

Secondary sodium-ion batteries (SIBs), as alternative to lithium-ion batteries (LIBs), have attracted the battery researchers entire the world because sodium is more abundant and cheaper than lithium.

SIB and LIB have many technical similarities in electrodes and electrolytes. One of the similarities is anode material of $Na_2Ti_3O_7$, or called NTO in SIBs, has the same form with $Li_4Ti_5O_{12}$ (LTO) in LIBs [4].

Ti-based oxides have been considered as potential electrode material for their small structural expansion and applicable operating voltage [5]. As a typical insertion Ti-base oxides, $Na_2Ti_3O_7$ is a promising negative electrode for rechargeable batteries since it has a very low insertion voltage (0.6 V *vs* Li^+/Li, or 0.3 V *vs* Na/Na^+) and high specific capacity (178 mAh/g) [6, 7]. This materials, $Na_2Ti_3O_7$, is very low cost and has a good cycling stability.

In this preliminary study, we will synthesize NTO powder via solid state reaction with dry and wet mixing, determine phase via XRD, and study intercalation process via cyclic voltammetry. Like in LiBs, anode act as host for ion while intercalating and de-intercalating process so $Na_2Ti_3O_7$ also act as host for Na ion [8, 9]. We will study intercalating process in $Na_2Ti_3O_7$ powder with Li ion because Li^+ radius is smaller than Na^+. Smaller ion will have more mobile while inserting/de-inserting process [10].

2. Experimental method

The NTO phase can be prepared easily by solid state reaction from Na_2CO_3 and TiO_2 by following equation [11].

$$Na_2CO_3 + 3\ TiO_2 \rightarrow Na_2Ti_3O_7 + CO_2 \quad (1)$$

NTO powder was synthesized via a ceramic route using TiO_2 (Sigma Aldrich) anatase as Ti sources and Na_2CO_3 (Sigma Aldrich) as Na sources. Raw materials were mixed stoichiometrycaly with ethanol and without ethanol-assisted or wet and dry mixing respectively to become precursor. Both precursor were tested by Simultaneous Thermal Analysis (STA) merck Linseis/PT1600 to determine the sintering temperature. The mixture was then heated in air at up to 900 °C for 4 h. The structural characterization of product was done by X-ray diffraction (XRD) using Rigaku. Particle size of powder was measured by Particle Size Analyzer (PSA) merck cilas 1190 using water as dispersant medium.

The electrode was prepared with common way from slurry containing 85 wt% of $Na_2Ti_3O_7$ powder, 5wt% of acetylene black as conductive agent, and 10% of polyvinylidene fluoride (PVDF) as binder. All materials were dissolved in N,N Dimethylacetamide (DMAC) solution instead of using NMP because PVDF more soluble in DMAC than in NMP solution [12]. The paste was coated on a Cu foil by the screen printing method. The working electrode was dried at 80 °C in vacuum oven for 30 minutes and then pressed into the final thickness of 40-60 um. After they were completely dried, then the electrode sheet was cut to a 14 mm diameter. The electrodes were assembled with lithium metal, separator and electrolyte in glove box become half-cell. After that, the half cells were measured by the cyclic voltammetry using battery cycler WBCS3000.

3. Results and discussion

Figure 1 shows the results of thermal analysis using STA of precursor $Na_2Ti_3O_7$ via (a) dry and (b) wet milling. Both of patterns show identical pattern. From DSC curve, there is small endothermic peak at the range of 25-100 °C for all samples and it is associated with evaporation water content adhering on raw materials surface. The difference of both sample is the peak of second sample (wet mixing) is wider than first sample, which indicates that second sample absorb more energy because there is ethanol content in the sample. From TGA curve at the range 25-100 °C, There is no significant mass change and it indicates that there is no crystallization or formation of a new phase. In that range, relative mass increases of about 7% and 3% occur for precursor (a) and (b) respectively, and it indicates that there is oxidizing process. From DSC curve at the range 800-900 °C, there is high and sharp endothermic peak for all sample and it associated with crystallization and formation of new phases. This endothermic process is accompanied by a drastic mass change. After 900 °C, there is no endothermic process and mass change, and it means that the crystallization proses is over.

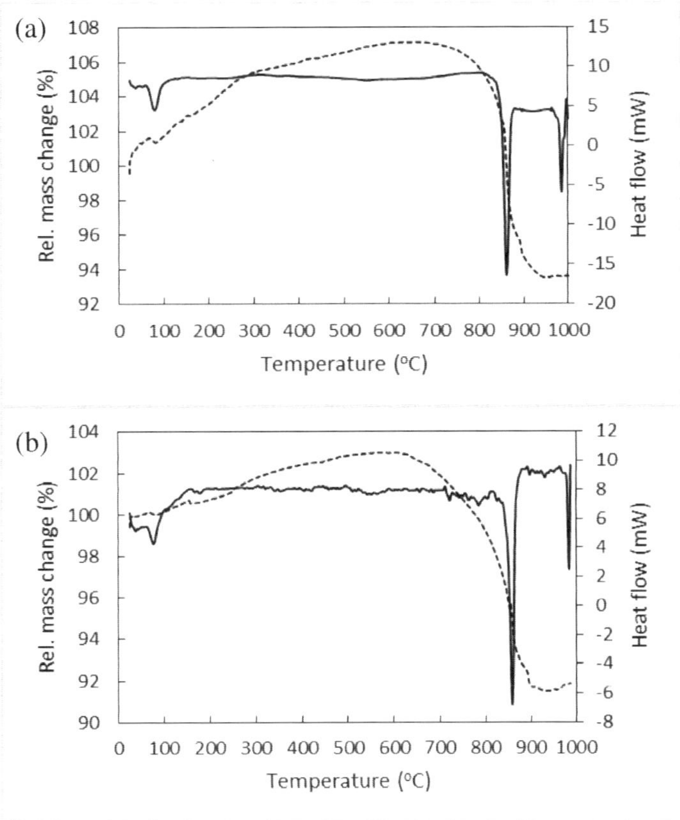

Figure 1. Thermal analysis using simultaneous thermal analysis (STA) of precursor $Na_2Ti_3O_7$ (a) dry mixing (b) wet mixing.

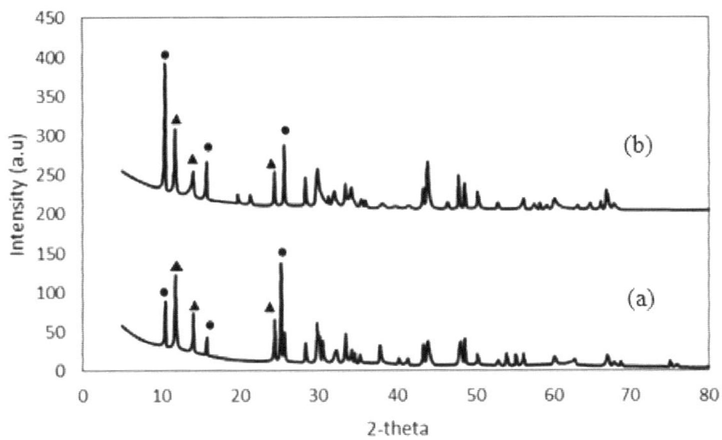

Figure 2. X-ray diffraction patterns of all samples of $Na_2Ti_3O_7$ (a) dry mixing (b) wet mixing.

Figure 2 shows XRD pattern of NTO with and without ethanol assistance as showed by (a) and (b) sign, respectively. The XRD patterns show that there are two dominant phases such as $Na_2Ti_3O_7$ and $Na_2Ti_6O_{16}$ which are denoted by solid circle and triangle marks, respectively [13]. Both of patterns show good XRD diagrams with high peak, sharp and narrow which indicated that these material have high crystallinity. The existence of $Na_2Ti_6O_{16}$ phase indicated that the reaction does not occur perfectly and material need more heat or longer time in sintering process. However, ethanol-assisted process before sintering can increase the $Na_2Ti_3O_7$ and decrease the $Na_2Ti_6O_{16}$ phase, it indicates that ethanol-assisted help mixing process and the precursor become more homogenous than the sample without ethanol assistance.

Table 1. Particle size of samples.

Measurement	Dry mixing (μm)	Wet mixing (μm)
1	6.12	3.53
2	5.62	3.70
3	6.36	3.33
Mean	6.05	3.52

Table 1 shows that samples prepared by wet mixing have smaller particles than samples prepared by dry mixing. Sample b has particle size almost half time bigger than other because ethanol could dissolve and pulverize raw material to smaller sizes.

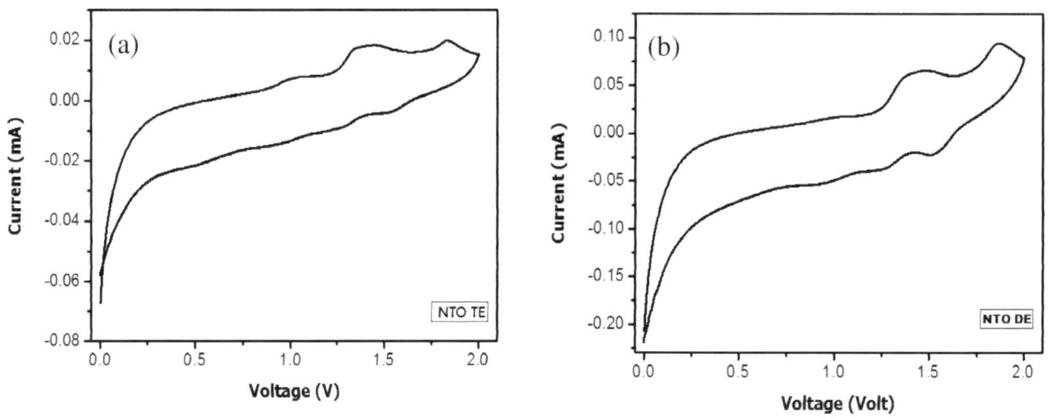

Figure 3. Cyclic voltammogram patterns of all samples of $Na_2Ti_3O_7$ (a) dry mixing (b) wet mixing.

Figure 3 shows the CV of $Na_2Ti_3O_7$ both (a) dry mixing and wet mixing in the potential range of 0.0-2.0 V (*vs* Li/Li^+) at a scan rate of 0,1 $mV.s^{-1}$. All samples show identical pattern with small pairs of oxidizing and reducing peak, which indicates that sample can be charge-discharge slowly. The second sample has a higher intensity than the first sample because it is attributed from the phase and the particle size. The second sample has higher crystallinity and smaller particle than the first sample, and therefore this material will be more active in electrochemical process. In the first cathodic scan, a weak peak at around 1.3 V could be attributed to insertion of lithium ion to the $Na_2Ti_3O_7$ structure. The small peak emerging at 1.8V could be associated with $Na_2Ti_6O_{16}$ phases [13], while the large peak ranged from 0.4 to 0.0 V correspond to the de-insertion of Li^+ and the formation of solid electrolyte interface (SEI) layer [5]. The oxidizing peak appearing at 1.3 V is similar with the result of an experiment that used Na reference and Na-electrolyte [8]. It indicates that Li environment can be used to analyze intercalating process in NTO powder.

4. Conclusion

$Na_2Ti_3O_7$ powder was prepared by solid state reaction with dry mixing and wet mixing. Thermal analysis shows that both sample must be sintered at 900 °C, while ethanol assistance just decreased the amount of energy. Phase analysis showed that there are co-existence of $Na_2Ti_3O_7$ and $Na_2Ti_6O_{16}$ phases but ethanol-assisted procession before sintering can increase the $Na_2Ti_3O_7$ phase. Ethanol assistance also can decrease particle size from 6 micrometer to 3 micrometer on average while in sintering process. Cyclic voltammetry shows that all samples have a pair of unbalanced redox peak, show two oxidizing peaks and show formation SEI layer.

Acknowledgment

The author would like to acknowledge the financial and experimental tools support from the Ministry of Research, Technology and Higher Education (Menristek-Dikti); and Research Center for Physics – Indonesian Institute of Sciences (LIPI). The author also would like to thank the Battery Research Group in Research Center for Physics – Indonesian Institute of Sciences (LIPI) for helping the battery fabrication and analysis.

References

[1] Yang Z Zhang J Kintner-Meyer M C W Lu X Choi D Lemmon J P and Liu J 2011*Chem. Rev.* **111** 3577-613
[2] Dunn B Kamath H and Tarascon J M 2011 *Science* **334** 928-35
[3] Leiden D 2002 *Hanbook of Batteries* McGraw-Hill
[4] Sun J Lee H W Pasta M Sun Y Liu W Li Y Lee H R Liu N Cui Y 2016 *Energy Storage Materials* **4** 130-6
[5] Cheng C C Ning Z Chang L Y Jing W Y Jun C 2016 *Acta Phys. Chim. Sin.* **32(1)** 349-55
[6] Rousse G Arroyo-de Dompablo M E Senguttuvan P Ponrouch A Tarascon J M Palacin M R 2013 *Chem. Mater* **25** 4946-56
[7] Zarrabeitia M Nobili F Munoz M A Rojo T Cabanas M C 2016 *Journal of Power Sources* **330** 78-83
[8] Ko J S Nguyen V V T D Kim H S Muller G A Serino A C Weiss P S Dunn B S 2017 *ACS Appl. Mater. Interfaces* **9** 1416-24
[9] Yu L Wang L P Liao H Wang J Feng Z Lev O Loo J S C Saugrati M T Xu Z J 2018 *Small* **170** 3338-60
[10] Rahm R Hoffmann R Ashcroft N W 2016 *Chem. Eur. J.* **22** 14625-32
[11] Baliteau S Sauvet A L Lopez C Fabri P 2007 *Solid State Ionics* **178** 1517-22
[12] Liu F Hashim N A Liu Y Abed M R Li K 2011 *Journal of Membrane Science* **375** 1-27
[13] Rudola A Sharma N Balaya P 2015 *Electrochemistry Communication* **61** 10-3

IOP Conf. Series: Materials Science and Engineering **432** (2018) 012059 doi:10.1088/1757-899X/432/1/012059

The improving conductivity of $LiFePO_4$ by optimizing the calendaring process

Deswita[1], Sudaryanto[1], H Djodi[1] and E Kartini[1]

[1] Centre for Science and Technology Advanced Materials, National Nuclear Energy Agency Build. 71 Puspiptek Area Serpong, Tangerang – Indonesia 15313

deswita@batan.go.id

Abstract. Lithium ion batteries have important applications in various electronic devices. The use of battery components such as $LiFePO_4$ cathodes determines battery performance. However, the electronic conductivity of $LiFePO_4$ is low; therefore, increasing the conductivity of the $LiFePO_4$ cathode sheet is very important. The aim of this research is to obtain the conductivity of $LiFePO_4$ cathode sheet which can be applied as lithium ion battery cathode through calendaring process optimization. $LiFePO_4$ cathode sheet preparation begins with $LiFePO_4$ slurry ($LiFePO_4$ + Carbon Black and PVDF / NMP) slurry, followed by coating it on Al foil surface, resulting in a sample denoted as C0. The calendaring process was then carried out to the cathode sheet, at 100°C by repeating once, twice, and three times, producing samples hereinafter referred to as C1, C2, and C3 respectively. $LiFePO_4$ cathode sheet products are characterized with X-ray diffractometer (XRD) and scanning electron microscopy (SEM) combined with energy-dispersive spectrophotometer (EDS) and LCR-meter. The XRD pattern shows that the $LiFePO_4$ cathode sheet consists of $LiFePO_4$ and carbon crystal phases. The cross-sectional microstructure of the $LiFePO_4$ cathode sheet shows that the sheets becomes denser and thinner as the number of calendaring increases. The thickness of the $LiFePO_4$ cathode sheet decreases with increasing calendaring process. The $LiFePO_4$ cathode sheet for the C0 sample was about 160μm in thick, and decreased gradually to 140, 130 and 120μm for the C1, C2 and C3 samples, respectively. The conductivity of the sample was improved from 9.9×10^{-5} to 2.9×10^{-5}, 3.0×10^{-4}, and 5.5×10^{-4} $S.cm^{-1}$. The density also was improved from 2.359 to 2.516 and to 2.638 $g.cm^{-1}$ with increasing calendaring process. It can be concluded that the optimizing calendaring process of the $LiFePO_4$ cathode sheet with the $LiFePO_4$ cathode sheet is important to improve the performance of batteries.

1. Introduction

Currently, the development of technology is running rapidly along with the growth and development of electronic equipment, especially cell phones and laptops. This electronic equipment requires the battery as its power source because it is used portably. To meet the growing need for portable power in the future, batteries are needed to provide large amounts of energy and power, and they need to be light, cheap, safe, and environmentally friendly [1–5]. In 1991 Sony first commercially produced lithium-ion battery; the production of lithium-ion batteries has increased significantly because it has created an electronic revolution in the world. Currently, Japan is the largest battery manufacturer, chiefly by Sony,

Panasonic, and Toshiba. Lithium-ion batteries are also the leader of battery products that control 46% or about 4 billion US dollars market share in 2007 [5, 6]. The lithium ion battery itself consists of anode, separator, electrolyte, and cathode. The cathode material is the most expensive material and one of the hardest to produce components of lithium ion batteries, and therefore intensive research is needed to focus on the cathode. The first cathode material used in lithium ion batteries is $LiCoO_2$. Then other cathode materials such as $LiMO_2$ oxide compounds (M are metals which can be Co, Ni, Mn,), LiM_2O_4 spinal compound (M = Mn), and olivine $LiMPO_4$ (M = Fe, Mn, Ni, Co). $LiFePO_4$ is extensively developed as a cathode material because it is stable, cheap, and environmentally friendly. Efforts to improve electrical conductivity can be done in several ways, adding doping when synthesizing cathode material, and mixing cathode materials with metal powders or coating cathode materials with conductive carbon [7, 8]. In previous research, the standardization of $LiFePO_4$ cathode standard materials by adding carbon black has been performed [9–11], but has not yet obtained optimal conductivity. In this research, we produced $LiFePO_4$ cathode sheet by adding carbon black (CB) and polyvinylidene fluoride (PVDF) polymer in N-methylpyrrolidone (NMP) solution and variation of calendaring process [12–14]. Thus far, no other previous work has concerned the influence of the calendaring process. Therefore, in this study it is expected that through this calendaring process can produce the optimum $LiFePO_4$ material conductivity to be used as battery cathode material. The quality of the resulting product is characterized using the XRD to determine the phase of the $LiFePO_4$. The density was calculated manually from measured mass and the volume of the sample. SEM-EDS to observe the surface morphology and its elements content. LCR meter to measure electrical conductivity.

2. Method

2. 1 $LiFePO_4$ cathode sheet preparation

The materials used in this study include commercial materials such as polyvinylidene fluoride (PVDF) as binder, $LiFePO_4$ powder (LFP MTI), carbon black (CB) as additive, N-methylpyrrolidone (NMP) as solvent [5], and aluminum sheet (Al) as a current collector.

$LiFePO_4$ and CB powder of 85:10 (in weight percent ratio) [6] was mixed in agate mortar and then dried in oven at 80 °C for eight hours. The PVDF polymer of 5% by weight which only acts as a binder is dissolved in the MNP solvent and fed into mixer vacuum vial. Add the $LiFePO_4$/ CB mixture to the PVDF solution stirred with the vacuum mixer for 2 hours until a $LiFePO_4$-shaped cathode of slurry is formed. $LiFePO_4$ cathode slurry is then superimposed (casting) on aluminum (Al) sheet foil using doctor blade, to form a $LiFePO_4$ cathode sheet. Then dried in a vacuum oven at 80 °C for 24 hours to obtain $LiFePO_4$ (one layer) cathode sheet. The slurry of the $LiFePO_4$ cathode sheet double layer is done in the same manner and composition. Furthermore, the slurry is coated on an uncoated Al foil sheet so that the aluminum sheet is centered between two $LiFePO_4$ slurries, then dried in a vacuum oven until a double layer $LiFePO_4$ cathode sheet is obtained. This sample is called a C0 sample that is ready for characterization. In a portion of the $LiFePO_4$ cathode sheet, a calendaring process was repeating for one, two and three times using a hot rolling machine [12] at a temperature about 100 °C, which are called C1, C2 and C3 samples. The detail cathode preparation has been described elsewhere [10].

2.2 $LiFePO_4$ cathode characterization

The phases inside the $LiFePO_4$ cathode sheet was measured using an X-ray diffractometer (XRD, PAN-analytical Empyrean), using a Cu-K irradiation source (λ = 1.5406Å) in the range 2θ = 10 - 80°, with a step size 0.5° and step time 10 seconds. The morphology and cross-sectional surface of the samples were observed with the Scanning Electron Microscope (SEM, JEOL - JSM 6510) equipment coupled with the energy-dispersive spectrophotometer (EDS) equipment. The conductivity of the $LiFePO_4$ cathode sheet was measured using impedance conductance resistance equipment (LCR meter, HIOKI LCR Hit ESTER 3532-50 for EIS). Silver paste as ion-blocking electrode and current collector was used on both sides of the samples. The impedance spectrum is measured in atmospheric environment and at room temperature using a voltage of 1 Volt and a frequency range of 100 Hz - 5MHz. The conductivity of the sample was calculated from measured impedance value and the dimension of the sample. The density was calculated manually from measured mass and the volume of the sample.

3. Results and discussion

3.1 X-ray diffraction pattern (XRD)

The X-ray diffraction pattern has been measured using a CuKα source (λ = 1.5406 Å), 2θ = 10 to 80°. Figure 1 shows the XRD pattern of $LiFePO_4$ cathode sheet samples before and after the calendaring process. The result shows that the sample has formed crystals well which is characterized by high and sharp diffraction patterns. The highest diffraction pattern lies at 2θ = 35.8° corresponding to the open data crystallography (COD). The material is lithium iron phosphate with $LiFePO_4$ chemical formula, Pnma space group (62), and orthorhombic crystal structure. The lattice parameter a = 10.330Å, b = 6.003Å, c = 4.697Å and atomic density (ρ) about 3.595 gram.cm^{-3}. The results are in agreement with the experiments performed by Wang Jiajun *et al.* (2015) [15] and Wang W. L *et al.* (2012) [9]. Yiu *et al.* [16] get results the lattice parameter from a $LiFePO_4$ without calendaring process of a = 10.332Å, b = 6.010Å, c = 4.692Å. It is seen that there is no significant change in the crystals structure.

The diffraction pattern of the carbon phase is observed at 2θ = 26.78° as indicated by the sign (x) in Figure 1. Carbon or graphite with the formula C structured hexagonal crystals, group space P 63/mm (194), with parameters Lattice a = b = 2.464Å, c = 6.711Å and density of 2.261 gram.cm^{-3}. The similarity of XRD patterns from all samples showed no change in crystal structure due to the calendaring process.

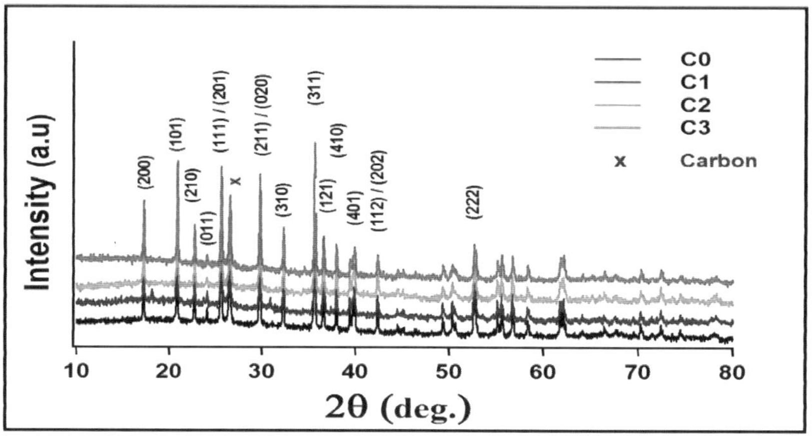

Figure 1. XRD pattern of $LiFePO_4$ cathode sheet for C0, C1, C2 and C3.

3.2 Scanning electron microscope (SEM)-energy disperse spectrophotometer (EDS)

Figure 2 is the microcrystalline sheet structure of $LiFePO_4$ cathode prior to calendaring (sample C0). The resulting particles ranged from 0.4 to 1.0 μm, with the size measured on particles having clear grain boundaries. On the surface of the sample looking uneven and having many cavities, the polymer of PVDF appears to bind to the $LiFePO_4$ particles in the form of webs, as shown in Figure 2C0. Figure 2C1 is the microcrystalline sheet structure of $LiFePO_4$ after one round of the calendaring process (sample C1), the sample surface has not been uniform even though the calendaring process has been conducted and also still agglomerated. This is because after the calendaring process, the sample surface is covered by a PVDF polymer binder.

Figure 2(C2) is the microstructure of the calendaring C2 sample. Visible, the surface is more evenly distributed and the PVDF polymer integrates with the $LiFePO_4$ matrix and slightly cavities between one particle and the other evenly. The cavity in this cathode material is necessary to facilitate the movement of Li ions. Figure 2C3 is the microstructure of the sample after calendaring three times (C3). Can be seen more evenly distributed sample surface with fewer cavities and solid, also seen some micro-cracks (micro cracks).

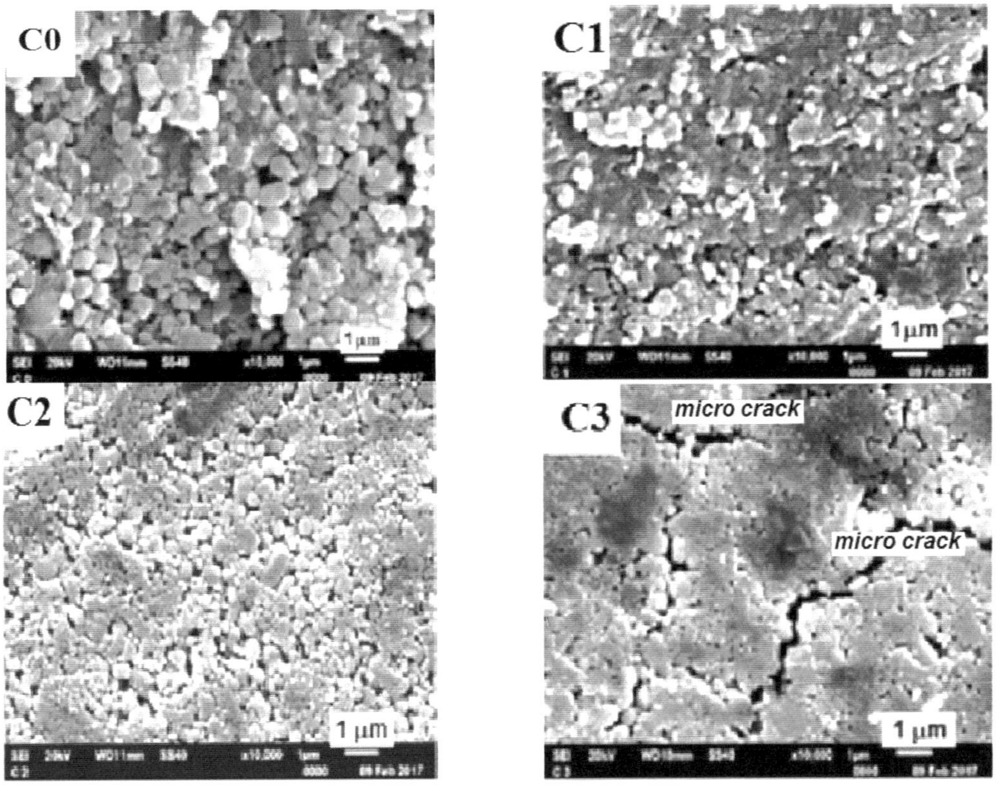

Figure 2. The microstructure of $LiFePO_4$ cathode sheet for C0, C1, C2 and C3.

Figure 3 shows the cross-sectional view of the $LiFePO_4$ cathode sheet before and after the calendaring process. The influence of calendaring process to the decrease of the $LiFePO_4$ cathode sheet thickness is described in Figure 4. Based on Figure 3 and Figure 4, $LiFePO_4$ cathode sheet thickness was obtained about 160μm (Figure 3C0). After several times calendaring process at a temperature of 100 °C, the thickness of the $LiFePO_4$ cathode sheet decreased to 140 μm (12.50%), 130 μm (18.75%), and 120 μm (18.00%) respectively.

In the Figure 3(C0), the $LiFePO_4$ cathode material has not been perfectly attached to Aluminium (Al) foil as a current collector seen as a thin cavity above Al foil. In $LiFePO_4$ cathode material there is also a transverse cavity up to about 10μm in size. After the calendaring process for samples of C1, C2, and C3 $LiFePO_4$ cathodes with Al foil are well attached, they appear to be also denser with decreasing sample thickness, as shown in Figure 3C1 to Figure 3C3. The thickness vs. calendaring process is described in Figure 4.

In the calendaring C3 process, again appears the existence of micro-sized cracks, as shown in Figure 3C3. To determine the conductivity of the effect of the calendaring process, the sample was then measured with LCR meter equipment as shown in Figure 9.

IOP Conf. Series: Materials Science and Engineering **432** (2018) 012059 doi:10.1088/1757-899X/432/1/012059

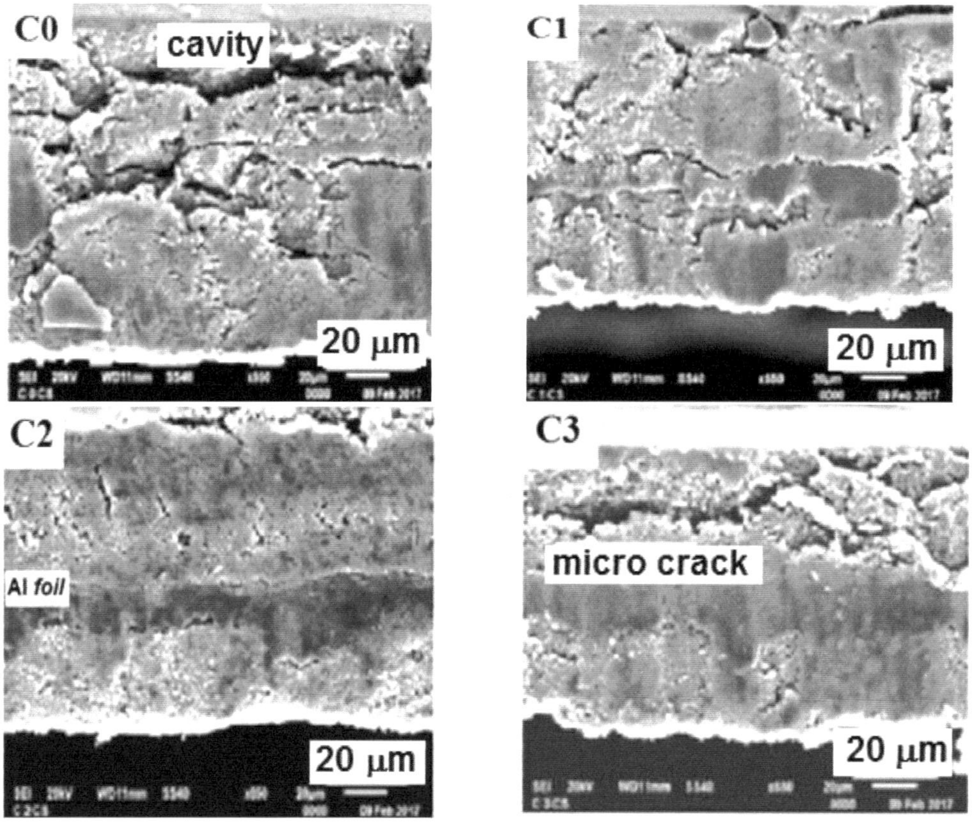

Figure 3. The cross section of $LiFePO_4$ cathode sheet for C0, C1,C2 and C3.

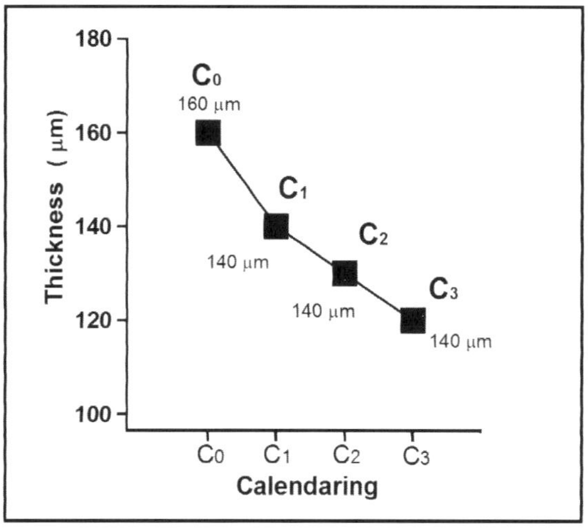

Figure 4. The influence of calendaring process to the thickness of $LiFePO_4$ cathode sheets.

The SEM-EDS is only capable for detecting elements with an atomic number of more than 5, so Li cannot be detected because its atomic number is 3 [16]. Figure 5 shows the percent mass curve of the calendaring process. From Figure 5 it appears that the elements of Fe and P representing elements of $LiFePO_4$ increase with the increase of calendaring process. The F value representing the PVDF polymer seen in C1 is an increase in the percent mass of F. This shows that the PVDF polymer binder covers the $LiFePO_4$ surface, as shown in the microstructure of Figure 2 of C1 sample. With the increase in calendaring processes (samples C2 and C3) the F element decreases, which shows the $LiFePO_4$ surface not covered by the PVDF polymer binder seen in the microstructure Figure 2 samples C2 and C3.

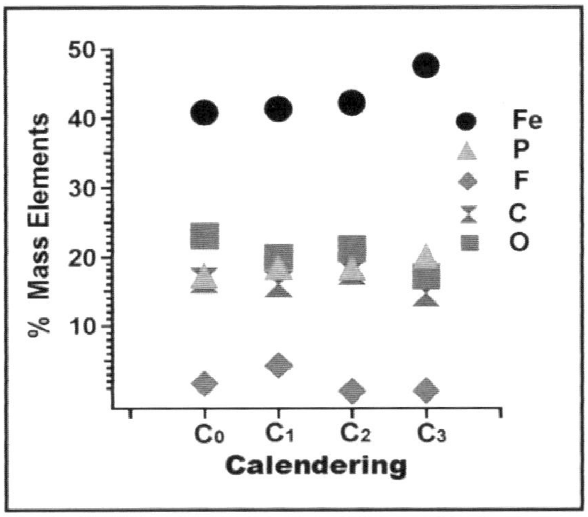

Figure 5. EDS counting result (in mass percent) of the element for C0, C1, C2 and C3.

3.3 Impedance conductance resistance (LCR meter)

The conductivity (σ) of the electrode is calculated from the measured impedance value on LCR meters and electrolytic dimensions such as the thickness and surface area of the sample. Figure 6 shows the conductivity change (σ) of the $LiFePO_4$ cathode sheet against the frequency measured at room temperature, while Figure 7 shows the influence of calendaring process to the change of the $LiFePO_4$ cathode sheet conductivity. From Figure 6 and Figure 7, the conductivity value of C0 is about 9.8×10^{-5} $S.cm^{-1}$. The conductivity of the C sample is obtained as about 2.9×10^{-5} $S.cm^{-1}$. The decrease in conduc-

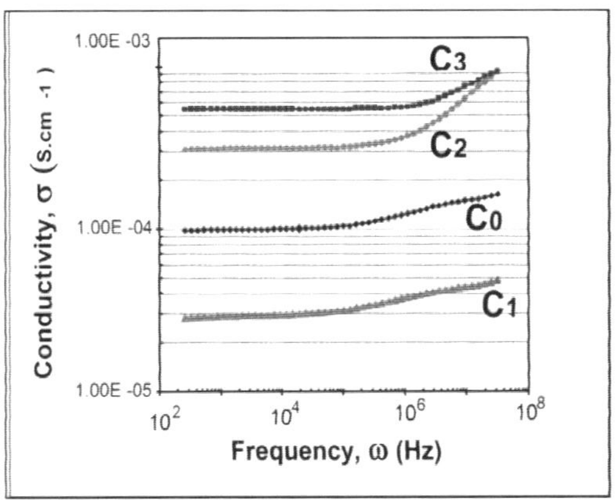

Figure 6. Conductivity curve vs. Frequency of $LiFePO_4$ cathode sheet for C0, C1, C2, and C3.

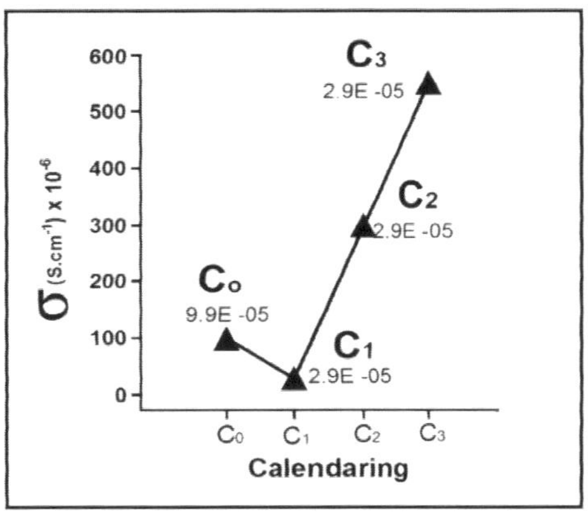

Figure 7. Conductivity curve vs. calendaring of $LiFePO_4$ cathode sheet for C0, C1, C2, and C3.

tivity of the C1 sample may be due to the closed surface of the sample by the PVDF polymer acting as a binder. In the calendaring process, the C2 conductivity sample increased significantly to 3.0×10^{-4} S.cm^{-1}. The increase of significant conductivity values from sample C1 to C2 may the shrinking of the PVDF polymer binder from the sample surface, thus increasing the electrical contact and increasing the concentration of mobile ions. Then the conductivity on the C3 sample increased to 5.5×10^{-4} S.cm^{-1} after calendaring 3 times. This increase is not very significant, because it is still one order with a sample of C2. Synthesis of $LiFePO_4$ cathode sheets by hydrothermal process has been carried out with the conductivity obtained being about 5.6×10^{-5} S.cm^{-1} [17].

3.4 Density

The density of the sample was calculated manually from measured mass and the volume of the sample. Figure 8 shows the density curve of the $LiFePO_4$ cathode sheet through calendaring process. It is seen that the density of cathode sheets of $LiFePO_4$ sample C1 increases the calendering process until it reaches the highest value in calendering process C2. This means that the density of the sample is getting denser and thinner with increasing calendering process. The density of the $LiFePO_4$ cathode sheet then decreases with the continuation of the calendering process up to C3, as described in Figure 8. The decrease in the C3 sample density is due to the formation of micro-size cracks formed during the calendering process, as shown by SEM morphology in Figure 2 and Figure 3. This causes C3 samples to be impossible to apply as $LiFePO_4$ cathode sheets on lithium ion batteries.

From Figure 8, it is seen that through the calendaring process, the density of the $LiFePO_4$ cathode sheet is more denser. When the density of $LiFePO_4$ cathode sheet increases, then the conductivity also rises, as shown in Figure 7 and Figure 8. It is clear that the calendaring process can solidify the $LiFePO_4$ cathode sheets causing the microstructure to be better organized with a much smoother surface, thereby enhancing electrical contact and increasing concentration Mobile ions that occur through the Li^+ ion leap mechanism. In addition, the PVDF polymer used to make the electrolytic membrane has a high enough mechanical strength to withstand the pressure between the cathode and anode, has a high degree of chemical stability and ionic conductivity, easy to make in thin size, has thermal stability, has dimensional stability or size, and able to form a nice interface with electrodes. As a result, the conductivity of the $LiFePO_4$ cathode sheet results in increased calendering process.

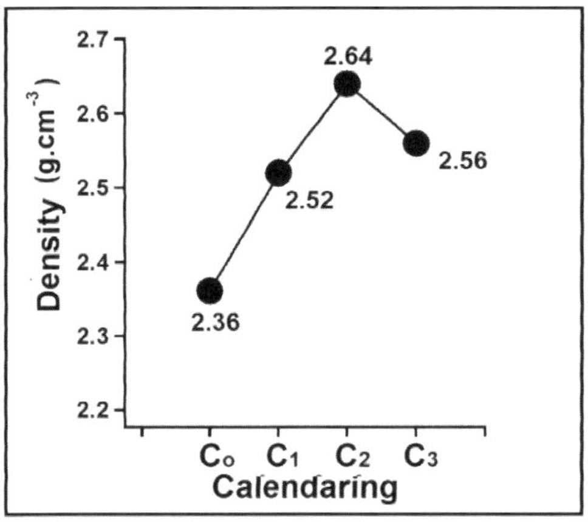

Figure 8. The density of $LiFePO_4$ cathode sheet vs. calendaring process for C0, C1, C2 and C3.

4. Conclusion

$LiFePO_4$ cathode sheets have been prepared from $LiFePO_4$, carbon black (CB), and polyvinylidene fluoride (PVDF/NMP) powder mixtures. The improvement of the $LiFePO_4$ cathode sheet conductivity has been successfully performed by calendaring process optimization. The XRD pattern shows that the $LiFePO_4$ cathode sheet consists of $LiFePO_4$ and carbon crystal phases. The cross-sectional microstructure of the $LiFePO_4$ cathode sheet shows the sheets getting denser and thinner as the number of calendaring increases. The thickness of the $LiFePO_4$ cathode sheet obtained by the calendaring process about 160μm then decreased about 12.5% to 140 μm, 18.75% to 130 μm, and 25.00% to 120 μm with the increase of calendaring process. The conductivity of the $LiFePO_4$ cathode sheet obtained is 9.9×10^{-5}; 2.9×10^{-5}, 3.0×10^{-4}, and 5.5×10^{-4} $S.cm^{-1}$ for samples C0, C1, C2, and C3, respectively. It can be concluded that the calendaring process can increase the conductivity of the $LiFePO_4$ cathode sheet with the $LiFePO_4$ cathode sheet.

Acknowledgment
This study is a part of the the Battery Group research Programs in the Center for Advanced Materials Science and Technology, and funded by DIPA PSTBM-BATAN for Fiscal Year 2017.

References

[1] Kartini E Honggowiranto W Jodi H and Jahya A K Adams S Kawamura J 2014 *(Eds) Proc. the14th Asian Conf. Solid State Ionics* Singapore Research Publishing 163–73
[2] Jodi H Zulfia A Deswita and Kartini E 2016 *Int. J. Technol.* **8** 1117–26
[3] Johnson I D Blagovidova E Dingwall P A Brett D J Shearing P R and Darr J A 2016 *J. Power Sources*, **326** 476–81
[4] Yurwendra A O and Noerochim L 2014 *J. POMITS Tech.* **3(2)** F229–31
[5] Nur I P A Kartini E Dwi P L Faisal M and Supardi 2016 *Int. J. Ionics* **22(7)** 1051–7
[6] Scipioni R Jurgensen P S Ngo D T Simonsen S B Liu Z Yakal-Kremski K J Wang H Hjelm J Norby P Barnett S A and Jensen S H 2016 *J. Power Sources* **307** 259–69
[7] Halim A Widiastuti W Setyawan H Machmudah S Nurtono T and Winardi S 2014 *AIP Conf. Proc.* **173(1586)** 173–8
[8] Triwibowo J Lestariningsih T Priyono S Purawiardi R I and Daulay L 2015 *J. Mater. Energy Indones.* **5(2)** 1–7
[9] Li-Wang W Jin E M and Gu H B 2012 *Trans. Electr. Electron. Mater.* **13(3)** 121–4
[10] W Honggowiranto and Gunawan I 2015 *Proc. Nat. Sem. XXIV Chem. in Indust. and Envi.* 57–64
[11] Deswita and Gunawan I 2016 *Proc. Nat. Semin. XIX Chem. in Develop.* 69–74
[12] Haselrieder W Ivanov S Christen D K Bockholt H and Kwade A 2013 *ECS Trans.* **50(26)** 59–70
[13] Smekens J Gopalakrishnan R Van den Steen N Omar N Hegazy O Hubins A and Van Mierlo J 2016 *Energies* **104(9)** 2–12
[14] Van-Bommel A and Divigalpitiya R 2012 *J. Electrochem. Soc.***159(11)** A1791–5
[15] Wang J and Sun X 2015 *Energy Environ. Sci.* **8** 1110–38
[16] Yiu Y M Yang S Wang D Sun X and Sham T K 2013 *J. Phys. Conf. Ser.* **430(1)** 2014
[17] Ding Z Gu K Zheng H Liu C Liu L Yan L X and Guo 2014 *Electrochim. Acta* **146** 585–59

IOP Conf. Series: Materials Science and Engineering **432** (2018) 012060 doi:10.1088/1757-899X/432/1/012060

Carotenoid pigment as sensitizers for applications of dye-sensitized solar cell (DSSC)

A Supriyanto[1, 2, a], **F Nurosyid**[1, 2], **and A H Ahliha**[1]

[1] Physics Department of Graduate Program, Sebelas Maret University, Indonesia
[2] Department of Physics, Faculty of Mathematics and Natural Science, Sebelas Maret University, Indonesia

[a]agusf22@staff.uns.ac.id

Abstract. A dye-sensitized solar cell is the third generation of solar cells that can convert solar energy into electrical energy using photo-electrochemical principles. Natural dye extracted from carotenoid pigments such as orange fruits, carrots, and tomatoes. The absorbance of dye and TiO_2 were characterized using UV-Vis spectrophotometer. The chemical bond contained TiO_2/dye were characterized using FTIR spectrophotometer. The efficiency of DSSC was calculated using *I-V* meter Keithley. The absorbance spectra results show that carotenoid dyes have peaked in the wavelength range of 400 nm-510 nm. The highest efficiency was obtained from tomatoes dye at 0.03% due to the interaction between TiO_2 and carboxyl –OH group. Adsorption of carotenoid dyes on the surface of TiO_2 can be used as the photosensitizer for DSSC.

1. Introduction

Dye-sensitized solar cell (DSSC) technology is a third-generation solar cell technology based on the dye that can convert solar energy into electrical energy using photo-electrochemical principles [1-5]. Components for assembling the DSSC consists of a transparent conductive glass substrate, a semiconductor having a wide band gap, photosensitizers, electrolyte, and a counter electrode [6]. The best semiconductor material often used in DSSC fabrication is TiO_2 [7]. The performance of photosensitizers adsorbed on the surface of TiO_2 is one of the important factors to determine the efficiency of the dye-sensitized solar cell [8, 9]. Natural dyes is one component that plays an important role in improving the performance of DSSC because it can be that convert solar energy into electrical energy. The natural dye can be fabricated in a simple, very cheap and environment-friendly.

Some fruits, plants, flowers, and leaves displayed various colours and contained several pigments, which can be simply extracted and then used as sensitizer because most of yellow, red, and orange plants contained a lot of carotenoid, which help in absorbing photon from sunlight in the visible region at the wavelength range of 400 nm-510 nm. Gomez-Ortiz *et al.* carotenoid have carboxyl group enable to bind to the interface of TiO_2 [10] because carboxyl groups at the ends of the chain of compounds which may be binding to Ti from TiO_2. The entire carotenoid is poly-isoprenoid have a system with single and double bonds conjugated [11] as has been reported [12], carotenoid molecules are highly reactive and degrades more quickly due to the small number of oxygen. Extraction of carotenoids is very difficult because there is no optimal solvent for preventing the degradation process. Gao *et al.* used a carotenoid dye with an efficiency of 34% on the stability of the light irradiation for 1 hour [13]. *Daucus carota* as a carotenoid dyes for DSSC resulted the absorbance at the wavelength peak in 415 nm and 508 nm. The performance of DSSC based on *daucus carota* dye showed that the maximum voltage (V_{max}) of 23.9 x 10^{-2} V and the maximum current (I_{sc}) of 3.3 x 10^{-5} A, and conversion efficiency was very low is 12.5 x 10^{-4}% [14]. In this work, a natural dye is extracted from carotenoid pigments such as orange fruits, carrots, and tomatoes. The aim of this work is to examine the chemical bond of a carotenoid dye on TiO_2 for the efficiency of the dye-sensitized solar cell.

2. Experimental

TiO_2 pastes were prepared using TiO_2 nano-powder in ethanol. The TiO_2 solution was stirred at 300 rpm for 30 minutes. TiO_2 pastes that have been formed are inserted into the bottle and stored in an area free from direct sunlight. The TiO_2 paste is deposited onto the FTO substrate glass with the active cell area 1 cm x 1 cm by spin coating technique. TiO_2 films were annealed in a furnace at 400 ^{0}C for 10 minutes.

The material used is natural dyes extracted from orange fruits, carrots, and tomatoes, and then cut into small cubes. Carotenoid dyes were dissolved in n-hexane solvent. It was kept for 24 hours. It were filtered using filter paper. The dye was evaporated using a rotary evaporator to get the more concentrated solutions. The working electrode was soaked in the dye solution for 24 hours. The dye solution and TiO_2/dye layers were characterized using a UV-Visible spectrophotometer to determine its absorbance spectrum and TiO_2/dye was characterized using a FTIR spectrophotometer to know the chemical bonds contained in TiO_2/dye.

The electrode counter is made of transparent conductive glass (FTO) coated with a platinum catalyst. The working electrode and electrode counter are assembled, thus forming a sandwich structure (Figure 1). The electrolyte solution is dropped into the gap between the two electrodes. The DSSC was characterized using a *I-V* meter *Keithley* 2602A.

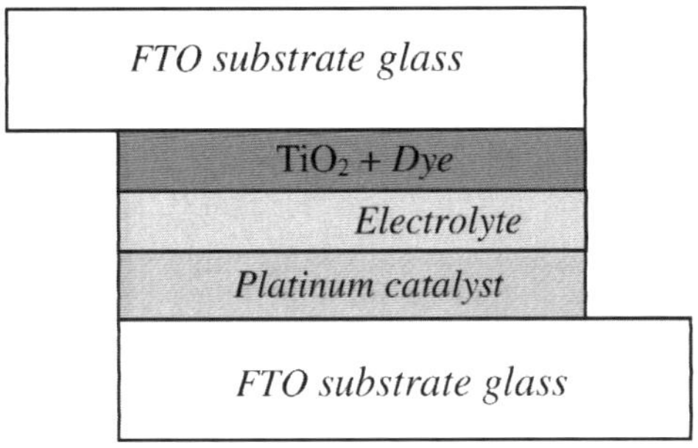

Figure 1. The structure of DSSC.

3. Results and discussions

3.1. Optical characteristics of carotenoid dyes

Characterization of the absorbance of the dye was examined using UV-Vis spectrophotometer. Figure 2 shows the absorbance peaks of orange fruit and tomatoes dye at 441 nm and 466.5 nm while the dye carrots at 433 nm and 472.5 nm. Absorbance peak difference is not significant because the three dye is in the wavelength range of 400-510 nm in the visible region which indicates the content of carotenoid pigments. High concentration of carotenoid dye causes the dye molecules are absorbed and reflected photons of light bulbs too much. The greater the wavelength used, the smaller the resulting absorbance values. It can be produced as white light at each wavelength can be selected more detail by the prism so that the spectrophotometer measurements imply the absorption of light energy away by a chemical system as a function of the wavelength of the radiation [15].

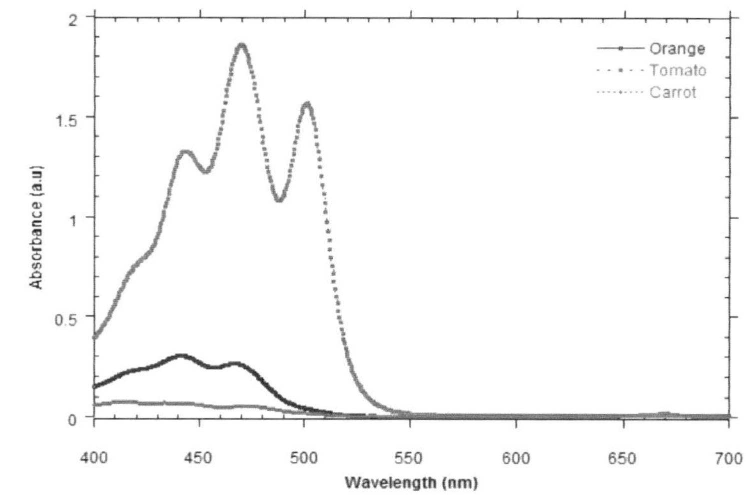

Figure 2. The absorbance spectra of carotenoid dyes.

3.2. Fourier transform infrared (FTIR) characterization of caroenoid dyes

Figure 3 shows the FTIR spectra of TiO_2 and the carotenoid dye. O-H stretching vibration of the carboxyl group appears on the three samples of tomatoes, carrots, and orange fruits. TiO_2/ dye of tomatoes at wavenumber 2395 cm^{-1} with transmittance of 84.8%, TiO_2/ dye of carrots at 2925 cm^{-1}, and 2856 cm^{-1} with transmittance value of 81.8% and 84.7%, while for TiO_2/ dye of orange fruits appeared at 2924 cm^{-1} and 2854 cm^{-1} with transmittance of 73.7% and 78.5%. The two union monomers of carboxylic acids are capable of forming hydrogen bonds on two carboxyl groups. Therefore, the vibration of O-H in the carboxylic acid appears as a series of bands that are distributed over a very wide range (3300-2500 cm^{-1}). This resilient vibration occurs due to compression of the bond.

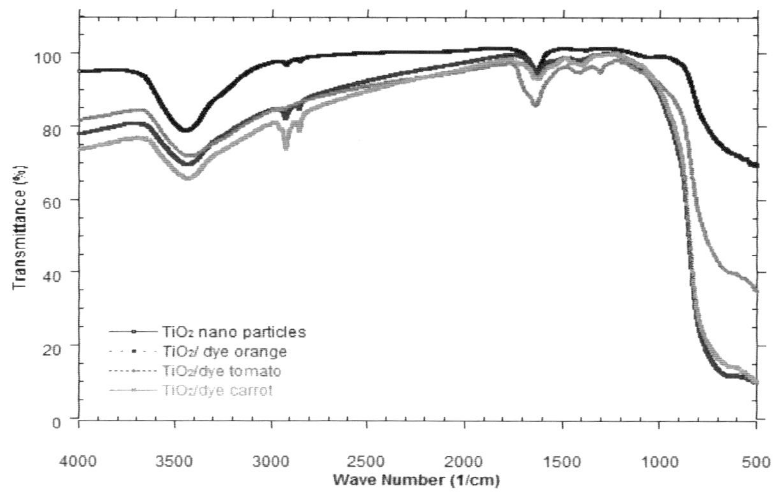

Figure 3. The FTIR spectra of carotenoid dyes.

TiO_2/ dye of tomatoes are capable of absorbing very large radiation due to the widened frequency in the lower direction. This is due to the tomato dye being absorbed into the more TiO_2 cavities. Vibrations that can absorb more radiation are caused by changes in the dipole moment in the molecule. The greater of the dipole moment, the resulting absorption is also more intense. -CH_3 and C=C that appears in the infrared spectrum is estimated to come from carotenoid compounds. Carotenoid have carboxyl groups at the end of the chain of compounds that enables to bind to the TiO_2 [10]. The entire carotenoid is polyisoprenoid have a system with a single bond and conjugated double bonds [11].

3.3. I-V characterization of DSSC

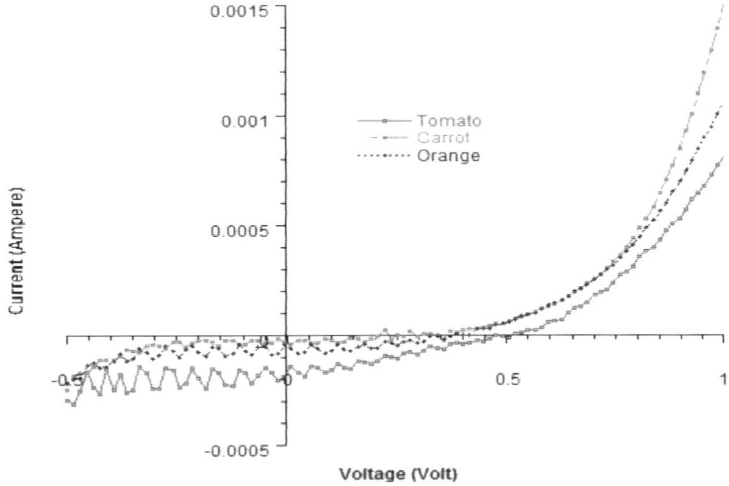

Figure 4. The I-V curve of the DSSC with carotenoid dyes.

Current-voltage characteristics of the DSSC fabricated using a light source 1000W/m^2. Figure 4 shows sampled current-voltage characteristics of the DSSCs. Table 1 shows the results of current-voltage measurement of each dye extraction such as I_{sc}, V_{oc}, FF, and η. Table 1 shows the DSSC of tomato dye having an open-circuit (V_{oc}) voltage value higher than 0.51 *V* compared to the V_{oc} extract of orange fruits 0.37 *V* and carrots 0.36 *V*. The highest current density obtained by DSSC dye tomatoes of 0.00014 *A*. This current density value is greater than the current density of the orange fruits extract of 0.00006 *A* and carrots 0,00004 *A*. The differences that occur in the density of the cell flow may be due to the structural differences in the dye molecule and how quickly the charge is injected into in the TiO_2 conduction band [16].

Table 1. Photoelectrochemical parameters of DSSCs based on carotenoid dyes.

Dye	I_{sc} (mA/cm^2)	V_{oc} (V)	FF	η (%)
Tomato	0.51	0.14	0.37	0.03
Orange fruit	0.37	0.06	0.58	0.02
Carrot	0.36	0.04	0.64	0.009

The efficiency of DSSC from each extract of tomato, orange fruits, and carrots was 0.03%, 0.02%, and 0.009%. The efficiency of tomato DSSC has a higher value compared to other DSSCs. This result is supported by a characteristic result from FTIR TiO_2/ dye of tomato which has a value of transmittance of 84.8% in the range of 3500-2500 cm^{-1} wavenumbers. The FTIR analysis results support the efficiency of the tomato DSSC sample, although its efficiency value is still low probably because the dye molecules adsorbed on the TiO_2 surface are still low. The nanostructure of TiO_2 provides a pathway for electrons in order to maintain a high surface area during dye absorption process. Therefore, the unavailable bond between the dye and TiO_2 molecules during the electron transfer process causes a low-efficiency value [17, 18].

4. Conclusion

Based on research that the highest efficiency obtained by tomatoes dye was 0.03% in carotenoid pigment. This efficiency occurs due to the tomatoes dye binding to TiO_2 which is supported with the FTIR transmittance at wavenumber 3500-2500 cm^{-1} of 84.8%, resulting in the carboxyl O-H group stretching. The interaction between TiO_2 and dye causes the transfer of electrons from dye molecules to the TiO_2 conduction band. The amount of dye absorbed into the TiO_2 cavities is large, so the amount of light absorbed also rise.

Acknowledgements

This work was supported by Institute for Research and Community services, Sebelas Maret University by Excellence Research Grant -UNS (Research Scheme) (contract number. 623/UN27.21/PP/2017).

References

[1] O'Regan B and Gratzel M 1991 *Nature* **353** 737-40
[2] Dhamodharan P Manoharan C Bououdina M Venkadachalapathy R and Ramalingan S 2017 *Solar Energy* **141** 127-44
[3] Ahliha A H Nurosyid F and Supriyanto A 2017 *AIP Conf. Proc.* **1868** 060001-6
[4] Ahliha A H Nurosyid F Supriyanto A and Kusumaningsih T 2018 *IOP Conf. Series: Mater. Sci. and Engin.* **333** 012018 1-5
[5] Sengupta D Mondal B and Mukherjee K 2015 *Spectrochim. Acta Part A: Molecul. and Biomolecul. Spectros.* **148** 85-92
[6] Cakar S Guy M Ozacar M and Findik F 2016 *Electrochim. Acta* **209** 407-22
[7] Hemamali G G G M N and Kumara G R A 2013 *Int. J. of Sci. and Research Pub.* **3** 1-3
[8] Gratzel M 2003 *Rev. J. of Photochem. and Photobiol. C: Photochem. Rev.* **4** 145-53
[9] Tu L Pan H Xie H Yu A Xu M Chai Q Cui Y and Zhou X 2012 *Solid State Sci.* **14** 616–21
[10] Ananth S Vivek P Solaiyammaland T and Murugakoothan P 2015 *Optik* **126** 1027–31
[11] Orti'z N M G Maldonado I A V Espadas A R P Rejo'n G J M Barrios J A A and Oskama G 2009 *J. of Solar Energy Mater. & Solar Cells* **94** 40-4
[12] Hemalatha K V Karthick S N Raj C J Hong N Y Kim S K and Kim H J 2012 *J. of Electrochim. Acta part A: Molecul. and Biomolecul. Spectros.* **96** 305-9
[13] Haugan J A Aakermann T and Liaaen-Jensen S 1992 *Academic Press* 231-45
[14] Gao F G Bard A J and Kispert L D 2000 *J. of Photochem. and Photobiol. A: Chemistry* **130** 49-56
[15] Suryana R Khoiruddin and Supriyanto A 2013 *Mater. Sci. Forum* **737** 15-9 ISSN: 1662-9752
[16] Underwood 1986 *Quantitative Chemical Analysis* Jakarta: Erlangga (In Indonesia)
[17] Nwanya A C Ugwuoke P E Ejikeme P M Oparaku O U and Ezema F I 2012 *Int. J. Electrochem. Sci.* **7** 11219-35
[18] Maabong K Muiva C M Monowe P Sathiaraj T S Hopkins M Nguyen L Malungwe K and Thobega M 2015 *Int. J. of Renew. Energy Research* **5** 54-60

Comparative study on the ionic conductivities and redox properties of $LiPF_6$ and LiTFSI electrolytes and the characteristics of their rechargeable lithium ion batteries

A R Septiana[1], W Honggowiranto[2], Sudaryanto[2], E Kartini[2,a], R Hidayat[1,b]

[1] Physics Program Study, Faculty of Mathematics and Natural Sciences, Institut Teknologi Bandung, Jl. Ganesha 10, 40132, West Java, Indonesia
[2] Center for Science and Technology of Advanced Material, National Nuclear Energy Agency of Indonesia, Kawasan PUSPIPTEK Serpong, Tangerang Selatan, 15314, Banten, Indonesia

[a]kartini@batan.go.id
[b]rahmat@fi.itb.ac.id

Abstract. The ionic conductivities of two different electrolytes, namely lithium hexafluorophosphate ($LiPF_6$) and lithium bis (trifluoromethylsulfonyl) imide (LiTFSI), in carbonate-based solvents have been investigated. The ionic conductivity of LiTFSI electrolyte is slightly larger than the $LiPF_6$ electrolyte, namely 2.7 mS/cm *vs.* 2.4 mS/cm. The results of cyclic voltammetry and electrochemical impedance spectroscopy measurements show that LiTFSI electrolyte exhibit a better reversible redox reaction. Therefore, in this work, the full-cell battery using LiTFSI electrolyte exhibited higher specific capacity than the battery cell using $LiPF_6$ electrolyte, namely 83.1 mAh/g and 101.5 mAh/g for the $LiPF_6$ and LiTFSI electrolytes, respectively. Higher capacity in LiTFSI battery is thus related to better ionic conductivity and reversible redox reaction of LiTFSI electrolyte.

1. Introduction

For more than two decades, lithium-ion batteries (LIBs) have become an important energy storage for portable electronics, such as mobile phones, communication devices, laptops, digital audio players etc. Currently, LIBs have been also developed as high-power electric storages for renewable energy and electric vehicles applications due to its superior performance, flexibility in design, high energy, and excellent power density properties [1–3]. However, such applications require more chemically and thermally stable electrolytes, where the electrolyte composition plays an important role. The electrolytes used in LIB are commonly comprised of a lithium salt, a mixture of organic solvents and additives [4]. Commercially available LIB mostly contain lithium hexafluorophosphate ($LiPF_6$) in mixtures of organic carbonate solvents as the electrolyte due to its high conductivity [5–10]. However, $LiPF_6$ is thermally unstable and may decompose to LiF and PF_5. Moreover, PF_5 is highly reactive to water and hydrolize to form HF and PF_3O, where HF is highly reactive acid and corrosive. It may attack both negative and positive electrodes, resulting in the battery performance reduction [11, 12]. The detailed mechanism of reaction between water and $LiPF_6$ in organic solvents has been reported in previous research [10, 12, 13]. Due to this drawback, the use of other lithium salts should also be seriously considered [8, 9, 14, 15]. Among some alternatives, lithium bis(trifluoromethylsulfonyl) imide (LiTFSI) have been considered as one of promising alternative because of less moisture reactivity, better ionic dissociation, high thermal and high electrochemical stability in comparison to $LiPF_6$ [16, 17].

In this paper, we reported the results of our comparative study on LiTFSI and $LiPF_6$ electrolytes on their ionic conductivity and their electrochemical properties as well as the charge-discharge capacity of LIB using these electrolytes.

2. Experimental

In this research, 1M LiTFSI in EC:DEC (1:1) and 1M $LiPF_6$ in EC:DEC (1:1) solutions were used as the electrolytes, where EC is ethylene carbonate and DEC is diethylene carbonate. The solutions were prepared in a glove box filled with an inert gas atmosphere. The ionic conductivity of LiTFSI and $LiPF_6$ electrolytes were determined from the results of the Electrochemical Impedance Spectroscopy (EIS) measurements, using a Hioki 3532-50 LCR Hi-Tester with the frequency range of 42 Hz up to 1 MHz. The electrolyte was put inside a CR2016 coin-type cell in between of a pair of two stainless steel electrodes (stainless steel | electrolyte | stainless steel). With known cell dimensions, the conductivity σ can be obtained from the following formula [18]:

$$\sigma = \frac{t}{A}\frac{1}{R_b} \qquad (1)$$

where t is the thickness of the electrolyte layer (cm), A is the effective contact area (cm^2) and R_b is the bulk resistance (Ω).

For the cyclic voltammetry (CV) measurements, the coin cell used for the measurement consists of a lithium metal as an anode and $LiFePO_4$ as a cathode. The cell was swept at 50μV/s scan rate and voltage range of 2-4 V. For the charge-discharge characterizations, the coin cells were consisting of a graphite as the negative electrode and $LiFePO_4$ as the positive electrode, which was separated by a polypropylene microporous separator (celgard). The electrodes were heated at 80°C under vacuum for 12h before assembled. The charge-discharge characterizations were conducted using an 8-Channel Battery Analyzer at 0.1 C rate.

3. Results and discussions

3.1. Electrolyte conductivity

The conductivity of the electrolyte plays a key role because the internal resistance of the battery cell is mostly due to the electrical resistivity of the electrolyte solution [19]. The measured frequency-dependent ionic conductivity of these electrolytes are shown in Figure 1. As shown in this figure, the ionic conductivity spectra consist of two regions in the frequency range, where at the low-frequency dispersive region is due to the electrode polarization effects and the frequency independent plateau region corresponds to the dc conductivity of the electrolyte. This is a typical characteristic of Li salt electrolytes as reported elsewhere [20]. The ionic conductivities were then calculated using equation (1), where the R_b values were taken from their plateau region, resulting in the ionic conductivity values of 2.7 mS/cm and 2.4 mS/cm for LiTFSI and $LiPF_6$ electrolytes, respectively. The measured ionic conductivities of these salts here are slightly lower than for the result as reported by Murmann *et al.*, which may be affected by the presence of membrane separator used in the cells here [21].

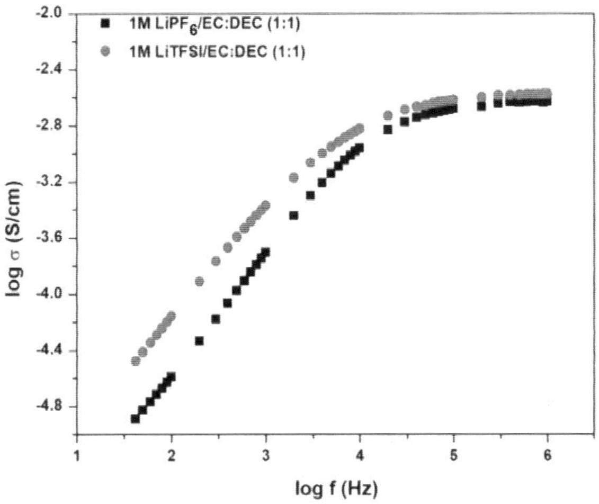

Figure 1. The frequency-dependent ionic conductivity spectra of the LiTFSI and $LiPF_6$ electrolytes.

3.2. Cyclic voltammetry characteristics

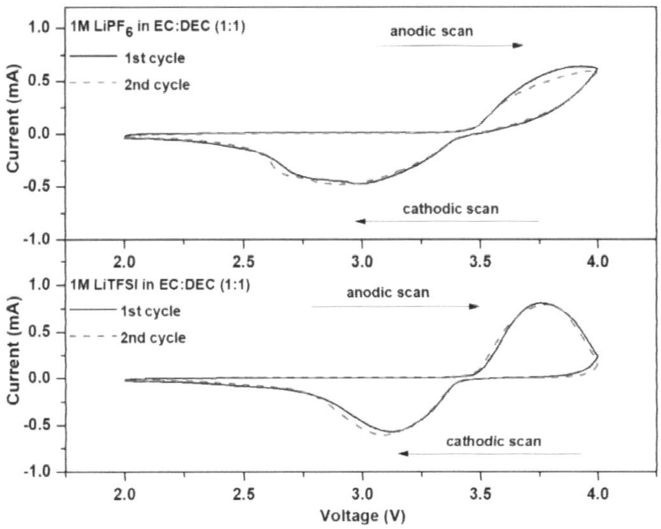

Figure 2. The CV curves of the $Li/LiFePO_4$ full cell batteries with different electrolytes, namely $LiPF_6$ and LiTFSI, taken at 50μV/s scan rate.

Figure 2 shows the CV curves of the $Li/LiFePO_4$ full cell batteries using LiTFSI and $LiPF_6$ electrolytes. For the $LiPF_6$ electrolyte, the redox peaks can be clearly seen at around 3 V and 3.9 V *vs.* Li/Li^+. For the LiTFSI electrolyte, the redox peaks are at around 3.1 V and 3.75 V vs. Li/Li^+. For a comparison, Wang *et al.* reported that the oxidation and reduction peaks for a Li/$LiFePO_4$ cells are at about 3.6 V and 3.3 V [22]. In general, CV shape with more symmetry and narrower redox peaks would indicate a better electrochemical activity. The figure thus indicates that the cell using LiTFSI electrolyte shows a better electrochemical activity in comparison to that using $LiPF_6$ electrolyte. Moreover, the areas of the oxidation and reduction peaks for the battery using LiTFSI electrolyte are almost the same, indicating highly reversible redox reaction. However, the battery using $LiPF_6$ based electrolyte show unequal oxidation and reduction peaks, indicating less reversible redox reaction. The midpoint of the anodic and cathodic peaks is about 3.425 V for the cell using LiTFSI electrolyte and 3.450 V for the cell using $LiPF_6$ electrolyte, which corresponds to the open-circuit voltage (OCV) of the battery [23].

3.3. Charge-discharge and EIS characteristics

The electrochemical process on the electrodes during charging process is:

Cathode : $LiFePO_4 \rightarrow FePO_4 + Li^+ + e^-$ [24]

Anode : $Li^+ + e^- \rightarrow Li$

The electrochemical process on the electrodes during discharging process is:

Cathode : $FePO_4 + Li^+ + e^- \rightarrow LiFePO_4$

Anode : $Li \rightarrow Li^+ + e^-$

Figure 3 shows the measured charge-discharge curves of these batteries taken at the charge-discharge rate of 0.1 C. Shown in the figure is the first cycle and the fifth cycle only, drawn in the voltage range of 2.5-3.8V for their vertical axis. It can be seen that there is a plateau region with the potential at around 3.3-3.4 V. The specific charge and discharge capacity at the first cycle for the battery using $LiPF_6$ electrolyte is 89.7 mAh/g and 83.1 mAh/g, while it is larger, namely 105.7 mAh/g and 101.5 mAh/g, for the battery using LiTFSI electrolyte. Capacity loss upon the first cycle is mainly due to the loss of lithium resulting from the formation of the Solid Electrolyte Interphase (SEI) at the surface of the graphite negative electrode as reported by Castro *et al.* [24]. The charge and discharge capacity seems to decrease slightly with the number of cycles.

In order to understand the observed difference in their charge-discharge capacity, the EIS measurements have also been carried out after the first and fifth cycles. EIS is a versatile tool for investigating the impedance changes due to characteristics changes in the electrodes and electrolyte caused by any irreversible or undesired reactions inside the cell, such as excessive SEI formation [25, 26].

Figure 4 shows the Nyquist plots obtained from the EIS measurements of these battery cells at a dc bias voltage of 0V taken after the first and fifth charge-discharge cycles. It seems that the shape of those Nyquist plots for both batteries are almost similar. However, it can be clearly seen that the semicircle for the cell using $LiPF_6$ electrolyte becomes much larger after the fifth charge-discharge cycle. This may indicate the increase in the resistance or/and the decrease in capacitance. In general, the semicircle part at the high-middle frequency region may correspond to the charge transfer resistance (R_{ct}). The inclined line at the low-frequency region may be related to the Warburg impedance (W), that reflects the diffusion of Li^+ in the active material [26, 27].

In order to better understand on the difference in their impedance characteristics, those Nyquist plots were fitted by an equivalent circuit as shown in Figure 4 [28–31]. The equivalent circuit consists of a resistor (R_1) describing the bulk resistance of the cell and two R|CPE elements connected in series. The first RC element (R_2 in parallel with CPE_1) is related to various layers formed as the electrode surface (SEI formation). The constant phase element (CPE) is related to the non-ideal capacitor behavior, which is quantified by the n value in between 0 and 1. If n is equal to 1, the CPE then behaves as an ideal capacitor.

A non-ideal capacitor behavior may arise from the presence of SEI layer. The second RC element is attributed to the interfacial processes (charge transfer) controlled by the lithium ion diffusion into graphite,

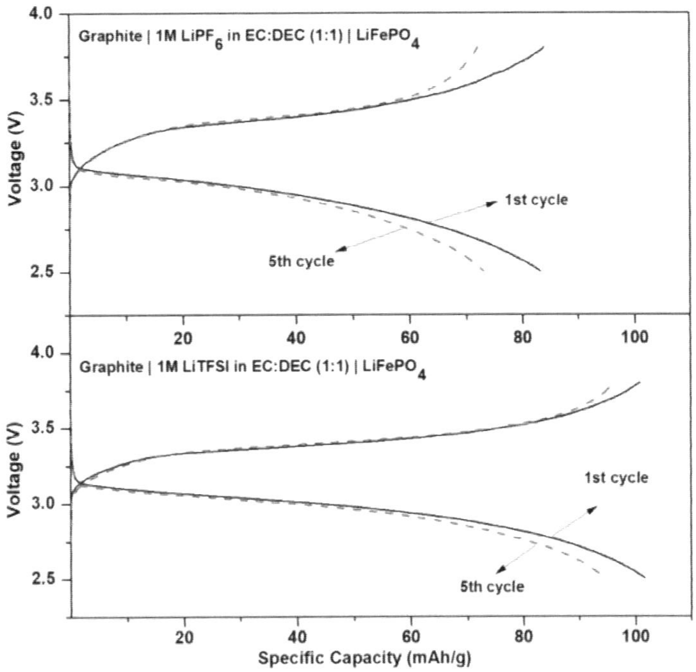

Figure 3. Charge-discharge curves of full-cell (Graphite |electrolyte |$LiFePO_4$) batteries using $LiPF_6$ and LiTFSI electrolytes taken at 0.1 C rate.

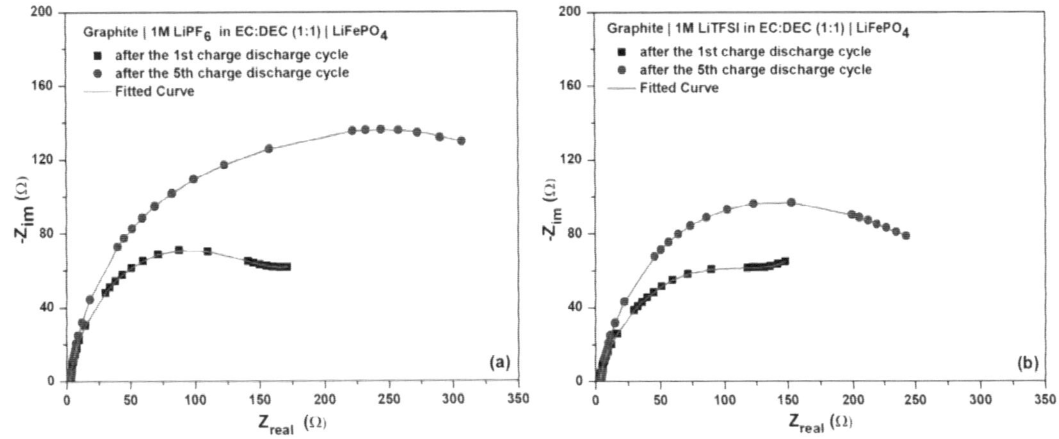

Figure 4. The Nyquist plots obtained from the EIS measurements of the (Graphite |electrolyte |$LiFePO_4$) battery cells using $LiPF_6$ and LiTFSI electrolytes taken after the first and fifth charge-discharge cycles. The thin lines represent the fitting curves.

which is modeled with a Warburg impedance [32]. The combination of R_3 and Zw is commonly called as the Faradic impedance. Low R_3 value generally corresponds to a fast kinetics of the faradic reaction [28].

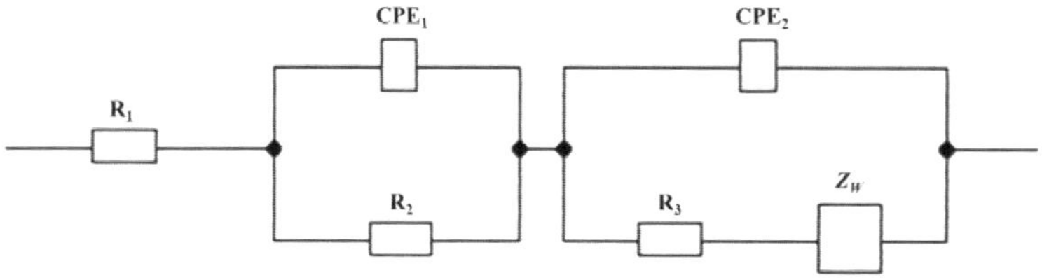

Figure 5. The equivalent circuit used in the curve fitting of the Nyquist plots in Figure 4.

Table 1. The obtained fitting parameters of the equivalent circuit model for the Nyquist plots in Figure 4.

Parameter	1M $LiPF_6$ in EC:DEC (1:1)		1M LiTFSI in EC:DEC (1:1)	
	1st cycle	5th cycle	1st cycle	5th cycle
R_1 (Ω)	2.27	2.83	2.90	3.94
R_2 (Ω)	122.17	253.73	108.44	162.25
R_3 (Ω)	33.63	120.10	8.71	73.93
Aw ($\Omega s^{0.5}$)	584.40	439.52	734.30	562.88
CPE_1 (μF)	17.241	28.561	21.824	23.201
n_1	0.722	0.811	0.809	0.788
CPE_2 (μF)	6.312	3.739	5.880	3.658
n_2	0.912	0.955	1.000	0.987

The fitting results are tabulated in Table 1. Based on these fitting results, taken after the fifth charge-discharge cycle, the first RC element ($R_2|CPE_1$) of the cells slightly change. However, the cell using LiTFSI electrolyte shows a much smaller increase in its R_2 and CPE_1 values in comparison to the cell using $LiPF_6$ electrolyte. This may indicate less continues SEI layer formation in the cell using LiTFSI. This is in agreement with a smaller capacity reduction in the fifth charge-discharge cycles for the cell using LiTFSI electrolyte. In addition, for the second RC element, the higher R_3 values in the cell using $LiPF_6$ electrolyte may indicate a slower kinetics of redox reaction and it thus leads to a smaller battery capacity. This is consistent with the observed charge-discharge capacity above, where the cell using $LiPF_6$ electrolyte shows a smaller battery capacity in comparison to the cell using LiTFSI electrolyte.

4. Conclusion

LiTFSI in EC:DEC electrolyte shows a slightly larger ionic conductivity in comparison to $LiPF_6$ in EC:DEC electrolyte. In addition, the LiTFSI electrolyte also shows a more reversible cyclic voltammetry shape, indicating more reversible redox of Li^+ ions in the battery. The charge-discharge capacity of the battery cell using LiTFSI electrolyte also show higher capacity and smaller increasing with number of cycles. Therefore, these experimental results show that the LiTFSI electrolyte has more superior characteristics compared to the $LiPF_6$ electrolyte.

Acknowledgments

This work was partly supported by Program Molina, LPDP – ITB and partly supported by Program INSINAS Riset PRATAMA Konsorsium 2016.

References

[1] Ryou M Lee J Lee D Kim W Jeong W K Choi J W Park J Lee Y M 2012 *Electrochim Acta* **83** 259–63
[2] Alias N and Mohamad AA 2015 *J. Power Sources* **274** 237–51
[3] Zhang R Chen Y and Montazami R 2015 *Materials* **8** 2735–48
[4] Zhang B Zhou Y Li X Ren X Nian H Shen Y Yun Q 2013 *Int. J. Electrochem. Sci.* **8** 12735–40
[5] Yang H and Zhuang G Jr PR 2006 *J. Power Sources* **161** 573–9
[6] Gnanaraj JS Thompson RW Dicarlo JF and Abraham KM 2007 *J. Electrochem. Soc.* **154** A185–91
[7] Zhang Z Chen X Li F Lai Y Li J Liu P Wang X 2010 *J. Power Sources* **195** 7397–402
[8] Dougassa YR Tessier C El-Ouatani L Anouti M Jacquemin J 2013 *J. Chem. Thermodyn.* **61** 32–44
[9] Zhang L Chai L Zhang L Shen M Zhang X Battaglia VS Stephenson T Zheng H 2014 *Electrochim. Acta* **127** 39–44
[10] Kawamura T Okada S and Yamaki J 2006 *J. Power Sources* **156** 547–54
[11] Chen Z Lu WQ Liu J and Amine K 2006 *Electrochim. Acta* **51** 3322–6
[12] Lux SF Lucas IT Pollak E Passerini S Winter M Kostecki R 2012 *Electrochem. Commun.* **14** 47–50
[13] Sloop SE Pugh JK Wang S Kerr JB Kinoshita K 2001 *Electrochem. Solid-State Lett.* **4** A42–4
[14] Zhang SS Xu K and Jow TR 2006 *J. Power Sources* **156** 629–33
[15] Shieh DT Hsieh PH and Yang MH 2007 *J. Power Sources* **174** 663–7
[16] Bolloli M Alloin F Kalhoff J Bresser D Passerini S Judeinstein Lepretre JC Sanchez JY 2015 *Electrochim. Acta* **161** 159–70
[17] Yang B Li C Zhou J Liu J Zhang Q 2014 *Electrochim. Acta* **148** 39–45
[18] Osman Z Mohd Ghazali MI Othman L Md and Isa KB 2012 *Results Phys.* **2** 1–4
[19] Dahbi M Ghamouss F Tran-Van F Lemordlant D Anouti M 2011 *J. Power Sources* **196** 9743–50
[20] Vijaya N Selvasekarapandian S Malathi J Iwai Y Nithya H Kawamura J 2012 *Proc. of the 13th Asian Conf. on Solid State Ionics* 342–33
[21] Murmann P Niehoff P Schmitz R Nowak S Gores H Ignatiev N Sartori P Winter M Schmitz R 2013 *Electrochim. Acta* **114** 658–66
[22] Wang WL Jin EM and Gu H-B 2012 *Trans. Electr. Electron. Mater.* **13** 121–4
[23] Yu DYW Fietzek C Weydanz W Donoue K Inoue T Kurokawa H Fujitani S 2007 *J. Electrochem. Soc.* **154** A253–7
[24] Castro L Dedryvère R Ledeuil J-B Breger J Tessier C Gonbeau D 2012 *J. Electrochem. Soc.* **159** A357–63
[25] Aurbach D 2000 *J. Power Sources* **89** 206–18
[26] Gao P Zhang C and Wen G 2015 *J. Power Sources* **294** 67–74
[27] Wang L Zhao J He X Gao J Li J Chunrong W and Changyin J 2012 *Int. J. Electrochem. Sci.* **7** 345–3
[28] Zhang SS Xu K and Jow TR 2004 *Electrochim. Acta* **49** 1057–61
[29] Zhang SS Xu K and Jow TR 2006 *Electrochim. Acta* **51** 1636–40
[30] Liao L Zuo P Ma Y Chen X An Y Gao Y Yin G 2012 *Electrochim. Acta* **60** 269–73
[31] Li X Luo D Zhang X and Zhang Z 2015 *J. Power Sources* **291** 75–84
[32] Sharova V Moretti A Diemant T Varzi A Behm RJ Passerini S 2018 *J. Power Sources* **375** 43–52

Synthesis of Li4Ti5O12-Sn by ultrasonic method as anode materials for lithium ion battery

Y Purwamargapratala[1], S Sudaryanto[1], E Kartini[1] and M Manawan[2,3]

[1] Centre of Advanced Materials Science and Technology, National Nuclear Energy Agency, South Tangerang 15314, Indonesia
[2] Energy Engineering, Politeknik Negeri Jakarta, Depok, Indonesia
[3] Integrated Laboratory, Esa Unggul University, Jakarta, Indonesia

pratala_yustinus@yahoo.com

Abstract. Research on synthesis of $Li_4Ti_5O_{12}$ has been carried out using ultrasonic method with the objective to study the influence of the Sn addition to the materials conductivity and structure of lithium titanate. The starting materials used were LiOH and TiO_2, while Sn was used as an additive with percentages of 0, 5%, 10%, 15% and 20%. Lithium hydroxide, titanium dioxide, and Sn were mixed into an aquabidest media and stirred for two hours at a rate of 300 rpm. Then it was reacted with the help of ultrasound for two hours, filtered and washed with distilled water and then rinsed with acetone. After drying overnight at room temperature, the resulting powder material is compacted using a hydraulic press at a pressure of 4000 psi and the obtained pellets were sintered in the furnace at 800 °C for two hours. Characterization was performed using LCR meter and X-ray diffraction (XRD) to measure the conductivity and the crystal structure, respectively. The SEM-EDS are used to observe the morphology and the composition of the material. The results of the Rietveld analysis showed that the un-doped sample correspond to cubic crystal structure with space group of Fd-3m that belong to $Li_4Ti_5O_{12}$ and monoclinic crystal structure with space group of C12/c1 that belong to Li_2TiO_3. It turned out that the increase of Sn; did not change in the ratio between $Li_4Ti_5O_{12}$ and Li_2TiO_3, as well as in the lattices constant for all three phases. The optimum conductivity of 6.57 x 10^{-6} S/cm was obtained for 5 % Sn addition to $Li_4Ti_5O_{12}$. A more homogeneous particle distribution was observed by SEM, due to ultrasonic method. It is concluded that the Sn addition to $Li_4Ti_5O_{12}$ has improved structural and electrical performance of the anode materials for lithium ion battery.

1. Introduction

The battery research is one of fast growing researches in developing country such as in Indonesia [1]. A battery is a device that can convert the chemical energy in an active material into electrical energy, a process known by the electrochemical process present in oxidation and reduction reactions. A battery consists of three parts, namely the anode, cathode and electrolyte. The anode is defined as the negative electrode where oxidation takes place and the cathode is the positive electrode where reduction process takes place.

Battery anodes used in lithium batteries are generally composed of graphite material. The advantages of graphite is its high capacity, nevertheless it also has limitations which are not suitable for high discharge rates due to formation of lithium dendrites, and is susceptible to the occurrence of short circuits in the battery which could cause an explosion, so it is less safe to use. There are other materials that have a high enough voltage difference against Li/Li that therefore ensures the formation of the lithiation phenomenon on the electrode surface [2]. One that has been useful is the lithium titanate ($Li_4Ti_5O_{12}$) ceramic material.

$Li_4Ti_5O_{12}$ is a ceramic lithium-titanium oxide (LTO), better known as lithium titanate, which has a spinel structure. The main properties of the ceramic material is the ability of the crystalline structure or the crystalline symmetry to undergo a transformation during a Li^+ ion insertion. Ariyoshi *et al.* [3], reported 0002 A and 0006 A discharge lattice shrinkage in subsequent discharge processes observed using a high precision synchroton XRD to measure the minute changes in the crystal lattice.

Lithium titanate is one of the most promising anode material for lithium batteries despite having a lower specific capacity of 175 $mAhg^{-1}$. In comparison graphite has a capacity of 372 $mAhg^{-1}$ [4]. Lithium titanate is a good anode material for applications that have battery power capability and long life cycle. LTO superiority lies in its power density and chemical stability, and LTO-based batteries has a higher voltage: 2.5 V compared to LiCoO [5] and LFP-based 1.9 V batteries [6]. Lower operating voltage brings significant advantages in terms of security. These batteries can also be recharged quickly. Data show that LTO-based batteries can be safely charged at a higher rate at 10 ^0C, and can be charged in less than 10 minutes. LTO-based batteries also have a wider operating temperature range and their coulombic efficiency is exceeding 98%. The energy density is lower compared to other lithium ion batteries, but still higher than the lead-acid (accu) and NiCd batteries [7].

In various studies, researchers add additives to lithium titanate. Some additives have been added to the original material synthesized with only LiOH and TiO starting materials. Various additives such as chromium (Cr) [8], silicon (Si) [9], magnesium (Mg) [10], graphene [11], aluminum (Al) [12], and carbon (C) [13] were added to LTO. Synthesis of lithium titanate ($Li_4Ti_5O_{12}$) by addition of excess lithium carbonate (Li_2CO_3) in titanium dioxide (TiO_2) xerogel was done by Syahrial *et al.* [14].

The aim of this research is to synthesis $Li_4Ti_5O_{12}$ with addition of tin (Sn) to increase conductivity. The use of ultrasonic process is expected to produce a homogeneous composite LTO-Sn particles and to further increase the conductivity. It is necessary to optimize the concentration of Sn during the formation of LTO-Sn composite.

2. Experimental method

Lithium hydroxide (LiOH) was mixed with 60 mL aquabidest and stirred for 30 minutes by a magnetic stirrer, followed by the addition of titanium dioxide (TiO_2). The batches were stirred for 15 minutes and then SnO was added with various concentrations of 0%, 5%, 10%, 15% and 20%. The mixture was stirred for another 15 minutes.

The mixture in the form of suspension was treated by ultrasonic method for 2 hours at a frequency of 50 Hz. Then it was filtered using a vacuum filter, washed with acetone and finally the powder products were dried for 15 hours at room temperature. The powder samples was compacted under pressure of 4000 psi for 1 minute with hydraulic presser to form pellets. The five samples in the form of pellets were then sintered in a furnace at a temperature of 800 °C for two hours. Detailed experiment has been described elsewhere [15].

Characterizations were done by XRD (X-ray diffractometer), LCR meter, and SEM-EDS (Scanning Electron Microscope Energy Dispersive Spectroscopy) to determine the crystal structure, the conductivity, and the morphological observation.

3. Results and discussion

The X-ray diffraction results is shown in Figure 1. The data was analyzed by Rietveld method using Profex-BGMN that employed Fundamental Parameter Approach (FPA) to determine the instrument profile [16]. The results of the Rietveld analysis was shown in Table 1 where the un-doped sample correspond to cubic crystal structure with space group of Fd-3m that belong to $Li_4Ti_5O_{12}$

[5] and monoclinic crystal structure with space group of C12/c1 that belong to Li_2TiO_3. It turned out that the addition of Sn; oxidized and became SnO_2 (P42/mnm), because there is no diffraction peaks of pure Sn (Fm-3m). The expected *R*-factor (R_{exp}) and weighted profile *R*-factor (R_{wp}) for all patterns is less than 10% indicated a good data quality and a good agreement between simulated pattern and experimental data, as shown in Figure 1 [17]. The goodness of fit (χ^2) has an average value of 1.5, indicated the successful performed of the Rietveld refinements.

There is no change in the ratio between $Li_4Ti_5O_{12}$ and Li_2TiO_3 as the addition of Sn. There is also no significant change in the lattices constant for all three phases as the Sn concentration increase. The use of ultrasonic is intended to produce a homogeneous particle distribution with a small grain size data before ultrasonic. It has been very helpful in optimizing the calcination process. Figure 1 shows that the addition of 5% SnO causes diffraction peaks to appear at 2θ = 17.36° which is a characteristic of Sn peak. This peak also appears at other concentration as shown in Figure 1(c-e). Other peaks at 2θ = 35° and 43°, also appear for addition of SnO into LTO. The rest of the peaks are similar to the diffraction of LTO sample with no Sn addition.

This means that the LTO-Sn composite can be obtained upon the addition of 5% to 20% SnO. Addition of 20% SnO give another peak at 33.9°. The addition of Sn exceed 10% could result in a new phase at the 2θ angular positions of 18.37°; 35.60°; 43.27°; 62.85°; 82.33° and denoted as the SnO phase. This shows that there is an excess of SnO which can no longer usefully interact with $Li_4Ti_5O_{12}$. The refinement results of the x-ray diffraction data are shown in Table 1.

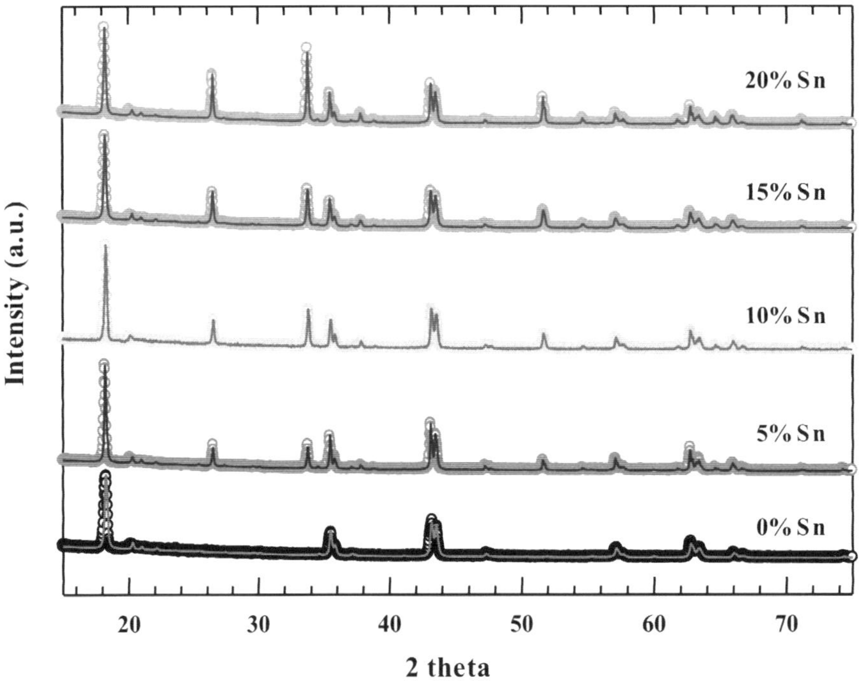

Figure 1. X-ray diffraction pattern of the sample with additives $Li_4Ti_5O_{12}$ - Sn: 0 % Sn, 5% Sn, 10% Sn, 15% Sn, and 20% Sn.

It shows that as the Sn concentrations increase, the crystallite size of $Li_4Ti_5O_{12}$ and SnO_2 decrease up to Sn=15% and then increase at Sn=20%. The opposite effect happened to strain where the %strain of $Li_4Ti_5O_{12}$ and SnO_2 decrease as the Sn concentration increase up to Sn=15% and then decrease at

Sn=20%. The crystallite size of Li_2TiO_3 not significantly change as the addition of Sn, but the %strain pattern follow the $Li_4Ti_5O_{12}$.

Table 1. X-ray Refinement results of $Li_4Ti_5O_{12}$- Sn with various concentrations.

Sn	Chemical Formula	Space group	Crystallite Size (nm)	Strain (%)	Lattice Parameters :			QPA	R-factor		
					a (Å)	b (Å)	c (Å)	wt%	R_{wp} (%)	R_{exp} (%)	X^2
0 %	Li4Ti5O 12	F d -3 m	79.3 ± 6.2	0.000143	8.3545 (4)			62.32 (68)	7.48	5.08	1.47
	Li2TiO3	C 1 2/c 1	69.5 ± 3.1	0.000403	5.0733 (1)	8.7734 (2)	9.7421 (1)	37.68 (68)			
	SnO2	P 42/m n m	-	-	-	-	-	-			
5 %	Li4Ti5O 12	F d -3 m	139.9 ± 8.0	0.0000206	8.3586 (2)			59.64 (57)	7.46	5.07	1.47
	Li2TiO3	C 1 2/c 1	66. 6 ± 3.1	0.000015	5.0709 (6)	8.7788 (1)	9.7550 (1)	35.49 (57)			
	SnO2	P 42/m n m	97..4 ± 3.4	0.0000436	4.7380 (2)	-	3.1865 (2)	5.81 (10)			
10 %	Li4Ti5O 12	F d -3 m	128.8 ± 7.5	0.000084	8.3578 (2)			57.32 (5)	6.94	4.84	1.43
	Li2TiO3	C 1 2/c 1	63.7 ± 3.0	0.000065	5.0718 (1)	8.7768 (1)	9.7540 (1)	34.11 ((5)			
	SnO2	P 42/m n m	90.3 ±4.5	0.000047	4.7377 (0)	-	3.1854 (0)	9.47 (12)			
15 %	Li4Ti5O 12	F d -3 m	109.1 ± 6.6	0.000094	8.3586 (0)			53.95 (64)	6.86	4.81	1.43
	Li2TiO3	C 1 2/c 1	61.8 ± 3.2	0.000076	5.0736 (1)	8.7751 (1)	9.7562 (1)	32.10 (65)			
	SnO2	P 42/m n m	83.2 ± 3.9	0.000071	4.7380 (0)	-	3.1857 (0)	14.79 (14)			
20 %	Li4Ti5O 12	F d -3 m	135.7 ± 7.5	0.0000314	8.3590 (0)			50.57 (62)	6.74	4.66	1.45
	Li2TiO3	C 1 2/c 1	66.3 ± 3.5	0.000031	5.0705 (1)	8.7805 (2)	9.7536 (2)	30.10 (66)			
	SnO2	P 42/m n m	110.2 ± 4.4	0.0000126	4.7383 (0)	-	3.1863 (0)	20.13 (17)			

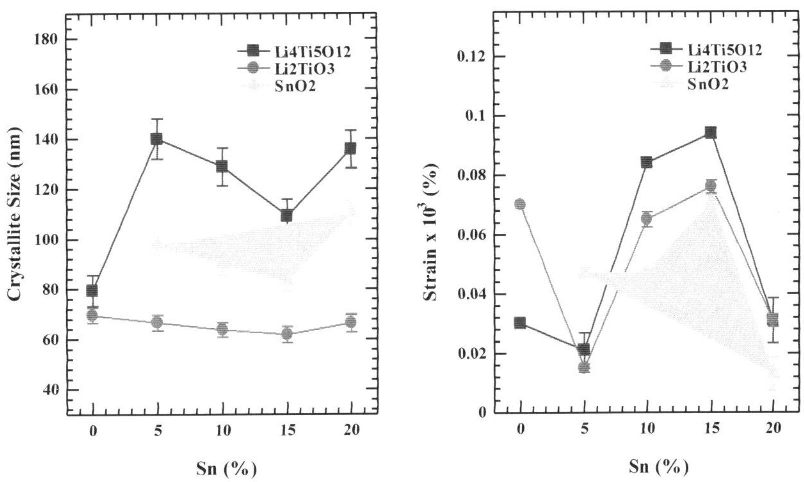

Figure 2. Mean crystallite size and strain of $Li_4Ti_5O_{12}$-Sn compound as a function of 0 Sn addition.

Observations using optical microscopy showed the morphology of Sn distribution in the sample. The SEM observation results in Figure 3(b-e) show that the addition of SnO into $Li_4Ti_5O_{12}$ are well distributed and homogenous. In comparation to LTO with 0% Sn addition in Figure 3(a), they all show similarity to each other, meaning that the use of ultrasonic does not change the microstructure of LTO. Sn particles with a concentration of 5% Sn seemed to be more homogeneously spread than in samples with higher Sn concentrations. This shows that the use of Sn as an additive with a concentration up to 5% is effective. This is confirmed by the EDS (energy dispersive spectrometer) measurement results in the range of 0-20 keV and counting rate of 2449 cps, as shown in Figure 3, which shows the appearance of the Sn peak in the EDS spectrogram. The atomic% of Ti,O and Sn for various Sn concentration is shown in Table 2.

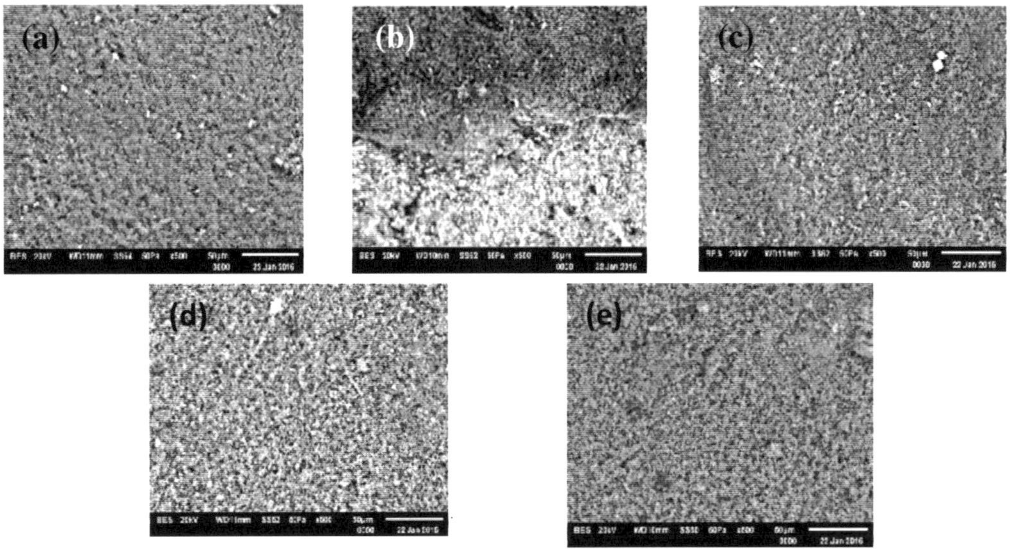

Figure 3. SEM Observation of morphology of samples $Li_4Ti_5O_{12}$-Sn (a) 0% Sn (b) 5% Sn; (c) 10% Sn; (d) 15% Sn; (e) Sn 20%.

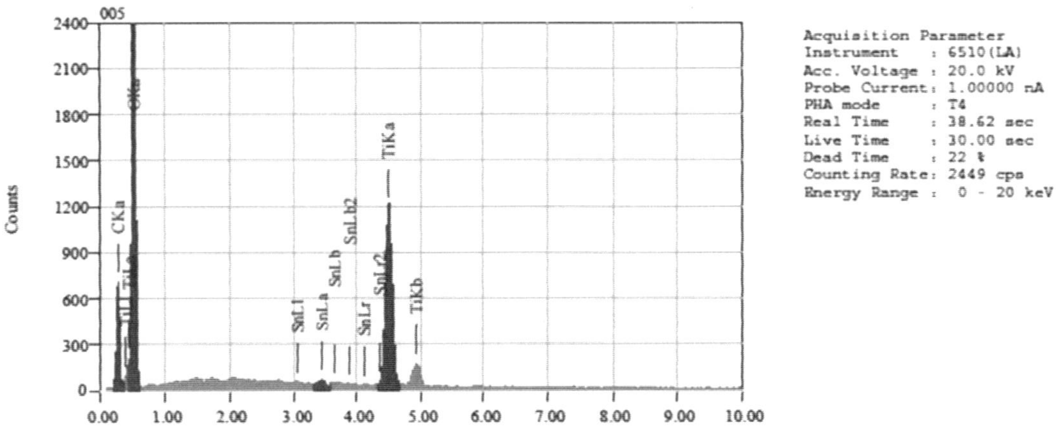

Figure 4. Results of EDS measurements of samples Li4Ti5O12-Sn (5% Sn).

Table 2. Atom% of the components containing in $Li_4Ti_5O_{12}$-Sn with variation of % Sn.

No	%Sn	Atom %					
		C 0.277 keV	O 0.277 keV	Ti 0.277 keV	Sn 0.277 keV	Al 1.486 keV	Fe 6.398 keV
1	0		68.98	31.02			
2	5		51.72	44.78	3.50		
3	10	66.38	17.42	0.94	0.09	14.40	0.76
4	15		70.84	28.47	0.69		
5	20	18.58	53.19	19.22	2.22	6.89	

The distribution of Sn in the $Li_4Ti_5O_{12}$ sample will affect the value of ionic conductivity. A smoother distribution of Sn in a material would increase the conductivity. Ions will flow evenly throughout the surface of the material. Inversely, an uneven or coarse distribution of Sn in the material would cause the value of the conductivity of the material to decrease, the ions will only flow through dots or lines formed in some parts of the sample. The obtained conductivity values sample of LCR meter measurement are shown in Table 3.

Table 3. Conductivities of $Li_4Ti_5O_{12}$-Sn with variation of % Sn.

No.	Sn (%)	Conductivity (S/cm)
1.	0	2.61×10^{-5}
2.	5	6.57×10^{-6}
3.	10	3.31×10^{-5}
4.	15	3.03×10^{-6}
5.	20	8.41×10^{-7}

Conductivity values of $Li_4Ti_5O_{12}$ changes with addition of Sn in not linear trend. The maximum conductivity of 6.57×10^{-6} is found in the sample with 5% Sn, while the lowest conductivity of 8.41×10^{-7} is found in the sample with 20% Sn. The $Li_4Ti_5O_{12}$ conductivity is influenced by the distribution of Sn in the sample. In this case the optimum addition is less than 15% Sn, further increasing will reduce the conductivity.

4. Conclusion

Synthesis of $Li_4Ti_5O_{12}$-Sn using LiOH, TiO_2 and Sn as additive using ultrasonic method has been successfully performed. The composite materials consisted of three phases, a cubic crystal structure with space group of Fd-3m that belong to $Li_4Ti_5O_{12}$ and monoclinic crystal structure with space group of C12/c1 that belong to Li_2TiO_3 and SnO. The best $Li_4Ti_5O_{12}$-5% Sn is a good candidate for anode material in lithium ion battery. The ultrasonic method is found to be good method to synthesis the LTO with various additives.

Acknowledgement
This work was supported by the 2017 DIPA PSTBM BATAN and partially funded by INSINAS grant (No. IRPK-2017-076) Ministry of Research, Technology and Higher Education of the Republic of Indonesia (KEMENRISTEK DIKTI).

IOP Conf. Series: Materials Science and Engineering **432** (2018) 012062 doi:10.1088/1757-899X/432/1/012062

References

[1] Kartini E 2015 *AIP Conf. Proc.* **1697** 100001-2
[2] Fu R Zhou X Fan H Blaisdell D Jagadale A Zhang X Xiong R 2017 *Energies* **10** 2174
[3] Ariyoshi K Yamato R Ohzuku T 2005 *Electrochim. Acta* **51** 1125-9
[4] Jiang X Zhang M Shi S He G Song X Sun Z 2014 *Nanoscale Research Lett.* **9** 219
[5] Tian B Xiang H Zhang L Li Z Wang H 2010 *Electrochim. Acta* **55(19)** 5453–8
[6] Zaghib K Mauger A Groult H Goodenough JB Julien CM 2013 *Materials* **6** 1028-49
[7] Yang SB Feng XL Mullen K 2011 *Sandwich Like. Adv. Mater* **13** 3575-9
[8] Akinyele D Belikov J Levron Y 2017 *Energies* **10** 1760
[9] Kumar P Kennady CJ 2015 *Int. J. Thin. Fil. Sci. Tec.* **4(2)** 147-53
[10] Si Q Mori D Takeda Y Yamamoto O Imanishi N 2017 *Energies* **10** 1803
[11] Tao H Feng Z Liu H Kan X Chen P 2011 *The Open Mater. Sci. J.* **5** 204-14
[12] Ying S Lei W Cheng HM 2011 *J. of Powder Source* **196** 8610-7
[13] Sun H Hanlon D Dinh D Boland JB 2018 *2D Mater.* **5** 015024
[14] Syahrial AZ 2016 *Int. J. of Tech.* **3** 392-400
[15] Purwamargapratala Y Ginting J 2015 *Proc. of 2nd Nuclear Energy Tech. Sem.* Denpasar 2015
[16] Doebelina N Kleebergc R 2015 *J. of App. Crystallo.* **48** 1573-80
[17] Toby BH 2006 *Powder Diffr.* **21(1)** 67–70

IOP Conf. Series: Materials Science and Engineering **432** (2018) 012063 doi:10.1088/1757-899X/432/1/012063

Indonesian consortium of lithium ion battery for solar street lamp

E Kartini[1], A Purwanto[2], S Sudaryanto[1], W Honggowiranto[1], R N S Rofika[3]

[1] Science and Technology Center for Advanced Materials, National Nuclear Energy Agency, Indonesia
Puspiptek Serpong, South Tangerang 15314, Banten, Indonesia
[2] Univeristy of Sebelas Maret (UNS), Solo, Indonesia
[3] Department of Physics, Bandung Institute of Technology, Bandung, Indonesia

kartini@batan.go.id

Abstract. Indonesia's renewable energy systems are reducing the world's dependence on fossil fuels by providing constant energy sources such as lithium ion battery (LIB) for application on solar street lamps. There are advantages of using LIB that will improve its performance and life cycle, and also reduce the maintenance cost. Due to these reasons, a National Consortium on Lithium Ion Battery for Solar Street Lamp was formed in 2016 and supported by the National Innovation System (INSINAS) from the Ministry Research Technology and Higher Education. The project was divided into four work breakdown structures (WBS), with various activities and targets. The goal of this consortium is to produce a lithium ion battery module with the specification 120 Watt hour, so as to run the 10 Watt street lamp. The module consists of 40 cylinder cells 18650. The first year's target is to produce cylinder cells 18650 with specific output of 3.2 Volt and 1 Ah. This paper will review the activities of the consortium, WBS, and the characteristic of the cylinder cells.

1. Introduction

With issues such as global warming and carbon emissions now, the demands for renewable energy technologies are continuing to grow. Solar, Hybrid, and Small Electric Vehicles such as Golf carts *etc.* require deep cycle batteries that are capable of supplying large amounts of power for both short and long periods of time [1]. Due to the future global prospect of increasing renewable energy, the research on lithium ion battery for application in solar cells of, for example, street lamps, must be considered [2]. Therefore, the government design a new strategic planning of lithium ion battery program under the National Innovation System (INSINAS). The national consortium on lithium ion battery for solar street lamp, consisting of several institutions, universities and industry, was formed in 2016. The program was run by the grant from the Ministry Research Technology and Higher Education. Its main idea is to gather the researchers who are working on battery research into one renewable energy program, with the main goal to provide a solution for energy storage based on lithium ion battery made by The Indonesian consortium. Furthermore, the goal is to build the National-Battery Research Institute (N-BRI), that can provide solutions for battery science and technology from research straight into application. The N-BRI will be started from the combination of industries, institutions and universities. Our starting point is the consortium that has been running well from 2016 to 2018, which consists already of several institutions, universities and industries [2].

In order to achieve this goal, based on various background, capacity and capabilities, the program is divided into four work break down structures (WBS), with one coordinator for every WBS (Figure1), furthermore we set the target for each WBS. The road map of Consortium is also the most important so as to determine the goal every year. This consortium is a good example of joint collaborations between multiple parties. Though its activities have been around for several years, there is no article that review about the consortium itself and their agenda. Therefore, the aim of this paper is to review the research activities of consortium, work break down structure (WBS), the road map and progress that has been achieved in the past year.

2. Methodology

As shown in diagram of WBS in Figure 1, there are several activities performed and linkage to produce the battery module. At first the WBS1 produced battery components such as the cathode $LiFePO_4$ and the anode graphite (C). The fabrication was located at the Laboratory of UNS, Solo, as described [3]. In WBS2, these componnents were assembled into cylinder cell 18650 at the Integrated Battery Laboratory, BATAN [4]. The WBS3 performed testing and standardization of cells using charge discharge instruments at BATAN. The cell grading and assembly for module was done by WBS4 at ASTRA and B4T, respectively. In addition, WBS0 is designed to work, with a separate program, to produce the raw materials for the battery components.

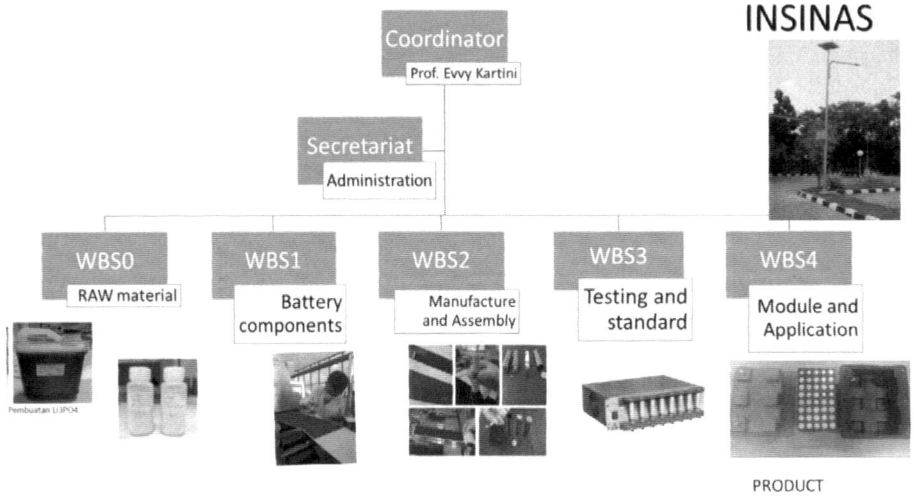

Figure 1. Diagram of the Work Break Down Structure (WBS).

The road map of the consortium is shown in Figure 2. In 2016 the target is to produce battery components, including cathode and anode materials, based on local resources. The next target is to produce the cylinder cell 18650 with the specification of 1 Ah and 3.2Volt. In 2017, a battery module will be designed to provide 40 cells with the specification 120 Watt hours.

3. Results and discussion

As shown in Figure 1, the target of every WBS is different, depending on the various activities and targets. The WBS1 has activities consisting of mainly research and development of battery components. The WBS2 has a target to assembled the battery components into cylindrical cells 18650. The WBS3 has the duty to perform battery testing and standardization of the lithium ion battery (LIB) cells. The design of battery module and application in a solar street lamp will be performed by WBS4. The activities were performed conjointly among the scientists from various universities, institutions and industry, such as Institut Teknologi Sepuluh November (ITS), Surabaya; Bandung Institute of Technology (ITB), Bandung; Universitas Sebelas Maret (UNS), Solo; University of Indonesia (UI), Jakarta; Politeknik Negeri Jakarta (PNJ), Jakarta; National Nuclear Energy Agency (BATAN); Agency for Assessment and Application Technology (BPPT); National Standardization Agency (BSN); B4T, Ministry of Industry; Indonesian Institute of Science (LIPI) and industry EDC-ASTRA. As shown in Figure 3, all the laboratories have different job descriptions, depending on the availability of laboratory and the main focus of the research.

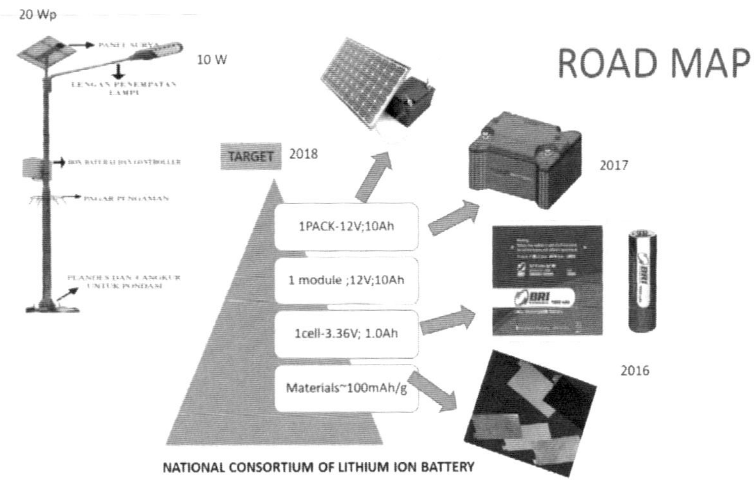

Figure 2. Road map of the national consortium of lithium ion battery.

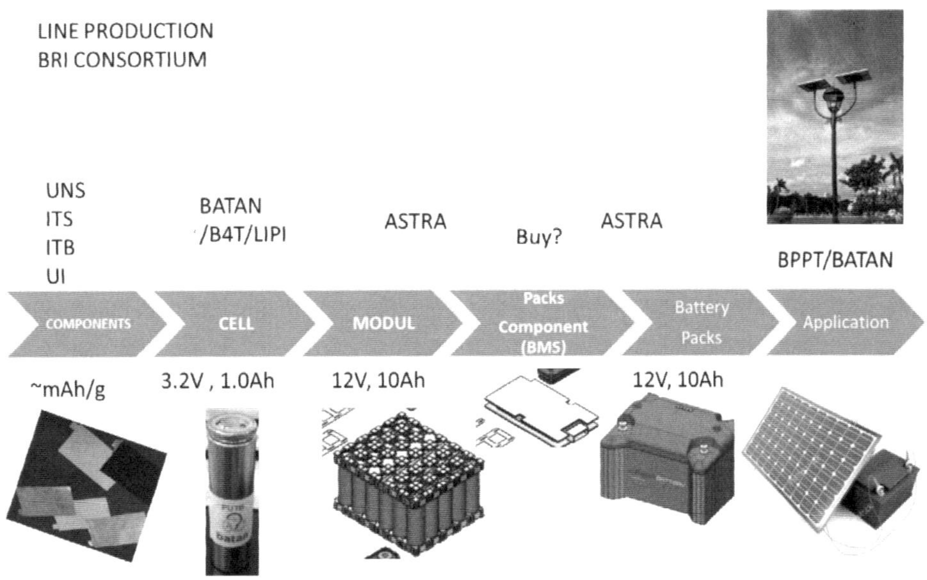

Figure 3. The line production of consortium activities.

In the first year, the consortium has decided to apply $LiFePO_4$ and Graphite as the cathode and anode materials respectively for the battery cell [5, 6, 7]. The choice of these electrodes is due to increased stability during charge and discharge, simplicity in preparation and their environmentally friendliness [4, 5, 8]. The battery components such as cathode and anode sheets were prepared at the Battery Laboratory, UNS. The battery assembly into cylinder cell 18650 was done at the Integrated Battery laboratory, National Nuclear Energy Agency (BATAN), Indonesia. Prior to the fabrication, the cathode and anode materials were characterized and compared with the commercial components. The cylinder 18650 testing was performed at BATAN by using charge-discharge analyzer.

The cathode $LiFePO_4$ and anode graphite sheets were cut using a slitting machine with the width of 65 mm and 58 mm respectively. The lengths of the cathode and the anode were optimized to achieve higher capacity, then the sheets were assembled into cylindrical cell 18650, named there after as empty cell length was 650 mm, and the diameter was 18 mm. The polymer cellgard and $LiPf_6$ were used as separator and liquid electrolyte, respectively. The assembly was done in a glove box with flowing argon containing 0.01 ppm O_2. The use of the glove box is a requirement to avoid the absortion of moisture that can destroy the cell. The testing of the cylinder cell was performed to determine initialization, charge-discharge and cycling performance. Some of prototypes cylinder cell 18650 is shown in Figure 4.

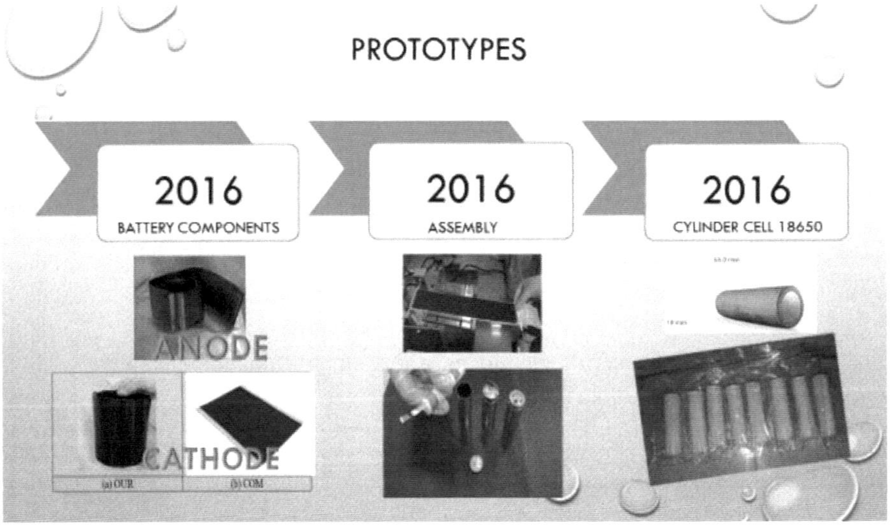

Figure 4. Prototypes of cathode, anode, and cell 18650.

The targets of consortium in the first year (2016) are to produce good quality battery components and to produce the cylinder cell 18650 that delivers the capacity as designed. Furthermore, to produce several cylinders repeatedly and reproduceably, in order to obtain similar output of cylinder cells. Figure 4 shows the battery components of $LiFePO_4$ cathode and graphite anode that were ready for fabrication into cylinder cells. The products of cylinder cell 18650 are also shown in the Figure 4.

4. Battery testing

Figure 5 shows the initial charging for the first 5 cylces of the cylinder cell 18650 produced by the consortium. As shown in Figure 5, the input charge and discharge current was 250 mAh, with higher and lower limit voltages at 3.7 Volt and 2.5 Volt respectively. The output voltage for the cell is 3.2 Volt, in agreement with the $LiFePO_4/C$ battery. The discharge capacity and efficiency (%) curves are shown in Figure 6. The capacity during the first cycle is still 1.3 Ah, but is increased to 1.5 Ah for the second cycle up to 5th cycle. The continuing cycles are now in progress, until 200 cycles to observe the cycle ability and stability of the battery. The output of the cell, has met not only requirement as designed at the beginning, which was 1 Ah, but also shows better capacity. The reproduceability was shown by production of another 100 cells, with the same condition of preparation and testing.

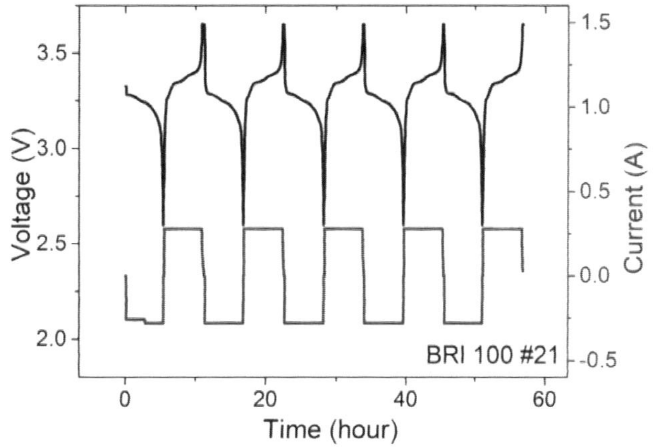

Figure 5. Battery testing of cylinder cell 18650.

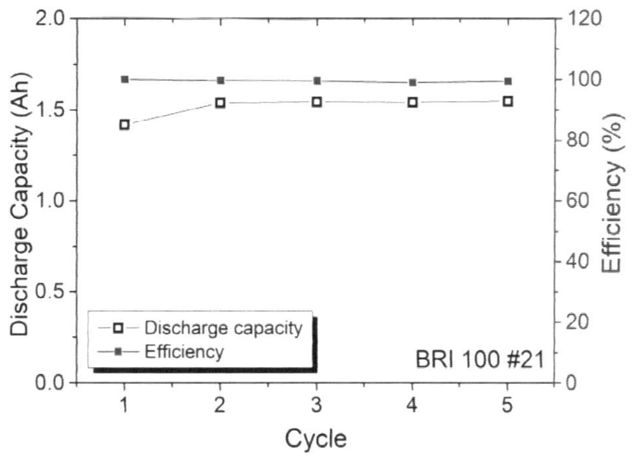

Figure 6. Discharge capacity and efficiency of cylinder cell 18650.

In addition, WBS1's researches are to develop new materials for battery components such as: Preparation of lithium iron phosphate $LiFePO_4$ added polypyrrole PPy as cathode materials for li-ion battery [9]; The improving conductivity of $LiFePO_4$ by optimizing the calendaring process [10]; Synthesis of $Li_4Ti_5O_{12}$-Sn by ultrasonic method as anode materials for lithium ion battery [11]; Effects of annealing temperature on the electrochemical characteristics of ZnO microrods as anode materials of lithium ion battery using chemical bath deposition [12]; Comparative study on the ionic conductivities and redox properties of $LiPF_6$ and LiTFSI electrolytes and the characteristics of their rechargeable lithium ion [13]; The effect of acetonitrile as an additive on the Ionic conductivity of imidazolium-based ionic liquid as the electrolyte for li-ion battery [14]; and Development of battery performance data acquisiton system for monitoring battery performance inside solar cell system [15].

5. Conclusion

The consortium has been performed well for the last two years. It can be shown from the results achieved, which coincide with the target at the beginning. Several international publications related to the battery components produced have been published. The prototypes of cylinder cell 18650 have shown a good performance and can be further applied for the battery module. The cooperation among institutions plays an important role in this consortium.

Acknowledgement
This work was funded by INSINAS grant (No. RT-2016-0650 and IRPK-2017-076) of the Ministry of Research, Technology and Higher Education of the Republic of Indonesia (KEMENRISTEK DIKTI).

References

[1] Scrosati B Garche J 2010 *J. Power Sources* **195** 2419–30
[2] Kartini E 2015 *AIP Proc. 5th Int. Conf. on Women in Phys. Waterloo, Canada, 5-8 August 2015* **1697** 100001–2
[3] Honggowiranto W Sudaryanto S Kartini E Purwanto A 2016 *AIP Proc."6th Nanosci. and Nanotech. Symp. (NNS2015)* 030042–1–7
[4] Honggowiran W Kartini E 2016 *AIP Proc.* **1710** 030045
[5] Liu J Wang Z Zhang G Liu Y Yu A 2013 *Int. J. Electrochem. Sci.* 2378–87
[6] Hu B Wang X Wei Q Shu H Yang X Bai Y Wu H Song Y Liu L 2013 *J. Alloys Compd.* **579** 18–26
[7] Kim H Kim S W Park K Y Kim J Jeon S Kang K 2012 *Carbon N. Y.* **50(5)** 1966–71
[8] Hu C Yi H Fang H Yang B Yao Y Ma W Dai Y 2011 *Mater. Lett.* **65** 1323–26
[9] Gunawan I Ginting J Purwamargapratala Y Sudaryanto S 2018 *IOP Proc. 1st MRS-INA C&C 2017.* in press
[10] Deswita Sudaryanto S Kartini E 2018 *IOP Proc. 1st MRS-INA C&C 2017.* in press
[11] Purwamargapratala Y Sudaryanto S Kartini E Manawan M 2018 *IOP Proc. 1st MRS-INA C&C 2017.* in press
[12] Pambudi Y D S Setiabudy R Yuwono AH Kartini E Lee JK Hudaya C 2018 *Ionics.* in press
[13] Septiana A R Honggowiranto W Sudaryanto S Kartini E Hidayat R 2018 *IOP Proc. 1st MRS-INA C&C 2017.* in press
[14] Rofika R N S Septiana A R Honggowiranto W Sudaryanto S Jodi H Kartini E Hidayat R 2018 *Ionics,* in press
[15] Mulyono A E Mustika T Sulaikan H P Kartini E 2018 *IOP Proc. 1st MRS-INA C&C 2017.* in press

Effect of Solution Treatment Temperature on The Microstructure and Mechanical Properties of Al-5.1Zn-1.9Mg Alloy Produced by Squeeze Casting

Dimas Pratamawansyah Putra and **Bondan Tiara Sofyan**[1]

[1]Department of Metallurgy and Materials Engineering, Faculty of Engineering Universitas Indonesia, Kampus UI, Depok 16424, Indonesia

Email: bondan@eng.ui.ac.id

Abstract. Aluminum alloys are widely used in aviation industries, especially for the body and wings of aircraft. The heat treatment for 7xxx series aluminum alloys is precipitation hardening which consists of solution treatment, quenching, and ageing. One key for successful ageing process is the amount of solute elements and vacancies dissolve in the matrix during solution treatment. Therefore, this research is aimed to study the effects of solution treatment temperature on the hardness and microstructure of Al-5.1Zn-1.9Mg (wt. %) alloy which produced by squeeze casting. Solution treatment temperatures were varied to 220, 420, and 490 °C for 60 minutes, followed by quenching. The samples were then stored at -10 °C to prevent natural ageing. Characterization included hardness testing and microstructural observation by using OM and SEM-EDS, XRD, and STA. Ageing was conducted at 130 °C for 48 h followed by hardness testing and microstructural observation.The results showed that increasing solution treatment temperature induced enhancement of second phase dissolution and the amount of trapped vacancies in the matrix. However dissolution of second phase was hardly detected at solution treatment temperature of 220 °C. It was shown by the volume fraction of the second phase found after homogenizing was 7.13 % and decreased to 7.06, 4.80, and 4.19 % after solution treatment at temperatures 220, 420, and 490 °C respectively. Therefore the increase in hardness after ageing at 130 °C for 48 hours was 11.54, 42,1, and 66.7 HRB for solution treatment temperatures of 220, 420, and 490 °C respectively. This can be concluded that inceasing solution treatment temperature enhanced the hardness of alloy after ageing.

Keywords : Al-Zn-Mg, Solution Treatment, Precipitation Hardening, Fraction Volume

1. Introduction

Aluminum 7XXX series with zinc and magnesium as alloying element are commonly used in aviation industries due to a very high strength to weight ratio and their mechanical properties can be enhanced by heat treatment. Maximum solubility of zinc in aluminum is 83.1 wt. %, which will strengthen the alloy through solid solution [1]. Sofyan, *et.al.* [2] found that the increase in zinc also refines the dendrite structure in Al-xZn-6Mg composites. Magnesium addition in the Al-Zn alloy enhances formation of $MgZn_2$ semi coherent precipitate and increases hardness [3]. Addition of other alloying elements, such as Cu, may even increase the hardness. However, when Zn and Mg content of the alloy is already high, addition of Cu will not give significant increase to the hardness [4]. The mechanical properties of these alloys increase by formation of precipitates which evenly disperse in the matrix through precipitation hardening and hinder the movement of dislocations [5].

The stages of precipitation hardening are solution treatment, quenching, and ageing. During solution treatment, matrix will be richer in alloying elements coming from dissolution of second phase particles [6]. The main purpose of solution treatment is to dissolve most of second phase particles, so that maximum precipitation hardening would be obtained during ageing. The solid solubility of atom increases with the rise of temperature, but at the same time, high temperature will create more vacancies that will be trapped during quenching. Therefore, solution treatment temperature should be carefully determined to obtain optimum hardness. Based on Figure 1, at the composition of Al-5.1Zn-1.9Mg (wt. %), the Mg_5Al_8 dan $Mg_3Zn_3Al_2$ second phase begin to dissolve at 277 °C (550 K). Therefore, conducting solution treatment at above 277 °C is expected to dissolve more solutes to the α-matrix. It is also supported by Li, *et.al.* [7], in Al-5.48Zn-2.02Mg (wt. %) alloy with solution treatment at 460 °C for 1 h, second phase already dissolved and the solubility increases with rise of temperature.

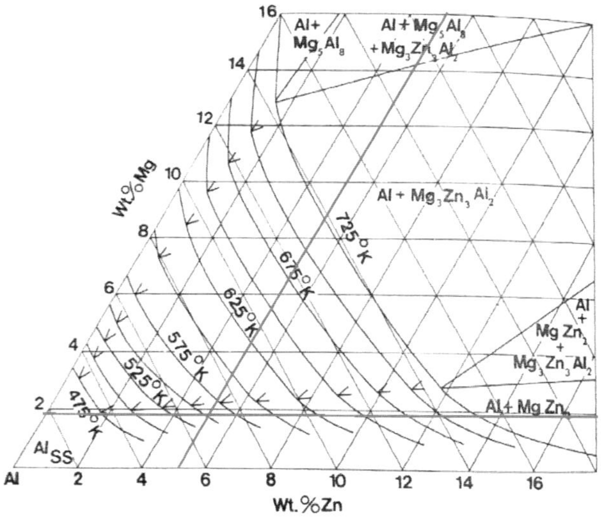

Figure 1. Liquidus projection of the ternary diagram of Al-Zn-Mg [8]. Red line shows the composition of Al-5.1Zn-1.9Mg alloy.

The number of vacancies at a certain temperature can be calculated by using equation (1) and (2) [5].

$$N_D = N \times \exp(-Q_D / kT) \quad (1)$$

$$N = \rho \times N_A/A_r \quad (2)$$

Where Q_D is the activation energy, k is the Boltzmann constant (8.62 x 10^{-5} eV /atom K), T is temperature, ρ is density, N_A is the Avogadro number (6.02 x 10^{23} atom/mol), and Ar is the relative atom. The calculation shows that the number of vacancies at 220, 420, and 490 °C is 0.13x10^{16}, 21.23x10^{16}, 67.17x10^{16} respectively. The exponential increase at 490 °C is potential to be harmful to the mechanical properties of the alloy.

Studies of precipitation hardening have been conducted by many researchers. Among all the parameters, the temperature and time of ageing process have been massively studied. However, few research investigated the effect of solution treatment temperature. Therefore, this paper investigated the effects of solution treatment temperatures of 220, 420, and 490 °C on the microstructures and hardness of Al-5.1Zn-1.9Mg (wt. %) alloy. It was intended to learn how solution treatment temperature affected the dissolution of second phases into the matrix and the amount of vacancy in aluminum alloy, that then influenced the precipitation processes.

2. Experimental Method

The alloy was made by squeeze casting process with pure Al, Zn and Mg ingots as starting materials. The ingots were melted in an electrical furnace at 850 °C and degassed by argon as well as stirred for 30 second for proper mixing. The metal mould with dimension of 17x17x1.5 mm^3 was preheated to 300 °C, then the molten alloy was poured into mould. Semi solid alloy was then squeezed at 76 MPa for 10 min. The chemical composition of the as-cast alloy was shown in Table 1. The as-cast alloy was then homogenized at 400 °C for 4 h in a muffle furnace followed by air cooling. The samples were solution treated at 220, 420, 490 °C for 1 h, followed by water quenching and ageing at 130 °C for 48 h.

Table 1. The chemical composition of the Al-5.1Zn-1.9Mg (wt. %) alloy.

Zn	Mg	Fe	Cr	Si	Ni	Al
5.140	**1.938**	0.267	0.048	0.044	0.027	Bal

Microstructures were observed by using optical microscope and Scanning Electron Microscope (SEM). The samples were polished and etched by Keller's reagent (2.5ml HNO_3 + 1 ml HCl + 1.5 ml HF + 95 ml distilled water). The as-homogenized and as-quenched microstructures were further quantitatively analyzed by using MagniSci. Brinell hardness test was performed according to ASTM E10 and five indentations were made for each measurement. In addition, STA test was conducted on STA 6000 Parkin Elmer at as-quenched samples to examine phase transformation, with scanning rate of 10 °C/min.

3. Results And Discussion

3.1 Microstructure and Hardness Analysis

Microstructures of Al-5.1Zn-1.9Mg alloy in various conditions are shown in Figure 2. All of them possess dendritic shape and black interdendritic phases. The black interdendritic phases consist of T ($Mg_3Zn_3Al_2$) and β (Mg_5Al_8) in accordance to ternary phase diagram in Figure 1. The as-cast microstructure, (Figure 2.(a)), has wide interdendritic area and measured to be 10.58 %, see Figure 3.

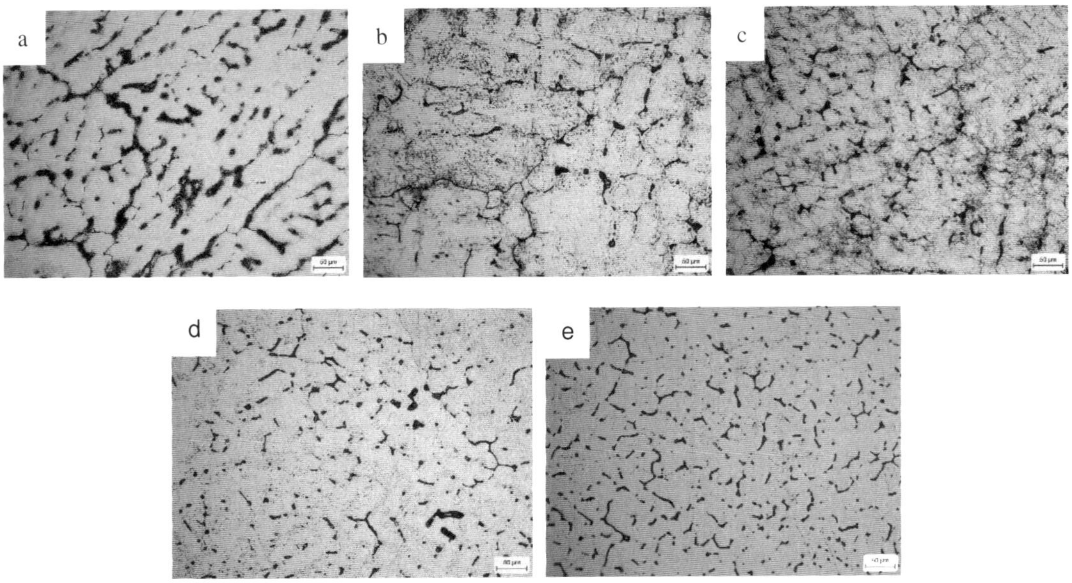

Figure 2. Microstructures of Al-5.1Zn-1.9 Mg: (a) as-cast, (b) as- homogenized, and solution treated at: (c) 220, (d) 420, (e) 490 °C.

Homogenization seemed to dissolve some of the interdendritic phase, that reduced the volume to 7.13 % (Figure 2(b)). This is in line with the purpose of homogenization to reduce the dendritic structure and segregation of alloying elements, as well as to uniform grain size [9]. It also led to a decrease in hardness from 53.5 HRB in as-cast to 47.25 HRB in as-homogenized (Figure 3).

Microstructure of as-quenched sample after solution treatment at 220 °C shows no significant different with that of as-homogenized condition. The volume fraction of the dendritic phase is 7.06 % (Figure 3). The hardness also shows the same number with as-homogenized, that is 42.96 HRB. This indicates that at solution treatment temperature of 220 °C, the second phase dissolution has not occurred as predicted by the liquidus projection of the ternary diagram of Al-Zn-Mg alloy in Figure 1.

Meanwhile, solution treatment at 420 °C rounded the dendritic structure and significantly decreased the interdendritic phase to 4.80 %, (Figure 2.(d)). It is also supported by the hardness number, which is out of range with HRB unit, so that it was tested using Rockwell E units and showed the value of 38.5 HRE. The low hardness maybe caused by the trapping of Zn and Mg atoms in the α-Al matrix that produced unstable state [5]. This indicates that at 420 °C the interdendritic phase already dissolves into the matrix. Further increase of solution treatment temperature to 490 °C reduced the second phase fraction to 4.19 % and hardness to 34.4 HRE (Figure 3) as well as the trapping of vacancies within the matrix [5]. More detailed observation by using SEM is provided ini Figure 4. Reduction of second phase at higher solution treatment temperature is confirmed in this figure. This supports the notion that higher solution treatment temperature leads to enhanced dissolution of the second phase [6].

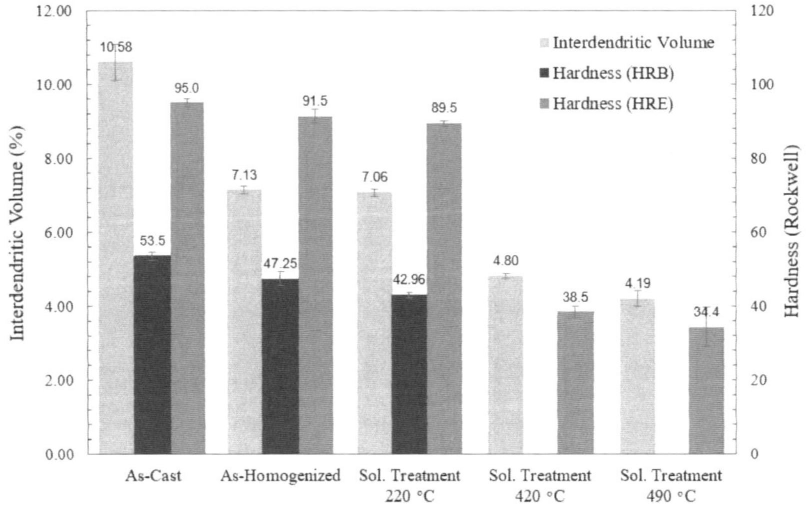

Figure 3. The effect of treatment on the percentage of interdendritic volumes and the hardness of Al-5.1Zn-1.9Mg alloy.

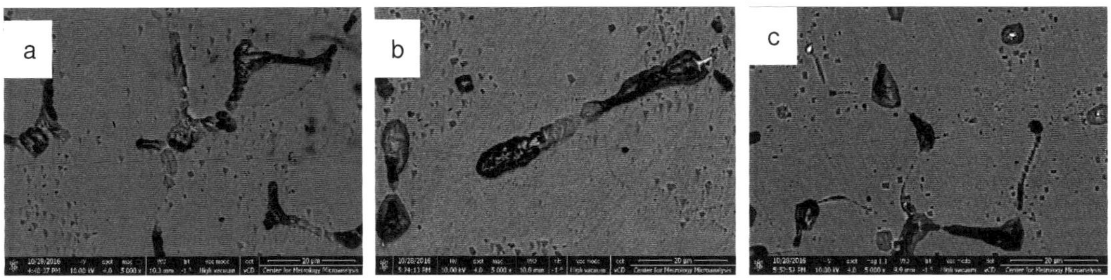

Figure 4. SEM images of of Al-51Zn-1.9Mg alloy after solution treatment at temperature of : (a) 220, (b) 420, and (c) 490 °C followed by water quenching.

Figure 5 provides the microstructures of Al-5.1Zn-1.9Mg alloy after solution treatment at varied temperature followed by 130 °C for 48 h. Figure 5.(a) shows the ageing microstructure with solution treatment at 220 °C. The second phase, which is black, did not dissolve and still visible, so that supposed to be not many precipitates formed [8]. It is indicated by a slightly increase of hardness from 42.96 HRB in as-quenched to 54.5 HRB after 48 h ageing (Figure 6).

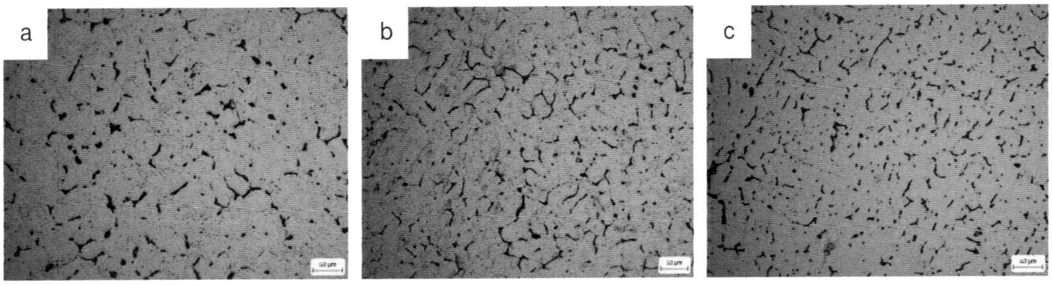

Figure 5. Microstructure of Al-51Zn-1.9Mg (wt. %) after ageing at 130 °C for 48 h: (a) solution treatment 220 °C, (b) solution treatment 420 °C, (c) solution treatment 490 °C.

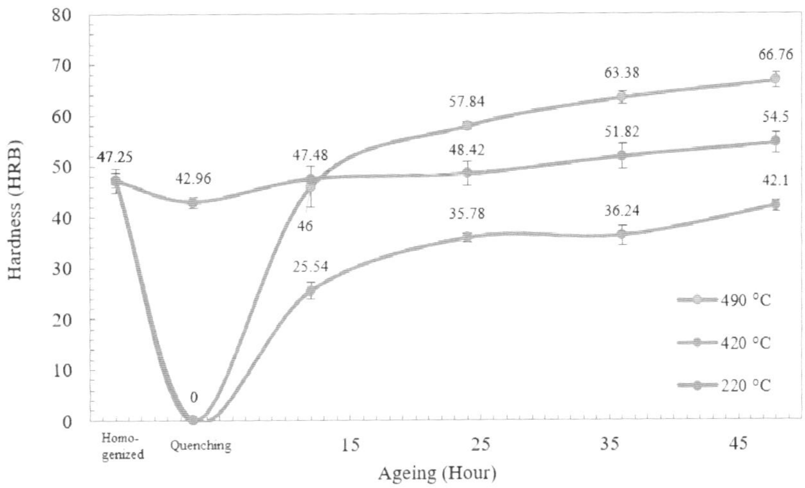

Figure 6. The effect of solution treatment to the hardness of Al-5.1Zn-1.9Mg (wt. %) alloy after ageing at 130 °C.

The ageing microstructure with solution treatment at 420 °C is shown in Figure **4**(b). Precipitates may have formed and characterized by significant increase in hardness to 42.1 HRB. However, the precipitates are not visible by optical microscopy because they are very small and evenly dispersed. Interestingly, samples that solution treated at 490 °C, achieved the highest hardness after ageing, compared to other samples. It is in contrast with their lowest as-quenched hardness. The solution treatment at 490 °C reduced the interdendritic volume to 4.19 % as well created 67.17×10^{16} vacancies. Both the solute atom and vacancies promote formation of precipitates that contribute to the highest hardness after ageing [4].

3.2 Simultaneous Thermal Analysis

Figure 7 illustrates the STA results on Al-5.1Zn-1.9Mg alloy after solution treatment at 220, 420, 490 °C, followed by quenching. The summary of temperatures of phase formation and dissolution is provided in Table 2 and Table 3. Figure 7.(a) shows that there is one endothermic peak at 286.6 °C, indicating the dissolution β (Mg_5Al_8) and T ($Mg_3Zn_3Al_2$) as the only reaction and no precipitates formed. While at 420 and 490 °C (Figure 7.(b) and (c)), three exothermic and four endothermic peaks were detected, in connection with the formation of GP zone, η', η precipitates and the dissolution of GP zone, η', η, as well as the second phase. In general, the peaks of 490 °C solution treated samples are slightly lower than those of 420 °C. This is because higher temperature of solution treatment created more vacancies, so that precipitation is easier to occur.

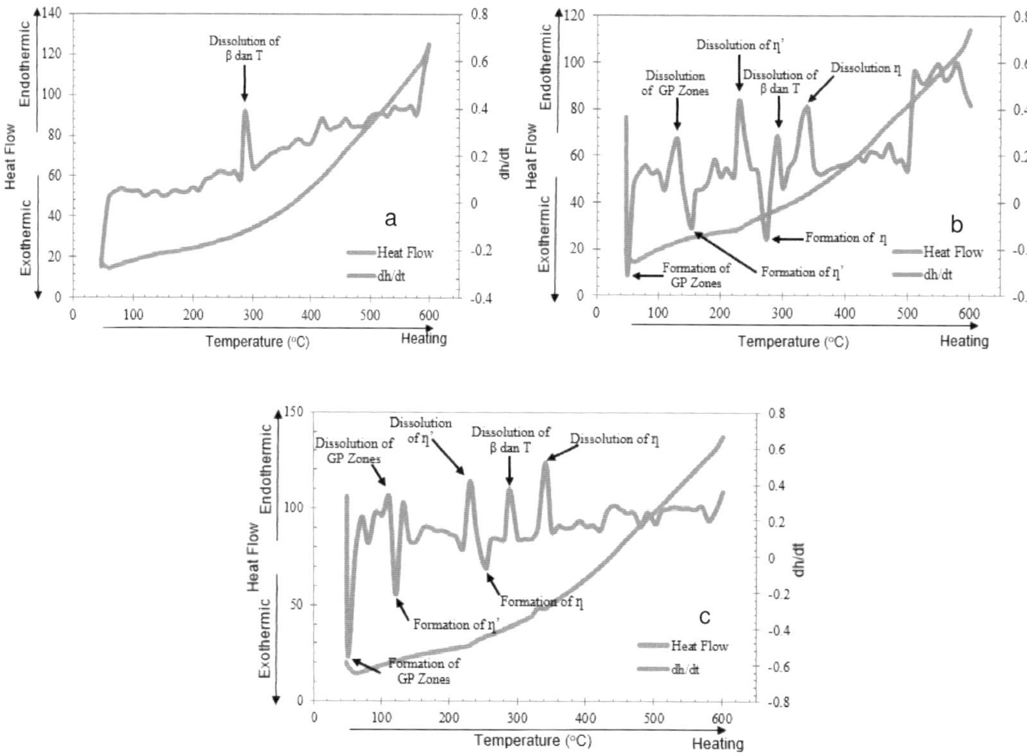

Figure 7. Heat flow and derivatives to temperature of Al-5.1Zn-1.9Mg (wt. %) alloy with solution treatment at: (a) 220, (b) 420, (c) 490 °C.

Table 2. Temperature of phase formation of Al-5.1Zn-1.9Mg alloy

Solution Treatment	Formation Temperature (°C)		
	GP zone	η'	η
220 °C		None	
420 °C	50.14	153.98	273.90
490 °C	50.02	120.02	253.52

Table 3. Temperature of phase dissolution of Al-5.1Zn-1.9Mg alloy

Solution Treatment	Dissolution Temperature (°C)			
	GP zone	η'	β & T Second Phases	η
220 °C	None	None	286.6	None
420 °C	131.04	230.11	291.74	340.10
490 °C	110.50	230.46	287.84	341.11

According to Krishna *et al.*[10], in Al-4Zn-2Mg (% wt.) alloys, the phase forming intervals indicated by exothermic reaction are GP zone at temperature of 20-120 °C, η' at 120-250 °C, and η at 150-300 °C. While the dissolution interval of each phase through endothermic reaction are GP zone at temperature of 50-150 °C, η' at 200-250 °C, and η at 300-350 °C. All exothermic and endothermic peaks in Figure 7.(b) and (c) are in the same temperature intervals. Therefore, it suggests that the precipitation occurs in the studied alloy is GP zones→ η' → η ($MgZn_2$) [11].

4. Conclusion

The increase in solution treatment temperature on Al-5.1Zn-1.9Mg (wt. %) enhanced the dissolution of second phase β (Mg_5Al_8) and T ($Mg_3Zn_3Al_2$) and decreased the as-quenched hardness. It is shown by interdendritic phase volumes in the as-cast, as-homogenized, solution treated at 220, 420, 490 °C were 10.58, 7.13, 7.06, 4.80, and 4.19 %, respectively. While the hardnesses were 53.5 HRB, 47.25 HRB, 42.96 HRB, 38.5 HRE, 34.4 HRE, respectively. The solution treatment at 490 °C led to the highest hardness after ageing due to the highest dissolved solute atoms and the highest amount of vacancies trapped in the matrix that promoted formation of precipitates. Therefore vacancies in precipitation hardening has an important role as a place for precipitate to grow. The phase transformation occurred in the Al-5.1Zn-1.9Mg alloy during ageing was GP zone → η' → η ($MgZn_2$).

Acknowledgment

The authors are grateful with the provision of research funding through *Hibah Penelitian Unggulan Perguruan Tinggi* (*PUPT*) 2017 from Universitas Indonesia.

References

[1] G.E. Totten, MacKenzie, D.Scoot, *Handbook of Aluminum: Physical Metallurgy and Processes, Vol.1*, New York: Marcel Dekker, Inc, 2003.
[2] B.T. Sofyan, D. Rahmalina, E.S. Siradj, H. Mochtadi, *Effect of addition Zn alloying element to the ballistic performance of Al-Zn-6Mg matrix composite reinforced by silicon carbide particles*, Prosiding Seminar Nasional InSINas, (2012) 141 - 145.
[3] W.F. Smith and N.J. Grant, *Effect of chromium and copper on precipitation in Al Zn Mg alloys*, Metallurgical Transactions, 2(5), (1971) 1333-1340.
[4] A. Syakuura, B.T. Sofyan, S.P. Ringer, *Mechanical properties and precipitation hardening of Al-Zn-Mg alloys with variation in Cu content during ageing at 120 °C*, Majalah Ilmu dan Teknologi – Metalurgi, 27(2), (2012) 85-94.
[5] W.D. Callister Jr, D.G. Rethwisch, *Materials Science and Engineering: An Introduction, 8th.ed*, New York: John Wiley & Sons, Inc, 2010.
[6] G. Peng, K. Chen, S. Chen, H. Fang, *Evolution of the second phase particles during the heating-up process of solution treatment of Al–Zn–Mg–Cu alloy*, Materials Science & Engineering, A 641, (2015) 237–241.
[7] B. Li, Q. Pan, C. Chen, H. Wu, Z. Yin. *Effects of solution treatment on microstructural and mechanical properties of Al-Zn-Mg alloy by microalloying with Sc and Zr*, Journal of Alloys and Compounds, 664, (2016) 553-564
[8] L.F. Moldofo, *Aluminium Alloys: Structure and Properties*, Boston: Butterworth&Co Ltd, 1976.
[9] ASM Internasional, *ASM Handbook Volume 02: Nonferrous Alloys and Special-Purpose Materials*, Ohio: ASM International Handbook Committee, 1990.
[10] K.G. Krishna, K. Sivaprasad, K. Venkateswarlu, K.C.H. Kumar, *Microstructural evolution and aging behavior of cryorolled Al–4Zn–2Mg alloy*, Materials Science and Engineering, A 535, (2012) 129-135.
[11] S. Skolianos, T. Griortsidis, Xatzifotiou, *Effect of applied pressure on the microstructure and mechanical properties of squeeze cast aluminum AA6061 alloy*, Materials Science and Engineering, A 231, (1997) 17-24.

AUTHOR INDEX

AUTHOR INDEX